Biotechnological Approaches *for* Sustainable Environment Management

Editors

Sudhir Pratap Singh
Department of Industrial Biotechnology
Gujarat Biotechnology University
GIFT City, Gandhinagar, Gujarat, India

Jyoti Singh Jadaun
Department of Botany, Dayanand Girls PG College
Kanpur, UP, India

Amit Kumar Rai
National Agri-Food and Biomanufacturing Institute (NABI)
Mohali, Punjab, India

CRC Press
Taylor & Francis Group
Boca Raton London New York

CRC Press is an imprint of the
Taylor & Francis Group, an **informa** business
A SCIENCE PUBLISHERS BOOK

This book benefits graduates, research scholars, and academicians interested in the literature related to green technologies of environment management. This book will reinforce our understanding of biotechnological avenues applicable to combat the problem of waste management for keeping a viable environment. All the editors would like to sincerely appreciate the significant contribution of authors from different parts of the globe for the book. We would also acknowledge the team of CRC Press for their excellent support during the publication process.

Sudhir Pratap Singh
Jyoti Singh Jadaun
Amit Kumar Rai

Biotechnological Approaches *for* Sustainable Environment Management

Editors

Sudhir Pratap Singh
Department of Industrial Biotechnology
Gujarat Biotechnology University
GIFT City, Gandhinagar, Gujarat, India

Jyoti Singh Jadaun
Department of Botany, Dayanand Girls PG College
Kanpur, UP, India

Amit Kumar Rai
National Agri-Food and Biomanufacturing Institute (NABI)
Mohali, Punjab, India

CRC Press
Taylor & Francis Group
Boca Raton London New York

CRC Press is an imprint of the
Taylor & Francis Group, an **informa** business
A SCIENCE PUBLISHERS BOOK

Cover Credit: **Cover designed by the editors.**

First edition published 2025
by CRC Press
2385 NW Executive Center Drive, Suite 320, Boca Raton FL 33431

and by CRC Press
4 Park Square, Milton Park, Abingdon, Oxon, OX14 4RN

CRC Press is an imprint of Taylor & Francis Group, LLC

Library of Congress Cataloging-in-Publication Data (applied for)

ISBN: 978-1-032-35839-0 (hbk)
ISBN: 978-1-032-35840-6 (pbk)
ISBN: 978-1-003-32892-6 (ebk)

DOI: 10.1201/9781003328926

Typeset in Times New Roman
by Prime Publishing Services

Preface

Global industrial demands in the various spheres of technology and manufacturing generate several kinds of waste and release toxic chemicals and gases into the environment. This leads to the gradual deterioration of the green status of the environment. Though several conventional strategies are available for pollution abatement, more is needed to handle the excessive environmental pollution. Numerous scientists and researchers are developing new bio-techniques for detecting and degrading environmental pollutants worldwide. Microbial biotechnology provides an extended platform for utilising omnipresent microbial resources to curtail hazardous substances to protect the environment. Indeed, it is an alarming stage to develop a sustainable environment, which is not only required for us but also for the wellbeing of our future generations. This book is compiled in such a manner to show diversified applications of microbial agents to combat a variety of wastes generated from agro-industrial practices.

The book deals with the sustainable management of wastes from fruits, vegetable processing industry, dairy industry, sugar and starch industry, animal husbandry (poultry) and slaughterhouses, and management of biomedical wastes. The chapters provide insight into the production of value-added products from agricultural wastes and conversion strategies of waste vegetable oils into biofuels. Chapters also focus on the application of microbes for curtailing the waste generated from petroleum and biorefinery industries. The exponentially expanding plastic burden on Earth significantly threatens the world's ability to maintain a sustainable environment. In the last few years, several microbes have been reported for the degradation of different types of plastic materials. Further, the book explores innovative strategies for addressing plastic waste management. Another critical issue of environmental concern is the tremendous production of e-waste; as technology is evolving rapidly, people are used to digital devices equipped with the latest updated version, but meanwhile, old stocks of used devices are left this. E-waste has increased massively in the last two decades, so we must consider it seriously. This book also covers the applications of biosensors in environmental monitoring and provides in-depth knowledge about the multidimensional applications of nanotechnology for a healthy environment.

This book benefits graduates, research scholars, and academicians interested in the literature related to green technologies of environment management. This book will reinforce our understanding of biotechnological avenues applicable to combat the problem of waste management for keeping a viable environment. All the editors would like to sincerely appreciate the significant contribution of authors from different parts of the globe for the book. We would also acknowledge the team of CRC Press for their excellent support during the publication process.

Sudhir Pratap Singh
Jyoti Singh Jadaun
Amit Kumar Rai

Contents

Chapter **1**

Biological Valorization of Dairy Wastes for Value-added Products

Varsha Sharma[1,2], Anu Radha[1,2], Vinod Kumar[1], and Vivek Narisetty*[3]

[1]Fermentation and Microbial Biotechnology Division, CSIR-Indian Institute of Integrative Medicine (CSIR-IIIM), Jammu, 180001, India.

[2]Academy of Scientific and Innovative Research (AcSIR), Ghaziabad, 201002, India.

[3]School of Water, Energy and Environment, Cranfield University, Cranfield MK43 0AL, UK.

INTRODUCTION

One of the main industries in the country that stimulates rural economies is the dairy sector, but the processing and cleansing steps involves generation of significant liquid and solid wastes. These wastes typically get disposed of. One of the major concerns that tends to be addressed in major sustainable policies is industrial waste management, in particular dairy waste which consists of high organic matter and high nutrients available from milk and other dairy products' residues. The traditional disposal of these dairy wastes into the local water bodies results in increased biological oxygen demand and chemical oxygen demand,

*For Correspondence: viveknarisetty@gmail.com

thereby polluting the environment, aquatic habitat, and other societal concerns. Although it's classified as hazard for the environment, it can be a huge opportunity for the biorefineries (Adesra et al. 2021, Usmani et al. 2022, Asunis et al. 2020). Hence, these wastes need to be treated or valorised into value added products, so that this management does not pose concerns on the safety of the environment and the human beings.

In the year 2022, over 544 million metric tonnes of cow milk, over 22 million metric tonnes of cheese and over 20 million metric tonnes of other dairy products were produced worldwide [Dairy products: worldwide production by category 2022 | Statista]. During the production of one liter of dairy milk and other products, approximately 6-10 liters of wastewater is produced. (Awasthi et al. 2022), further year-on-year CAGR of 0.7–0.9% growth was estimated. The biorefineries should thus take into account the massive volume of wastewater produced as a sustainable feedstock for value-adding, given the rapid growth rate of the dairy sector (kasmi 2018).

Biological valorisation of the dairy wastes results in value-added products like bioplastics, bioethanol, biohydrogen, carboxylic acids, flavour compounds, single cell protein, polyhydroxyalkanoates, organic fertilizer, and many more (Awasthi et al. 2022). The present chapter discusses the state of the art of processing techniques, valorisation of dairy wastes into value-added biobased products (Figure 1.1).

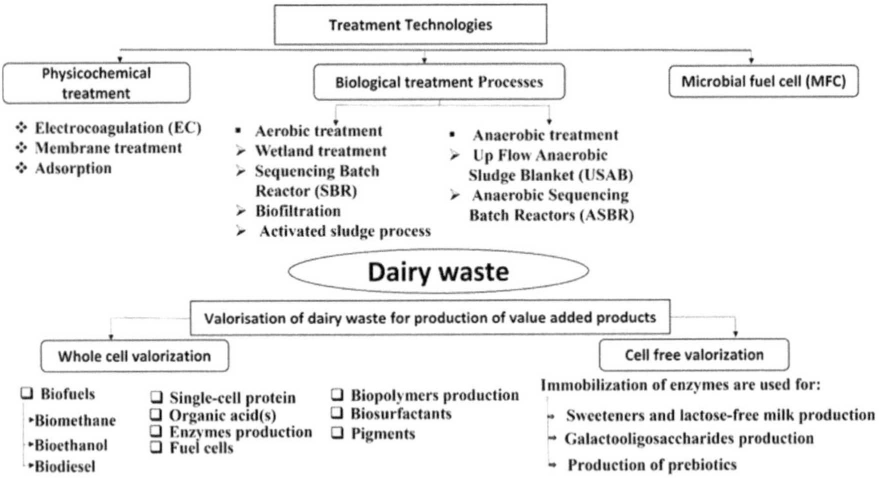

Figure 1.1 Schematic representation of dairy waste treatment and valorisation for value-addition.

COMPOSITION OF DAIRY WASTE

Components of dairy effluents includes soluble proteins, dissolved solids, D-Lactose, lipids, salts, chlorides, and sulphates. Therefore, it has a high BOD5/COD ratio of between 0.4 and 0.8 and a high Chemical Oxygen Demand (COD) that ranges from 0.1 kg/m^3 to 100 kg/m^3 (Ahmad et al. 2019a). The high concentration of phosphate and nitrogen in dairy effluent may cause eutrophication, endangering aquatic life. Dairy effluent also contains lipids ranges from 0.1 to 10.6 kg/m^3 and nutrients including phosphate and nitrogen (Yonar et al. 2018).

Dairy waste streams degrade biologically more quickly than sewage treatment plants because they are warmer than municipal wastewater (10–20°C) on an annual basis temperature (17–25°C). Industrial dairy effluents typically fluctuate in temperature from 17–18°C in the winter to 22–25°C in the summer (Mishra et al. 2021, Li et al. 2021). Biodegradation and oxygen consumption rates are estimated to be 1.5 times high in the summer than in the winter using the Arrhenius equation. Due to the use of hot water for equipment washing and cleaning, the design winter temperature of 15°C is chosen for this type of wastewater.

Dairy effluent must have a pH range between 6 and 9 in order to be treated biologically. The effluents from milk and butter factories show active reactions that are close to neutral (pH = 6.8–7.4). When making cheese, sweet whey has a pH range of 5.9 to 6.6, whereas mineral acid coagulation produces an acidic whey with a pH range of 4.3 to 4.6 (Li et al. 2021). The discharge of alkaline cleansers may cause a rapid rise in the short-term pH of the whole flow, reaching up to 10–10.5. In the sewer system with sumps, prolonged exposure to anaerobic conditions results in liquid acidification from lactic acid fermentation, which lowers pH.

During the cheese-making process, the whey produced is unusually high in both organic as well as volumetric load and that is the major pollutant in dairy effluent. Because whey is a nutrient-rich effluent, when released into the environment without proper treatment, it can lead to eutrophication and severe oxygen depletion. Whey typically has a lactose concentration of 4–5%, proteins and lactic acid combined of 1%, less than 0.5% in lipids, and a range of 1.5–3% in salt. Because whey's BOD5/COD ratio is greater than 0.5, it can be treated biologically (Carvalho et al. 2013). However, the breakdown of proteins can produce a bad smell if biological processes used for whey treatment are not properly controlled.

Sewage pipes may clog due to the presence of higher settleable solids, which are mostly produced by the dairy industry's procedures for making butter and cheese. Additionally, regular cleaning with the required chemical solutions is necessary to prevent the creation of molecular proteins and fats deposition along the pipes' walls that transport dairy effluent. The expense of maintaining the conveyance system used to transport dairy wastewater goes up as a result. Because of the presence of amino acids and inorganic phosphates in milk, varying concentrations of total nitrogen (TN) and total phosphorus (TP) are also seen in various types of dairy effluent (Britz et al. 2006). However, total nitrogen and total phosphorus values depends on biological processes, consequently METs (Microbial Electrolysis Cells) can also be used to clean dairy effluent while concurrently recovering valuables (Choudhury et al. 2021).

ENVIRONMENTAL IMPACT OF DAIRY WASTES

Impacts on Environment

Dairy manufacturing is one of the most polluting sectors, producing an average of 2.5 L of wastewater per litre of processed milk. This can range from 0.2 to 10 L of effluent. (Shete and Shinkar 2013). The volume, concentration, and composition of dairy effluents vary depending on various factors, including the kind of product processed, production strategy, operating procedures, processing plant design, water management strategies, and water conservation efforts. Sweet whey's biochemical contents, which include lactose, proteins, phosphates, nitrates, and nitrogen, render it 60 to 80 times more polluting than domestic wastewater.

In addition to the detergents and sanitizers used in cleaning, dairy effluent contains considerable levels of milk components such as casein and inorganic salts. Their high chemical and biological oxygen demand (BOD and COD) is mainly produced by all of these components, which is much greater than the limitations set by the Indian Standard Institute (ISI), now known as the Bureau of Indian Standard (BIS), for industrial effluent discharge (Burke et al. 2018). These wastes are often dumped into nearby streams or land without previous treatment, which has been linked to major environmental concerns.

Dairy effluents degrade fast, instantly reducing the dissolved oxygen content of the receiving streams, resulting in anaerobic conditions and the release of intense unpleasant odors due to irritation. The receiving water becomes a breeding ground for mosquitos and flies, which carry illnesses such as dengue fever, yellow fever, and chikungunya, in addition to malaria. Furthermore, it has been discovered that some fish and algae species are toxic when dairy manure is present in higher amounts (Kolhe et al. 2009). Dairy waste products has been shown to be toxic to fish owing to casein precipitation, which further decomposes into a foul-smelling, black sludge at certain dilutions. The dairy effluent contains soluble organics, suspended solids, and trace organics. They produce flavor and odor, emit gas, add color or turbidity, and contribute to eutrophication. They also reduce the concentration of dissolved oxygen.

Pollution of the water, air, and biodiversity are the key environmental issues associated with milk production. They typically create a rise in bacteria and algae, depleting the oxygen supply in the water and eventually suffocating rivers, causing fish population to progressively diminish. As a result, several techniques must be used to treat dairy effluents (Deshpande et al. 2012).

Effects on the Atmosphere

Ammonia and methane emissions may be discharged into the environment as a result of the breakdown of animal faeces in the dairy. Methane produces greenhouse gases that can result in global warming, while ammonia can irritate the respiratory system because it forms fine airborne particulates. Dust generated by animal activities also contributes to respiratory illnesses like asthma.

Gaseous Emissions

Industrial processes contribute to a large number of emissions into the environment. Carbon dioxide, sulphur oxides, and nitrogen oxides are released into the environment by boiler stacks. Nitrous oxide (N_2O) is released from the soil at wastewater irrigation sites and methane may be released from anaerobic waste treatment systems (Deshpande et al. 2012). Future studies would likely need to consider the consequences of emissions of three important greenhouse gases: carbon dioxide, methane, and nitrous oxide.

Dust/Odours

From boiler stacks, powder dryers, and other sources, particulate pollutants can be released. When particulate emissions are excessive, dust and powder are coated on the nearby buildings, which is not only adverse but can also be corrosive. Another type of visual pollution that comes from factories is their smoke and steam plumes. At industrial processing locations, the release of offensive odours must be taken into account. Bad odours might be produced by a lot of waste treatment conveniences.

Effects on Water

For every living species, water is the most valuable natural resource. One of the greatest contributors to wastewater is the dairy sector. The treatment of dairy wastewater is a major issue since it releases an enormous amount of nutrients, organic and inorganic particles, and increases COD and BOD levels. (Cristian et al. 2010). Such contaminated water pollutes water bodies and significantly affects our ecology and biodiversity if inadequately cleaned. Turbidity and offensive odour in the water are caused by the dairy effluents. The aquatic habitat develops toxicity due to the organic content present in the wastewater. One of the main compounds generated by the dairy industry is casein. It degrades into extremely dreadful-smelling, dark-coloured waste when exposed to environmental factors. The demise of aquatic species may result from this. The wastewater may disrupt human and marine hormonal balance when combined with freshwater sources.

River Oxygen Levels and BOD

Rivers that have been fully aerated have oxygen concentrations of at least 8 g/m^3 when the water is between 15 and 25°C. The oxygen concentration in discharges to rivers must consequently be kept at or above 6 g/m^3. In order to achieve this, the discharge to the river must not raise its BOD by more than 3 g/m^3 (depending on the river's reaeration properties). Dairy processing effluent contains a high percentage of biodegradable organic materials. Bacteria in streams will consume the waste's organic components. Eq. (1) states that oxygen is used up during the biodegradation process in rivers.

$$\text{Organic material} + O_2 = CO_2 + H_2O + \text{Bacteria} \qquad (1)$$

Colour and Turbidity

Highly coloured wastewaters tend to change the colour of receiving water. Waste products from dairy factories probably contain little soluble color, however color may appear following different types of treatment. Light is reflected back to the spectator by the colloidal and particle components of the waste material. Large quantities of chemicals found in milk waste will make discharges turbid.

Sewage Fungus

The formation of some filamentous slimes in waterways is aided by low molecular weight chemical substances. Sewage fungus is the name given to all these bacterial colonies. *Sphaerotilusnatans* is the type of bacteria that is most prevalent in this group. One of the primary ingredients in wastewater from the dairy industry is lactose, a lower molecular weight sugar that is known to promote the growth of sewage fungus.

Inorganic Components (Mainly Nitrogen and Phosphorus)

One of the industry's main objectives is to recover the waste's protein, or the organic nitrogen component, and transform it into products that can be marketed. The effluent from dairy factories consequently contains a significant amount of nitrogen. Protein will be lost to the waste streams to some extent. Proteins include nitrogen, which bacteria convert to inorganic forms such as ammonia and the ions nitrite, nitrate, and ammonium (Cameron et al. 2002). These many inorganic nitrogen types all have different environmental implications. Both people and livestock are toxic from high amounts of nitrate ions. Newborns may convert nitrate to nitrite, bloodstream absorption, and met-haemoglobinization.

Hemoglobin is not capable of carrying oxygen. Infants less than six months are particularly vulnerable to met-hemoglobinemia because they do not have the enzyme necessary to convert methemoglobin back into hemoglobin. Hence, hemoglobin is not capable of carrying oxygen. The typical limit for nitrate-nitrogen in drinking water sources is 10 g/m^3 to defend human. Met-hemoglobinemia can affect livestock as well. Due to their more neutral stomach pH and rumen bacteria that convert nitrate to nitrite, ruminants are susceptible to met-hemoglobinemia-related deaths. Although it has been suggested that stock drinking water should include no more than 30 g.m3 of nitrate-nitrogen, this usually comes from consuming nitrate-rich feed. Nitrate, nitrite, and ammonium ions, as well as inorganic phosphates, serve as plant nutrients in rivers (Kothari et al. 2017). A minimum of approximately 30 to 100 mg m^3 of total inorganic nitrogen or 15 to 30 mg m^3 of dissolved reactive phosphorus (inorganic phosphorus) should be maintained in receiving waters, respectively.

Effects on Soil

Dairy waste application to land has long been a recognised method of waste disposal; however, there are a few research which have examined the long-term consequences of dairy wastewater irrigation on soil.

Nutrients (Mainly Nitrogen and Phosphorus)

In soil-based treatment systems, the main methods for nutrient removal include:

(i) Plant absorption and assimilation into animal products
(ii) Adsorption and soil immobilisation
(iii) Environmental losses
(iv) Losses by leaching to groundwater.

Plants take up to 500 kg/ha of nitrogen each year, and contains around 30 kg of phosphorus in total. Almost 90% of the nitrogen and phosphorus are restored to the grassland as animals consume the grass (Cameron et al. 2002). Ammonia from dung and urine volatilizes into the atmosphere, and denitrification also contributes to nitrogen losses to the atmosphere. Denitrification is a process in which microorganisms convert nitrate to nitrous oxide or dinitrogen gas. This happens when there is enough organic carbon source for energy and anoxic conditions (a lack of oxygen).

When wastewater is used as irrigation, denitrification rate can be relatively high. At some sites of irrigation, nitrogen losses (mostly in the form of nitrate) to groundwater may occur depending on the levels of nitrogen removed through other methods. Nitrate poisoning of groundwaters that are then utilized as livestock or human drinking water sources is the primary reason that nitrogen-containing wastes are normally not disposed of in soils. In such cases, conventional drinking water approvals are anticipated to be followed. Due to the high retention and immobilization of phosphates in soils, phosphorus seldom causes a problem when leached into ground water.

Sodium and Other Minerals

Sodium, potassium, calcium, and magnesium are all insoluble in soil because they are immobilized and occupy cation exchange sites on soil colloid and clays. The structure of the soil to which irrigation water is applied can be affected by sodium in the water. When irrigation water contains higher salt concentrations than calcium and magnesium, clays can swell and diffuse, which reduces hydraulic conductivity and soil infiltration rates (Balks et al. 1998). These modifications have an impact on gas permeability also, which in turn may reduce the soil's suitability for growing crops and pasture. The fact that soil having higher sodium concentrations may first act to be in good condition but could eventually deteriorate when irrigation or precipitation with lesser conductivity is applied is a complicating factor (Halliwell et al. 2001).

CURRENT APPROACHES FOR DAIRY WASTE TREATMENT

Physicochemical Methods

According to Nadais et al. (2010), the physicochemical treatment is mostly used to remove milk lipids and protein colloids from dairy effluent. Physicochemical treatment involves coagulation and flocculation to assist in the treatment of industrial wastewater (Ghernaout et al. 2019). Prior to sedimentation or filtration, flocculation-coagulation is used to gather wastewater-associated particles into settleable or filterable flocs. The process of particle aggregation is called flocculation (Vigneswaran et al. 2005). To lower hardness and phosphate content, aluminium sulphate ($Al_2\ (SO_4)^{3-}$) is used as a coagulant (Khan et al. 2017a). Examples of coagulant involves *Moringa oleifera* (natural) or alum and ferrous sulphate (chemical). In comparison to natural coagulant, chemical coagulant is more efficient at lowering BOD, COD, and pH (Khan et al. 2017a). Chemical coagulants have the capacity to aggregate the insoluble and dissolved organic matter found in DWW into larger aggregates, allowing for its removal by sedimentation, floatation, and filtering (Kushwaha et al. 2011). Coagulation method treats DWW and eliminates 40% of the organic matter and nitrogen content, using coagulants such as calcium hydroxide ($Ca(OH)_2$), aluminium sulphate ($Al_2\ (SO_4)^{3-}$), and iron chloride ($FeCl_3$). According to Hamdani et al.'s (2005) research, calcium hydroxide eliminates 89% of all phosphorus and 94% of suspended particles. The removal of COD from dairy effluent may also be accomplished using chitosan as an adsorbent. The application of crab shell chitosan as an adsorbent for the treatment of dairy waste water was examined in the study by Geetha Devi et al. (2012). Using batch adsorption methods, experiments showed that chitosan performed best when circumstances were ideal, reducing 79% COD, 93% turbidity, and 73% TSS. Adsorption is the most effective approach for removing organic materials out of all the physical and chemical techniques that are accessible. Many researchers use inexpensive adsorbents including coconut coir, rice husk, fly ash, etc. Additionally, absorbent used to treat DWW is the powdered activated carbon (Kushwaha et al. 2011). Coagulation and flocculation help to diminish turbidity in wastewater caused due to suspended and colloidal materials. Additionally, coagulation and flocculation helps to remove the organic materials that cause rise in COD and BOD levels. When coagulant is added to wastewater, the particulate matter existing in it is disrupted, and this is followed by particle collision which results in flotation and sedimentation in the end (Sarkar et al. 2006b).

Electrocoagulation (EC)

An alternate method of treating dairy effluent is electrocoagulation or flotation, a technique which may effectively remove suspended particles, oil and grease from various industrial effluents. EC starts by the electric current flow and combines flotation and coagulation (Şengil and Ozacar 2006). The fundamental idea behind this procedure is the dissolving of the metal electrode into the metal ions. The metal ions created by this electrode can also go towards spontaneous generation of hydroxides especially if aluminium metal electrodes are employed (Deepak 2014). It is a highly successful and economically beneficial method. This procedure uses less sludge and requires fewer chemicals since the equipment is straightforward and simple to use. The COD value is reduced by 85–87% on average when the EC method is used. In an additional study, the concentration of COD and oil-grease was observed to characterise dairy wastewater (Şengil and Ozacar 2006), the initial COD value during this EC procedure was close to 18,200, and the elimination rate of COD by this technique was 95–97%. Approximately, 97–99% of the oil-grease concentration was reduced.

Membrane Treatment

A selective barrier, the membrane process is a technique for removing or separating particle and colloidal materials from wastewater. In order to remove all particulate and colloidal materials from wastewater, membrane methods such as ultra-filtration, microfiltration, reverse osmosis, nano-filtration, and electrodialysis are applicable ("Technological Approaches for Novel Applications in Dairy Processing" 2018). It is also known as pressure-driven technology, which works by applying external pressure to allow molecules to move from lower-concentration to higher-concentration. In nano-filtration, salts and organic chemicals are retained by its 0.4–2.0 nm pore size. When it comes to eliminating monovalent salts and low molecular weight organic substances, it is largely permeable (Kumar Dasarahally-Huligowda et al. 2019). Reverse osmosis produces water of higher quality than nano-filtration since the latter does not provide superior permeate flux. According to Vourch et al. (2008), RO technology has a 99.8%, 96.3%, and 99.5% removal efficiency for TOC, TKN, and lactose, respectively. Physical-chemical processes are appropriate because they effectively remove organic materials. This approach has the benefit of flexible operations, easy maintenance, and simple procedures. It creates a lot of chemical sludge and is ineffective in removing dissolved materials. Additionally, pathogens and other toxic substances cannot be removed by it.

Adsorption

Since adsorption has been shown to be effective for removing organic components from wastewaters, activated carbon is frequently used in the treatment of wastewaters. However, other adsorbents, such as coal fly ash and rice husk ash, may also be utilized to treat wastewater streams and are inexpensive (Rao et al. 2002). For the treatment of wastewater resulting from dairy waste, Rao et al. (2002) combined powdered activated carbon (PAC) with other inexpensive adsorbents. PAC has superior efficiency in lowering the number of total solids when compared to other adsorbents including fly ash, straw dust, bagasse and coconut coir. Chitosan and other inorganic coagulants were utilised by Sarkar et al. (2006) as pretreatment procedures before the dairy effluent went through membrane separation procedures.

Biological Processes

Since organic waste makes up the majority of dairy effluent, biological degradation is one of the best methods for getting rid of it. However, the sludge produced, particularly during aerobic biodegradation processes, can provide major and costly disposal challenges. The ability of sludge to adsorb certain organic components, particularly toxic heavy metals, may worsen this. The advantages of biological treatment include the microbial modifications of organic chemicals and the possibility for heavy metal adsorption by bacteria. (Sengil and Ozacar 2006).

Aerobic Treatment

BOD, phosphorus, and nitrogen are removed from wastewater using the aerobic treatment process (Janet Joshiba et al. 2019). According to Janet Joshiba et al. (2019), this procedure involves subjecting industrial wastewater to an oxidation reaction in the presence of oxygen, which reduces the hazardous bacteria that are present in DWW. This highly difficult procedure can only be carried out in the presence of oxygen, which microbes utilise to adapt organic materials. These materials are subsequently converted into biomass, CO_2, and H_2O (Janet Joshiba et al. 2019, Shete and Shinkar 2013). Suspended growth and attached growth, sometimes known as biofilm processes, are the two primary biological treatment methods. The organic matter is broken down by microbes during suspended growth in the presence of oxygen (Wang et al. 2009). Due to its operational flexibility, increased reaction rate, and need for less area, the treatment of wastewater using biofilm offers a number of advantages over the suspended growth technique (Andersson 2009). A challenging step in bio-filtration is the development and production of biofilm on the surface of the bio carrier. The features of the media surface, the environmental factors, the parameters of the water, and the type of microorganisms all affect attachment strength. Up to the point at which biomass is lost due to sloughing and degradation, biomass will continue to expand. When the substrate is rich, the biofilm develops and accumulates more quickly than the rate of detachment, but when the substrate is poor, sloughing counteracts the biofilm's rapid growth (Holá et al. 2006).

Wetland Treatment

Microbial communities for wastewater processing are a part of a natural process of wastewater treatment (Vymazal 2005a, Slavov 2017). Aquatic and terrestrial plants are set up in an artificial system. According to Ahmad et al. (2019b), this treatment is regarded as a sustainable wastewater treatment since it accomplishes the same task as conventional treatment (efficient wastewater treatment). To remove toxins from industrial wastewater, these plants' roots are in touch with water over an extended length of time (Schwantes et al. 2019). Hence, these plants have phytoremediation property and helps to reduce damage caused by metals in the environment (Schwantes et al. 2019, Ding et al. 2015). As reported by

Ibekwe et al. (2003), the wetland process removed 25% of the total nitrogen and reduced ammonium by 16%.

The DWW (full form) is treated aerobically in a wetland system. The removal effectiveness of BOD5 in the aerobic pond was 85% at 20°C, and facultative wetlands are used for high load wastewater treatment. BOD and COD are removed with high removal efficiencies in the wetland process, whereas phosphorus and nitrogen are removed at lesser rates. Based on Verhoeven et al. (1999) study, nitrogen removal is often greater than 50% while phosphorus removal is typically low. When wastewater retention time was increased from 2 to 7 days during the treatment process, total nitrogen levels decreased in the unplanted wetlands by 12 to 14% and in the planted wetlands by 48 to 75%. According to Tanner et al. (1995), the overall rise from 12–36% of phosphorus removal efficiency in unplanted wetlands and from 37–74% in planted wetlands have been reported. The *Typha domingensis* was chosen for a study because of its high production and effectiveness in removing nutrients (Schierano et al. 2017). A study by Slavov (2017) reported that the effluents from the aerobic wetland process are of a good quality and remove COD, TSS, BOD5, organic N, and total coliforms with an efficiency of more than 90%.

Wetland treatment systems employ straightforward procedures that include self-sustaining microbial populations, affordable treatment, minimal initial investment, and low operating and maintenance costs (Ibekwe et al. 2003). It also uses less energy and does not require skilled labour, according to a study on vertical flow constructed wetlands. Disadvantages of this system involves the necessity for a sizable surface installation as well as the possibility of ground water contamination. The removal of NH_3 from anaerobic wetland is constrained, while the amount of phosphorus from PO_4^{3-} is constrained in aerobic processes (Slavov 2017).

Constructed wetlands is another term for green treatment technology. Municipal, industrial, home, and agricultural wastewater have all been successfully treated using this approach. Pathogens, nutrients, and organic materials are all effectively removed by this procedure (Wu et al. 2015). A built wetland's main goal is to eliminate microbiological pollutants. According to the literature, faecal and total coliforms were removed in 95–99% of the 60 created wetlands with macrophyte, and faecal streptococci were removed in 80–90% of the cases (Vymazal 2005b).

Surface flow wetlands or free water surface constructed wetlands which comprises of small basins and a device that regulates the depth of the water. The roots of macrophytes require 20 to 40 cm of soil to maintain them. They successfully get rid of suspended solids. Filtration and sedimentation are used to efficiently remove suspended particles, while nitrification and denitrification are used to efficiently remove nitrogen. In aerobic zones, nitrifying bacteria oxidises ammonium, while in anoxic zone denitrifying bacteria convert nitrate into free nitrogen or nitrous oxide. As a result of the limited interaction between the water column and the soil during this phase, the removal of phosphorus is somewhat sluggish (Vymazal 2005a). A small number of wetland species can be found in subsurface flow produced wetland because water flows through the porous media generated by sand and gravel (Miklas 2011). However, a larger surface area is provided by macrophytes including yellow flag and canna lilies for the associated development. They avoid clogging due to a vertical flow system, and they guard against ice during the winter. They help stabilise the surface of wetland beds and improve the conditions for physical filtration (Brix 1994).

Sequencing Batch Reactor (SBR)

The sequencing batch reactor (SBR), which works on the principle of fill-and-draw activated sludge for the removal of unwanted components, appears to be the most promising technology for treating dairy wastewater. The cleared supernatant is removed from the tank after the mixed liquor has undergone the necessary treatment and been allowed to settle. The primary difference between the SBR and the classic continuous flow activated sludge system is that the former executes activities such as equalization, aeration, and sedimentation over time rather than in a spatial sequence (Bandpi et al. 2004, Mahvi et al. 2008). Single-batch reactor could be used to accomplish equalisation, clarity and aeration. Clarifiers and other equipment can therefore be eliminated to reduce overall costs (Loo et al. 1990). SBR treatment has a cheap investment cost since it is frequently run at a higher TS (Neczaj et al. 2008). Operating factors that impact the effectiveness of SBR treatment include hydraulic retention time (HRT), organic loading, pH, dissolved oxygen concentration, temperature, mixed liquor-suspended solid (MLSS), and wastewater strength (Rao et al. 2002). An industrial milk factory's effluent was treated using a bench-scale aerobic Sequencing Batch Reactor (SBR). A tiny bubble air diffuser provided oxygen to the reactor. While the COD content ranged from 400 to 2500 mg/l, the removal effectiveness of COD was over 90%. MLVSS was around 3000 mg/l, whereas the reactor's ideal dissolved oxygen was between 2 and 3 mg/l (Bandpi et al. 2004). Neczaj et al. (2008) used two SBR to treat dairy wastewaters with a starting COD content of 400–7500 mg/L to study the impact of operational parameter on treatment efficiency. It took 19 hours of aeration time and 2 hours of anoxic phase to remove 98.6% COD and 80.1% TKN. Reduced HRT or increased organic loading both result in lower removal efficiency.

Biofiltration

The microorganisms are attached to a porous substrate during the biofiltration process, which aids in the breakdown of foreign contaminants found in wastewater. By forming a biofilm, the bacteria grow on the medium surface which engulf medium sized particles and float on the liquid phase (Srivastava et al. 2008). The media in the filter bed is made up of relatively inert materials that offer a lot of surface area for attaching biomass and supplying nutrients. Since a steady supply of nutrients and organic materials are necessary for the bio filtration method to function effectively and consistently, microorganisms play a major role in determining how effective it is. Additionally, some chemoautotrophic microbes are used as an inorganic source of energy. The major factors involved in the effectiveness of biofiltration include temperature, pH, the starting concentration of harmful pollutants, and oxygen concentration (Van Loosdrecht et al. 1990).

Activated Sludge Process

It is recommended for the reduction of nitrogen, ammonia, and carbon compounds in dairy wastewater (Renou et al. 2008). Activated sludge process (ASP) is a very effective technology for treating dairy wastewater (Umejewska K 2017). Protozoa and microorganisms, which aid in the degradation of organic materials and nutrients found in wastewater, are the key components of ASP. According to Tocchi et al. (2012), these bacteria may eat the organic materials that are suspended in dairy effluent and turn them into activated sludge, which is then recycled and removed. The wastewater flows through an aerated tank in this process, and the generated activated sludge is taken out by using clarifier from wastewater (Patel et al. 2016). According to Tocchi et al. (2012), this technique may also be used to extract complex compounds such as lactose, oils, lipids, and proteins. However, aerobic sludge process is less effective than the aerobic granular sludge process. Aerobic granular sludge process has increased settle ability, high biomass recovery, high shock resistance, and lower toxicity, that is more effective than the aerobic sludge process (Vashi et al. 2018). According to Lateef et al. (2013), COD and BOD were removed 96% effectively in just 5 days.

Anaerobic Treatment

Anaerobic treatment mostly uses biological processes to handle high strength wastewater. Anaerobic digestion is a process that uses bacteria to stabilise organic materials by converting them to biogas (composed of methane and CO_2), biomass, and inorganic products in the absence of oxygen. However, 95% of organic load in the waste water stream is reportedly capable of being converted into biogas (methane and carbon dioxide). Methane can be used as a source of heat or electricity when dairy waste is processed anaerobically. Another benefit of this procedure is that less sludge is produced, which lessens disposal issues with sludge, requires less space, uses less energy, and produces no unpleasant odours. Due to lesser energy requirement, anaerobic system is more effective than aerobic systems. Hence, they are typically considered to be more cost-effective for the biological stabilisation of dairy wastes (Rao et al. 2002).

Up Flow Anaerobic Sludge Blanket (USAB)

The treatment of dairy and food wastes has seen the largest application of UASB reactors (Gavala et al. 1999). Sludge blankets, gas-solid separators, influent-distribution systems, and effluent-withdrawal systems make up the fundamental parts of a typical UASB reactor. The influent in the UASB reactor is dispersed at the bottom and moves in an upward flow (Tchobanoglous et al. 2003). According to Gavala et al. (Gavala et al. 1999), steady-state operation utilising wastewater with a COD influent of 2050 mg/L resulted in a 90% COD reduction at an 0.031 kg COD/m^{3}d organic loading rate. Fats are present in dairy effluent, and their inhibitory effects on anaerobic treatment prevent their efficient removal from occurring quickly (Vidal et al. 2000). According to some writers, this issue could be resolved by pre-treating lipids with enzymatic hydrolysis. According to Cammarota et al.'s research (2001), treating dairy wastewaters in a UASB reactor with high amounts of fat (868 mg/L) removes more than 50% of the COD and up to 944 mg/L of volatile suspended solids (VSS). However, when treated in the same UASB reactor, pre-treatment with lipases from *Penicillium restrictum* demonstrated a 90% higher COD removal efficiency. Numerous authors have noted that the buildup of organic matter in continuous UASB reactors has caused a reduction in efficiency. Monali et al.'s study (2009) evaluated the UASB reactor's effectiveness in treating dairy wastewater and estimated its biogas production efficiency. According to their findings, the removal efficiencies of 87.06% COD, 94.50% BOD, and 56.54% TSS, and the 0.28 to 0.43 VFA/Alkalinity ratio 6.9 to 7.1 was discovered to be the pH range of the reactor. Methane accounts for 75% of the daily average biogas output, which is 179.35 m^3/day.

Hybrid Anaerobic Digesters

For the treatment of dairy wastewater, a lab-scale mesophilic hybrid anaerobic digester that combines a UASB and an anaerobic filter design was employed. A 0.354 m^3 CH_4/kg methane production and removal of COD at 1.7 days of HRT was produced by the anaerobic digester (Calli et al. 2002). A lab-scale upflow hybrid reactor used to treat acidic cheese whey produced removal efficiencies of over 95% after two days of HRT and an OLR of around 11 kg COD/(m^{3}day) (Koshta et al. 2010). With two reactors (A and B), Koshta et al. (2010) assessed the efficiency of the anaerobic fixed film reactor for the dairy effluent. At 3 days HRT and 2 days HRT, respectively, reactors A and B reached their optimal COD removal efficiencies of 87.69 and 89.42, respectively. 20.46, 28.63, and 28.75 l/day of biogas were produced during HRT of 5, 3, and 2 days in reactor A, and 20.00, 28.83, and 30.83 l/day in reactor B. At a 3 day HRT, reactors A and B produced the highest levels of biogas and methane, respectively, at 0.73 and 0.43 l/g COD removal and 0.80 and 0.52 l/g COD removal. TDS removal was concurrently 58.56 and 60.09 percent in reactors A and B, respectively.

Anaerobic Sequencing Batch Reactors (ASBR)

The excellent quality of dairy effluents is also obtained with ASBRs. The lab-scale ASBR system investigation was conducted to remove 62% COD and 75% BOD, using a substrate comprising non-fat dry milk at 58°C during a 6 h hydraulic retention time (HRT) (Dugba and Zhang 1999). In a different lab-scale investigation, volatile organic substances are removed by thermophilic ASBR and mesophilic ASBR systems from dairy effluent between 26 and 44% and 26 and 50%, respectively (Nadais et al. 2005). High mass transfer rates, little clogging, high load handling capacity, high biomass, and minimal space requirements are all benefits of AFBR (Burghate et al. 2013, Shete and Shinkar 2013). According to Janet Joshiba et al. (2019), the greatest COD elimination effectiveness was 84%.

Sludge Treatment

Each stage in treatment process produces sludge, which is collected in thickening tanks and chemicals are added to aid in the further solid particle's aggregation. Sludge is uniform, virtually odourless, and black in colour. Sludge needs to further decompose since it includes a lot of organic material to lessen the bad smell. The organic materials are finally poured into a digester, where they are converted into carbon dioxide and methane, a very low quantity of ammonia, hydrogen gas and hydrogen sulphide under anaerobic circumstances. The basic ingredients of a digester, which may be used as fuel for heating, are methane and carbon dioxide. Sludge has a lot of moisture—between 94 and 73%. Therefore, the centrifuge decanter is used to discharge nearly 1/8th of the initial volume solid phase and is the most effective method for dewatering it. Sludge that has been dewatered can subsequently be dumped as garbage or used as fertiliser (Abdulrazak 2007).

Integrated Technologies

Bio-electrochemical systems (BES) combine biological and electrochemical processes to clean waste and generate power, hydrogen, and other useful products at the same time. As a result, BES is crucial to the circular economy (Jung et al. 2020). ''Power'' and ''Pollution'', respectively, are two of the most significant limitations of electrochemical and biological processes that can be addressed by this system. Electrons are transferred from the anode to the cathode as a result of electrochemically active microorganisms reducing organic materials at the anode (Khan et al. 2017b, Do et al. 2020, Mishra et al. 2020, Sambavi et al. 2020). The quantity of electrons collected and transported to the anode affects output energy efficiency (Ivase et al. 2020). Microbial Fuel cells (MFC), a form of BES in which bacteria in the anode section of the cell create electricity by oxidising organic substances in the waste, are the most sophisticated and extensively researched BES. Microbes produce both protons and electrons during this process. Bacteria move electrons from the cathode to the anode. The generated protons should be transferred *via* the proton exchange membrane from the anode chamber to the cathode chamber, where they should combine with oxygen, electrons, and other elements to form water (Kondaveeti et al. 2019, Mishra et al. 2020, Do et al. 2020, Yadav et al. 2020).

WHOLE CELL OR CELL FREE APPROACHES FOR VALORIZATION OF DAIRY WASTE

A crucial answer in the present day for waste to wealth is the production of value-added products from milk, whey, and other dairy products waste. Even though dairy byproducts are polluting, they are a rich source of lactose and other nutrients that can promote microbial growth and the production of bio-products.

Biofuels

Biomethane

By-products from the dairy sector with high COD levels can be utilised as a potential feedstock for anaerobic digestion (AD) to produce biogas (Kozowski et al. 2019). According to Wainaina et al. (2019), AD primarily consists of three stages: (1) hydrolysis of the complex molecule into simpler soluble materials by different bacteria including *Peptococcus*, *Acetivibrio* and *Bacillus*; (2) synthesis of hydrogen and volatile fatty acids (VFAs) via *Propionibacterium*, *Clostridium,* and *Butyribacterium*; (3) synthesis of biogas from hydrogen and VFAs by certain archaea and methanogenic bacteria (Wainaina et al. 2019). In a circular economy, produced biogas might be used as a thermal energy source for steam generation throughout the cheese-making process. The wastewater treatment unit can utilise this biogas as a source of electrical power (Comino et al. 2009). Additionally, the anaerobic digesters' solid and liquid fractions can be used as soil fertilisers (Kavacik and Topaloglu 2010). Hence, the anaerobic digesters reduce energy consumption and also control wastewater pollution caused by the dairy sector (Asunis et al. 2020, Hublin et al. 2012).

Whey's high COD, low pH, and lack of alkalinity make it difficult to handle anaerobically (Kavacik and Topaloglu 2010). A classic continuous stirred tank reactor (CSTR)-type biogas plant may not be able to anaerobically digest dairy wastes having low COD values including milk and yoghurt (Karadag et al. 2015). However, the UASB reactor system has the highest COD removal efficiency in comparison to several other reactors. Additionally, the dairy effluents' carbohydrates promote the development of bacteria that produce acids but suppress the methanogens that lead to reactor instability. However, co-digesting dairy effluents with other wastes, such as manure, cow dung, goat straw bedding, poultry, or animal wastes, has addressed this issue (Lovato et al. 2019, Szaja and Montusiewicz 2019, Fernández-Rodrguez et al. 2021). Research has demonstrated that these combinations can stimulate the growth of methanogens and microbial synergism while keeping the carbon/nitrogen (C/N) ratio and biogas generation in a healthy state (Gelegenis et al. 2007, Comino et al. 2009, 2012). To reduce instability and increase biomethane production, the acetogenesis and methanogenesis processes can be separated into a two-stage anaerobic digestion that requires a shorter hydraulic retention time (HRT) than standard anaerobic digestion (Asunis et al. 2020).

Bioethanol

Compared to lignocellulosic feedstocks, lactose-containing whey and expired milk have an advantage since they don't require a specific pretreatment (Akbas et al. 2014 Pescuma et al. 2015, Ryan and Walsh 2016, Okamoto et al. 2019), which is currently encouraged for the production of bioethanol. Microorganisms that can naturally digest lactose can be employed to produce ethanol from lactose-containing dairy waste, or galactosidase can be used to perform lactose hydrolysis (Guimares et al. 2010). Recently, a β-galactosidase functional in a wide pH range of acidic to neutral pH has been characterized (Thakur et al. 2022). Acid active β-galactosidase is useful in hydrolysing lactose present in acidic whey samples. Examples of yeast that can produce ethanol by lactose fermentation include *Kluyveromyces marxianus* and *K. lactis* (You et al. 2017, Sampaio et al. 2020, Yamahata et al. 2020, Tesfaw et al. 2021).

Biodiesel

Methyl ethyl esters, often known as biodiesel, are one of the substitutes for fossil fuels and are obtained from plants, microalgae, or oily fungal biomass (Mathew et al. 2021, Kumar et al. 2020a, b). Growth of fungal (*Mortierella isabellina, Thamnidium elegans, Mucor* sp., and *Fusarium*) and microalgal (*Chlorella, Anabaena, Scenedesmus, Chlamydomonas,* and *Acutodesmus*, among others) species in dairy wastes can promote wastewater treatment, oily biomass production, and bioenergy generation (Chokshi et al. 2016). According to study conducted by Espinoza-Gonzalez et al. (2014) on *Chlorella protothecoides* strain that produces the most lipids from dairy by-product stream, when fermented in batches produces 42% lipids and 20.5% lipids in fed-batch. Other oleaginous microbes that may manufacture lipids from whey include *Apiotrichumcurvatum, Scenedesmus sp.*, and *Cryptococcus laurentii* strains (Pescuma et al. 2015, Borges et al. 2016). Acid-catalyzed transesterification and esterification are the two major steps that may be evaluated for the manufacture of biodiesel from oily biomass that contains oleic and palmitic acids (Vicente et al. 2009, Chan et al. 2018).

Production of Organic Acid(s)

Important microbiological compounds with acidic characteristics are organic acids including acetic, oxalic, propionic, succinic, lactic, citric, valeric, and capric acids which are the most prevalent organic acids (Pandey et al. 2016, Louasté and Eloutassi 2020). Recently, *Lactobacillus buchneri*, an acid-tolerant lactic acid bacteria, was immobilised for the production of acetic acid from whey powder (Veeravalli and Mathews 2018). The analysis revealed that acetate and propylene glycol were produced from 57% of the lactose in the whey powder. According to Veeravalli and Mathews (2018), *L. buchneri* has the capacity to produce acetic acid industrially and might benefit from metabolic or evolutionary engineering.

Yalcin et al. (2009) reported that *Y. lipolytica* NBRC 1658 and *Y. lipolytica* 57 yeast strains produced highest quantity of citric acid, 32.65 g/L and 49.23 g/L, respectively from whey supplemented with fructose. Hence, the citric acid made by the dairy sector may be utilised again to make dairy products, which might lower overall costs in the cheese manufacturing unit and can also be used during ice cream preparation as an acidifier, emulsifier, and an antioxidant (Soccol et al. 2006, Yalcin et al. 2009). Some attempts have been made to produce propionic acid, an organic acid which is used as a herbicide and also as a cellulosic polymer (Ahmadi et al. 2017) from lactose-containing whey using continuous *Propionibacterium acidipropionici* ATCC 4875 fermentation in anaerobic conditions (Gupta and Srivastava 2001). The effluent from the dairy sector was combined with *Propionibacterium acidipropionici* to create propionic acid together with other VFAs (acetic and butyric) (Atasoy et al. 2020a, Atasoy and Cetecioglu 2021). The use of the anaerobic sequencing batch reactor (ASBR) has proven to be more effective for the production of bioenergy and VFA from cheese whey. Additionally, Atasoy et al. (2020b) reported that nearly 80% propionic acid as a major organic acid was produced by *Desulfovibrionaceae* and *Synergistaceae* from cheese processing wastewater. The result of anaerobic metabolism in the Krebs cycle is succinic acid. There is a large demand for this acid's production since it is widely employed in many different sectors (detergent, cosmetics, herbicides, fungicides, etc.) (Zeikus et al. 1999). Succinic acid is produced chemically, which is not only costly but also increases greenhouse gas emissions. This is the major justification for the current consideration of whey valorisation as a green method of producing this acid. The study reported higher production of succinic acid (0.58 g h^{-1} L^{-1}) at pH 6.80 from 50 gL^{-1} cheese whey (Wan et al. 2008). The microbial synthesis of succinic acid from whey under anaerobic circumstances was investigated by Louasté and Eloutassi (2020). *Actinobacillus succinogenes* growth on whey had a succinic acid output and productivity in batch fermentation of 62.1% and 0.81 g/L/h, respectively. The succinate production may be increased by using an immobilisation system with batch, repeated-batch, and continuous batches (0.89–1.09 g h^{-1} L^{-1}) (Uysal and Hamamc 2021). Additionally, acetic acids and formic acids were byproducts that were found at 48 hours of whey fermentation (Wan et al. 2008).

Enzymes Production

By-products from the dairy sector are one of the inexpensive sources for the synthesis of enzymes (Ryan and Walsh 2016). Furthermore, developing enzymes from these substrates might be a long-term solution to the dairy industry's pollution concern. Enzymes such as β-galactosidase, α-amylase, manganese peroxidase, protease, penicillin acylase, penicillin amidase, polygalacturonase, cutinase, inulinase, and lipase can be produced by some yeasts and bacteria using lactose waste (Knob et al. 2019, Phukon et al. 2020, 2022, Kumari et al. 2021, Padhi et al. 2023). As an alternative, a novel *Paracoccus marcusii* strain that can make β-galactosidase from lactose has been discovered (Kalathinathan and Muthukaliannan 2021). *Anoxybacillus beppuensis* manufacture a thermostable α-amylase from agro-dairy wastes (combination of whey and sugarcane bagasse) (Jabeen et al. 2019). When *Serratia marcescens* was cultivated on whey, it was examined if the different strains generated extracellular metalloprotease (Romero et al. 2001, Panesar and Kennedy 2012). Protease enzyme can be produced by *Aspergillus* and *Mucor* fungi by growing them in whey (El-Shora and Metwally 2008).

IMMOBILIZATION OF ENZYMES FOR VALUE ADDED PRODUCTS

Sweeteners and Lactose-free Milk Production

Lactose is not a particularly useful sweetener, although it may be converted to one by an enzymatic process. Recently, immobilised *β*-D-galactosidase was used to hydrolyse lactose and ultra-high temperature pasteurised (UHT) milk lactose for synthesis of modified Arabic gum-based hydrogel (Wolf et al. 2018). The ability of a bi-enzymatic system, which consists of glucose isomerase from *Streptomyces rubiginosus* and galactosidase from *Kluyveromyces lactis*, to increase lactose's sweetness was examined (Lorenzen et al. 2013). In the study reported, more than 70% of the lactose in cheese whey was hydrolysed, along with a reduction in protein leaching, using a cross-linked gelatin and ca-alginate encapsulated spheres of *β*-galactosidase from *K. lactis* (Mörschbächer et al. 2016).

Galacto-oligosaccharides Production

The intriguing prebiotic component galacto-oligosaccharides (GOS) can be produced from lactose in dairy waste streams. Eskandarloo and Abbaspourrad (2018) reported that covalently linked *β*-galactosidase onto 3-aminopropyl tri-ethoxysilane-functionalized glass beads produced GOS from whey permeates in a continuous packed-bed bioreactor. The greatest GOS yield recorded was 39.3%, and the highest lactose conversion rate was 56.4% after the second cycle of reusability, according to the results. The improved transglycosidation activity of *A. oryzae β*-galactosidase immobilised on silica nanoparticles was evaluated by Banjanac et al. (2016). However, significant increase from free enzyme 30 $gL^{-1}h^{-1}$ to 90 $gL^{-1}h^{-1}$ assisted by galactosidase nanoparticles have been reported.

Production of Prebiotics

Lactulose disaccharide, a prebiotic consisting of fructose and galactose monosaccharides, is considered to work as a therapeutic agent in the treatment of constipation and encephalopathy. Immobilization technology may affect and improve the final yields of lactulose. However, immobilization technology used to produce lactulose from whey lactose has proven to improve yield of the disaccharide, as reported in *K. lactis* immobilized *β*-galactosidase on silica gel (Song et al. 2013a, b). De Albuquerque et al. (2018) studied biosynthesis of lactulose using cheese whey by immobilized *K. lactis* β-galactosidase enzyme on glutaraldehyde-activated chitosan using cheese whey.

A trisaccharide of glucose, galactose, and fructose—lactosucrose prebiotic—was reported to be synthesised using covalently linked β-galactosidase to chitosan microspheres. Additionally, immobilization process increases thermal stability of enzyme which hence helps in the production of lactosucrose due to higher operation temperature condition around 50°C and 64°C. The study showed no loss of lactosucrose yield using microspheres-insolubilized biocatalyst up to 30 repeated cycles. Up to 30 cycles of use, the biocatalyst without microspheres showed no reduction in lactosucrose output. These findings show the adaptability of the engineered galactosidase enzyme system in multiple cycles of reuse to synthesise a variety of high-value products depending on the processing conditions. Further, an enzymatic process for the transformation of lactose into 4-galactosyl kojibiose has been developed by employing dextransucrose (Lata et al. 2018, Chourasia et al. 2022).

Single-cell Protein

Numerous studies have concentrated on the microbial fermentation for the production of single-cell proteins (SCPs), which can be used as human foods or animal feeds because dairy wastes are suitable substrates for bioprocessing (Ryan and Walsh 2016). *K. fragilis*, a lactose-using yeast that is generally recognised as GRAS, was employed in the majority of trials (Ghaly and Kamal 2004, Ghaly et al. 2005, Schultz et al. 2006). Three yeasts, *K. marxianusvar. marxianus, K. marxianus var. lactis*, and *Candida pintolopepsii*, were utilised in the 1950s in France to produce SCP from deproteinized whey (Panesar and Kennedy 2012, Pescuma et al. 2015). *Debaryomyces robertsiae*, also known as *Wingearobertsii*, was one of the yeast strains used to produce SCP; it had a maximum output of 89% and a greater protein content (Sandhu and Waraich 1983). Baker's yeast (*S. cerevisiae*) may also be grown from whey in a batch fermentation system with the addition of nutrients and lactose hydrolysis by immobilised lactase (Castillo 1990). Microorganisms that are useful for producing protein-rich biomass include filamentous fungus. The most popular fungal genera, known as GRAS (*Aspergillus, Rhizopus, Neurospora, Monascus,* and *Fusarium*), are employed primarily to produce fungal biomass from various substrate types (Souza Filho et al. 2019, Sar et al. 2020a, b, Sar et al. 2021). Through the cultivation of *Aspergillus oryzae* and *Neurospora* intermedia strains, the dairy wastes potential as a substrate was used for the production of fungal biomass and reported 11 g/L of fungal biomass generation of *A. oryzae* and *Neurospora*, both detected from expired milk (Thunuguntla et al. 2018). The high fat materials including cream bioconversion is challenging. Hence, two-step fermentation system is used for this, with *Neurospora* intermedia (a lactose consumer) coming after *A. oryzae* (a fat-degrader) in the first reactor. According to certain reports (Mahboubi et al. 2017a, b), the biomass generated by this fungus includes 30–40% protein and may be utilised as animal feed. The growth of diverse fungal species can also produce oil-rich biomass from dairy wastes like whey (Kakkad et al. 2015, Chan et al. 2018, Ibarruri and Hernández 2019, Chan et al. 2020). However, biodiesel and nutraceuticals production can be done using 24–32% oily fungal biomasses (Chan et al. 2018, 2020).

Biopolymers Production

The plastic and mechanical characteristics of some biopolymers, such PLA and polyhydroxyalkanoates (PHAs), are equivalent to those of plastics derived from fossil fuels (Gironi and Piemonte 2011). For the bioconversion of dairy industry wastes to PHA, the following three potential methods were suggested: (a) Few microorganisms, including *Thermus thermophilus*, *Pseudomonas hydrogenovora*, *Hydrogenophagapseudoflava*, *Bacillus megaterium*, or engineered strains bearing lactose-degrading genes (e.g., engineered *Cupravidus necator*), directly convert lactose to PHA; (b) Bioconversion of whey through enzymatic or chemical route to break lactose into galactose and glucose; (c) PHA are produced through two-step bioconversion which involves fermentation of lactose into lactic acid and then PHA production from lactic acid through PHA-producing microorganisms (Mollea et al. 2013, Ryan and Walsh 2016). Use of mixed microbial cultures (MMC) has been reported to increase the yield of PHAs from complicated substrates (Zikmanis et al. 2020). Exopolysaccharides (EPS) can also be created by bioprocessing using dairy waste. These extracellular polymeric ingredients are frequently employed to enhance food texture in food items (Iliev et al. 2001, Abedin et al. 2023, Phukon et al. 2024).

Fuel Cells

Surprisingly, trichloroethene (TCE), a hazardous solvent, may dissolve when exposed to dairy wastewater, cheese whey and cheese wastes/wastewaters (Sekar et al. 2019, Veeramani et al. 2020). This is because these substances have the capacity to operate as an electron donor. The solubility of TCE was reported to be improved by using 10% w/w whey powder by factor of three in laboratory and six in the field trials when added to the prepared microcosms. It shows that dairy wastes may be used extensively for a variety of tasks (Macbeth et al. 2006). Using microbial fuel cells (MFCs) to produce power is one of the other uses for dairy wastes in this regard. Microorganisms in the anode of an MFC process organic wastes anaerobically, generating electrons that are transmitted to the external circuit. Microorganisms in the anode of the MFC anaerobically oxidise organic wastes, producing electrons that are then carried to the external circuit (Tremouli et al. 2013, Ghasemi et al. 2017). The potential for producing power from diluted cheese whey in a two-chamber mediator-less microbial fuel cell (MFC) was examined in this work. By normalizing to the anodic electrode's geometric area of 13.8 cm^2, the greatest power density achieved with diluted cheese whey was 18.4 mW/m^2, which translates to an MFC voltage of 0.23 V and a current density of 80 mA/m^2 (Antonopoulou et al. 2010). Additionally, Ghasemi et al. (2017) used two distinct techniques to handle concentrated whey and cheese whey. Immobilized *L. bulgaricus* cells were used to ferment whey in an immobilised cell reactor (ICR), and produced 10.7 gL^{-1} lactic acid which is the highest concentration reported with 0.125/h dilution rate with 50 gL^{-1} lactose. The COD removal rates for whey-fed and concentrated whey-fed MFCs were 95% and 86%, respectively. Additionally, 288.12 mW/m^2 power density was reported for concentrated whey-fed

MFC (Ghasemi et al. 2017). However, similar study using reactor effluent done by Wenzel et al. (2017) showed 439 mW/m^2 power density that was 1000 times higher than raw cheese whey. Enzymatic fuel cells (EFC) may also be utilised to produce electricity from whey in addition to MFC. Hence, this system based on redox-active enzymes is considered as a cutting-edge technique for producing green energy. In a research, cheese whey was used to generate electricity using an EFC that had cellobiose dehydrogenase immobilised. This enzyme is recognised as a special anodic enzyme that converts lactose, cellodextrins, or cellobiose into bioelectrical energy and has a strong ability to transport electrons. When cheese whey was exhibited in the EFC under ideal circumstances, a 1.839 W/cm^2 maximum power density was attained (Choi et al. 2020). Hence, studies showed that contemporary biological therapies are more effective than conventional ones at recovering and turning wastes into goods and energy sources.

Pigments

Whey is a cheap raw material that can be used to make pigments, which are coloured substances that are present in microbial cells, plant and animal. Carotenoids pigments are employed in the food, feed, pharmaceutical, and cosmetic sectors as antioxidants, anticancer agents, colouring, and flavouring additives. Carotenoid synthesis in whey ultrafiltrate was achieved by co-cultivating lactose-positive yeasts or bacteria with carotenoid producers (Panesar and Kennedy 2012, Frengova et al. 2004). The study conducted reported the highest yield of carotenoid from deproteinized whey by a filamentous fungi *Blakesleatrispora*, in the shake flask, airlift bioreactor, and bubble column bioreactor (Varzakakou et al. 2010a, b, Roukas et al. 2015). *Desmodesmus sp.* L2B Bold, a freshwater microalgal strain, produced 0.5 g/mL carotenoids when cultivated on cheese whey (Bonett et al. 2020). Additionally, *Cellulosimicrobium* bacteria demonstrated the capacity to produce carotenoids from whey. However, when 60% w/v amount was raised of whey in the medium, this strain generated 17.5 mg/L carotenoids (Bakhtiyari et al. 2020).

Biosurfactants

Biosurfactants are biological substitutes for surfactants that exhibit more selective and focused action in challenging environments. Surfactants, which are used in the detergent, oil, food, pharmaceutical, and cosmetic sectors, are often created chemically. *Bacillus*, *Acinetobacter*, and *Pseudomonas* create the majority of known biosurfactants, including emulsan, surfactin, and rhamnolipids (Decesaro et al. 2020). Recently, Decesaro et al. (2020) synthesised biosurfactants from whey by growing *Bacillus methylotrophicus* and *Bacillus pumilus* on the ultrafiltered whey permeate.

CONCLUSION AND FUTURE PERSPECTIVES

With the current SDGs and low carbon or zero carbon processing strategies, dairy waste has considerably untapped potential as the biogenic waste carbon feedstocks for the production of various chemicals and other valuable products. To recycle and treat huge amounts of biogenic waste on an industrial scale, significant research has to be taken into account from supply chain aspects, technologies for valorisation, production chains and designs, and finally consumer perspective. The dairy waste fits efficiently into circular biobased economy model as the potential feedstock rather than eliminating it as waste side stream. Although decades of research is accessible through scientific literature, significant additional efforts are needed to overcome the technical challenges in various unit procedures, like initial pretreatments, recovery of fermentable sugars, identifying the potent microbial strain for production of desired products, scale-up strategies, and finally techno-economic and life cycle assessment to understand and develop an economically and commercially feasible biorefinery business model. In this context, diary waste side streams are an efficient feedstock for resource recovery and an important aspect for sustainability.

REFERENCES

Abdulrazak, A. 2007. Dairy industry effluent treatment: anaerobic treatment of whey in stirred batch reactor. Scribd.

Abedin, M.M., Chourasia, R., Phukon, L.C., Sarkar, P., Ray, R.C., Singh, S.P., et al. 2023. Lactic acid bacteria in the functional food industry: biotechnological properties and potential applications. Critical Reviews in Food Science and Nutrition. 1–19. DOI: 10.1080/10408398.2023.2227896.

Adesra, A., Srivastava, V.K. and Varjani, S. 2021. Valorization of dairy wastes: integrative approaches for value-added products. Indian Journal of Microbiology. 61(3): 270–278.

Ahmad, T., Aadil, R.M., Ahmed, H., Ur Rahman, U., Soares, B.C.V., Souza, S.L.Q., et al. 2019a. Treatment and utilization of dairy industrial waste: a review. Trends in Food Science and Technology. 88: 361–372. https://doi.org/10.1016/j.tifs.2019.04.003

Ahmad, M., Yang, Q., Zhang, Y., Ling, J., Sajjad, W., Qi, S., et al. 2019b. The distinct response of phenanthrene enriched bacterial consortia to different PAHs and their degradation potential: a mangrove sediment microcosm study. Journal of Hazardous Materials. 380: 120863. doi: 10.1016/j.jhazmat.2019.120863

Ahmadi, N., Khosravi-Darani, K. and Mortazavian, A.M. 2017. An overview of biotechnological production of propionic acid: from upstream to downstream processes. Electronic Journal of Biotechnology. 28: 67–75. https://doi.org/10.1016/j.ejbt.2017.04.004.

Akbas, M.Y., Sar, T. and Ozcelik, B. 2014. Improved ethanol production from cheese whey, whey powder, and sugar beet molasses by "*Vitreoscilla* hemoglobin expressing". *Escherichia coli*. Bioscience, Biotechnology, and Biochemistry. 78(4): 687–694. https://doi.org/10.1080/09168451.2014.896734.

Andersson, S. 2009. Characterization of bacterial biofilms for wastewater treatment. Technology. Retrieved from http://kth.diva-portal.org/smash/get/diva2:209486/FULLTEXT01.

Antonopoulou, G., Stamatelatou, K., Bebelis, S. and Lyberatos, G. 2010. Electricity generation from synthetic substrates and cheese whey using a two-chamber microbial fuel cell. Biochemical Engineering Journal. 50(1): 10–15. https://doi.org/10.1016/j.bej.2010.02.008.

Asunis, F., De Gioannis, G., Dessì, P., Isipato, M., Lens, P.N.L., Muntoni, A., et al. 2020. The dairy biorefinery: integrating treatment processes for cheese whey valorisation. Journal of Environmental Management. 276: 111240. https://doi.org/10.1016/j.jenvman.2020.111240.

Atasoy, M., Eyice, Ö. and Cetecioglu, Z. 2020a. Volatile fatty acid production from semi-synthetic milk processing wastewater under alkali pH: the pearls and pitfalls of microbial culture. Bioresource Technology. 297: 122415. https://doi.org/10.1016/j.biortech.2019.122415.

Atasoy, M., Eyice, O. and Cetecioglu, Z. 2020b. A comprehensive study of volatile fatty acids production from batch reactor to anaerobic sequencing batch reactor by using cheese processing wastewater. Bioresource Technology. 311: 123529. https://doi.org/10.1016/j.biortech.2020.123529.

Atasoy, M. and Cetecioglu, Z. 2021. Bioaugmentation as a strategy for tailor-made volatile fatty acid production. Journal of Environmental Management. 295: 113093. https://doi.org/10.1016/j.jenvman.2021.113093.

Awasthi, M.K., Paul, A., Kumar, V., Sar, T., Kumar, D., Sarsaiya, S., et al. 2022. Recent trends and developments on integrated biochemical conversion process for valorization of dairy waste to value-added bioproducts: a review. Bioresource Technology. 344: 126193.

Bakhtiyari, A.S., Etemadifar, Z. and Borhani, M.S. 2020. Use of response surface methodology to enhance carotenoid pigment production from *Cellulosimicrobium* strain AZ. SN Applied Sciences. 2(12): 2096. https://doi.org/10.1007/s42452-020-03549-6.

Balks, M.R., Bond, W.J. and Smith, C.J. 1998. Effects of sodium accumulation on soil physical properties under an effluent-irrigated plantation. Australian Journal of Soil Research. 36(5): 821–830.

Bandpi, A.M. and Bazari, H. 2004. Biological treatment of dairy wastewater by sequencing batch reactor. Iranian Journal of Environmental Health Science and Engineering. 1: 65–69.

Banjanac, K., Carević, M., Ćorović, M., Milivojević, A., Prlainović, N., Marinković, A., et al. 2016. Novel β-galactosidase nanobiocatalyst systems for application in the synthesis of bioactive galactosides. RSC Advances. 6(99): 97216–97225.

Barnett, J.W., Robertson, S.L. and Russell, J.M. 2007. Environmental issues in dairy processing. Environment Portfolio, New Zealand Dairy Research Institute, Private Bag 11029, Palmerston North.

Bilal, M. and Iqbal, H.M.N. 2019. Sustainable bioconversion of food waste into high-value products by immobilized enzymes to meet bio-economy challenges and opportunities–a review. Food Research International. 123: 226–240. https://doi.org/10.1016/j.foodres.2019.04.066.

Bonett, J.E.A., de Sousa Geraldino, P., Cardoso, P.G., de Freitas Coelho, F., Duarte, W.F. and Souza, S.L.Q. 2020. Isolation of freshwater microalgae and outdoor cultivation using cheese whey as substrate. Biocatalysis and Agricultural Biotechnology. 29: 101799. https://doi.org/10.1016/j.bcab.2020.101799.

Britz, J.T., van Schalwyk, C. and Hung, Y.T. 2006. Treatment of dairy processing wastewaters. pp. 1–28. *In*: Wang, L.K., Hung, Y.T., Lo, H.H. and Yapijakis, C. (eds). Waste Treatment in the Food Processing Industry. Boca Raton, FL, USA: CRC Press.

Brix, H. 1994. Functions of macrophytes in wetlands. Water Science and Technology. 29(4): 71–78.

Bruzzese, E., Volpicelli, M., Squaglia, M., Tartaglione, A. and Guarino, A. 2006. Impact of prebiotics on human health. Digestive and Liver Disease. 38: S283–S287.

Burke, N., Zacharski, K.A., Southern, M., Hogan, P., Ryan, M.P. and Adley, C.C. 2018. The dairy industry: process, monitoring, standards, and quality. pp. 1–25. *In*: Díaz, A.V. and García-Gimeno, R.M. (eds). Descriptive Food Science. IntechOpen: Rijeka, Croatia.

Calli, B. and Yukselen, M.A. 2002. Anaerobic treatment by a hybrid reactor. Environmental Engineering Science. 19: 143–150.

Cameron, K.C., Di, H.J., Reijnen, B.P.A., Li, Z., Russell, J.M. and Barnett, J.W. 2002. Fate of nitrogen in dairy factory effluent irrigated onto land. New Zealand Journal of Agricultural Research. 45(2): 207–216.

Cammarota, M.C., Teixeira, G.A. and Freire, D.M. 2001. Enzymatic pre-hydrolysis and anaerobic degradation of wastewaters with high-fat contents. Biotechnology Letters. 23(20): 1591–1595.

Carvalho, F., Prazeres, A.R. and Rivas, J. 2013. Cheese whey wastewater: characterization and treatment. Science of the Total Environment. 445–446: 385–396. https://doi.org/10.1016/j.scitotenv.2012.12.038

Castillo, F.J. 1990. Lactose metabolism by yeasts. Yeast Biotechnology and Biocatalysis. 297–320.

Chan, L.G., Cohen, J.L., Ozturk, G., Hennebelle, M., Taha, A.Y., de Moura Bell, L.N., et al. 2018. Bioconversion of cheese whey permeate into fungal oil by *Mucor circinelloides*. Journal of Biological Engineering. 12(1): 25. https://doi.org/10.1186/s13036-018-0116-5.

Chan, L.G., Dias, F.F.G., Saarni, A., Cohen, J., Block, D., Taha, A.Y., et al. 2020. Scaling up the bioconversion of cheese whey permeate into fungal oil by *Mucor circinelloides*. Journal of the American Oil Chemists' Society. 97(7): 703–716. https://doi.org/10.1002/aocs.12372.

Choi, H.S., Yang, X., Kim, D.S., Yang, J.H., Han, S.O., Park, C., et al. 2020. Power generation from cheese whey using enzymatic fuel cell. Journal of Cleaner Production. 254. 120181. https://doi.org/10.1016/j.jclepro.2020.120181.

Chokshi, K. Pancha, I. Ghosh, A. and Mishra, S. 2016. Microalgal biomass generation by phycoremediation of dairy industry wastewater: An integrated approach towards sustainable biofuel production. Bioresource Technology. 221: 455–460, 10.1016/j.biortech.2016.09.070

Choudhury, P., Ray, R.N., Bandyopadhyay, T.K., Basak, B., Muthuraj, M. and Bhunia, B. 2021. Process engineering for stable power recovery from dairy wastewater using microbial fuel cell. International Journal of Hydrogen Energy. 46(5): 3171–3182.

Chourasia, R., Phukon, L.C., Abedin, M.M., Padhi, S., Singh, S.P. and Rai, A.K. 2022. Whey valorization by microbial and enzymatic bioprocesses for the production of nutraceuticals and value-added products. Bioresource Technology Reports. 19: 101144. https://doi.org/10.1016/j.biteb.2022.101144.

Cohen, B. 1931. The bacterial culture as an electrical half-cell. Journal of Bacteriology. 21(1): 18–19.

Comino, E., Rosso, M. and Riggio, V. 2009. Development of a pilot scale anaerobic digester for biogas production from cow manure and whey mix. Bioresource Technology. 100(21): 5072–5078. https://doi.org/10.1016/j.biortech.2009.05.059.

Comino, E., Riggio, V.A. and Rosso, M. 2012. Biogas production by anaerobic co-digestion of cattle slurry and cheese whey. Bioresource Technology. 114: 46–53. https://doi.org/10.1016/j.biortech.2012.02.090.

Cristian, O. 2010. Characteristics of the untreated wastewater produced by food industry. Analele Universitatii din Oradea, Fascicula Protectia Mediului. 15: 709–714.

de Albuquerque, T.L., Gomes, S.D.L., D'Almeida, A.P., Fernandez-Lafuente, R., Gonçalves, L.R.B. and Rocha, M.V.P. 2018. Immobilization of β-galactosidase in glutaraldehyde-chitosan and its application to the synthesis of lactulose using cheese whey as feedstock. Process Biochemistry. 73: 65–73.

Decesaro, A., Machado, T.S., Cappellaro, Â.C., Rempel, A., Margarites, A.C., Reinehr, C.O., et al. 2020. Biosurfactants production using permeate from whey ultrafiltration and bioproduct recovery by membrane separation process. Journal of Surfactants and Detergents. 23(3): 539–551. https://doi.org/10.1002/jsde.12399.

Deepak, S. 2014. Treatment of dairy wastewater by electrocoagulation using aluminum electrodes and settling, filtration studies. International Journal of ChemTech Research. 6(1): 591–599.

Deshpande, D.P., Patil, P.J. and Anekar, S.V. 2012. Biomethanation of dairy waste. Research Journal of Chemical Sciences. 2(4): 35–39.

Ding, J., Zhao, F., Cao, Y., Xing, L., Liu, W., Mei, S., et al. 2015. Cultivation of microalgae in dairy farm wastewater without sterilization. International Journal of Phytoremediation. 17(3): 222–227. https://doi.org/10.1080/15226514.2013.876970.

Do, M.H., Ngo, H.H., Guo, W., Chang, S.W., Nguyen, D.D., Varjani, S., et al. 2020. Microbial fuel cell-based biosensor for online monitoring wastewater quality: a critical review. Science of the Total Environment. https://doi.org/10.1016/j.scitotenv.2019.135612.

Duarte, L.S., da Natividade Schöffer, J., Lorenzoni, A.S.G., Rodrigues, R.C., Rodrigues, E. and Hertz, P.F. 2017. A new bioprocess for the production of prebiotic lactosucrose by an immobilized β-galactosidase. Process Biochemistry. 55: 96–103.

Dugba, P. and Zhang, R. 1999. Treatment of dairy wastewater with two-stage anaerobic sequencing batch reactor systems: thermophilic versus mesophilic operations. Bioresource Technology. 68: 225–233.

El-Shora, H.M. and Metwally, M.A.A. 2008. Production, purification, and characterization of proteases from whey by some fungi. Annals of Microbiology. 58(3): 495–502. https://doi.org/10.1007/BF03175548.

Eskandarloo, H. and Abbaspourrad, A. 2018. Production of galacto-oligosaccharides from whey permeate using *β*-galactosidase immobilized on functionalized glass beads. Food Chemistry. 251: 115–124.

Espinosa-Gonzalez, I., Parashar, A. and Bressler, D.C. 2014. Heterotrophic growth and lipid accumulation of *Chlorella protothecoides* in whey permeate, a dairy by-product stream, for biofuel production. Bioresource Technology. 155: 170–176. https://doi.org/10.1016/j.biortech.2013.12.028.

Frengova, G., Simova, E. and Beshkova, D. 2004. Use of whey ultrafiltrate as a substrate for production of carotenoids by the yeast *Rhodotorula rubra*. Applied Biochemistry and Biotechnology. 112(3): 133–141. https://doi.org/10.1385/abab:112:3:133.

Gavala, N., Kopsinis, H., Skiadas, I.V., Stamatelatou, K. and Lyberatos, G. 1999. Treatment of dairy wastewater using an upflow anaerobic sludge blanket reactor. Journal of Agricultural Engineering Research. 73: 59–63.

Geetha Devi, M., Joefel, J.D. and Feroz. S. 2012. Dairy wastewater treatment using low molecular weight crab shell chitosan. Journal of the Institution of Engineers (India): Series E. 93(1): 9–14. https://doi.org/10.1007/s40034-012-0005-2.

Gelegenis, J., Georgakakis, D., Angelidaki, I. and Mavris, V. 2007. Optimization of biogas production by co-digesting whey with diluted poultry manure. Renewable Energy. 32(13): 2147–2160. https://doi.org/10.1016/j.renene.2006.11.015.

Ghasemi, M., Ahmad, A., Jafary, T., Azad, A.K., Kakooei, S., Wan Daud, W.R., et al. 2017. Assessment of immobilized cell reactor and microbial fuel cell for simultaneous cheese whey treatment and lactic acid/electricity production. International Journal of Hydrogen Energy. 42(14): 9107–9115. https://doi.org/10.1016/j.ijhydene.2016.04.136.

Gironi, F. and Piemonte, V. 2011. Bioplastics and petroleum-based plastics: strengths and weaknesses. Energy Sources Part A. 33(21): 1949–1959.

Gottschall, N., Boutin, C., Crolla, A., Kinsley, C. and Champagne, P. 2007. The role of plants in the removal of nutrients at a constructed wetland treating agricultural (dairy) wastewater, Ontario, Canada. Ecological Engineering. 29(2): 154–163. https://doi.org/10.1016/J.ECOLENG.2006.06.004.

Guimarães, P.M.R., Teixeira, J.A. and Domingues, L. 2010. Fermentation of lactose to bio-ethanol by yeasts as part of integrated solutions for the valorization of cheese whey. Biotechnology Advances. 28(3): 375–384. https://doi.org/10.1016/j.biotechadv.2010.02.002.

Halliwell, D.J., Barlow, K.M. and Nash, D.M. 2001. A review of the effects of wastewater sodium on soil physical properties and their implications for irrigation systems. Australian Journal of Soil Research. 39(6): 1259–1267.

Hernández-Hernández, O., Muthaiyan, A., Moreno, F.J., Montilla, A., Sanz, M.L. and Ricke, S.C. 2012. Effect of prebiotic carbohydrates on the growth and tolerance of Lactobacillus. Food Microbiology. 30(2): 355–361.

Holá, V., Růžička, F. and Votava, M. 2006. The dynamics of staphylococcus epidermis biofilm formation in relation to nutrition, temperature and time. Scripta Medica Facultatis Medicae Universitatis Brunensis Masarykianae. 79(3): 169–174.

Hublin, A., Zokić, T.I. and Zelić, B. 2012. Optimization of biogas production from co-digestion of whey and cow manure. Biotechnology and Bioprocess Engineering. 17(6): 1284–1293. https://doi.org/10.1007/s12257-012-0044-z.

Ibarruri, J. and Hernández, I. 2019. Valorization of cheese whey and orange molasses for fungal biomass production by submerged fermentation with *Rhizopus* sp. Bioprocess and Biosystems Engineering. 42(8): 1285–1300. https://doi.org/10.1007/s00449-019-02127-4.

Ibekwe, A.M., Grieve, C.M. and Lyon, S.R. 2003. Characterization of microbial community in constructed dairy wetland. Applied and Environmental Microbiology. 69(9): 5060–5069. https://doi.org/10.1128/AEM.69.9.5060.

Iliev, I., Radoilska, E., Ivanova, I. and Enikova, R. 2001. Biosynthesis of exopolysaccharides by two strains of *Lactobacillus bulgaricus* in whey-based media. Mededelingen (Rijksuniversiteit te Gent. Fakulteit van de Landbouwkundige en Toegepaste Biologische Wetenschappen) 66(3b): 511–516.

Ivase, T.J., Nyakuma, B.B., Oladokun, O., Abu, P.T. and Hassan, M.N. 2020. Review of the principal mechanisms, prospects, and challenges of bioelectrochemical systems.

Environmental Progress and Sustainable Energy. 39(1): 1–9. https://doi.org/10.1002/ep.13298.

Jabeen, F., Hussain, A., Younis, T., Manzoor, M. and Samiullah, K. 2019. Isolation of thermophilic *Anoxybacillus beppuensis* JF84 and production of thermostable amylase utilizing agro-dairy wastes. Environmental Progress and Sustainable Energy. 38(2): 417–423. https://doi.org/10.1002/ep.12991.

Janet Joshiba, G., Senthil Kumar, P., Femina, C.C., Jayashree, E., Racchana, R. and Sivanesan, S. 2019. Critical review on biological treatment strategies of dairy wastewater. Desalination and Water Treatment. 160: 94–109. https://doi.org/10.5004/dwt.2019.24194.

Jung, S., Lee, J., Park, Y. and Kwon, E.E. 2020. Bioelectrochemical systems for a circular bioeconomy. Bioresource Technology. 300: 122748. https://doi.org/10.1016/j.biortech.2020.122748.

Kadier, A., Jain, P., Lai, B., Kalil, M.S., Kondaveeti, S., Alabbosh, K.F.S., et al. 2020. Biorefinery perspectives of microbial electrolysis cells (MECs) for hydrogen and valuable chemicals production through wastewater treatment. Biofuel Research Journal. 7: 1128–1142. https://doi.org/10.18331/BRJ2020.7.3.1.

Kakkad, H., Khot, M., Zinjarde, S., Ravi Kumar, A., Ravi Kumar, V. and Kulkarni, B.D. 2015. Conversion of dried *Aspergillus candidus* mycelia grown on waste whey to biodiesel by *in situ* acid transesterification. Bioresource Technology. 197: 502–507. https://doi.org/10.1016/j.biortech.2015.07.118.

Kalathinathan, P. and Muthukaliannan, G.K. 2021. Characterization of a potential probiotic strain Paracoccus marcusii KGP and its application in whey bioremediation. Folia Microbiologica. 1–12. https://doi.org/10.1007/s12223-021-00886-w.

Karadag, D., Köroğlu, O.E., Ozkaya, B. and Cakmakci, M. 2015. A review on anaerobic biofilm reactors for the treatment of dairy industry wastewater. Process Biochemistry. 50(2): 262–271. https://doi.org/10.1016/j.procbio.2014.11.005.

Kasmi, M. 2018. Biological processes as a promoting way for both treatment and valorization of dairy industry effluents. Waste and Biomass Valorization. 9(2): 195–209.

Kavacik, B. and Topaloglu, B. 2010. Biogas production from co-digestion of a mixture of cheese whey and dairy manure. Biomass and Bioenergy. 34(9): 1321–1329. https://doi.org/10.1016/j.biombioe.2010.04.006.

Khan, M.H., Hugar, S.C. and Khatmode, N.P. 2017a. Removal of dairy wastewater characteristics by using natural and chemical coagulant. Asian Journal of Convergence in Technology. 3(1): 2–4.

Khan, N., Danish, K.M., Khan Sultana, S., Khan, M.Z. and Ahmad, A. 2017b. Bioelectrochemical systems for transforming waste to energy nishat. pp. 111–128. *In*: Oves, D.M., Khan, M.Z. and Ismail, I.M.I. (eds). Modern Age Environmental Problem and their Remediation. Springer, Cham. https://doi.org/10.1007/978-3-319-64501-8.

Knob, A., Izidoro, S.C., Lacerda, L.T., Rodrigues, A. and de Lima, V.A. 2020. A novel lipolytic yeast *Meyerozyma guilliermondii*: efficient and low-cost production of acid and promising feed lipase using cheese whey. Biocatalysis and Agricultural Biotechnology. 24: 101565. https://doi.org/10.1016/j.bcab.2020.101565.

Kolhe, A.S., Ingale, S.R. and Bhole, R.V. 2009. Effluent of Dairy Technology. Shodh, Samiksha aur Mulyankan (International Research Journal)—ISSN-0974-2832 Vol. II, Issue-5.

Kondaveeti, S., Mohanakrishna, G., Lee, J-K. and Kalia, V.C. 2019. Methane as a substrate for energy generation using microbial fuel cells. Indian Journal of Microbiology. 59: 121–124. https://doi.org/10.1007/s12088-018-0765-6.

Koshta, V. 2010. Study on performance evaluation for anaerobic fixed film reactor for treatment of dairy effluent using commercial packing media. M.Tech Thesis. Anand Agricultural University, Anand. 96–99.

Kothari, R., Kumar, V., Pathak, V.V. and Tyagi, V.V. 2017. Sequential hydrogen and methane production with simultaneous treatment of dairy industry wastewater: bioenergy profit approach. International Journal of Hydrogen Energy. 42(8): 4870–4879.

Kozłowski, K., Pietrzykowski, M., Czekała, W., Dach, J., Kowalczyk-Juśko, A., Jóźwiakowski, K., et al. 2019. Energetic and economic analysis of biogas plant with using the dairy industry waste. Energy. 183: 1023–1031. https://doi.org/10.1016/j.energy.2019.06.179.

Kumar Dasarahally-Huligowda, L., Goyal, M.R. and Ansar Rasul Suleria, H. (Eds). 2019. Nanotechnology Applications in Dairy Science: Packaging, Processing, and Preservation. Apple Academic Press. https://doi.org/10.1201/9780429425370

Kumar, A.K., Sharma, S., Dixit, G., Shah, E. and Patel, A. 2020a. Techno-economic analysis of microalgae production with simultaneous dairy effluent treatment using a pilot-scale high volume V-shape pond system. Renewable Energy. 145: 1620–1632.

Kumar, M., Rathour, R., Gupta, J., Pandey, A., Gnansounou, E. and Thakur, I.S. 2020b. Bacterial production of fatty acid and biodiesel: opportunity and challenges. pp. 21–49. *In*: Kumar, R.P., Gnansounou, E., Raman, J.K. and Baskar, G. (eds). Refining Biomass Residues for Sustainable Energy and Bioproducts. Academic Press. https://doi.org/10.1016/B978-0-12-818996-2.00002-8.

Kumari, M., Padhi, S., Sharmas S., Phukon, L.C., Singh, S.P. and Rai, A.K. 2021. Biotechnological potential of psychrophilic microorganisms as the source of cold-active enzymes in food processing applications. 3 Biotech. 11(11): 479.

Kushwaha, J.P., Srivastava, V.C. and Mall, I.D. 2011. An overview of various technologies for the treatment of dairy wastewaters. Critical Reviews in Food Science and Nutrition. 51(5): 442–452. https://doi.org/10.1080/10408391003663879.

Lata, K., Sharma, M., Patel, S.N., Sangwan, R.S. and Singh, S.P. 2018. An integrated bioprocess for production of functional biomolecules utilizing raw and by-products from dairy and sugarcane industries. Bioprocess and Biosystems Engineering. 41(8): 1121–1131. https://doi.org/10.1007/s00449-018-1941-0.

Li, Y., Qi, C., Zhang, Y., Li, Y., Wang, Y., Li, G., et al. 2021. Anaerobic digestion of agricultural wastes from liquid to solid state: performance and environ-economic comparison. Bioresource Technology. 332: 125080.

Loo, K.V. and Liao, P.H. 1990. Treatment of poultry processing wastewater using sequencing batch reactors. Canadian Agricultural Engineering. 32: 299–305.

Lorenzen, P.C., Breiter, J., Clawin-Rädecker, I. and Dau, A. 2013. A novel bi-enzymatic system for lactose conversion. International Journal of Food Science and Technology. 48(7): 1396–1403.

Louasté, B. and Eloutassi, N. 2020. Succinic acid production from whey and lactose by *Actinobacillus succinogenes* 130Z in batch fermentation. Biotechnology Reports. 27: e00481. https://doi.org/10.1016/j.btre.2020.e00481.

Lovato, G., Albanez, R., Triveloni, M., Ratusznei, S.M. and Rodrigues, J.A.D. 2019. Methane production by co-digesting vinasse and whey in an AnSBBR: effect of mixture ratio and feed strategy. Applied Biochemistry and Biotechnology. 187(1): 28–46. https://doi.org/10.1007/s12010-018-2802-7.

Macbeth, T.W., Nelson, L., Rothermel, J.S., Wymore, R.A. and Sorenson, K.S. 2006. Evaluation of whey for bioremediation of trichloroethene source zones. Bioremediation Journal. 10(3): 115–128. https://doi.org/10.1080/10889860600952339.

Mahboubi, A., Ferreira, J.A., Taherzadeh, M.J. and Lennartsson, P.R. 2017a. Production of fungal. Fermentation. 3(4): 48. https://doi.org/10.3390/fermentation3040048.

Mahboubi, A., Ferreira, J.A., Taherzadeh, M.J. and Lennartsson, P.R. 2017b. Value-added products from dairy waste using edible fungi. Waste Management. 59: 518–525. https://doi.org/10.1016/j.wasman.2016.11.017.

Mahvi, A.H. 2008. Sequential batch reactor: a promising technology in wastewater treatment. Iranian Journal of Environmental Health Science and Engineering. 5: 79–90.

Mathew, G.M., Raina, D., Narisetty, V., Kumar, V., Saran, S., Pugazhendi, A., et al. 2021. Recent advances in biodiesel production: challenges and solutions. Science of the Total Environment. 794: 148751. https://doi.org/10.1016/j.scitotenv.2021.148751.

Miklas, S. 2011. Green Energy and Technology. Wetland Systems Storm :Water Management Control.

Mishra, A., Kumar, M., Bolan, N.S., Kapley, A., Kumar, R. and Singh, L. 2021. Multidimensional approaches of biogas production and up-gradation: opportunities and challenges. Bioresource Technology. 338: 125514.

Mishra, B., Varjani, S., Kumar, G., Awasthi, M.K., Awasthi, S.K. and Sindhu, R. 2020. Microbial approaches for remediation of pollutants: innovations, future outlook, and challenges. Energy and Environmental. 1–19. https://doi.org/10.1177/0958305X19896781.

Mollea, C., Marmo, L. and Bosco, F. 2013. Valorisation of cheese whey, a by-product from the dairy industry. Food Industry. IntechOpen. doi:10.5772/53159.

Monali, G., Dhoble, R.M. and Pittule, A.P. 2009. Biomethanation of dairy waste water through UASB at mesophilic temperature range. International Journal of Advanced Engineering Science and Technology. 8: 1–9.

Mörschbächer, A.P., Volpato, G. and Souza, C.F.V.D. 2016. *Kluyveromyces lactis* β-galactosidase immobilization in calcium alginate spheres and gelatin for hydrolysis of cheese whey lactose. Ciência Rural. 46(5): 921–926.

Nadais, H., Capela, I., Arroja, L. and Duarte, A. 2005. Optimum cycle time for intermittent UASB reactors treating dairy wastewater. Water Research. 39(8): 1511–1518.

Nanotechnology Applications in Dairy Science: Packaging, Processing, and ... - Google Books." n.d. https://books.google.com.eg/books? hl=en&lr=&id=XU-fDwAAQBAJ&oi=fnd&pg=PT17&dq=polyunsaturated+fatty+acids+omega+3+nanocapsules+for+ruminants&ots=vZABkya5rh&sig=LN10atyS8gOpWPx3h8k18GN-r3o&redir_esc=y#v=onepage&q&f=false.

Neczaj, E., Kacprzak, M., Kamizela, T., Lach, J. and Okoniewska, E. 2008. Sequencing batch reactor system for the co-treatment of landfill leachate and dairy wastewater. Desalination. 222: 404–409.

Okamoto, K., Nakagawa, S., Kanawaku, R. and Kawamura, S. 2019. Ethanol production from cheese whey and expired milk by the brown rot fungus *Neolentinus lepideus*. Fermentation. 5(2). https://doi.org/10.3390/fermentation5020049.

Overcash, M.R. and Pal, D. 1979. Design of land treatment systems for industrial wastes—theory and practice. Ann Arbor Science Publishers. Michigan. 702.

Pachpute, A.A., Kankal, S.B. and Jadhav, M.V. 2014. Use of artificial wetland for treatment of dairy industry wastewater for analysis of BOD and COD. International Journal of Scientific Engineering and Research (IJSER). 2(6): 38–41.

Padhi, S., Sarkar, P., Abedin, M.M., Sirohi, R., Singh, S.P. and Rai, A.K. 2023. Microbial enzymes and metabolites in health care: an overview. pp. 3–23. *In*: Sirohi, R. Rai, A.K., Vandenberghe, L.P.deS. and Parameswaran, B. (eds). Microbial Enzymes and Metabolites for Health and Well-Being. CRC Press. https://doi.org/10.1201/9781003369295.

Pandey, A., Negi, S. and Soccol, C.R. 2016. Current Developments in Biotechnology and Bioengineering: Production, Isolation and Purification of Industrial Products. Elsevier Science.

Panesar, P.S. and Kennedy, J.F. 2012. Biotechnological approaches for the value addition of whey. Critical Reviews in Biotechnology. 32(4): 327–348. https://doi.org/10.3109/07388551.

Pescuma, M., de Valdez, G.F. and Mozzi, F. 2015. Whey-derived valuable products obtained by microbial fermentation. Applied Microbiology and Biotechnology. 99(15): 6183–6196. https://doi.org/10.1007/s00253-015-6766-z.

Phukon, L.C., Chourasia, R., Kumari, M., Godan, T.K., Sahoo, D., Parameswaran, B., et al. 2020. Production and characterisation of lipase for application in detergent industry from a novel *Pseudomonas helmanticensis* HS6. Bioresource Technology. 309: 123352.

Phukon, L.C., Chourasia, R., Padhi, S., Abedin, M.M., Godan, T.K., Parameswaran, B., et al. 2022. Cold-adaptive traits identified by comparative genomic analysis of a lipase-producing *Pseudomonas* sp. HS6 isolated from snow-covered soil of Sikkim Himalaya and molecular simulation of lipase for wide substrate specificity. Current Genetics. 68(3–4): 375–391.

Phukon, L.C., Chourasia, R., Abedin, M.M., Sharma, S., Singh, S.P. and Rai, A.K. 2024. Exopolysaccharides from extremophiles and their potential applications. pp. 139–163. *In*: Bhatia, S.K., Panesar, P.S. and Mehariya, S. (eds). Microbial Exopolysaccharides. CRC Press.

Rao, M. and Bhole, A.G. 2002. Removal of organic matter from dairy industry wastewater using low-cost adsorbents. Journal of Indian Chemical Engineering, Section A. 44: 25–28.

Romero, F.J., Salas, J.A., Quirós, L.M., García, L.A. and Dıáz, M. 2001. Production, purification and partial characterization of two extracellular proteases from *Serratia marcescens* grown in whey. Process Biochemistry. 36(6): 507–515. https://doi.org/10.1016/S0032-9592(00)00221-1.

Roukas, T., Varzakakou, M. and Kotzekidou, P. 2015. From cheese whey to carotenes by *Blakeslea trispora* in a bubble column reactor. Applied Biochemistry and Biotechnology. 175(1): 182–193. https://doi.org/10.1007/s12010-014-1260-0.

Ryan, M.P. and Walsh, G. 2016. The biotechnological potential of whey. Reviews in Environmental Science and Biotechnology. 15(3): 479–498. https://doi.org/10.1007/s11157-016-9402-1.

Sambavi, S.M., Vishali, S., Varjani, S. and Mullai, P. 2020. Electricity generation in a microbial fuel cell using iron oxide nanoparticles. Indian Journal of Experimental Biology. 58: 571–577.

Sampaio, F.C., de Faria, J.T., da Silva, M.F., de Souza Oliveira, R.P. and Converti, A. 2020. Cheese whey permeate fermentation by *Kluyveromyces lactis*: a combined approach to wastewater treatment and bioethanol production. Environmental Technology. 41(24): 3210–3218. https://doi.org/10.1080/09593330.2019.1604813.

Sandhu, D.K. and Waraich, M.K. 1983. Conversion of cheese whey to single-cell protein. Biotechnology and Bioengineering. 25(3): 797–808. https://doi.org/10.1002/bit.260250315.

Sar, T., Ferreira, J.A. and Taherzadeh, M.J. 2020a. Bioprocessing strategies to increase the protein fraction of *Rhizopus oryzae* biomass using fish industry side streams. Waste Management. 113: 261–269. https://doi.org/10.1016/j.wasman.2020.06.005.

Sar, T., Ozturk, M., Taherzadeh, M.J. and Ferreira, J.A. 2020b. New insights on protein recovery from olive oil mill wastewater through bioconversion with edible filamentous fungi. Processes. 8(10). https://doi.org/10.3390/pr8101210.

Sar, T., Ferreira, J.A. and Taherzadeh, M.J. 2021. Conversion of fish processing wastewater into fish feed ingredients through submerged cultivation of *Aspergillus oryzae*. Systematic Microbiology and Biomanufacturing. 1(1): 100–110. https://doi.org/10.1007/s43393-020-00009-5.

Sar, T., Harirchi, S., Ramezani, M., Bulkan, G., Akbas, A., Pandey, A., et al. 2022. Potential utilization of dairy industries by-products and wastes through microbial processes: a critical review. Science of the Total Environment. 810. https://doi.org/10.1016/j.scitotenv.2021.152253.

Sarkar, B., Chakrabarti, P.P., Vijaykumar, A. and Kale, V. 2006. Wastewater treatment in dairy industries—possibility of reuse. Desalination. 195: 141–152.

Schierano, M.C., Maine, M.A. and Panigatti, M.C. 2017. Dairy farm wastewater treatment using horizontal subsurface flow wetlands with typha domingensis and different substrates. Environmental Technology. 38(2): 192–198. https://doi.org/10.1080/09593330.2016.1231228.

Schultz, N., Chang, L., Hauck, A., Reuss, M. and Syldatk, C. 2006. Microbial production of single-cell protein from deproteinized whey concentrates. Applied Microbiology and Biotechnology. 69(5): 515–520. https://doi.org/10.1007/s00253-005-0012-z.

Schumann, C. 2002. Medical, nutritional and technological properties of lactulose. An update. European Journal of Nutrition. 41(1): i17–i25.

Schwantes, D., Gonçalves, A.C., Schiller, A.d.P., Manfrin, J., Campagnolo, M.A. and Veiga, T.G. 2019. *Salvinia auriculata* in post-treatment of dairy industry wastewater. International Journal of Phytoremediation. International Journal of Phytoremediation, 21(13): 1368–1374. https://doi.org/10.1080/15226514.2019.1633260.

Sekar, A.D., Jayabalan, T., Muthukumar, H., Chandrasekaran, N.I., Mohamed, S.N. and Matheswaran, M. 2019. Enhancing power generation and treatment of dairy wastewater in microbial fuel cell using Cu-doped iron oxide nanoparticles decorated anode. Energy. 172: 173–180. https://doi.org/10.1016/j.energy.2019.01.102.

Sengil, A. and Ozacar, M. 2006. Treatment of dairy wastewaters by electrocoagulation using mild steel electrodes. Journal of Hazardous Materials. 137(2): 1197–1205.

Shete, B. and Shinkar, N.P. 2013. Dairy industry wastewater sources, characteristics and its effects on environment. International Journal of Current Engineering and Technology. 3(5): 1611–1615. http://inpressco.com/category/ijcet.

Silvério, S.C., Macedo, E.A., Teixeira, J.A. and Rodrigues, L.R. 2015. Perspectives on the biotechnological production and potential applications of lactosucrose: a review. Journal of Functional Foods. 19: 74–90.

Slavov, A.K. 2017. General characteristics and treatment possibilities of dairy wastewater—a review. Food Technology and Biotechnology. 55(1): 14–28. https://doi.org/10.17113/ftb.55.01.17.4520.

Soccol, C.R., Vandenberghe, L.P.S., Rodrigues, C. and Pandey, A. 2006. New perspectives for citric acid production and application. Food Technology and Biotechnology. 44(2).

Song, Y.S., Lee, H.U., Park, C. and Kim, S.W. 2013a. Batch and continuous synthesis of lactulose from whey lactose by immobilized β-galactosidase. Food Chemistry. 136(2): 689–694.

Song, Y.S., Lee, H.U., Park, C. and Kim, S.W. 2013b. Optimization of lactulose synthesis from whey lactose by immobilized β-galactosidase and glucose isomerase. Carbohydrate Research. 369: 1–5.

Souza Filho, P.F., Andersson, D., Ferreira, J.A. and Taherzadeh, M.J. 2019. Mycoprotein: environmental impact and health aspects. World Journal of Microbiology and Biotechnology. 35(10): 147. https://doi.org/10.1007/s11274-019-2723-9.

Szaja, A. and Montusiewicz, A. 2019. Enhancing the co-digestion efficiency of sewage sludge and cheese whey using brewery spent grain as an additional substrate. Bioresource Technology. 291: 121863. https://doi.org/10.1016/j.biortech.2019.121863.

Szczodrak, J. 2000. Hydrolysis of lactose in whey permeate by immobilized β-galactosidase from *Kluyveromyces fragilis*. Journal of Molecular Catalysis B: Enzymatic. 10(6): 631–637.

Tanner, C.C., Clayton, J.S. and Upsdell, M.P. 1995. Effect of loading rate and planting on treatment of dairy farm wastewaters in constructed wetlands—II. Removal of Nitrogen and Phosphorus. Water Research. 29(1): 27–34. https://doi.org/10.1016/0043-1354(94)00140-3.

Tchobanoglous, G., Burton, F.L. and Stensel, H.D. 2003. Waste Water Engineering: Treatment and Reuse. Fourth Edition. Metcalf and Eddy Inc., Tata McGraw-Hill Publishing Company Limited, New Delhi, India. 39–41.

Technological Approaches for Novel Applications in Dairy Processing. 2018. Technological Approaches for Novel Applications in Dairy Processing. https://doi.org/10.5772/68053.

Tesfaw, A., Oner, E.T. and Assefa, F. 2021. Evaluating crude whey for bioethanol production using non-*Saccharomyces* yeast, *Kluyveromyces marxianus*. SN Applied Sciences. 3(1): 42. https://doi.org/10.1007/s42452-020-03996-1.

Thakur, M., Rai, A.K. and Singh, S.P. 2022. An acid-tolerant and cold-active β-galactosidase potentially suitable to process milk and whey samples. Applied Microbiology and Biotechnology. 106: 3599-3610. doi: 10.1007/s00253-022-11970-7.

Thunuguntla, R., Mahboubi, A., Ferreira, J.A. and Taherzadeh, M.J. 2018. Integration of membrane bioreactors with edible filamentous fungi for valorization of expired milk. Sustainability. 10(6). https://doi.org/10.3390/su10061940.

Tremouli, A., Antonopoulou, G., Bebelis, S. and Lyberatos, G. 2013. Operation and characterization of a microbial fuel cell fed with pretreated cheese whey at different organic loads. Bioresource Technology. 131: 380–389. https://doi.org/10.1016/j.biortech.2012.12.173.

Umiejewska, K. 2017. Biological anaerobic-aerobic treatment of dairy wastewater. *In*: Proceedings of the 15th International Conference on Environmental Science and Technology. 5 p. Rhodes, Greece, 31 August–2 September 2017. p5.

Usmani, Z., Sharma, M., Gaffey, J., Sharma, M., Dewhurst, R.J., Moreau, B., et al. 2022. Valorization of dairy waste and by-products through microbial bioprocesses. Bioresource Technology. 346: 126444.

Uysal, U. and Hamamcı, H. 2021. Succinic acid production from cheese whey via fermentation by using alginate immobilized *Actinobacillus succinogenes*. Bioresource Technology Reports. 16: 100829. https://doi.org/10.1016/j.biteb.2021.100829.

Varzakakou, M., Roukas, T., Kotzekidou, P. and Giamoustaris, A. 2010a. Effect of non-ionic surfactants and beta-ionone on the morphology of *Blakeslea trispora* and carotenoids production from cheese whey in submerged aerobic growth: a statistical approach. Food Biotechnology. 24(2): 197–214. https://doi.org/10.1080/08905436.2010.482455.

Varzakakou, M., Roukas, T., Papaioannou, E., Kotzekidou, P. and Liakopoulou-Kyriakides, M. 2010b. Autolysis of *Blakeslea trispora* during carotene production from cheese whey in an airlift reactor. Preparative Biochemistry and Biotechnology. 41(1): 7–21. https://doi.org/10.1080/10826068.2010.525436.

Veeramani, V., Rajangam, K. and Nagendran, J. 2020. Performance of cobalt oxide/carbon cloth composite electrode in energy generation from dairy wastewater using microbial fuel cells. Sustainable Environmental Research. 30(1): 16. https://doi.org/10.1186/s42834-020-00058-4.

Veeravalli, S.S. and Mathews, A.P. 2018. Exploitation of acid-tolerant microbial species for the utilization of low-cost whey in the production of acetic acid and propylene glycol. Applied Microbiology and Biotechnology. 102(18): 8023–8033. https://doi.org/10.1007/s00253-018-9174-3.

Verhoeven, J.T.A. and Meuleman, A.F.M. 1999. Wetlands for wastewater treatment: opportunities and limitations. Ecological Engineering. 12: 5–12.

Vicente, G., Bautista, L.F., Rodríguez, R., Gutiérrez, F.J., Sádaba, I., Ruiz-Vázquez, R.M., et al. 2009. Biodiesel production from biomass of an oleaginous fungus. Biochemical Engineering Journal. 48(1): 2227. https://doi.org/10.1016/j.bej.2009.07.014.

Vidal, G., Carvalho, A., Mendez, R. and Lema, J.M. 2000. Influence of the content in fats and proteins on the anaerobic biodegradability of dairy wastewaters. Bioresource Technology. 74: 231–239.

Vigneswaran, S., Ngo, H.H., Chaudhary, D.S. and Hung, Y.T. 2005. Physicochemical treatment processes for water reuse. pp. 635–676. *In*: Wang, L.K., Hung, YT. and Shammas, N.K. (eds). Physicochemical Treatment Processes. Handbook of Environmental Engineering, vol 3. Humana Press. https://doi.org/10.1385/1-59259-820-x:635

Vourch, M., Balannec, B., Chaufer, B. and Dorange, G. 2008. Treatment of dairy industry wastewater by reverse osmosis for water reuse. Desalination. 219(1–3): 190–202. https://doi.org/10.1016/J.DESAL.2007.05.013.

Vymazal, J. 2005a. Constructed wetlands for wastewater treatment. Ecological Engineering. 25(5): 475–477. https://doi.org/10.1016/j.ecoleng.2005.07.002.

Vymazal, J. 2005b. Removal of enteric bacteria in constructed treatment wetlands with emergent macrophytes: a review. Journal of Environmental Science and Health—Part A. Toxic/Hazardous Substances and Environmental Engineering. 40(6–7): 1355–1367. https://doi.org/10.1081/ESE-200055851.

Wainaina, S., Lukitawesa, Awasthi, M.K. and Taherzadeh, M.J. 2019. Bioengineering of anaerobic digestion for volatile fatty acids, hydrogen, or methane production: a critical review. Bioengineered. 10(1): 437–458. https://doi.org/10.1080/21655979.2019.1673937.

Wan, C., Li, Y., Shahbazi, A. and Xiu, S. 2008. Succinic acid production from cheese whey using *Actinobacillus succinogenes* 130 Z. Applied Biochemistry and Biotechnology. 145(1): 111–119. https://doi.org/10.1007/s12010-007-8031-0.

Wang, K., Shammas, K. and Hung, Y. 2009. Advanced Biological Treatment Processes. Google Books.

Watkins, M. and Nash, D. 2014. Dairy factory wastewaters, their use on land and possible environmental impacts—a mini review. Open Agriculture Journal. 4: 1–9. https://doi.org/10.2174/1874331501404010001.

Wenzel, J., Fuentes, L., Cabezas, A. and Etchebehere, C. 2017. Microbial fuel cell coupled to a biohydrogen reactor: a feasible technology to increase energy yield from cheese whey. Bioprocess and Biosystems Engineering. 40(6): 807–819. https://doi.org/10.1007/s00449-017-1746-6.

Wolf, M., Gasparin, B.C. and Paulino, A.T. 2018. Hydrolysis of lactose using β-d-galactosidase immobilized in a modified Arabic gum-based hydrogel for the production of lactose-free/low-lactose milk. International Journal of Biological Macromolecules. 115: 157–164.

Wu, H., Zhang, J., Ngo, H.H., Guo, W., Hu, Z., Liang, S., et al. 2015. A review on the sustainability of constructed wetlands for wastewater treatment: design and operation. Bioresource Technology. 175: 594–601. https://doi.org/10.1016/J.BIORTECH.2014.10.068.

Wu, L., Xu, C., Li, S., Liang, J., Xu, H. and Xu, Z. 2017. Efficient production of lactulose from whey powder by cellobiose 2-epimerase in an enzymatic membrane reactor. Bioresource Technology. 233, 305–312.

Yadav, B., Pandey, A., Kumar, L.R. and Tyagi, R.D. 2020. Bioconversion of waste (water)/residues to bioplastics—a circular bioeconomy approach. Bioresource Technology. 298: 1–51. https://doi.org/10.1016/j.biortech.2019.122584.

Yalcin, S.K., Bozdemir, M.T. and Ozbas, Z.Y. 2009. Utilization of whey and grape must for citric acid production by two *Yarrowia lipolytica* strains. Food Biotechnology. 23(3): 266–283. https://doi.org/10.1080/08905430903106860.

Yamahata, N., Toyotake, Y., Kunieda, S. and Wakayama, M. 2020. Application of multiple sensory evaluations to produce fermented beverages made from sole whey using *Kluyveromyces marxianus*. International Journal of Food Science and Technology. 55(4): 1698–1704. https://doi.org/10.1111/ijfs.14440.

Yonar, T., Sivrioğlu, Ö. and Özengin, N. 2018. Physico-chemical treatment of dairy industry wastewaters: a review. *In*: N. Koca (ed.). Technological Approaches for Novel Applications in Dairy Processing. IntechOpen: Rijeka, Croatia. https://doi.org/10.5772/intechopen.74606.

You, S., Chang, H., Yin, Q., Qi, W., Wang, M., Su, R., et al. 2017. Utilization of whey powder as a substrate for low-cost preparation of *β*-galactosidase as the main product, and ethanol as a by-product, by a liter-scale integrated process. Bioresource Technology. 245: 1271–1276. https://doi.org/10.1016/j.biortech.2017.08.092.

Zeikus, J.G., Jain, M.K. and Elankovan, P. 1999. Biotechnology of succinic acid production and markets for derived industrial products. Applied Microbiology and Biotechnology. 51(5): 545–552. https://doi.org/10.1007/s002530051431.

Zikmanis, P., Kolesovs, S. and Semjonovs, P. 2020. Production of biodegradable microbial polymers from whey. Bioresources and Bioprocessing. 7(1): 36. https://doi.org/10.1186/s40643-020-00326-6.

Chapter **2**

Advancement in Processing and Utilization of Mango (*Mangifera indica*) By-Products

Priyanka Prasad[1], Jatindra K. Sahu*[1] and Jyoti Singh Jadaun[2]

[1]Centre for Rural Development and Technology, Indian Institute of Technology Delhi, New Delhi-10016, India.

[2]Dayanand Girls Post Graduate College, Kanpur-208001, India.

INTRODUCTION

Mangifera indica (Mango) is a member of the family *Anacardiaceae* which consists of about 30 other species of tropical fruit trees. Mango trees attain a height of up to 10–40 m, having a crown radius of approximately 10 m, perennial with a symmetrical, rounded canopy. Mango trees have an umbrella-like form (Jahurul et al. 2015). Once ripened, the fruit produces a characteristic resiny sweet smell (Parvez 2016). Although India produces various kinds of fruits, India contributes ~40.48% of total world mango production. Its uniqueness makes it the national fruit of India and Philippines and is the national tree of Bangladesh (Parvez 2016). In terms of production, mango is the second highest produced tropical fruit after only bananas (Vijay et al. 2019). Although many varieties of mangoes are grown in India, only 3% of them find place and dominate in the trade and export. The commercially important varieties of mango include *totapuri*, alphonso, *dasheri*, *kesar*, *banganpalli*, *langra* and *chausa*.

* For Correspondence: jksahu@iitd.ac.in

The popularity of *Mangifera indica* is not only because of its peculiar flavour but also due to good nutritional value. The composition of the fruit changes from cultivar to cultivar (Asif et al. 2016). As per the nutrient report given by the United States Department of Agriculture (USDA), the ripened mango pulp contains 83.4 g/100 g of water for fresh fruit. Mature fruits of mango are the main source of sugars such as starch, pectin and other carbohydrates like glucose, fructose and sucrose. Mango pulp is devoid of lipids, although the seeds are considered as a good source of fatty acids (Jahurul et al. 2015). Mango contains 32.34 to 76.81% carbohydrate, 6–15% fat, 6–10% protein, 0.26% crude fibre, and 1.5 to 4% ash approximately on a dry weight basis. It is also a good source of vitamins like retinol (1082 IU), B-complex thiamine (0.028 mg), niacin (0.669 mg), Pantothenic acid (0.197 mg), riboflavin (0.038 mg), vitamin B6 (0.119 mg), folate (43 μg), carotene (640 μg), ascorbic acid (36.4%) and also have different minerals such as calcium (11 mg), magnesium (168 mg), potassium (168 mg), sodium (1 mg), phosphorous (14 mg) and iron (0.16 mg) (Lebaka et al. 2021). Due to enormous nutritional value, mango and its constituents have been used for ailments under the traditional medicinal therapies in south-east Asian and African countries. Recent study also reveals that mango peel is reservoir of several phytochemicals having good pharmaceutical activities such as antimicrobial, anti-inflammatory, anti-cancer and anti-diabetic etc. (Lebaka et al. 2021).

Generally, ripe mangoes have been finding applications in production of various processed products like frozen, canned, dehydrated products and beverages (Ediriweera et al. 2017). Other components of the fruit like seed and peel are important but they are devoid of direct consumption. Therefore, peels and seeds (stones) share major part of the total waste generated from mango processing units (Figure 2.1). This chapter covers different by-products of mango and their possible utilization for manufacturing value added products like 5-hydroxymethylfurfural (HMF), ethanol and enzyme production, as functional food ingredients etc.

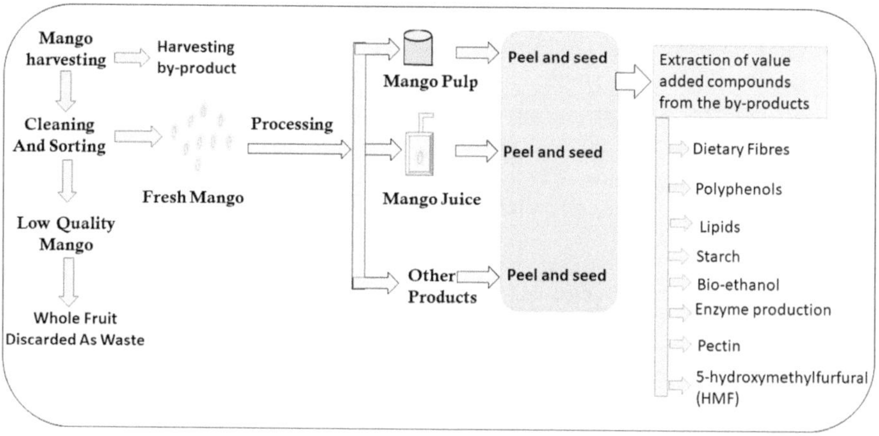

Figure 2.1 Schematic representation of by-products generated during mango processing and their utilization for production of various value added products.

BY-PRODUCTS OF MANGO AND THEIR NUTRITIONAL COMPOSITION

Generally, the term waste is considered for whole food or part of food which is fit to consume but discarded (Sagar et al. 2018). Unprocessed or processed mango generates peel and seed (stone) as the major waste products (Figure 2.1). 20–60% of the whole fruit weight is covered by seed while mango peel constitutes only 7–24%. Kernel inside the seed governed 45–75% total weight of the fruit (Kittiphoom 2012, Fowomola et al. 2010). Properties and uses of these by-products of mango are discussed in this section.

Mango Kernel

The mango fruit has one oblong seed that differs from cultivar to cultivar and has a hairy outer surface. Seed content varies from 7–24% by fruit weight of various mango cultivars. The seed coat is 1–2 mm thick and is 4–7 cm in length with 3–4 cm width. It covers the soft embryo/kernel. In the seed there is the seed nucleus, which accounts for 45–75% of the entire seed. Mango kernel majorly comprises of carbohydrate (32–77%), which is primarily present in the form of starch. It is also rich in lipids (6–15%), proteins (6–10%) and crude fibres (0.26–4.6%). Decent mineral content is indicated by the ash content of 1.46 to 3.71% on dry weight basis. The major amino acids are leucine, valine, and lysine. Apart from macronutrient and vitamins, mango kernel is also rich in minerals consisting of calcium (170 mg), magnesium (210 mg), and potassium (368 mg) per 100 g of kernel on dry basis (Kaur et al. 2022). However, the composition as determined by Nzikou et al. (2010) is much different as his study showed lesser potassium (158 mg), magnesium (22.34 mg), phosphorus (20 mg) calcium (10.21 mg), and sodium (2.7 mg) per 100 g mango kernel on dry basis. Mango kernel is rich in fat soluble vitamins and their precursors like beta-carotene (15.27 IU), tocopherol (1.30 mg/100 g dB), phylloquinone (0.59 mg/100 g dB), and ascorbic acid (0.56 mg/100 g dB). It also contains other vitamins like thiamine (0.08 mg), riboflavin (0.03 mg), biotin (0.19 mg) and cobalamine (0.12 mg) per 100 g kernel weight on dry basis (Fowomola 2010). Interestingly, mango seed kernels also contain phytochemicals like polyphenols (anthocyanins, isomangiferin, gallotannins, flavonols, mangiferin, homomangiferin, kaempferol, benzophenone derivatives, and quercetin) and phytosterols like campesterol, and tocopherols (Table 2.2).

Mango Kernel Oil (MKO)

Mango kernel contains about 8–15% oil which is generally termed as mango kernel oil. The content of oil in the kernel is like soybean and cottonseed which contains 18–20% oil content (Nadeem et al. 2016). In a study done by Sikdar et al. (2017), conditions for oil recovery from mango kernel was optimized by using Soxhlet extraction method and maximum oil yield was 17.3%. The kernel oil is solid at room temperature and has a melting point of 32–36°C with nutty odour. Appearance of pure oil was smooth and creamy in colour. The soft-yellow coloured MKO has comparatively stronger odour than butter. Just like olive oil, it also has a mild fragrance. Its lower melting point (23–27°C) makes it suitable for cosmetic applications. Generally, processing is required for commercial vegetable oils, but MKO can be used without much processing because of low content of free fatty acids. MKO contains 52–56% saturated and 42–44% unsaturated fatty acids (Zahoor et al. 2023). Chemical characterization of the oil reveals that it majorly contains oleic acid (44.89%) and stearic acid (36.57%) followed by minor amounts of palmitic (10.06%), linoleic (6.0%) and linolenic (2.48%) acids (Abdelaziz 2018). The content of fats and oils in mano kernel oil is reported as 1,3-distearoyl-2-oleoyl-glycerol (29.4 to 40%), 1-stearoyl-2,3-dioleoylglycerol (14.6 to 23%), 1-palmitoyl-2-oleoyl-3-stearoylglycerol 5.7–14.76% and 1,3-dipalmitoyl-2-oleoylglycerol (6.89–8.9%). The oil also possesses high antioxidant properties which are due to presence of its phenolic (9.87 mg/g) content. This is much higher than other common oils like chia and almond (Kittiphoom and Sutasinee 2013).

A mango variety form Pakistan revealed 15.5% free fatty acids. The melting point of the oil was found to be 35.2°C, whereas the iodine value and saponification value were reported to be 55.2 and 195, respectively. Low peroxide value (0.22 meqO_2/kg) indicated good stability (Nadeem et al. 2017). MKO finds its application in foods as it does not require partial hydrogenation. Numerous studies have also been conducted to check the efficacy of MKO as opposed to trans-fat options. Another reason why MKO finds its use in the food industry is because it is extremely resistant to auto-oxidation (Kittiphoom and Sutasinee 2013). The physico-chemical properties of MKO are given in Table 2.1.

Mango Peel

Mango peel contains various components which are not just only nutritionally valuable but also important for economical perspective. Mango peel is a potential source of bioactive compounds which is helpful in preventing lifestyle-related disorders. Usage of mango peels leads to not just only value addition but it will also curtail the waste generation. The mango peel contains approx. 7–24% of the weight of mango fruit. Total fibre content of mango peel accounts for 28 to 78% which has 14 to 50% of insoluble dietary fibre and 13–28% of soluble dietary fibre, neutral sugar and uronic acid. Dietary fibre possesses an important trait that it has high oil-holding capacity which is associated to the structure of the polysachharides and depends upon surface property, thickness, overall charge density, and the hydrophobicity of the fibre particle. Another advantage is that mango peel has low oil holding capacity due to which the dietary fibre extracted from mango peel can be used in fried products where its use can avoid greasy impression. The moisture content and ash content ranges from 8.13–9.30% and 1.38–1.57%, respectively.

The overall content of polyphenols in green peels of raw mango ranged from 90 to 110 mg/g peel; it is higher in ripe peels, ranging from 55 and 100 mg/g depending on the maturity stage. Similar to polyphenols, carotenoid content was

Table 2.1 Effect of different solvents on physicochemical properties of mango kernel oil.

Physicochemical property	Ethanol extract	Hexane extract	Petroleum ether extract	Methanol extract	Water extract
Total oil yield (g/g d.b)	6.96 ± 0.13	8.46 ± 0.01	8.04 ± 0.07	13.00	2.53
Acid value (mg KOH/g oil)	27.55 ± 0.55	0.10 ± 0.012	0.15 ± 0.083	5.35	—
Peroxide value (mg/g oil)	26.35 ± 2.1	8.72 ± 3.4	8.82 ± 2.8		—
Iodine value (mg/100 g oil)	59.17 ± 2.3	38.50 ± 3.9	37.25 ± 3.0	39.50	—
Saponification value (mg KOH/g oil)	206.00 ± 13.8	207.50 ± 14.2	190.20 ± 12.8	207.50	—
Total phenolic content (mg GAE/g)	53.50 ± 7.5	98.70 ± 8.8	77.00 ± 7.4	117.00	118.1 ± 0.2

Table 2.2 Phytochemical properties of compounds extracted from mango byproducts.

Compound	Molecular formula	Biological activities	References
Gallic acid	$C_7H_6O_5$	Radical scavenger, preventive and therapeutic effects for cardiovascular diseases, cancer, antifungal, neurodegenerative disorders and in aging.	Pal et al. 2018
m-digallic acid	$C_{14}H_{10}O_9$	Wound-healing, anti-depressant, anti-Parkinson	Nayeem et al. 2016
p-digallic acid	$C_{14}H_{10}O_9$	Antioxidant activity, anti-inflammatory, hepatoprotective	Onuh et al. 2017
Ellagic acid	$C_{14}H_6O_8$	Relaxation of blood vessels, oxygen free radical scavenging, lipid lowering effect, anti-inflammatory and anti-carcinogenic effect	Usta et al. 2013
Methyl gallate	$C_8H_8O_5$	Anti-proliferative, apoptosis	Lopez-Cobo et al. 2017
Mangiferin	$C_{19}H_{18}O_{11}$	Inhibitory activity, anti-cancer, antioxidant activity	Matkowski et al. 2013
Penta-galloyl glucose	$C_{41}H_{32}O_{26}$	Free radical sink, an anti-inflammatory agent, anti-diabetic agent, enzymatic resistant properties	Patnaik et al. 2019
Quercetin	$C_{15}H_{10}O_7$	Antioxidant, anti-inflammatory, antiviral and anticancer as well as the function to ease some cardiovascular diseases	Wang et al. 2016

also found higher in peels of ripened fruits and it ranges from 74 to 436 g. The amount of dietary fibre present in peel lies between 45 to 78% and was present in a higher amount in mature skins. The content of ascorbic acid and tocopherol in dry mango peel lies between 188–392 and 205–509 μg/g, respectively. It was observed that the content of these vitamins enhanced as the fruits ripened. Mango peels (raw and ripe) possess significant enzymatic activity and contains a variety of enzymes including proteases, peroxidases, polyphenol oxidases, xylanase and amylases among others. Due to the abundant content, the peel act as a great thickener, gelling and emulsifying agent which find application for manufacture of marmalades, beverages, dairy and meat products (del Pilar Sanchez-Camargo et al. 2019). It also has adequate amount of minerals like potassium, copper, zinc, manganese, iron, and selenium. The mango strips of Indian varieties Badami and Raspuri assortment are a great source of bioactive components like polyphenols, carotenoids, and dietary fibre. A study conducted on the peel of Irwin mangoes from Korea reported higher content of polyphenols and bioactive compounds than pulp, thereby resulting in potent antioxidant activity. The phenolic compounds extracted from peel are syringic acid, mangiferin, protocatechuic acid, ellagic acid, gallic acid, and quercitin (Kim et al. 2010). More bioactive compounds are found as carotenoids, violaxanthin, and luteine in the peel. Other important compounds reported in the mango peel are terpenoids (β-mircene, terpinene, α-guaiene, canfene, limonene, fenchene, trans-ocimene, α-humulene, cis-ocimene, α-pinene, β-pinene), sesquiterpenes, lactones, acids, esters, aliphatic alcohols, aldehydes, and other compounds which are involved in the synthesis of vitamins A, K, and E. Significant quantities of minerals like potassium (2910 mg/kg), manganese (950 mg/kg), calcium (4.445 mg/kg), iron (175 mg/kg) and zinc (32.5 mg/kg) are observed in the peel of Haden mango variety (de França et al. 1999).

Mango Pomace

Mango pomace is another by-product of the processing industry obtained after extracting juice from the pulp (Nagarajaiah and Prakash 2016, Quiles et al. 2018). Mango pomace represents about more than half of the total fruit weight and is composed of the peel and the kernel (Mahmoud et al. 2020). On analysis, the pomace was found to have a moisture content of 68.57%. The nutritional composition analysis was done on dry basis which revealed presence of high quantity of organic matter (97.24%), crude protein (10.59%) and crude fibre (6.14%) while fat content was found low (1.74%). Moreover, 78.77% carbohydrates and 2.76% ash was reported in the analysed samples. In a study carried out by Mahmoud et al. (2020), the effect of solid-state fermentation on cell wall constituents, chemical analysis, and nutritive values of mango pomace were determined. Moisture content for unfermented (UFMP) and fermented (FMP) mango pomace was 68.57% and 75.92%, respectively. Chemical analyses on dry matter basis (%) was carried out which showed that there is an increase in the values of crude protein, fibre, fat content, carbohydrate content and ash content with an exception of organic matter where the values dropped. Somewhat similar trend was observed in the energetic and nutritive values, thereby drawing a conclusion that fermentation (processing) enhances the nutritional quality of the mango pomace. Furthermore, neutral detergent fibre, acid detergent fibre, acid detergent lignin, hemicelluloses and celluloses represented 32.41%, 15.03%, 2.56%, 17.38% and 12.47%, respectively, while the energy composition presented the following values-gross energy (4286 kcal/kg dry mass), digestible energy (3257 kcal/kg dry mass), metabolizable energy (2671 kcal/kg dry mass) and net energy (1496 kcal/kg dry mass).

Chemistry of By-Products

Natural antioxidants provide protection against oxidative damage caused by reactive oxygen species and free radicals (Oktay et al. 2003, Hassan et al. 2011). Kim et al. (2007) conducted a study with *Mangifera indica* L. cv. Irwin extracts to determine the antioxidant content in mango peel extract. It was found that the mango peel extract exhibited stronger free radical scavenging ability than mango flesh extract on 1,1-diphenyl-2-picrylhydrazyl (DPPH) and alkyl radicals. Similarly, peel extract and flesh exhibited significant antiproliferative effect against all tested cancer cell lines in a dose-dependent manner due to presence of significant content phenolic and flavonoids. The high DPPH scavenging activity has potential in the counteraction of oxidative pressure related illnesses. Polyphenol rich parts of strip concentrate could be utilized as characteristic cell reinforcements and useful food/feed supplements (Ma et al. 2011). Peel strips mainly contain mangiferin (C-glucosyl xanthone), a heat-stable and medicinally potent phytochemical (Ajila et al. 2007, 2013). Through studies, it was concluded that mangiferin exerts anti-oxidant, anti-inflammation, immunomodulation and anti-apoptosis effects in the body. Mangiferin is a compound sparingly soluble in ethanol and insoluble in n-hexane and diethyl ether. It is sparingly soluble in water with a solubility only at 0.111 mg/mL.

A significant alternation was noticed in the antioxidant and polyphenolic content under variable processing techniques and solvent used for extraction. In a study carried out by Sogi et al. (2013), the effect of heat treatment on the total phenolic content of mango peels against standard ascorbic acid was determined. Results stated that heat treatment delivered a huge decrease ($p<0.05$) in tannin (76.44%), ascorbic corrosive (18.24%), and flavonoid (7.38%) polyphenol content (4.7%). The estimation of ascorbic acid content was in the range of 61.22–74.48 mg/100 g in mango peels.

PROCESSING AND UTILIZATION

This section covers the processing and utilization of mango by-products. The therapeutic applications of the products have also been discussed.

MKO

Lipid profiling and oil content varies in different parts of fruit such as fruit pulp, fruit peel, fruit pomace and nuts. Generally, oil is stored in seeds and it is utilized as energy and carbon sources when seeds sprout.

Processing

The integrity of essential bioactive components gets affected by the procedure used for processing. Different methods can be used for extraction of oil like cold and hot pressing, extraction using solvents or blend of both methods. These techniques can be made more efficient by amalgamating with other accessories like microwaves, enzymes and inert gases (Sultana and Ashraf 2019).

In order to extract mango seed oil, seeds are separated, washed and dried. After dehydration, the dried kernels are then roasted which ensures better mechanical separation of the hull. Separated kernels are crushed and ground into fine powder. This fine powder is then used for extraction of oil. Solvents like n-hexane can be used for extraction with good efficiency (Kittiphoom and Sutasinee 2013). Different steps involved in processing of MKO are:

1. **Drying and Grinding:** After separation of the peel and pulp of mango, the seeds are dried either in sun or in an oven to reduce the moisture content upto 15% which facilitates proper separation of the hull. Once optimum dehydration is achieved, the kernels are separated from the endocarp using drum pressing and then grinded by using different types of grinders. The powder is then used for immediate extraction. If not used immediately, the powder can be stored in tight jars at refrigerated temperatures.
2. **MKO extraction:** As already discussed, various methods can be used for extracting MKO, although generally the solvent extraction and hydraulic press extraction methods are used. A general set up of a soxhlet extractor includes a round bottom flask containing the solvent; in the neck of the flask, a glass siphon is fitted for extraction. On the upper part of the glass compartment a condenser is placed, which is cooled by continuous running water. The powder for oil extraction is kept in a thimble, which is made up of either glass or cellulose. Once the setup is arrranged, the solvent is heated which then converts into vapours. The process is repeated several times until complete extraction is ensured. This process remains uninterrupted until a constant supply of heat is there. Continuous washing of the sample allows the extracted material to collect in the flask along with the solvent. For MKO, solvents like n-hexane, chloroform, ethanol etc. can be used and have shown good efficiency (Karunanithi et al. 2015). Another used method is the extraction by pressing. Mechanical pressing is one of the oldest method used for extraction of oil from seeds. The benefit of this method is that it gives better quality of oil and cake residue in comparison to other methods like screw press or extraction by solvents (Babaria 2011).

Conditions and process parameters can be optimized to enhance the yield of MKO (Sultana and Ashraf 2019), one of which is usage of sealed containers where extraction takes place using high pressures. Such methods are called pressurised solvent extraction methods. The use of supercritical fluid extraction is an alternative technique. Carbon dioxide is one of the materials from which it is easy to obtain super critical fluid. It has a critical temperature and pressure of 31.3°C and 72.9 atm, respectively. Since these conditions can be readily attainable, making super critical carbon dioxide is easy. Supercritical CO_2 is used as a solvent for extraction and can dissolve certain organic compounds, which enables it to substitute a number of different solvents. Supercritical CO_2 has low toxicity, is cost effective, non-flammable and can be disposed of quickly. As soon as the extraction completes, the pressure returns back to atmospheric conditions and with this, the CO_2 instantaneously gets converted into a gas and escapes. The pure extracts are thus left behind which can be further used as desired. For extraction of MKO, mango kernel powder was inserted into the extraction vessel with pre-liquefied CO_2 supplied through CO_2 pump under controlled temperature and pressure (Kayathi et al. 2020).

In another study, it was found that use of microwaves prior to the extraction was quite beneficial for improving the oil yield; it also aids in reduction of extraction time by fivefold times (Kittiphoom 2012). Balacuit et al. (2021) also reported microwave assisted extraction as another advanced method which employs the use of high pressure and temperature (150°C). However, this method possessed a major limitation that some solvents such as the pure non-polar solvents (e.g. alkanes like hexane) are transparent to microwave radiation, and it hinders in production of desired results.

Another method employed for the extraction of MKO form the kernel is hydro distillation. This method is popularly used to extract essential oils from herbs and other medicinal plants in various medicinal systems. In this method, the kernel powder is placed in a tank and filled with generous amount of water. The tank is then heated from the bottom and converts water into vapour. This steam, while it rises, extracts the oil and carries it towards the condenser. On reaching the condenser, it is separated from the solvent. The powder-water mixture left behind in the tank gets separated into two layers using gravity separation method and collected, respectively. In a study, mango kernel seed oil extracted by using supercritical CO_2 and other conventional solvent (hexane, petroleum ether, ethanol and acetone) extraction techniques are compared (Awolu and Manohar 2019). Study revealed that the kernel oil extracted from conventional solvents had higher yields in comparison to supercritical CO_2 technique. The extracted oil obtained by using solvent extraction technique was rich in unsaturated fatty acids (UFA), whereas supercritical-CO_2 extracted oil had higher saturated fatty acid (SFA) content. Oleic acid and stearic acid are the predominant UFA and SFA.

Utilization

The seed find application in beauty and cosmetic industry. Seed powder and oil can be stored even for a period of more than 3–4 years if stored under refrigerated conditions. MKO can find application as a substitute of cocoa butter in chocolates and confectionaries. On being cold-pressed, mango kernel oil renders butter instead of oil which is an exceptional source of essential fatty acids, minerals, vitamins, hence acts as a good alternative of cocoa butter. Mango kernel butter shows valuable creaming properties for skin lotions and lubricants (Nadeem et al. 2016).

Application of MKO on skin and its ability to act as an antidepressant were studied by Nadeem et al. (2016). The study reported that MKO can be effectively used in the treatment of dry skin. On topical application, it provides softness and protection of the skin from frost bites. Rashes and itching are severe ailments in hot and humid weather conditions especially for infants and babies. MKO contains several phenolic compounds and other bioactive which possibly can play an important role in these conditions. The same compounds in MKO possess the power to eradicate dark spots on the skin and the phenolic compounds also fight against ageing. MKO also finds its use in sunscreens and shaving cream owing to its fatty acid composition and avoidance of formation and also removal stretch marks developed during pregnancy. Due to the presence of phenolics, MKO acts as an antidepressant and relieves muscle fatigue and cramps.

Since MKO possess very low quantities of free fatty acids and good amount of phenolic components, it leads to very low incubation period for MKO than any other commercially available vegetable oil. Hence, it can effectively be used to replace artificial antioxidants used to preserve fats and oils. The major phenolic compounds present in MKO are mangiferin, quercetin, caffeic acid and chlorogenic acid (Nadeem et al. 2016). Functional properties of MKO can be enhanced by using methods like fractionation, trans-esterification and inter-esterification, which can make it more suitable for industrial applications (Nadeem et al. 2016, 2017). Apart from its multiple benefits to the body, owing to its antioxidant and total phenol rich profile, another advantage of mango kernel oil is its utilization in the active packaging. Adilah et al. (2018) attempted to make use of by-products obtained from food industries. The authors used mango kernel extracts in various proportions (1–5%) as natural antioxidant in films, which were made using soy protein isolate and fish gelatin. The results stated that incorporation of MKO made the films thicker and more translucent. An enhancement in the antioxidant concentration was observed in both the films with an increase in extract concentration. During the process of refining, the medicinal efficacy of MKO is adversely affected (Adilah et al. 2018).

Mango Peel

Processing

The processing of mango peel (like the processing of other mango by products) involves the collection of the mango variety, which is to be processed, followed by removal of the by-products (seed, pulp, peels). The by-products are then subjected to drying in order to reduce the initial moisture content below 12–15%. This is followed by grinding of the by-products (here peels) to obtain a powdered form which can be further incorporated as per need (Deeksha and Sunita 2020).

As stated in the previous section, bioactive compounds like phenolic compounds and carotenoids are present in generous amounts in mango peel. The efficacy of phenolic compounds depends upon the efficiency of extraction. There are various methods for extraction of the compounds from mango peel involving the opening of the complex matrix and solubilizing the compounds using various solvents. Solvents like hexane, ethanol and acetone have been tested for mango peel extraction. This solvent assisted extraction if coupled with other techniques like microwave, giving better yield and efficiency (Rojas et al. 2015). Apart from this, super fluid extraction has also been tested and proved beneficial. Researchers have successfully extracted non-polar flavonoids and carotenoids using super critical carbon dioxide from mango peel (Garcia-Mendoza et al. 2015). Supercritical CO_2 extraction method and pressurized solvent processes can be used to extract by-products from mango peel. Similarly, del Pilar Sanchez-Camargo et al. (2019) showed higher content of crude proteins and phenolic compounds in MP displayed high contents of crude protein and phenolic compounds, and carotenoids in supercritical fluid extract bioactive from mango peel. For storage of mango peel, drying has been considered an effective process. Drying makes the final product shelf stable not only by reducing moisture content in the final product but also leads to inactivation of enzymes liable for the destruction of beneficial compounds and reduces the rate of microbial growth. However, since drying process involves raised temperatures, care must be taken to control the temperature as most of the bioactive compounds can attain significant damage from high temperature (Kittiphoom and Sustainee 2013). Several methods for drying mango peel were used in a study like cabinet, infra-red, vacuum, and freeze drying. Mango peel got dehydrated when kept 2 h in infrared, 4 h in cabinet, 7 h in vacuum and 11 h in freeze drying methods. The ascorbic acid content was observed highest and lowest in cabinet dried and vacuum dried mango peel powder, respectively. The carotenoid content was observed highest in freeze dried mango peel. The highest total phenolic content was reported in freeze dried peel due to avoidance of heat-related damage while in vacuum dried peel least phenolic content was observed (Sogi et al. 2013). However, research is needed on alternative drying methods for mango peeling, like heat pumps or refractory windows. These alternative methods are only used in research studies and their impact on the final phytochemical and nutrient composition is yet to be studied for mango peel.

In a new report, phenolic compound from mango strips were effectively removed utilizing subcritical water (Rodríguez De Luna et al. 2020).

Utilization

Mango peels can be used for the extraction of bioactive compounds (i.e. dietary fibre and polyphenols) which can be added in diverse food preparation such as bread, sponges, biscuits and other baking items (Thambi et al. 2016). These bioactive compounds are responsible for playing a major part in prevention of diseases. A study was performed to determine the importance of mango waste in the treatment/avoidance of diseases (Ashoush and Gadalla 2011). The research is focused on the influence of physical, rheological, antioxidant and sensory properties of biscuits having mango peels at various substitute levels (5%, 15% and 20%) and of mango kernels powders at 20%, 30%, 40%, and 50%. The findings showed that peel powder possessed good amount of ash, fibre and water while mango kernel powder was higher in fat and protein than mango peel powder. Farinographic results showed an improvement in water absorption capacity from 60.4 to 67.6% of wheat flour integrated with mango peel while it declined in case of kernel powder. The content of phenolic was also risen from 3.84 to 24.37 mg/g biscuits supplemented with different amounts of peel and kernel powder. On fortification of the biscuits with mango peel powder, the biscuits depicted enhanced free radical scavenging properties. Mango flavoured biscuits are satisfactorily produced by supplementation of mango peel and kernel powder up to 10% and 40%, respectively. Aslam et al. (2014) also attempted the incorporation of mango peel and kernel for enhancing antioxidant properties and dietary fibre content of biscuits. The use of kernel powder enhanced the content of total phenol, protein and ash in comparison to control and peel fortified biscuits. Best sensorial results were obtained by supplementing 10% peel powder and 5% kernel powder. Another important finding drawn from this study is that self-life of fortified biscuits is satisfactory. Quality was maintained up to 30 days on storage at room temperature.

Ajila et al. (2010) assessed quality of macaroni after incorporation of mango peel powder. The authors studied quality parameters like cooking parameters, firmness, bioactive composition and sensorial traits. As per expectation, the dietary fibre (increased by more than 50%), polyphenol (more than triple increase) and carotenoid (84 μg/g) contents were enhanced significantly on supplementation with mango peel powder.

Mango peel is also reservoir of distinct enzymes like proteases, peroxidases, amylases, xylanases and polyphenol oxidases. Proteases are well known for their industrial and biotechnological applications (Padhi et al. 2023, Sharif et al. 2023). Mango peel contains serine proteases and it can be purified simply as compared to other proteases. Furthermore, amylases extracted from mango peel find application in various industries like the fruit and beverage, confectionary, detergent and other biotechnological applications. Encapsulation has recently been used as a novel technique for conservation and delivery of these enzymes (Amid et al. 2011).

Yang and Li (2013) showed a new biological path for the production of silver nanoparticles as a derivative of mango peel extract. They have been shown to have an outstanding antibacterial activity in non-woven fabrics lined with silver nanoparticles. They suggested that these nanoparticles can be used for various medicinal applications.

Mango Pomace

Processing

For extraction of fibre, fruit pomaces are exposed to a series of steps, comprising of desugaring, grinding and bleaching (Quiles et al. 2018). As the name suggests, the desugaring step reduces the sugar content from the pulp. In this step, the free monosaccharides are removed from the pulp which gives a concentrated mixture of dietary fibre. After this, washing is performed slowly so that soluble fibres are not lost. Afterwards, mixture is subjected to filtration through a 100 μm nylon filter to segregate the insoluble residues. The above given process is repeated to ensure better extraction. To acquire fibre concentrate, desugared fruit residues are dried for 7 h in hot air oven at a temperature of 70°C (Rabetafika et al. 2014, Sancho et al. 2015). Ultrafine, fine, coarse fibre concentrate powder are obtained after grinding the fibre concentrate followed by bleaching. The categories are divided based on the particle size segregated by passing through sieves of different mesh sizes (Hu et al. 2018). Bleaching is performed by using hydrogen peroxide to separate coloured molecules like phenols and tannins. In this process, dried desugared powder is mixed with 2.5 L of 3% hydrogen peroxide and the pH is adjusted to 11.5 using 7M sodium hydroxide for a duration of 60 minutes. The insoluble portion is then neutralised using 7M hydrochloric acid and separated by filtering through a metallic sieve. The fraction so obtained is washed thrice, oven dried and ground in a mill. The powder so obtained is called the bleached fibre concentrate.

Utilization

As stated earlier, pomace processing is majorly carried out to extract the fibre rich polysaccharides and bioactive compounds present and utilize them for human benefit. Therefore, the beneficial properties of pomace are due to the presence of polysaccharides and bioactive compounds. Plant polysaccharides (here obtained from mango) have attracted considerable attention due to their antioxidant, anti-tumour, anti-diabetic, anti-cancer causing, anti-viral and immune-modulating properties. Hu et al. (2018) carried out a study wherein ultrafine extraction method was used to extract the polysaccharides and for anti-cancer activities. Ultrasonic extraction can be used to extract polysaccharides from mango pomace. Best results (3.89% yield) are obtained at the following conditions-temperature (74 °C), ultrasonic power (170 W), time (100 minutes), and raw material-to-water ratio (1:40 mg/mL). The use of celluloses from DEAE-52 and Sephadex G-100 column chromatography purified three novel polysaccharide fractions (MG-1, MG-2, and MG-3) from crude polysaccharide extract. The molecular weights of MG-1, MG-2, and MG-3 are determined as 1.78 T1104, 3.26 T104, and 2.79 TN106 Da, respectively, through HPGPC. Monosaccharide composition of MG-1, MG-2 and MG-3 is constituted of mangosis, glucuronic acid, rhamnozyme, galcturonic acid, glucose, xylose, galactose, arabine and fucose. In addition, FT-IR, 1H NMR and SEM have been performed for tentative structural characterizations of MGP.

NEW INSIGHTS IN BIOTECHNOLOGICAL VALORISATION OF MANGO BY-PRODUCTS

The bio-refinery approach helps in the separation and extraction of value-added compounds from mango by-products and encourages its subsequent use in production of biofuels, bio-chemicals, and biomaterials. Valorisation of mango waste by applying various biotechnological routes can help in gaining the economic benefits from fruit processing industries in a sustainable way. The pathway showing the valorisation of mango by-products is presented in Figure 2.1.

Production of 5-hydroxymethylfurfural (HMF)

Mango fruit by-product and residues are reported for the manufacturing of furan compound like 5-hydroxymethylfurfural (HMF), due to the presence of adequate amount of hexose sugars. Mangoes mainly consist of sucrose, followed by fructose and glucose. Fructose and glucose can be easily transformed into HMF. In addition to this, sucrose can be used as source of hexose sugars. Hydroxymethyl furfurals are highly efficient chemicals which can be subsequently utilised and transformed in to other high end compounds like fuels, polymers, and other platform chemicals. In a study, for the production of HMF, two separate feed stocks—one consisting of blend of mango peel and pulp (1:1) and other consisting of combination of mango pulp, peel, and seeds (2:1:2)—are used. One gram of each of above mentioned feedstocks was dissolved in distilled water at three different concentrations for six different reaction mixtures. In each reaction mixture, sulfuric acid was added as a catalyst with the pH between 2.9 and 3.2. The mixtures were given microwave treatment under different temperatures and time, which resulted in the formation of HMF at varying temperature and time combinations (Muñiz-Valencia et al. 2020).

Production of Bioethanol

Bioethanol production from mango fruit waste especially mango peel is used due to availability of low-cost free sugars in mango waste biomass. The high content of reducing sugars presents in dry and fresh mango peel results in their utilization as a cheap raw material for the production of bioethanol. Moreover, mango peel fermentation for the production of bioethanol with concurrent recovery of other bioactive compounds is a cost effective biorefinery approach (Kaur et al. 2022). In a study, the ethanol production was achieved as 5.13% w/v after fermentation of only mango peel. In addition to this, the production of bioethanol was increased by 45% when mango peel was fermented with the addition of yeast extract, peptone and wheat bran (Varakumar et al. 2011). Production of bioethanol by using mango peels is an eco-friendly process and it has the potential of being a substitute of petrol (Walia et al. 2013).

Production of Enzymes

Utilization of agro-industrial wastes for microbial production of enzymes is gaining worldwide attention due to their enormous availability, high carbon content, and renewability. The lignocellulosic waste biomass is a cost effective and valuable substrate for enzyme production. Lipases have been produced from mango seeds and peels with the use of *Yarrowia lipolytica*. In a study performed by Pereira et al. (2019), a significant increment in lipase production was achieved using mango peel as a substrate and supplementation of several nitrogen sources. Extracellular lipase production was achieved around 3500 U/L after supplementation of yeast extract (Pereira et al. 2019). In another study, mango kernel was utilized as a substrate for production of alpha-amylase using *Fusarium soloni*. Production of alpha-amylase was maximally achieved by maintaining following conditions, a substrate concentration of 5% (w/v), temperature 30°C, pH-4 after 9 days of incubation (Kumar et al. 2013b). High microbial production of pectinases was achieved when dried mango by-products were utilised as substrate in a study conducted by Purnachandra Reddy and Saritha (2015). In this study, pectin obtained after post isolation of pectinolytic fungi was proved as a good substrate for pectinase production. The highest activity of pectinase was found in submerged fermentation by using mango waste as carbon source (Purnachandra Reddy and Saritha 2015).

PHARMACOLOGICAL PROPERTIES OF BY-PRODUCTS OF MANGO

Mango by-products are known for their pharmacological properties. These find applications in treatment of ailments like cancer, cardiovascular diseases and allergy (Table 2.3). This section deals with illustration of the pharmacological properties of mango by products.

Antiviral Activity

The antiviral activity of mangiferin have been tested against Herpes simplex virus type 2 (HSV-2) and the results showed that it doesn't directly affect the HSV-2 but instead it inhibits the later events caused by HSV-2 replication. It also inhibits multiplication of Herpes simplex virus type 1 within the cell. It is also effective against the cytopathic effects of HIV (Hannan et al. 2013).

Anthelminthic and Anti-allergenic Activity

Anthelminthic and antiallergenic effects of mango stem bark constituent, mangiferin, were examined in mice tainted with nematode, *Trichinella spiralis*. The investigation revealed that the compound was able to inhibit the growth of nematodes. Mast cells are primarily responsible for the type I allergic response, which is triggered by the interaction of the surface receptors with distinct molecules such as antigen bound with IgE.

These interactions set in motion a chain of biochemical reactions that result in the release of active mediators which are responsible for caused allergic reactions (Che Rahim 2019). Other cells, especially eosinophils, basophils, neutrophils and B and Th2 lymphocytes, play a role in the allergenic responses (Rummun et al. 2021). Mangniferin is responsible for changes in mast cells. Mangiferin was effective against C-glucosylxanthone, making it especially effective against hypersensitive models. The study thus proposed that mango extract can be effective against hypersensitive issues (Hamzah et al. 2018).

Immunomodulatory Effect

Mangiferin has also been identified as a key constituent in mango which showed anti-inflammatory effect. The compound mediated downregulation and suppression of NF-κB otherwise activated by inflammatory agents like tumour nuclear factor. Mangiferin also increases the intracellular content of the natural antioxidant glutathione and possibly might play a role in chemotherapeutic agent-mediated cell death (Rajendran et al, 2013). Immunomodulatory effect of mangiferin was assessed on thioglycollate-elicited mouse macrophages that were stimulated with lipopolysaccharide and gamma interferon (IFN-γ). Mangiferin (10 μM) was found to inhibit NF-κB mediated signalling pathways. It also inhibited toll like receptor and a number of pro-inflammatory cytokines such as IL-1, IL-1α, IL-6, IL-12,

Table 2.3 Pharmacological properties of mango waste products.

Health disorder	Biochemical property and name of part used	Name of the compound	Outcomes	References
Antioxidant activity	Peel	3,4-Dihydroxybenzoic acid (protocatechuic acid)	Mangiferin is effective in gut protection against gastric injury due to its antisecretory and antioxidant mechanism	Barreto et al. 2008
	Fruit pulp	Kaempferol Linalool Mangiferin Quercetin β-Carotene		
	Peel	β-Carotene		
Anti-inflammatory	Peel	5-(11Z-Heptadecenyl)-resorcinol and 5-(8,Z,11,Z-Heptadecadienyl)-resorcinol	Inhibits inflammatory cellular activities, regulates expression of genes responsible for inflammation, increase in cellular resistance against injuries caused by inflammation	Knödler et al. 2007, Khare and Shanker 2016, Telang et al. 2013
	Fruit pulp	Kaempferol Mangiferin Shikimic acid		
Antimicrobial	Fruit pulp	Kaempferol Mangiferin Quercetin	Proven antibacterial activity of both gram-positive and negative bacteria and yeast *Candida albicans.*	Moskaug et al. 2004
	Peel	3,4-Dihydroxybenzoic acid(protocatechuic acid)		
Anti-diabetic	Mango peel	Polyphenols Carotenoids Mangiferin β-Carotene	A significant increase in fasting blood glucose, triglycerides, LDL cholesterol in diabetic rats	Gondi et al. 2015

Health disorder	Biochemical property and name of part used	Name of the compound	Outcomes	References
Cytotoxic and apoptotic	Fruit pulp	Kaempferol Linalool Mangiferin Quercetin β-Carotene	It is effective against cyclophosphamide-induced immune depression, like lymphoid organ atrophy, poor cellular response, low antigen-specific immunoglobulin M, it reduces lipid peroxidation, and decreases superoxide dismutase activities.	Schieber et al. 2003
	Peel	3,4-Dihydroxybenzoic acid(protocatechuic acid)		
	Seed	Mangiferin Gallic acid		
Neuroprotective	Fruit pulp	Linalool	Stimulates cell propagation and induces an increase in the level of nerve growth factor and tumor necrosis factor (TNF)-α. Mangiferin is effective in enhancing recognition memory via a mechanism which might lead to enhanced neurotrophin and cytokine level.	Park et al. 2012
		Mangiferin		
Cardio-protective	Fruit pulp	Mangiferin	Decreases production of lipid peroxides and retains the myocardial marker enzyme activity at near normal level.	Soong and Barlow 2006
Anticoagulant	Fruit pulp	Shikimic acid	Reduced yeast-induced hyperpyrexia	Wall-Medrano et al. 2020
Blood pressure lowering activity	Fruit pulp	Quercetin	Significantly reduced total serum cholesterol, triglycerides, low density lipoprotein, very low density lipoprotein and enhanced high density lipoproteins in rats.	Park et al. 2012
Gastro-protective	Seed	Mangiferin	Mangiferin significantly improved gastro-intestinal tract movement at oral doses of 30 mg/kg and 100 mg/kg by 89 and 93%, respectively.	Sandhu and Lim 2008

TNF-α, granulocyte and macrophage colony-stimulating factors (Leiro et al. 2004, Singh et al. 2009).

Moreover, besides controlling the reactive oxygen species-scavenging, mangiferin has also been found responsible for modulating the expression of many genes, which are found to be critical for the regulating apoptosis, viral multiplication, development and propagation of tumour, inflammation and many autoimmune conditions. The studies propose that mangiferin can act as a crucial component not just for treating but also for preventing cancer, inflammatory diseases and may provide protection against oxidative damage and mutagenesis (Ediriweera et al. 2017).

Anti-inflammatory and Anti-cancer Effect

The impact of vimang and mangniferin in vascular smooth muscle cells of mesenteric normaltensive and spontaneously hypertensive arteries (WCY) on expression of COX-2 and iNOS and noradrenaline-induced vasoconstriction was examined (SHR) in rats with and without interleukin-1 (1 ng/ml; 24 h) stimulation. In the absence of IL-1, Vimang (0.5 mg/ml) and mangiferin (0.025 mg/ml) had no effect on iNos or COX-2 vascular expression in either strain; however, they may prevent 2 IL-1β-induced enzymes, which suggest a strong anti-inflammatory function. The elimination of contractions of noradrenalin (0,1–30 μM), both in WKY and SHR rats, Vimang (1 mg/ml, 05 mg/ml and 25 mg/ml), but not mangiferin (1 mg/ml and 05 mg/ml) indicates that various compounds will mediate the inhibitor effect of Vimang on vasoconstrictors' reactions and on COX-2 and iNOS expression (Mirza et al. 2020).

Antidiabetic Activity

Mangiferin is very effective for streptozotocin-induced diabetic rats (Aswal et al. 2020). Diabetic experimental rats were given mangiferin and the diabetic control rats were supplied without or with insulin (positive control) each day for a period of 28 days. As predicted, streptozotocin treatment resulted in significant decline in catalase and superoxide dismutase function in the kidney. The functions increased in the heart muscles probably due to some alternative compensatory mechanism and was unchanged in the red blood corpuscles. It also enhanced the MDA, glycosylated Hb, creatine phosphokinase, glucose, tuberculosis, total cholesterol, low density lipoproteins, and diminished high-density lipoproteins in both the tissues (Muruganandan et al. 2005, Duang et al. 2011).

Mangiferin also offers antidiabetic effect through alternative pathways other than pancreatic ß-cell insulin release/secretion. The mechanisms may include stimulating the utilization of glucose by the peripheral tissues, enhancing glycogenesis and glycolysis pathways and reduction in glycemic response by inhibiting glucose intake (Kumar et al. 2013a, Aswal et al. 2020, Kumar et al. 2021). Owing to these properties, it can be said that mango by products are highly beneficial in numerous ways. However, their benefits are yet to be exploited for the use of mankind.

CONCLUSION

Mango is a major fruit crop known for its peculiar flavour. It is one of the highest grown fruits in the world after banana. The mango pulp is nutritionally rich, therefore utilized in several food preparations. The seed and peel are considered as the by-products of mango based industries. The nutritional content and phytochemical profile of these is no less. The scientific evidence regarding the commercial and food industry-based utilization of mango by-products has been summarized in this chapter. The functional characteristics of discarded by-products of mango processing units/industries uncovers novel and unexploited area with a vast scope of waste valorisation. The peel is rich in nutrients mainly fibre and carotenoids whereas the seed contains oil which is highly useful in cosmetic industry due to its characteristic like cocoa butter. Besides huge potential, there are limited applications of mango by-products due to lack of sustainable and active packaging for the products. Therefore, research is required regarding customizing the packing to preserve the beneficial features of packed products. Another limitation in revalorization of mango by-products in food industry is the lack of sensorial analysis. There is also need to characterize the compounds responsible for off-flavours. More research is required on extraction methods and utilization of mango by-products to find out novel products from these by products.

REFERENCES

Abdelaziz, S.A. 2018. Physico-chemical characteristics of mango kernel oil and meal. Middle East Journal of Applied Sciences. 8(1): 01–06.

Adilah, Z.M., Jamilah, B. and Hanani, Z.N. 2018. Functional and antioxidant properties of protein-based films incorporated with mango kernel extract for active packaging. Food Hydrocolloids. 74: 207–218.

Ajila, C.M., Naidu, K.A., Bhat, S.G. and Rao, U.P. 2007. Bioactive compounds and antioxidant potential of mango peel extract. Food Chemistry. 105(3): 982–988.

Ajila, C.M., Aalami, M., Leelavathi, K. and Rao, U.P. 2010. Mango peel powder: a potential source of antioxidant and dietary fiber in macaroni preparations. Innovative Food Science and Emerging Technologies. 11(1): 219–224.

Ajila, C.M. and Rao, U.P. 2013. Mango peel dietary fibre: composition and associated bound phenolics. Journal of Functional Foods. 5(1): 444–450.

Amid, M., Tan, C.P., Mirhosseini, H., Aziz, N.A. and Ling, T.C. 2011. Optimisation of serine protease extraction from mango peel (Mangifera Indica Cv. Chokanan. Food Chemistry. 124(2): 666–671.

Ashoush, I.S. and Gadallah, M.G.E. 2011. Utilization of mango peels and seed kernels powders as sources of phytochemicals in biscuit. World Journal of Dairy and Food Sciences. 6(1): 35–42.

Asif, A., Farooq, U., Akram, K., Hayat, Z., Shafi, A., Sarfraz, F., et al. 2016. Therapeutic potentials of bioactive compounds from mango fruit wastes. Trends in Food Science and Technology. 53: 102–112.

Aslam, H.K.W., Raheem, M.I.U., Ramzan, R., Shakeel, A., Shoaib, M. and Sakandar, H.A. 2014. Utilization of mango waste material (peel, kernel) to enhance dietary fiber content and antioxidant properties of biscuit. Journal of Global Innovations in Agricultural and Social Sciences. 2(2): 76–81.

Aswal, S., Kumar, A., Chauhan, A., Semwal, R.B., Kumar, A. and Semwal, D.K. 2020. A molecular approach on the protective effects of mangiferin against diabetes and diabetes-related complications. Current Diabetes Reviews. 16(7): 690–698.

Awolu, O.O. and Manohar, B. 2019. Quantitative and qualitative characterization of mango kernel seed oil extracted using supercritical CO_2 and solvent extraction techniques. Heliyon. 5(12).

Babaria, P.M. and Vashney, A.K. 2011. Extraction of Oil from Mango Kernel by Hydraulic Pressing. Lap Lambert Academic Publishing GmbH KG.

Balacuit, J.N.G., Guillermo, J.D.A., Buenafe, R.J.Q. and Soriano, A.N. 2021. Comparison of Microwave-assisted extraction to soxhlet extraction of mango seed kernel oil using ethanol and n-hexane as solvents. ASEAN Journal of Chemical Engineering. 21(2): 158–169.

Barreto, J.C., Trevisan, M.T., Hull, W.E., Erben, G., De Brito, E.S., Pfundstein, B., et al. 2008. Characterization and quantitation of polyphenolic compounds in bark, kernel, leaves, and peel of mango (*Mangifera indica* L.). Journal of Agricultural and Food Chemistry. 56(14): 5599–5610.

Che Rahim, A. 2019. Anti-proliferative effect of methyl gallate isolated from Mangifera pajang kosterman in selected cancer cell lines (Doctoral dissertation. Universiti Tun Hussein Onn Malaysia).

de França, L.F., Reber, G., Meireles, M.A.A., Machado, N.T. and Brunner, G. 1999. Supercritical extraction of carotenoids and lipids from buriti (*Mauritia flexuosa*), a fruit from the Amazon region. The Journal of Supercritical Fluids. 14(3): 247–256.

Deeksha, K. and Sunita, M. 2020. Utilization of mango and its by-products by different processing methods. Asian Journal of Science and Technology. 9. 8896–8910.

del Pilar Sanchez-Camargo, A., Gutierrez, L.F., Vargas, S.M., Martinez-Correa, H.A., Parada-Alfonso, F. and Narvaez-Cuenca, C.E. 2019. Valorisation of mango peel: Proximate composition, supercritical fluid extraction of carotenoids, and application as an antioxidant additive for an edible oil. The Journal of Supercritical Fluids. 152. 104574.

Duang, X.Y., Wang, Q., Zhou, X.D. and Huang, D.M. 2011. Mangiferin: a possible strategy for periodontal disease to therapy. Medical Hypotheses. 76(4): 486–488.

Ediriweera, M.K., Tennekoon, K.H. and Samarakoon, S.R. 2017. A review on ethnopharmacological applications, pharmacological activities, and bioactive compounds of *Mangifera indica* (Mango). Evidence-Based Complementary and Alternative Medicine. 2017: 6949835. doi: 10.1155/2017/6949835.

Fowomola, M.A. 2010. Some nutrients and antinutrients contents of mango (*Magnifera indica*) seed. African Journal of Food Science. 4(8): 472–476.

Garcia-Mendoza, M.P., Paula, J.T., Paviani, L.C., Cabral, F.A. and Martinez-Correa, H.A. 2015. Extracts from mango peel by-product obtained by supercritical CO_2 and pressurized solvent processes. LWT-Food Science and Technology. 62(1): 131–137.

Gondi, M., Basha, S.A., Bhaskar, J.J., Salimath, P.V. and Prasada Rao, U.J. 2015. Anti-diabetic effect of dietary mango (*Mangifera indica* L.) peel in streptozotocin-induced diabetic rats. Journal of the Science of Food and Agriculture. 95(5): 991–999.

Hamzah, N., Wan-Ishak, W.R. and Rahman, N.A. 2018. Nutritional and pharmacological properties of agro-industrial by-products from commonly consumed fruits. Journal of Food Science and Technology. 3(4): 396–416.

Hannan, A., Asghar, S., Naeem, T., Ullah, M.I., Ahmed, I., Aneela, S., et al. 2013. Antibacterial effect of mango (*Mangifera indica* Linn.) leaf extract against antibiotic-sensitive and multi-drug resistant *Salmonella typhi*. Pakistan Journal of Pharmaceutical Sciences. 26(4): 715–719.

Hassan, F.A., Ismail, A., Hamid, A.A., Azlan, A. and Al-sheraji, S.H. 2011. Characterisation of fibre-rich powder and antioxidant capacity of *Mangifera pajang* K. fruit peels. Food Chemistry. 126(1): 283–288.

Hu, H., Zhao, Q., Pang, Z., Xie, J., Lin, L. and Yao, Q. 2018. Optimization extraction, characterization and anticancer activities of polysaccharides from mango pomace. International Journal of Biological Macromolecules. 117. 1314–1325.

Jahurul, M.H.A., Zaidul, I.S.M., Ghafoor, K., Al-Juhaimi, F.Y., Nyam, K.L., Norulaini, N.A.N., et al. 2015. Mango (*Mangifera indica* L.) by-products and their valuable components: a review. Food Chemistry. 183. 173–180.

Karunanithi, B., Bogeshwaran, K., Tripuraneni, M. and Krishna Reddy, S. 2015. Extraction of mango seed oil from mango kernel. International Journal of Engineering Research and Development. 11(11): 32–41.

Kaur, B., Panesar, P.S., Anal, A.K. and Chu-Ky, S. 2022. Recent trends in the management of mango by-products. Food Reviews International. 1–21.

Kayathi, A., Chakrabarti, P.P., Bonfim-Rocha, L., Cardozo-Filho, L. and Jegatheesan, V. 2020. Selective extraction of polar lipids of mango kernel using supercritical carbon dioxide (SC–CO_2) extraction: process optimization of extract yield/phosphorous content and economic evaluation. Chemosphere. 260. 127639.

Khammuang, S. and Sarnthima, R. 2011. Antioxidant and antibacterial activities of selected varieties of Thai mango seed extract. Pakistan Journal of Pharmaceutical Sciences. 24(1): 37–42.

Khare, P. and Shanker, K. 2016. Mangiferin: A review of sources and interventions for biological activities. BioFactors. 42(5): 504–514.

Kim, Y., Brecht, J.K. and Talcott, S.T. 2007. Antioxidant phytochemical and fruit quality changes in mango (*Mangifera indica* L.) following hot water immersion and controlled atmosphere storage. Food Chemistry. 105(4): 1327–1334.

Kim, H., Moon, J.Y., Kim, H., Lee, D.S., Cho, M., Choi, H.K., et al. 2010. Antioxidant and antiproliferative activities of mango (*Mangifera indica* L.) flesh and peel. Food Chemistry. 121(2): 429–436.

Kittiphoom, S. 2012. Utilization of mango seed. International Food Research Journal. 19(4): 1325.

Kittiphoom, S. and Sutasinee, S. 2013. Mango seed kernel oil and its physicochemical properties. International Food Research Journal. 20(3): 1145.

Knödler, M., Wenzig, E.M., Bauer, R., Conrad, J., Beifuss, U., Carle, R., et al. 2007. Cyclooxygenase inhibitory 5-alkenylresorcinols isolated from mango (*Mangifera indica* L.) peels. Planta Medica. 73(9): 031.

Kumar, B.D., Krishnakumar, K., Jaganathan, S.K. and Mandal, M. 2013a. Effect of mangiferin and mahanimbine on glucose utilization in 3T3-L1 cells. Pharmacognosy Magazine. 9(33): 72.

Kumar, D., Yadav, K.K., Muthukumar, M. and Garg, N. 2013b. Production and characterization of [alpha]-amylase from mango kernel by *Fusarium solani* NAIMCC-F-02956 using submerged fermentation. Journal of Environmental Biology. 34(6): 1053.

Kumar, M., Saurabh, V., Tomar, M., Hasan, M., Changan, S., Sasi, M., et al. 2021. Mango (*Mangifera indica* L.) leaves: Nutritional composition, phytochemical profile, and health-promoting bioactivities. Antioxidants. 10(2): 299.

Lebaka, V.R., Wee, Y.J., Ye, W. and Korivi, M. 2021. Nutritional composition and bioactive compounds in three different parts of mango fruit. International Journal of Environmental Research and Public Health. 18(2): 741.

Leiro, J., Arranz, J.A., Yáñez, M., Ubeira, F.M., Sanmartín, M.L. and Orallo, F. 2004. Expression profiles of genes involved in the mouse nuclear factor-kappa B signal transduction pathway are modulated by mangiferin. International Immunopharmacology. 4(6): 763–778.

López-Cobo, A., Verado, V., Diaz-de-Cerio, E., Segura-Carretero, A., Fernandez-Gutiérrez, A. and Gomez-Caravaca, A.M. 2017. Use of HPLC- and GC–QTOF to determine hydrophilic and lipophilic phenols in mango (*Mangifera indica* L.) fruit and its by-products. Food Research International. 100: 423–434. doi: 10.1016/j.foodres. 2017.02.008.

Ma, X., Wu, H., Liu, L., Yao, Q., Wang, S., Zhan, R., et al. 2011. Polyphenolic compounds and antioxidant properties in mango fruits. Scientia Horticulturae. 129(1): 102–107.

Mahmoud, A.E.E.D., Omer, H.A.A.A., Mohammed, A.T. and Ali, M.M. 2020. Enhancement of chemical composition and nutritive value of some fruit pomace by solid-state fermentation. Egyptian Journal of Chemistry. 63(10): 3713–3720.

Matkowski, A., Kus, P., Goralska, E. and Wozniak, D. 2013. Mangiferin–a bioactive xanthonoid, not only from mango and not just antioxidant. Mini Reviews in Medicinal Chemistry. 13(3): 439–455.

Mirza, B., Croley, C.R., Ahmad, M., Pumarol, J., Das, N., Sethi, G., et al. 2020. Mango (*Mangifera indica* L.) a magnificent plant with cancer preventive and anticancer therapeutic potential. Critical Reviews in Food Science and Nutrition. 1: 27.

Moskaug, J.Ø., Carlsen, H., Myhrstad, M. and Blomhoff, R. 2004. Molecular imaging of the biological effects of quercetin and quercetin-rich foods. Mechanisms of Ageing and Development. 125(4): 315–324.

Muñiz-Valencia, R., Portillo-Pérez, G., Ceballos-Magaña, S.G., Cortés-Quintero, G.C., Nava-García, A.Y., Dumont, M.J., et al. 2020. Utilization of mango wastes as a potential feedstock for the production of HMF. Biomass Conversion and Biorefinery. 1–8.

Muruganandan, S., Srinivasan, K., Gupta, S., Gupta, P.K. and Lal, J. 2005. Effect of mangiferin on hyperglycemia and atherogenicity in streptozotocin diabetic rats. Journal of Ethnopharmacology. 97(3): 497–501.

Nadeem, M., Imran, M. and Khalique, A. 2016. Promising features of mango (*Mangifera indica* L.) kernel oil: a review. Journal of Food Science and Technology. 53(5): 2185–2195.

Nadeem, M., Imran, M., Iqbal, Z., Abbas, N. and Mahmud, A. 2017. Enhancement of the oxidative stability of butter oil by blending with mango (*Mangifera indica* L.) kernel oil in ambient and accelerated oxidation. Journal of Food Processing and Preservation. 41(3): e12957.

Nagarajaiah, S.B. and Prakash, J. 2016. Chemical composition and bioactivity of pomace from selected fruits. International Journal of Fruit Science. 16(4): 423–443.

Nayeem, N., Asdaq, S.M.B., Salem, H. and AHEl-Alfqy, S. 2016. Gallic acid: a promising lead molecule for drug development. Journal of Applied Pharmacy. 8(2): 1–4.

Nzikou, J.M., Kimbonguila, A., Matos, L., Loumouamou, B., Pambou-Tobi, N.P.G., Ndangui, C.B., et al. 2010. Extraction and characteristics of seed kernel oil from mango (*Mangifera indica*). Research Journal of Environmental and Earth Sciences. 2(1): 31–35.

Oktay, M., Culcin, I. and Kufrevioglu, O.I. 2003. Determination of *in vitro* antioxidant activity of fennel (*Foeniculum vulgare*) seed extracts. LWT–Food Science and Technology. 36: 263–271.

Onuh, J.O., Momoh, G., Egwujeh, S. and Onuh, F. 2017. Evaluation of the nutritional, phytochemical, and antioxidant properties of the peels of some selected mango varieties. American Journal of Food Science and Technology. 5(5): 176–181.

Padhi, S., Sarkar, P., Abedin, M.M., Sirohi, R., Singh, S.P. and Rai, A.K. 2023. Microbial enzymes and metabolites in health care: an overview. pp. 3–23. *In*: Sirohi, R., Rai A.K., de Souza Vandenberghe, L.P. and Parameswaram, B. (eds). Microbial Enzymes and Metabolites for Health and Well-Being. CRC Press,

Pal, S.M., Avneet, G. and Siddhraj, S.S. 2018. Gallic acid: Pharmacogical promising lead molecule: a review. International Journal of Pharmacognosy and Phytochemical Research. 10: 132–138.

Park, S.N., Lim, Y.K., Freire, M.O., Cho, E., Jin, D. and Kook, J.K. 2012. Antimicrobial effect of linalool and α-terpineol against periodontopathic and cariogenic bacteria. Anaerobe. 18(3): 369–372.

Parvez, G.M. 2016. Pharmacological activities of mango (*Mangifera indica*): a review. Journal of Pharmacognosy and Phytochemistry. 5(3): 01–07.

Patnaik, S.S., Simionescu, D.T., Goergen, C.J., Hoyt, K., Sirsi, S. and Finol, E.A. 2019. Pentagalloyl glucose and its functional role in vascular health: biomechanics and drug-delivery characteristics. Annals of Biomedical Engineering. 47(1): 39–59.

Pereira, A.D.S., Fontes-Sant'Ana, G.C. and Amaral, P.F. 2019. Mango agro-industrial wastes for lipase production from *Yarrowia lipolytica* and the potential of the fermented solid as a biocatalyst. Food and Bioproducts Processing. 115: 68–77.

Purnachandra Reddy, M. and Saritha, K.V. 2015. Bio-catalysis of mango industrial waste by newly isolated *Fusarium* sp. (PSTF1) for pectinase production. 3 Biotech. 5: 893–900.

Quiles, A., Campbell, G.M., Struck, S., Rohm, H. and Hernando, I. 2018. Fiber from fruit pomace: a review of applications in cereal-based products. Food Reviews International. 34(2): 162–181.

Rabetafika, H.N., Bchir, B., Blecker, C., Paquot, M. and Wathelet, B. 2014. Comparative study of alkaline extraction process of hemicelluloses from pear pomace. Biomass and Bioenergy. 61: 254–264.

Rajendran, P., Jayakumar, T., Nishigaki, I., Ekambaram, G., Nishigaki, Y., Vetriselvi, J., et al. 2013. Immunomodulatory effect of mangiferin in experimental animals with benzo(a)pyrene-induced lung carcinogenesis. International Journal of Biomedical Science: IJBS. 9(2): 68.

Rodríguez De Luna, S.L., Ramírez-Garza, R.E. and Serna Saldívar, S.O. 2020. Environmentally friendly methods for flavonoid extraction from plant material: impact of their operating conditions on yield and antioxidant properties. The Scientific World Journal. 2020.

Rojas, R., Contreras-Esquivel, J.C., Orozco-Esquivel, M.T., Muñoz, C., Aguirre-Joya, J.A. and Aguilar, C.N. 2015. Mango peel as source of antioxidants and pectin: microwave assisted extraction. Waste and Biomass Valorization. 6: 1095–1102.

Rummun, N., Pires, E., McCullagh, J., Claridge, T.W., Bahorun, T., Li, W.W., et al. 2021. Methyl gallate–rich fraction of *Syzygium coriaceum* leaf extract induced cancer cell cytotoxicity via oxidative stress. South African Journal of Botany. 137: 149–158.

Sagar, N.A., Pareek, S., Sharma, S., Yahia, E.M. and Lobo, M.G. 2018. Fruit and vegetable waste: bioactive compounds, their extraction, and possible utilization. Comprehensive Reviews in Food Science and Food Safety. 17(3): 512–531.

Sancho, S.D.O., da Silva, A.R.A., Dantas, A.N.D.S., Magalhães, T.A., Lopes, G.S., Rodrigues, S., et al. 2015. Characterization of the industrial residues of seven fruits and prospection of their potential application as food supplements. Journal of Chemistry, 2015: 264284. https://doi.org/10.1155/2015/264284.

Sandhu, K.S. and Lim, S.T. 2008. Structural characteristics and *in vitro* digestibility of mango kernel starches (*Mangifera indica* L.). Food Chemistry. 107(1): 92–97.

Schieber, A., Berardini, N. and Carle, R. 2003. Identification of flavonol and xanthone glycosides from mango (*Mangifera indica* L. Cv."Tommy Atkins") peels by high-performance liquid chromatography-electrospray ionization mass spectrometry. Journal of Agricultural and Food Chemistry. 51(17): 5006–5011.

Sharif, T., Bhatti, H.N., Bull, I.D. and Bilal, M. 2023. Recovery of high-value bioactive phytochemicals from agro-waste of mango (*Mangifera indica* L.) using enzyme-assisted ultrasound pretreated extraction. Biomass Conversion and Biorefinery. 13(8): 6591–6599.

Sikdar, D.C., Hegde, S., Swamynathan, V., Varsha, S. and Rakesh, R. 2017. Solvent extraction of mango (*Mangifera indica* L.) seed kernel oil and its characterization. International Journal of Technical Research and Applications. 5(4): 43–47.

Singh, S.K., Sharma, V.K., Kumar, Y., Kumar, S.S. and Sinha, S.K. 2009. Phytochemical and pharmacological investigations on mangiferin. Herba Polonica. 55(1): 126–139.

Sogi, D.S., Siddiq, M., Greiby, I. and Dolan, K.D. 2013. Total phenolics, antioxidant activity, and functional properties of 'Tommy Atkins' mango peel and kernel as affected by drying methods. Food Chemistry. 141(3): 2649–2655.

Soong, Y.Y. and Barlow, P.J. 2004. Antioxidant activity and phenolic content of selected fruit seeds. Food Chemistry. 88(3): 411–417.

Soong, Y.Y. and Barlow, P.J. 2006. Quantification of gallic acid and ellagic acid from longan (Dimocarpus longan Lour.) seed and mango (*Mangifera indica* L.) kernel and their effects on antioxidant activity. Food Chemistry. 97(3): 524–530.

Sultana, B. and Ashraf, R. 2019. Mango (*Mangifera indica* L.) seed oil. Fruit Oils: Chemistry and Functionality. 561–575.

Telang, M., Dhulap, S., Mandhare, A. and Hirwani, R. 2013. Therapeutic and cosmetic applications of mangiferin: a patent review. Expert Opinion on Therapeutic Patents. 23(12): 156.

Thambi, P.A., John, S., Lydia, E., Iyer, P. and Monica, S.J. 2016. Antimicrobial efficacy of mango peel powder and formulation of recipes using mango peel powder (*Mangifera indica* L.). International Journal of Home Science. 2(2): 155–161.

Usta, C., Ozdemir, S., Schiariti, M. and Puddu, P.E. 2013. The pharmacological use of ellagic acid-rich pomegranate fruit. International Journal of Food Sciences and Nutrition. 64(7): 907–913.

Varakumar, S., Kumar, Y.S. and Reddy, O.V.S. 2011. Carotenoid composition of mango (*Mangifera indica* L.) wine and its antioxidant activity. Journal of Food Biochemistry. 35(5): 1538–1547.

Vijay, K.V., Rahaman, S.M., Wadhwani, M.K., Kumari, S., Kumari, M., Kumar, S., et al. 2019. Growing mango: a profitable livelihood option for Bhagalpur farmers. Journal of Crop and Weed. 15(3): 25–27.

Walia, N.K., Bedi, S.S., Kundu, K. and Karmakar, R. 2013. Production of bioethanol from mango peel. International Journal of Engineering Research and Technology. 2(1).

Wall-Medrano, A., Olivas-Aguirre, F.J., Ayala-Zavala, J.F., Domínguez-Avila, J.A., Gonzalez-Aguilar, G.A., Herrera-Cazares, L.A., et al. 2020. Health benefits of mango by-products. Food Wastes and By-products: Nutraceutical and Health Potential. 159–191.

Wang, W., Sun, C., Mao, L., Ma, P., Liu, F., Yang, J., et al. 2016. The biological activities, chemical stability, metabolism and delivery systems of quercetin: a review. Trends in Food Science & Technology. 56: 21–38.

Yang, N. and Li, W.H. 2013. Mango peel extract mediated novel route for synthesis of silver nanoparticles and antibacterial application of silver nanoparticles loaded onto non-woven fabrics. Industrial Crops and Products. 48: 81–88.

Zahoor, S., Anwar, F., Qadir, R., Soufan, W. and Sakran, M. 2023. Physicochemical attributes and antioxidant potential of kernel oils from selected mango varieties. ACS Omega. 8(25): 22613–22622.

Chapter **3**

High Value Utilization of Sugar and Starch Industry Waste by Biotechnological Approaches

Hamed I. Hamouda[1,2,3,4], Hong Jiang*[1,2,3], Samah Shabana[5] and Xiangzhao Mao*[1,2,3]

[1]Qingdao Key Laboratory of Food Biotechnology, College of Food Science and Engineering, Ocean University of China, Qingdao, China.

[2]Laboratory for Marine Drugs and Bioproducts of Qingdao National Laboratory for Marine Science and Technology, Qingdao, China.

[3]Key Laboratory of Biological Processing of Aquatic Products, China National Light Industry, Qingdao, China.

[4]Processes Design and Development Department, Egyptian Petroleum Research Institute, Cairo 11727, Egypt.

[5]Department of Biochemistry and Molecular Biology, College of Marine Life Sciences, Ocean University of China, Qingdao 266003, PR China.

INTRODUCTION

Even though there have been several attempts to encourage "reuse, recycle, and reduce," or the sequential economy system, we have been accustomed to the "take, consume, and discard" way of life for generations. The world has been "stockpiling" wastes throughout time, and as a result, the soil, the forest, and the

* For Correspondence: jh@ouc.edu.cn; xzhmao@ouc.edu.cn

water are no longer capable of regenerating (Kee et al. 2022). This has encouraged policymakers, investors, governments, and NGOs to improve their strategy by steadily moving toward a circular economy system, a renewable economic system that promotes the use of waste, and uses fewer new raw materials to create innovative products (Kee et al. 2021). Agriculture-related feedstocks were found as excellent precursors for the synthesis of a wide range of value-added enzymes (Jain et al. 2022). Utilizing these renewable wastes as feedstock is particularly useful due to their ease of availability and abundant nitrogen and carbon content. At the industrial scale, there are several cost-effective technologies in use for recycling agro-industrial byproducts that promote microbial growth for the production of value-added products (Sadh et al. 2018). Based on a previous study, as the agricultural industry becomes more industrialized, there will be a rise in waste, posing an environmental hazard (Tripathi and Shrivastava 2019). The annual biomass production from farmland was estimated to be around 5 million tons. All of these factors prompted the researchers to consider the numerous methods for converting those cellulosic agricultural wastes into precursor products for different enzyme industries, hence reducing the environmental load caused by these residues (Figure 3.1).

Figure 3.1 Insights on sugar and starch by-product for production and applications of value-added products.

As the global population grows, so will the demand for enzymes, and recycling waste is one of the most effective ways to satisfy these worldwide demands (Raina et al. 2022). Agro-biomasses and a few other important feedstocks, such as sugar and starch by-product are already being utilized to produce a range of different goods, particularly vitamins, antioxidants, biofuels, etc., as shown in Figure 3.2 (Lin et al. 2021, Low et al. 2021). Hence, this study will focus on the value addition of sugar and starch by-products. Maize cob, oil cake, peel, pulp, seeds, lint, bagasse from sugarcane or sweet sorghum milling, and many other biomasses are commonly produced in the starch and sugar industries (Pandit et al. 2021). As revealed by numerous types of research, currently, 60% of industrial enzyme synthesis was from fungi, 24% from bacteria, 4% from yeast, 2% from *Streptomyces*, 6% from higher animals, and 4% from plants (Raina et al. 2022).

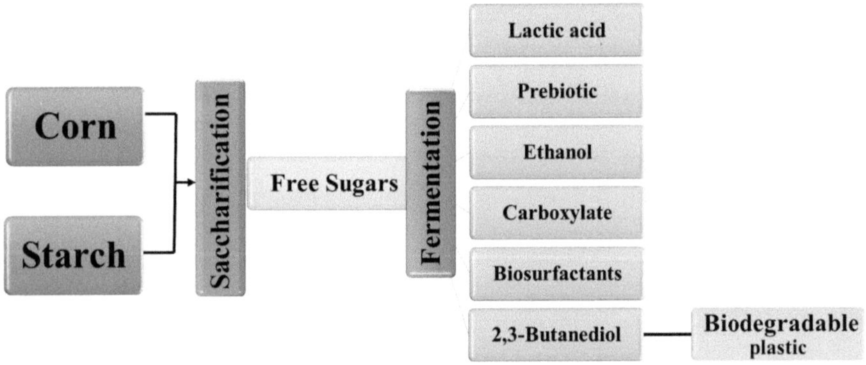

Figure 3.2 Value-added bioproducts from the low-cost sugar and starch lignocellulosic by-product.

Lactic bacteria may convert the sugars in the hydrolysates into lactic acid, which can then be catalytically reduced to propylene glycol. Microorganisms may also consume the sugars in the cell division process multiplication, and the obtained biomass can be utilized as a source of protein in fermentation, animal feed, nutrient extraction, and different industries (Modelska et al. 2017).

SUGAR INDUSTRY

Agricultural residues are wastes generated during the refining process of agricultural yield into products (Sadh et al. 2018). Agro-industrial wastes have a severe influence on the environment since wastewater is frequently discharged into the water supply and food and finally deposited in landfills. To develop a recycling industry, these waste substances might be converted into useful products or bioenergy (Najari et al. 2022). Due to the fast exhaustion of natural resources, waste accumulation, and the demand for better environmental and cost-effective treatment of waste have received attention (Arancon et al. 2017, Kee et al. 2022).

Sugar Plants

Sugar plants, like sweet sorghum (*Sorghum bicolor* L. Moench), sugar and energy beets (*Beta vulgaris*), and sugar and energy cane (*Saccharum officinarum*), are the most important crops globally with top producers being China, Brazil, India, Mexico, and Thailand (Formann et al. 2020). These are attractive, natural feedstocks due to their abundance, high sugar contents, and high carbohydrate yield per hectare (Sharma et al. 2021). Sucrose (-D-glucopyranosyl-(1 2)—D-fructofuranose) is widely used as refined sugar, and crystallized sucrose is predominantly obtained commercially from sugar beets (*Beta vulgaris*) and sugarcane (*Saccharum officinarum*) (Figure 3.2). Similar to several other chemical and food technology, the sugar and associated sugar-bioproduct sectors are presently encountering serious environmental issues (Eggleston and Lima 2015). The FAO has reported annual production of sugar cane of about 1.89 billion tons in over more than 100 countries (Formann et al. 2020). World sugar production reaches 165 million tons annually, with sugar cane contributing 80% of global sugar output, and sugar beets accounting for only 20% (Koo and Taylor 2014). During the milling process of the sugar industry, massive amounts of byproducts are generated.

Sugar beet is one of the world's most significant crops. Over 600 million tons of sugar are manufactured each year. Saccharose is the major product of sugar beet processing (Draycott 2008). However, following saccharose extraction, roughly 17 million tons of dry biomass or 68 million tons of wet sugar beet pulp (SBP) are produced globally (Kracher et al. 2014, Modelska et al. 2017). Sugar beet molasses is an available and low-cost by-product of sugar production. Molasses consists of complex lignocellulosic carbon-containing molecules that may be utilized for microbial growth as a low-cost carbon source and for the synthesis of valuable microbial metabolites such as amino acids, protein-enriched animal feed, enzymes, antibiotics, and organic acids (Mustafa et al. 2020).

While significant research has been conducted on conversion technologies of bioproducts into biofuels still customers want the highest feedstock quality and stability and reduced costs, and therefore, the focus nowadays is moving to establish long-term supply chains of biomass having good amount of sugar for the new, flexible bioenergy (Figure 3.3). All waste from sugar plants can be used as

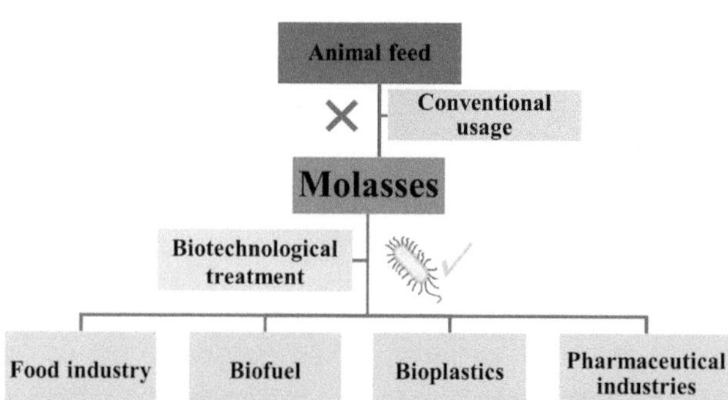

Figure 3.3 Value-added bioproducts and enzymes production from the low-cost sugar by-product.

feedstock. Electricity production of bioenergy from bagasse and leaf leftovers is becoming more widespread in more regions, and because of the high carbon content of bagasse and leaves, it may also be transformed into value-added substances such as biochar. Sugar plants are exceptional renewable sources for the synthesis of platform chemicals, which are used to produce a variety of end products such as chemicals, biomaterials, and bioplastics. Although new technologies for converting high-fiber sugar plant feedstock into biofuel have been established, large-scale commercialization has been limited (Eggleston and Lima 2015).

Major Sugar-Industrial By-Products

Molasses

Molasses is one of the most significant byproducts that might be utilized by microorganisms as a substrate to produce economic substances including methane and bioethanol (Lee et al. 2020). It is produced from the sequential operations of continuous centrifugation, evaporation, and crystallization of sugar juices during the process of table sugar production. Molasses is a black, thick liquid that contains sugars (~50% sucrose), salts, metal ions, amino acids, suspended colloids, and vitamins (Sharma et al. 2016). Due to the depletion of fossil fuels and other nonrenewable sources as well as the unstable market pricing for these feedstocks, there is an urgent demand for energy production from renewable feedstocks across the globe (Figure 3.3) (Lee et al. 2020, Wadhwa et al. 2016).

Sugar Beet Pulp

Sugar beet pulp was used for animal nutrition because of its high protein level, and because of its high sugar and low lignin concentration, it may be utilized to produce biogas or, in areas without animal cultivation, dumped in landfills (Tomaszewska et al. 2018). In the last few years, as a viable feedstock for enzymatic hydrolysis and subsequent conversion of biomass to value-added products, sugar beet pulp has become more popular.

Value-Added Sugar By-Products

Therapeutics

Glu (D-glucose), GalAc (D-galacturonic acid), and Ara (L-arabinose) are the most abundant monosaccharides of the sugar beet pulp; Ara is considered as a significant precursor material in pharmaceutical science (Foster et al. 2015). Sato and colleagues effectively recovered ferulic acid (FA) in the form of ferulate esters of oligosaccharides from sugar beet pulp via hydrothermal treatment and observed that >80% of arabinan concentration may be reclaimed as feruloylated arabinose oligosaccharides and arabinose oligosaccharides (Sato et al. 2013). Alkaline digestion was used to extract a high FA concentration from sugar beet pulp, and the extract was proven to have antioxidant properties (Aarabi et al. 2016).

Prebiotic Production

Sugar cane molasses as a by-product from the sugar industry was utilized as a cost-effective substrate, and sole carbon source for *Leuconostoc mesenteroides* MTCC 10508 growth, functional monosaccharide, and oligosaccharides biosynthesis. In this reaction, molasses was used as an acceptor by dextransucrase from *L. mesenteroides* to produce about 124 g oligosaccharides (DP3-DP6) per kg of molasses (Sharma et al. 2016). Prebiotics were defined as non-digestible short-chain carbohydrates (SCCs) that improve the host's health physiologically by regulating the probiotics' colonic growth and/or action following digestion (Bindels et al. 2015). Probiotic substances are resistant to stomach acidity and digestion by mammalian enzymes (Davani-Davari et al. 2019, Chourasia et al. 2021, Abedin et al. 2023). Its fermentation susceptibility by intestinal microbiota specifically promotes the vitality and activity of beneficial bacteria in the host's gut. Non-digestible oligosaccharides, such as fructooligosaccharides (FOS) and galactooligosaccharides (GOS), are frequently included in industrial prebiotics (Usmani et al. 2021b). The human digestive system cannot absorb prebiotic dietary oligosaccharides because they are highly resistant to enzymatic hydrolysis. Therefore, promoting the growth of friendly bacteria in the colon through probiotics has a significant chance of human health and preventing a variety of diet-related diseases (Gibson et al. 2004).

Antioxidative Compounds Production

Molasses is a potentially beneficial byproduct of the sugarcane processing sector that is high in antioxidants. It is a potential byproduct of sugarcane processing that is high in antioxidants. A previous study obtained neolignan glucosides as antioxidant agents and examined their capability to prevent DNA oxidation caused by free radicals of 2,2′-azobis (2-amidinopropane) dihydrochloride (AAPH)-peroxyl and hydrogen peroxide (H_2O_2)-hydroxyl. The result proves that the neolignan glucosides in sugarcane molasses scavenged the tested free radicals and prevented oxidative damage to the DNA (Asikin et al. 2016).

Lactate Production

Although lactate generation has not yet been a significant component of any sugarcane biorefinery, it is expected to be an electron source for chain extension and a preservation reagent for feedstock to keep it from spoiling while being stored. Also, lactate is considered as a monomeric unit of polylactate (PLA). Lactic acid, as a preservative and an acidulant, is added to different food products, such as cheese, beer, jellies, and meat. It is used in the pharmaceutical industry, as a flavor ingredient to improve the taste of oriental products, pickling processes, fermented foods, and dairy products (Formann et al. 2020, Chourasia et al. 2022, Chilakamarry et al. 2022, Abedin et al. 2023). Currently, in Thailand PLA is produced at a commercial scale from renewable agricultural feedstocks such as sugarcane juice via fermentation and polymerization (Morão and De Bie 2019). Rojan et al. (2005) explored the optimization of L-lactic acid production

Lactobacillus casei cultured on a combination of cassava starch and sugarcane bagasse. However, PLA production from sugarcane wastes is less developed, and with the growing use of 3D printing, PLA is becoming more and more popular (Nagarajan et al. 2016).

Cellulose Nanofibrils

High-strength polymeric fibrils produced from renewable agricultural feedstocks named cellulose nanofibrils (CNFs) have potential properties including cheap cost, biodegradability, good potential for recycling, and significant barrier and colloidal properties, which make them suitable raw materials for creating environment friendly technological devices. However, the industrial production of CNFs is mainly based on wood as renewable biomass; recently, sugarcane bagasse is also evaluated as viable biomass used for PLA production (Formann et al. 2020, Marcondes et al. 2020)

Biosurfactants and Viable Bioplastics

As most synthetic plastics used today are non-degradable, plastic waste is a significant source of pollutants. Thus, it is crucial to manage plastic trash internationally to minimize its negative impact on the ecosystem and natural resource depletion. The use of biodegradable polymers in place of petroleum-based, non-biodegradable plastic is an attractive approach to reduce the impact of plastic waste on the environment. A previous study provided insight into the utilization of molasses from sugarcane as an inexpensive carbon source for PHA and biosurfactant production. PHA is a type of bioplastic that is produced intracellularly by different kinds of bacteria like *Acinobacter* sp., *Aeromonas* sp., *Klebsiella* sp., *Bacillus* sp, recombinant *Escherichia coli*, and *Pseudomonas* sp (Cespedes et al. 2018, Yatim et al. 2017). Due to its perfect biodegradability, ecological acceptability, non-toxicity, and biocompatibility, it is widely recognized. Sugarcane utilization might promote the sustainable development goals of the circular economy (Kingsly et al. 2022).

Carboxylate Platform

High water-content sugarcane industry wastes like vinasse and sugarcane filter cake (SFC) are ideally suited for biotechnological applications, particularly those involving the use of microorganisms to produce bioenergy or chemicals. To enhance the moisture content of dryer biomass and to make them acceptable for biological breakdown processes, bagasse may be co-fermented with vinasse or waste streams to enhance their moisture levels. Anaerobic fermentation of sugarcane wastes by bacteria can be carried out to produce carboxylates. The synthesized carboxylates are useful basic compounds with a wide range of uses in the chemical industries or transformed into secondary products by biological processes, e.g., production of fuel, bioplastics, cosmetics, medicines, food and feed additives (Formann et al. 2020).

Furfural

Alkanes, tetrahydrofuran, and tetrahydrofurfuryl alcohol are all produced chemically from furfural, an organic molecule containing a furan ring and a carbonyl functional group (Yan et al. 2014). This substance is also a renewable key component for a variety of products, including plastics and food additives, making it an essential ingredient for agrochemicals like nematicides and fungicides, flavor enhancers, soil enhancers, the pharmaceutical industry, transportation fuel, energy, and books (Rodriguez et al. 2020). Furfural was produced from dehydration and hydrolysis reaction of pentoses sugars (Chandel et al. 2018).

Ethanol

Due to a lack of an efficient treatment system, the massive quantities of agro-waste produced have a severe impact on the environment. Agricultural biomass is very rich in nutrients and used for value-added compounds such as bioethanol (Figure 3.4) (Berlowska et al. 2017). Many countries are attempting to use the concepts of the circular economy to solve the global concerns of waste management and energy generation (Usmani et al. 2021a). Because of its cheap prices, lower octane number, and lower emissions, using bioethanol is both economical and environmentally good (Usmani et al. 2021a). Berlowska et al. (2017) studied bioethanol production using the released sugar from sugar beet pulp using a co-culture of *Scheffersomyces stipiti* and *Saccharomyces cerevisiae*. According to the previous study by (Kamzon et al. 2015), sugar beet may be converted into 37% ethanol, which could then be further distilled to get 95% concentration, to blend with gasoline appropriate to minimize GHG emissions. Lignocgellulosic biomass is degraded into free fermentable reducing sugars using microbial enzymes which can be fermented into ethanol. The following steps are involved in the bioconversion of lignocellulose biomass into ethanol: (i) Pretreatment step to breakdown the lignocellulosic biomass by microbial hydrolytic enzymes to liberate cellulosic compounds (Kamzon et al. 2015), (ii) Saccharification process, in which glucose is released from cellulose and hemicellulose due to action of cellulases and hemicellulases enzymes. Several fungi and bacteria have been reported to create cellulase and hemicellulase, fungi such as *Phanerochaete*, *Fusarium, Trichoderma,* etc., and bacteria such as *Thermonospora, Cellulomonas, Erwinia, Bacillus,* etc. (Nava-Cruz et al. 2016). Several physicochemical and biological factors including temperature, oxygen level, pH, types of microorganisms, substrate type and its concentrations, etc. influence the enzymatic hydrolysis of sugar beet pulp into value-added components (iii) fermentation of the liberated hexoses and pentoses into bioethanol efficiently using various microorganisms including *Zymomonas mobilis* and *Saccharomyces cerevisiae*. Some microorganisms are unable to utilize numerous hydrolysate substrates so, genetically modified strains are required that can utilize wide range of biomass and produce high yield of ethanol (Liu et al. 2018a).

Figure 3.4 Utilization of cane industry residues for the production of biofuel and other value added products (Madian et al. 2022).

Biohydrogen

Biohydrogen production from different renewable biomass such as molasses has gained huge interest due to the higher calorific value (4 and 4.5 times) than those from ethanol and coal, respectively. In contrast with fossil fuels, biohydrogen combustion did not release greenhouse gases (Carlozzi et al. 2019), established a continuous anaerobic fermentation bioreactor system (pilot-scale) for hydrogen production from sugarcane molasses, and the achieved hydrogen yield was about 5.57 m^3/m^3/day (Zhang et al. 2021b). It was clear that the most optimum technique for biohydrogen production was the combination of dark and photo fermentation. This two-step process has a maximum theoretical hydrogen production of 12 mol H_2/mol hexose sugar (Özgür et al. 2010). But when photosynthetic bacteria *Rhodobacter capsulatus* and thermophile *Caldicellulosiruptor saccharolyticus* were used for hydrogen production through the combining of photo and dark fermentation using the molasses as a carbon source, 13.7 mol H_2/mol sucrose was generated. However, the stability and consistency of this integrated two-step process should be examined further (Zhang et al. 2021b).

2,3-Butanediol Production

Butanediol was produced during a fermentation technique based on sugarcane molasses using *Bacillus subtilis* to achieve sustainable economic production. It has numerous potential applications such as the manufacture of perfumes, printing inks, moistening and softening agents, fumigants, lacquers, and resins (Deshmukh et al. 2016).

Biogas

There is a great demand for alternative energy supply due to the rising demand and cost of energy.

Biogas mostly consists of carbon dioxide, methane, and a minor amount of other gases such as hydrogen sulfide, nitrogen, and oxygen. Sugar beet pulp mainly contains cellulose, pectin, arabinose, and small quantities of lignin. It is an excellent substrate for the generation of bioenergy such as biomethane (Peighambardoust et al. 2021). Anaerobic digestion (AD) provides a great option for the treatment of sugarcane wastewater to produce biogas through a complicated anaerobic microbial fermentation by synergistic interactions between a variety of microbial species (Wei 2016). Biogas was utilized for heat and electricity generation by utilizing a combination of heat and power (CHP) or it can be converted to pure biomethane (about 98% purity) by different processes as reviewed before (Adnan et al. 2019, Formann et al. 2020). High-purity methane can be consumed for industrial and residential applications. It can be utilized as feedstock for the production of various chemicals or as fuel for light-duty vehicles and other fuels via CO_2 consumption (Verbeeck et al. 2018). Also, biomethane has the potential to significantly decrease emissions of greenhouse gases since it could replace huge volumes of natural gas. The by-products of the sugarcane bioethanol refinery were studied and optimized in detail from the biogas process (Fuess et al. 2018). Brooks et al. (2008) investigated the generation of biogas from sugar beet pulp *via* a fermentation process using anaerobic bacteria as an alternative to traditional fuel. In contrast, Felde (2008) reported that biogas generation by anaerobic digestion of sugar beet might produce 137% more fuel than sugar beet conversion to ethanol via fermentation (Usmani et al. 2021b).

CORN STARCH INDUSTRY

Corn is the most abundant starchy cereal crop, contributing to more than 85% of the global starch output. Several byproducts, characterized by production techniques such as steep liquor, corn bran, corn germ, and gluten, are generated primarily through maize starch production. They are cheap, nutrient-dense, and include a wide range of chemical components such as lipids, proteins, carbs, and minerals. In recent days, in an increasingly resource-constrained world, the use of these by-products for non-starch manufacturing is highly attractive, considering both nutritional value and economical aspects into account (Zhang et al. 2021a).

Corn (*Zea mays* L.) is a grass family monocotyledonous seasonal cereal crop. It seemed to have developed in Mesoamerica and is currently grown worldwide in tropical locations like Southeast Asia, South Asia, East Asia, the Americas, and the European Union, among others (Wrigley et al. 2015). On a dry basis (d.b.), corn kernels contain about 61–78% starch, around 10% non-starch polysaccharides, 6–12% protein, and 3–6% lipids (Sinha et al. 2011). It is also easily dried and movable. Corn is thus a near-ideal starch food crop for food products. Thus, it is the most important cereal crop on the globe. Global grain output reached 1136.3 MMT (million metric tons) in 2020–2021, and the United States produced approximately 31.7% (360.3 MMT). Corn, being the primary starch source, accounts for more than 85% of global starch production (Ranum et al. 2014). In the food sector, starches are commonly used as gelling agents, thickening, water retention agents, and bulking agents (Ai and Jane 2016). Moreover, around 40% of wet-milled corn starch is used in non-food sectors such as clothing, papermaking, adhesives, laundry, glass-fiber sizing, and bio-plastics (Lim 2012, Ranum et al. 2014). The seed coat is eliminated during corn starch manufacturing and is known as bran; however, it is high in polysaccharides like cellulose and hemicellulose. The tip cap, or the connecting point between the cob and the kernel, was also completely destroyed during the de-branning procedure (Wang et al. 2019). As a result, several by-products high in nutrients and organic components are formed throughout the starch production process, such as corn germ, maize bran, corn gluten, and so on.

Based on previous research, all the by-products from agriculture have incredible potential use and advantages (Dhar et al. 2017). Thus, the appropriate conversion and reuse of these feedstocks are critical to humans (Wang et al. 2016). The starch manufacturing sector plays an essential role in the industrial fields. Every year, the starch industry produced around 100 million tons of natural waste that can be utilized in a variety of non-starch industrial processes (Niles 2008). Corn steep liquor is widely employed in the fermenting field as an additive in the development of growth culture for nucleotides, glutamate, and antibiotics, as well as in the synthesis of phytic acid. Corn bran may be used to obtain edible oils, although it is most typically utilized to synthesize zeaxanthin, dietary fiber, and sterol (Ai and Jane 2016, Zhang et al. 2021a). Corn germ may be used to obtain edible oil and to prepare meals like biscuits and bread. Like corn gluten, it may be utilized as animal protein feed (Ai and Jane 2016). However, sustainable processing techniques based on cost decrease, eco-friendly manufacturing, and product acceptance have been identified as the most difficult difficulties. As a result, designing and developing by-products into creative goods is a critical development strategy for the food industries and other sectors (Robin et al. 2012).

Major Corn Starch By-Products

Milling steps to produce corn starch are categorized as either dry or wet according to their properties, as shown in Figure 3.5. Dry milling is a continuously abrasive procedure to produce dry-milled corn flour after eliminating the germ and bran (Wrigley et al. 2015). Recently, demand for the dry milling process has increased because corn products are usually used as industrial materials, feed, food, and carbohydrate sources for the synthesis of biofuels and other bioprocessing. Additionally, dry-milled corn is used to produce about 90% of the corn bioethanol (Anderson and Almeida 2019) Wet milling produces high-grade starch granules with the lowest mechanical damage and about 85% of industrial pure starch (Abu-Ghannam and Balboa 2018). Various components were isolated during this process, such as 68.0% starch, 12.0% corn bran, 7.5% corn germ, 6.5% steep liquor, 5.6% gluten, and 0.04% others (Rasby et al. 2003). These by-products contain redundant nutritional compounds and microelements, which are currently widely exploited and utilized by various industries.

Figure 3.5 An overview of dry and wet milling process (Zhang et al. 2021a).

For instance, corn germ was used as an edible oil supplement while bran and recovered protein meal can be utilized in the pharmaceutical and food industry (Xin et al. 2020).

In summary, various by-products of starch processing were generated. The chemical composition and variations in production technique between these by-products are briefly discussed in Table 3.1.

Table 3.1 Composition of the fatty acid in corn germ and bran.

Oil source	**Fatty acid composition (%)**							
	Palmitic acid (C16:0)	**Palmi-toleic acid (C16:1)**	**Stearic acid (C18:0)**	**Oleic acid (C18:1)**	**Linoleic acid (C18:2)**	**Linoleic acid (C18:2)**	**Arachidic acid (C20:0)**	Marinho et al. (2019)
Bran	12.2	0.1	2.3	22.3	54.5	4.2	0.6	
Germ	13.1	0.0	1.5	29.2	55.0	1.0	0.2	

Corn Steep Liquor

After 40 h of soaking the corn kernels at ~50°C in steep water, the germ cell walls become permeable, allowing organic substances to be released and separated from the endosperm or hull. At this moment, about 6.0% of the dry matter composition is liberated to the concentrated steep water, such as proteins, soluble sugars, minerals, amino acids, etc. (Sutivisedsak et al. 2013).

Corn Germ

Because of its high level of tocopherols and unsaturated fatty acids, corn germ oil can increase the shelf life and minimize the risk of serious illnesses (Sutivisedsak et al. 2013).

Corn Bran

Corn bran is another by-product of the dry milling process during the corn starch industries that are mostly produced from Pericarp (Benson and Pearce 1987). Corn bran is a lignocellulosic compound that contains various carbohydrate polymers, such as hemicellulose (30–50%), cellulose (20%), protein (10–13%), starch (9–23%), phenolic acids (4%), and crude oil (2–3%) (Robin et al. 2012). Purified bran was identified as an interesting supplementary dietary fiber for biscuits, bread, and pasta.

Corn Gluten

Corn gluten is derived mostly from corn bran and steep liquor, although it may also include destroyed corn kernels and a few microbes. Dry gluten is normally composed of 90% dry matter, with a significant amount of crude protein (19–24%). As a result, it is a great demand-supply for the protein product production line (Serna-Saldivar and Carrillo 2019).

Value-Added Corn Starch By-products

Therapeutic Enzymes

Pharmaceutical enzymes (prodrugs) have critical functions in disease diagnosis by enzyme-based immunoassays and enzyme-based biosensors (Vimal and Kumar 2019, Padhi et al. 2023). Although numerous pharmaceutical enzymes are produced from animal and plant sources, microorganisms produce the majority of them. Recently, antibiotics can be produced by agro-industrial by-products as substrates; Enzyme laccase EC 1.10.3.2 (p–benzenediol: oxygen oxidoreductase), multi-copper oxidases family (MCOs) were created by using sugarcane bagasse and wheat bran (Raina et al. 2022). Proteases have attracted biotechnological concern due to their numerous industrial purposes in the food, pharmaceutical, paper, textile industries, and leather as well as in peptide synthesis (Razzaq et al. 2019).

Several agricultural biomasses like maize oil cake, mustard oil, wheat bran, rice bran, orange peel, apple peel, and banana peel are utilized in the manufacturing of industrial protease by fermentations. Lipases (triacylglycerol acyl hydrolase, EC 3.1.1.3) have several pharmaceutical applications (Chandra et al. 2020). Lipases catalyze triacylglycerol degradation to glycerol and free fatty acids. Numerous biomasses have been investigated for lipase production, such as rice bran and sugarcane bagasse.

Extraction of Polysaccharides

Corn starch by-products as a whole have been detailed in earlier sections as human foods, feed for animals, and fermenting media for microorganisms without further extraction. However, many useful compounds and nutrients are valuable in extracting and separating from corn starch byproducts. Polysaccharides are the major constituent of corn hull as an edible plant substance that has biological activities and functional properties. It lowers blood pressure, blood sugar, and cholesterol while also preventing colon cancer and stimulating bowel movement (Cai et al. 2021, Zhang et al. 2021a).

Enzymes are usually used to extract polysaccharides from corn starch by-products; enzymatic extraction is widely applied with a high extraction yield and less polysaccharide structural damage (Philippini et al. 2021). Corn bran dietary fiber has received a lot of interest due to its hypolipidemic properties and anti-cancer. Dietary fiber can be produced by biological, chemical and physical methods. Chemical reagents such as alkalis and acids are used in the chemical technique to treat raw materials; it's very easy, and effectively removes protein and carbohydrates. Meanwhile, high temperature and high pressure, and ultrasonic are examples of physical therapy. But these procedures have considerable energy consumption and are frequently used in conjunction with other techniques. Fermentation may also be used to obtain dietary fiber, although its usage is limited due to the lengthy fermentation process. Recently, the enzymatic approach is the most often used for dietary fiber extraction (Hu et al. 2008).

To extract dietary fiber from corn feedstock, both enzymatic and chemical treatments can be used to degrade the glycosidic bonds. Enzymatic degradation was used, followed by alkali treatment at high pressure and high temperature. The obtained corn dietary fiber contains more hydroxyl groups and it has a larger specific surface area. Thus, it is an excellent dietary component for preventing hyperlipidemia by decreasing lipid contents and lowering cancer risk by absorbing dietary nitrites (Wang et al. 2015). Further processing of the obtained corn fiber can expand its industrial applicability by improving the mechanical characteristics of the membrane, enhancing emulsification stability, and so on. Nevertheless, the conversion of corn fiber extraction has to be explored further before it can be used in a variety of products.

The digestible starch can be processed further to increase its crystallinity and transform it into non-digestible prebiotic products. Starch contains a higher proportion of amylopectin, which is a branched structure. It can be treated with type 1 pullulanase to hydrolyze the α-1.6 linked branches of glucon chains. This significantly increases the amylose content in the starch. The liner glucan chains of amylose can be retrograded to produce a crystalline structure, type III resistant starch, which is resistant to gastric digestion (Thakur et al. 2021). The linere glucan chains can be subjected to further processing of complexation with fatty acid molecules, resulting in type V resistant starch (Thakur et al. 2023). The non-digestible resistant starch (type III and type V) products exhibit substantial prebiotic properties.

Lipids and their Derivatives

Oil has become a significant product of corn processing as the industry has grown. Lipid concentration of corn germ ranges from 18 to 41%, accounting for about 80% of the whole lipid in corn, and it's known commercially as corn germ oil. Corn germ oil is dietary vegetable oil with a small content of trans fatty acids (0.3%) (Table 3.1), and a high level of unsaturated fatty acids (80%), particularly linoleic acid (50%) [124, 125]. Linoleic acid, a polyunsaturated omega-6 fatty acid, is essential for our bodies that cannot be produced in the body (Espinosa-Pardo et al. 2020). Consumers prefer corn germ oil because of its mild aroma, high concentration of polyunsaturated fatty acids, significant oxidative stability, and preferable appearance (Shende and Sidhu 2016).

Corn germ oil was largely used as a frying oil due to the limitation of hydrogenated fat and its considerably oxidative stability. It is also used in bakery products, baking ingredients, and snack production (Hosseini et al. 2016). Linoleic acid can reduce cholesterol absorption and deposition in the artery wall by interacting with cholesterol in human blood. Generally, corn germ oil or bran oil is healthy vegetable oil with a high potential usage as a feedstock in pharmaceutical processing and applied in developing healthy foods, insecticides' production, paints, plastics, resins, soaps, textiles, biodiesel, and varnishes (Tang et al. 2016).

Protein Production

Corn protein content is around 7.8%. During the deep corn processing, around 30% of by-products such as corn barn, corn gluten, and corn germ are produced, which may be utilized as high-quality feed supplies. Corn endosperm contains around 8% protein, particularly zein which accounts for 35 to 60% of the endosperm protein. Zein mainly consists of hydrophobic amino acids (50%) with a minor concentration of charged basic and acidic amino acids, especially lysine and tryptophan. Therefore, the amino acid content of zein is unbalanced, leading to lower nutritive benefits. But it has distinguished properties in biologically active compounds, the preparation of nanomaterials for drug delivery systems, and edible materials (Tapia-Hernández et al. 2019). Corn germ has an excellent amino acid ratio; about 18 to 22% of corn germ mass is a protein that accounts for 29% of total corn protein. Corn germ is subjected to harsh conditions during corn starch processing, resulting in considerable denaturation of protein. Denatured proteins lose their functional characteristics and are mostly utilized in animal feed as an energy and protein supplement (Sun et al. 2015).

The protein content of defatted corn germ (CGM) ranges between 60 and 71%, and its necessary amino acid content is similar to that of the FAO/WHO model (Tapia-Hernández et al. 2019). To obtain excellent food qualities of corn gluten, and improve corn gluten products, we should hydrolysis macromolecular protein in corn to generate short peptides. Corn peptides have biological activity, such as antihypertensive (Guo et al. 2020), antioxidant (Jiang et al. 2018), antiproliferative (Li et al. 2013), and anti-inflammatory activity (Liang et al. 2018). Corn protein has a unique shape, molecular structure, and special solubility; thus, it can form a soft and transparent film.

Micro-functional Component Extraction

According to several studies, corn has a variable content of phytochemicals such as anthocyanins, carotenoids, and phytosterol phenolic acids. Corn bran contains several flavonoids, particularly anthocyanins and ferulic acid (Saulnier et al. 2001, Zhao et al. 2014). Ferulic acid binds to macromolecular polymers of the plant cell wall through an ester bond, thus corn bran is the best feedstock for ferulic acid extraction due to its high concentration (3%) compared to other crop's by-products (Saulnier et al. 2001). Ferulic acid may have a variety of therapeutic benefits, including anticancer, antioxidant, anti-inflammatory, and antimicrobial activity (Ou and Kwok 2004). Corn bran also contains a significant concentration (0.6–57.9 mg/kg) of carotenoids (zeaxanthin etc.) (Belitz et al. 2009). Corn yellow pigment (isoprene pigments) can enhance the body's immunity and protect vision and epithelial tissue. Corn oil has a higher content of sterols (phytosterols) about 8 mg/g oil compared to other vegetable oils.

Food Ingredients

As animals do not need high-quality feed compared to humans, using corn starch by-products in animal feed is a very important approach to utilize them as low-value by-products. So, several scientists are considering utilizing by-products as food ingredients directly in human nutrition to increase the application value of by-products. Corn starch byproducts that are high in antioxidants, protein, and fiber can be used as low-cost and low-calorie additives in food preparations to partially substitute sugar or fat (Grasso 2020). Corn bran has a high dietary fiber content (up to 90%), which can be utilized to improve the final fiber content of meals at a minimal insertion level.

In a study, de Sousa et al. (2019) examined the use of corn bran in a low-calorie snack in comparison to the composition without corn bran, and formulations with ~ 40% corn bran were highly appreciated by the experts. In another investigation, adding wheat flour with a maximum concentration of 20% and maize milling by-product to short-dough biscuits was very highly welcomed by the judges who participated in the sensory analysis (Paraskevopoulou et al. 2019). Adding defatted germ flour to wheat flour dramatically increases the protein level in the mixture. Furthermore, 30% of defatted germ could enhance the textural properties such as breaking energy, cutting energy, fracture force, breaking strength, cutting strength, and hardness (Barnwal et al. 2013). Defatted germ flour in different concentrations (5–25%) was supplemented with wheat flour to increase the hardness, fiber content, and yield of the bun; nevertheless, the volume was reduced (Arora and Saini 2016). Bread enrichment with corn germ protein hydrolysate (1–4%) might strengthen bread texture and decrease chewiness and hardness during storage (Karimi et al. 2021). A previous study revealed that chicken sausages with greater dietary fiber content, excellent acceptance, and storability were produced by substituting lean meat with 3% corn bran (Yadav et al. 2016). According to these reports, corn starch by-products can be employed to increase the nutritional content of meat items, biscuits, and bread for human food. Indeed, by-products as food additives imply that the end goods must conform to the food industry safety rules.

Livestock and Poultry Feed

Corn starch manufacturing by-products offer a high concentration of carbohydrates and protein, making them an excellent source for animal and poultry feed (Mulumpwa and Kang'ombe 2009). Germ meal, corn bran, and maize gluten meal gluten are all traded on a global scale, although corn fiber and pulp should be fermented or processed before being fed to animals (Mulumpwa and Kang'ombe 2009, Yang et al. 2021). The nutrient content and chemical composition of various corn starch production by-products vary significantly due to the nutrient redistribution in the processing; also, the digestion properties of various animals are extremely varied (Table 3.2). Corn bran is commonly utilized as a ruminant feed since it is cheaper and more abundant than grains and has higher nutritional value based on content and chemical composition. Furthermore, because of its rich dietary fiber content, corn bran isgreatly used in poultry feed to control the diversity of digestive tract bacteria and improve poultry growth (Yang et al. 2021). After adding a protein supplement to the corn brane, it can completely replace corn grain in cow feed without affecting milk content or causing digestion problems. For non-ruminants, corn bran does not influence the growth rate or digestibility, fermentation products, or gut hormones of weaned pigs. Meanwhile, adding 5% corn bran to the post-weaning diet increases diversity of microflora. This also promotes piglet anti-inflammatory properties (Liu et al. 2018b).

Table 3.2 Nutritional value of by-products as feed.

Items	Corn bran	Corn germ	Corn gluten feed	Corn gluten meal	Reference
Poultry nutritive values					
AMEn cockerel (MJ/kg DM)	8.8	9.8	8.7	16.8	Donkoh and Attoh-Kotoku (2009)
AMEn broiler (MJ/kg DM) – 8.5 – 16.6	–	–	8.5	16.6	Donkoh and Attoh-Kotoku (2009)
Ruminant nutritive values OM digestibility (%)	74.8	79.3	82.4	96.1	
Energy digestibility (%)	72.4	84.6	80.4	99.1	
DE ruminants (MJ/kg)	13.4	24.5	16.8	22.9	Urbano et al. (2014)
ME ruminants (MJ/kg)	11.0	20.3	14.9	16.6	
Nitrogen digestibility, ruminants (%)	–	–	74.4	81.0	
Pig nutritive values					
Energy digestibility, growing pig (%)	79.2	72.6	65.4	93.0	
DE growing pig (MJ/kg)	12.6	21.4	12.5	21.4	Liu et al. (2014)
MEn growing pig (MJ/kg)	12.88	13.9	11.6	19.8	
NE growing pig (MJ/kg)	7.3	9.2	7.7	12.7	
Nitrogen digestibility, growing pig (%)	45.5	71.0	60.2	86.2	

Notes: AMEn: Apparent Metabolizable Energy, Nitrogen-corrected; OM: Organic Matter; DE: Digestible Energy; ME: Metabolizable Energy; MEn: Metabolizable Energy, Nitrogen-corrected; NE: Net Energy.

Corn gluten is mainly composed of 75% protein, ~20% starch, 3% fat, 2% minerals, and 1% fiber. Corn gluten meal is commonly utilized as a suitable alternative for other animal or plant-based proteins due to its high protein concentration (Leeson 2008). Significant studies have been conducted in recent years on the use of corn gluten meals in the diets of dairy cows, broilers, and beef cattle, but less in pig feeding. In comparison to the previous technique, corn gluten meal is largely fermented to release soluble proteins, enhancing their consumption in cattle feed. The use of a 60:40 combination of ground corn gluten meal dramatically enhanced cow milk yield (Ribeiro Filho et al. 2009). Corn gluten has mostly been utilized in ruminant feed, with less focus on monogastric livestock feed. Various wet glutens were supplemented to the dairy cows' diets during the feeding production process. Cows fed 20 to 35% wet gluten feed provided higher milk than controls. Furthermore, milk lactose and protein yields rose with a reduced fat percentage, whereas fat yield remained stable. Moreover, milk protein and lactose yields were improved, while fat yield remained stable (VanBaale et al. 2001). The growth performance of the pigs was greatly enhanced when 15% corn gluten was introduced to the diet. However, increasing the percentage of corn gluten to 30% significantly reduced nutrient digestibility and growth performance. Thus, pigs' diets might include 15 to 30% more corn gluten feed.

Oil was firstly removed from the corn germ and the defatted germ was also regarded as a valuable supplement for all types of livestock (Schilling et al. 2017). It is rich in hemicellulose and digestible amino acids, has a pleasant taste and can be utilized as an energy and protein feed for ruminants, as well as a protein source for monogastric (Schilling et al. 2017). In China, Brazil, and Ethiopia, corn germ meals are frequently utilized in place of maize grain. Nevertheless, lamb's performance was adversely impacted by the corn germ meal. When 0.5% corn germ meal was added to the diet of hens, it improved the amount of omega-3 polyunsaturated fatty acids in egg yolk (Ehr et al. 2017).

Briefly, by-products from corn starch production demonstrated promising applications impacting the production of poultry and livestock. Nowadays, it is possible to significantly increase the feed value and application scope of corn starch production by-products by fermenting them or blending them with enzymes and/or other feed additives. Recently, the application range and feed value of corn starch by-products can be greatly enhanced by combining with enzymes or with fermentation. Corn pulp may be transformed into enzyme-hydrolyzed protein feed following enzyme treatment to enhance the health status and animal growth performance. An investigation showed that adding 10% corn pulp protein feed enhanced the rabbits' daily growth (Ortiz-Martinez et al. 2017, Zhang et al. 2021a).

SUBSTRATE RESOURCES OF MICROBIAL FERMENTATION

Byproducts of starch-based processing can be utilized not only as nutrition for animals and people but also as energy sources for microbial growth and substrates for biological fermentation. The biological fermentation field has developed to include amino acids, organic acids, starch sugar, polyols, yeast, enzyme preparations, paper, biological medicine, multipurpose biological products, new energy, and so on. It is a crucial step in resolving humanity's major problems (López-Prieto et al. 2021).

Biosurfactants

Biosurfactants are amphiphilic compounds that contain a mixture of lipopeptides with C16 and C18 including phenylalanine, valine, proline, histidine, cysteine, aspartic acid/asparagine, glycine, alanine, and leucine (El-Sheshtawy and Doheim 2014). A wide range of microorganisms produces biosurfactants extracellularly by using different substrates. Biosurfactants lower the surface tension between two liquids or between a liquid and a solid phase due to the presence of hydrophobic fatty acid and hydrophilic carbohydrates in their molecules. They have high biodegradability, surface activity, and low toxicity; thus, the utilization of industrial waste to synthesize high-value-added biomolecules is a potential approach to lowering total manufacturing costs. Biosurfactants have various applications in the dispersion of oil spills, also used for bioremediation to enhance contamination removal from water and soil (Kee et al. 2022). Citric acid's diverse uses have prompted ongoing research into expanding production processes and lowering substrate costs. So, corn starch by-product is a typical feedstock used in microbial fermentation (Zhang et al. 2021a).

Bioethanol Production

Bioethanol production was the first utilization of corn fermentation. Eventually, with the continuous development of many technologies and the rising need of consumers, several fermented goods, such as antibiotics, amino acids, organic acids, biosurfactants, and so on, started to be industrialized and manufactured on a large scale.

Corn steep liquor is extensively utilized as an alternative of expensive media ingredients. It is source of cheap C–, N–, trace element, or vitamins. Corn steep liquor is widely considered as a culture additive in the gentian, penicillin, erythromycin, and fermentation industries. Also, penicillin is a microbial secondary metabolite; its productivity was 20% enhanced after adding corn-steep liquor in the 1940s (Hofer et al. 2018).

FUTURE PROSPECTIVE

In the manufacture of the aforementioned products, a sustainable biorefinery model may be implemented utilizing sugar and starch by-products as renewable substrates. This is essential in establishing circular economy concepts and transforming the sugar and starch processing industry into current sustainability concepts. However, there are certain intrinsic obstacles to the creation of bioenergy, such as industrially scalable biotechnologies, preliminary deployment costs, and overall production profitability. Another problem, as previously indicated, is that microbes prefer glucose-rich sources, which limits their affinity for other kinds of feedstock. Common microorganisms, including yeast, have not been proven to digest galacturonic acid and arabinose present in a huge number of substrates. Furthermore, cellulase enzymes have inhibition feedback, decreasing the efficacy of bioconversion. To overcome this problem, simultaneous saccharification and fermentation (SSF) are highly recommended. wherein sugar is continually eliminated during the operation, preventing inhibitory effects by cellulase enzymes.

CONCLUSION

Sugar and starch industries generate a massive quantity of waste each year. To fulfill the ever-increasing global need, scientists have investigated using this economical and sustainable feedstock as a fermentation substrate for enzyme synthesis. Microbiological enzymes are now employed in a wide range of sectors, including medicines, food, detergents, textiles, biofuels, paper, etc. and utilizing these nutritious feedstocks also lowers production costs and helps to alleviate environmental challenges. Sugar and starch production steps include the grain processing chain. The growing amount of sugar and starch-producing crops, particularly in developing nations, has led to a significant number of residues that have attracted considerable interest. Innovative solutions for recycling these wastes are important for food manufacturing control. Many by-products are already utilized as feedstocks for value-added products in several countries around the world. During the past few years, investigation on the usage of sugar and starch production by-products has rapidly developed, specifically in the production and improvement of feed, nutritional supplements, bioactive substances, protein, oils, fuel, and other products. Most high-value studies on sugar and starch processing by-products have not yet been developed on an industrial scale, and extra application studies are needed in the future. This article summarized and discussed the currently use technique and study of sugar and corn starch industrial byproducts for animal food, a fermentation feedstock, carbohydrates, protein, animal fats, and so on. This insight may aid in the comprehension of previous discoveries as well as the creation of future studies on this topic.

REFERENCES

Aarabi, A., Honarvar, M., Mizani, M., Faghihian, H. and Gerami, A. 2016. Extraction and purification of ferulic acid as an antioxidant from sugar beet pulp by alkaline hydrolysis. Italian Journal of Food Science. 28: 362.

Abedin, M.M., Chourasia, R., Phukon, L.C., Sarkar, P., Ray, R.C., Singh, S.P., et al. 2023. Lactic acid bacteria in the functional food industry: biotechnological properties and potential applications. Critical Reviews in Food Science and Nutrition. 1–19. DOI: 10.1080/10408398.2023.2227896.

Abu-Ghannam, N. and Balboa, E. 2018. Biotechnological, food, and health care applications. Sustainable Recovery and Reutilization of Cereal Processing By-Products. Elsevier, 253–278.

Adnan, A.I., Ong, M.Y., Nomanbhay, S., Chew, K.W. and Show, P.L. 2019. Technologies for biogas upgrading to biomethane: a review. Bioengineering. 6: 92.

Ai, Y. and Jane, J.L. 2016. Macronutrients in corn and human nutrition. Comprehensive Reviews in Food Science and Food Safety. 15: 581–598.

Anderson, B. and Almeida, H. 2019. Corn dry milling: Processes, products, and applications. Corn. Elsevier. 405–433.

Arancon, R.A.D., Lin, C.S.K., Chan, K.M., Kwan, T.H. and Luque, R. 2017. Advances on waste valorization: new horizons for a more sustainable society. *In*: Waste Management and Valorization. Apple Academic Press. 23–66.

Arora, A. and Saini, C.S. 2016. Development of bun from wheat flour fortified with de-oiled maize germ. Cogent Food and Agriculture. 2: 1183252.

Asikin, Y., Takahashi, M., Mizu, M., Takara, K., Oku, H. and Wada, K. 2016. DNA damage protection against free radicals of two antioxidant neolignan glucosides from sugarcane molasses. Journal of the Science of Food and Agriculture. 96, 1209–1215.

Barnwal, P., Kore, P. and Sharma, A. 2013. Effect of partially de-oiled maize germ cake flour on physico-chemical and organoleptic properties of biscuits. Journal of Food Processing and Technology. 4: 1–4.

Belitz, H.-D., Grosch, W. and Schieberle, P. 2009. Chapter 15: Cereals and cereal products. pp. 670–745. *In*: Food Chemistry. Springer, Berlin, Heidelberg. https://doi.org/10.1007/978-3-540-69934-7_16.

Benson, G. and Pearce, R. 1987. Corn perspective and culture. pp. 1–29. *In*: S.A. Watson and P.E. Ramstad (eds). Corn: Chemistry and Technology. American Association of Cereal Chemists (AACC), St. Paul, MN, USA.

Berlowska, J., Pielech-Przybylska, K., Balcerek, M., Cieciura, W., Borowski, S. and Kregiel, D. 2017. Integrated bioethanol fermentation/anaerobic digestion for valorization of sugar beet pulp. Energies. 10: 1255.

Bindels, L.B., Delzenne, N.M., Cani, P.D. and Walter, J. 2015. Towards a more comprehensive concept for prebiotics. Nature Reviews Gastroenterology and Hepatology. 12: 303–310.

Brooks, L., Parravicini, V., Svardal, K., Kroiss, H. and Prendl, L. 2008. Biogas from sugar beet press pulp as a substitute for fossil fuel in sugar beet factories. Water Science and Technology. 58: 1497–1504.

Cai, Z., Wei, Y., Zhang, H., Rao, P. and Wang, Q. 2021. Holistic review of corn fiber gum: structure, properties, and potential applications. Trends in Food Science & Technology. 111: 756–770.

Carlozzi, P., Touloupakis, E., Di Lorenzo, T., Giovannelli, A., Seggiani, M. and Cinelli, P. 2019. Whey and molasses as inexpensive raw materials for parallel production of biohydrogen and polyesters via a two-stage bioprocess: new routes towards a circular bioeconomy. Journal of Biotechnology. 303: 37–45.

Cespedes, L.G., Nahat, R.A., Mendonça, T.T., Tavares, R.R., Oliveira-Filho, E.R. and Silva, L.F. 2018. A non-naturally-occurring P (3HB-co-3HAMCL) is produced by recombinant *Pseudomonas* sp. from an unrelated carbon source. International Journal of Biological Macromolecules. 114: 512–519.

Chandel, A.K., Garlapati, V.K., Singh, A.K., Antunes, F.A.F. and da Silva, S.S. 2018. The path forward for lignocellulose biorefineries: bottlenecks, solutions, and perspective on commercialization. Bioresource Technology. 264: 370–381.

Chandra, P., Singh, R. and Arora, P.K. 2020. Microbial lipases and their industrial applications: a comprehensive review. Microbial Cell Factories. 19: 1–42.

Chilakamarry, C.R., Sakinah, A.M., Zularisam, A., Sirohi, R., Khilji, I.A. and Ahmad, N. 2022. Advances in solid-state fermentation for the bioconversion of agricultural wastes to value-added products: opportunities and challenges. Bioresource Technology. 343: 126065.

Chourasia, R., Abedin, M.M., Phukon, L.C., Sahoo, D., Singh, S.P. and Rai, A.K. 2021. Biotechnological approaches for the production of designer cheese with improved functionality. Comprehensive Reviews in Food Science and Food Safety. 20: 960–979.

Chourasia, R., Phukon, L.C., Abedin, M.M., Padhi, S., Singh, S.P. and Rai, A.K. 2022. Whey valorization by microbial and enzymatic bioprocesses for the production of nutraceuticals and value-added products. Bioresource Technology Reports. 19: 101144. https://doi.org/10.1016/j.biteb.2022.101144.

Davani-Davari, D., Negahdaripour, M., Karimzadeh, I., Seifan, M., Mohkam, M. and Masoumi, S.J. 2019. Prebiotics: definition, types, sources, mechanisms, and clinical applications. Foods. 8: 92.

de Sousa, M.F., Guimarães, R.M., de Oliveira Araújo, M., Barcelos, K.R., Carneiro, N.S. and Lima, D.S. 2019. Characterization of corn (*Zea mays* L.) bran as a new food ingredient for snack bars. LWT. 101: 812–818.

Deshmukh, A.N., Nipanikar-Gokhale, P. and Jain, R. 2016. Engineering of Bacillus subtilis for the production of 2,3-butanediol from sugarcane molasses. Applied Biochemistry and Biotechnology. 179: 321–331.

Dhar, H., Kumar, S. and Kumar, R. 2017. A review of organic waste-to-energy systems in India. Bioresource Technology. 245: 1229–1237.

Donkoh, A. and Attoh-Kotoku, V. 2009. Nutritive value of feedstuffs for poultry in Ghana: chemical composition, apparent metabolizable energy, and ileal amino acid digestibility. Livestock Research for Rural Development. 21: 1–9.

Draycott, A.P. 2008. Sugar Beet.

Eggleston, G. and Lima, I. 2015. Sustainability issues and opportunities in the sugar and sugar-bioproduct industries. Sustainability. 7: 12209–12235.

Ehr, I., Persia, M. and Bobeck, E. 2017. Comparative omega-3 fatty acid enrichment of egg yolks from first-cycle laying hens fed flaxseed oil or ground flaxseed. Poultry Science. 96: 1791–1799.

El-Sheshtawy, H.S. and Doheim, M.M. 2014. Selection of Pseudomonas aeruginosa for biosurfactant production and studies of its antimicrobial activity. Egyptian Journal of Petroleum. 23: 1–6.

Espinosa-Pardo, F.A., Savoire, R., Subra-Paternault, P. and Harscoat-Schiavo, C. 2020. Oil and protein recovery from corn germ: extraction yield, composition, and protein functionality. Food and Bioproducts Processing. 120: 131–142.

Felde, A.v. 2008. Trends and developments in energy plant breeding-special features of sugarbeet. Sugar Industry/Zuckerindustrie. 133: 342–345.

Formann, S., Hahn, A., Janke, L., Stinner, W., Sträuber, H., Logroño, W. and Nikolausz, M. 2020. Beyond sugar and ethanol production: value generation opportunities through sugarcane residues. Frontiers in Energy Research. 8: 579577.

Foster, R.W., Tame, C.J., Bučar, D.K., Hailes, H.C. and Sheppard, T.D. 2015. Sustainable synthesis of chiral tetrahydrofurans through the selective dehydration of pentoses. Chemistry–A European Journal. 21: 15947–15950.

Fuess, L.T., Garcia, M.L. and Zaiat, M. 2018. Seasonal characterization of sugarcane vinasse: assessing environmental impacts from fertirrigation and the bioenergy recovery potential through biodigestion. Science of the Total Environment. 634: 29–40.

Gibson, G.R., Probert, H.M., Van Loo, J., Rastall, R.A. and Roberfroid, M.B. 2004. Dietary modulation of the human colonic microbiota: updating the concept of prebiotics. Nutrition Research Reviews. 17, 259–275.

Grasso, S. 2020. Extruded snacks from industrial by-products: a review. Trends in Food Science and Technology. 99: 284–294.

Guo, Y., Wang, K., Wu, B., Wu, P., Duan, Y. and Ma, H. 2020. Production of ACE inhibitory peptides from corn germ meal by an enzymatic membrane reactor with a novel gradient diafiltration feeding working-mode and *in vivo* evaluation of antihypertensive effect. Journal of Functional Foods. 64: 103584.

Hofer, A., Hauer, S., Kroll, P., Fricke, J. and Herwig, C. 2018. Indepth characterization of the raw material corn steep liquor and its bioavailability in bioprocesses of Penicillium chrysogenum. Process Biochemistry. 70: 20–28.

Hosseini, H., Ghorbani, M., Meshginfar, N. and Mahoonak, A.S. 2016. A review on frying: procedure, fat, deterioration progress and health hazards. Journal of the American Oil Chemists' Society. 93: 445–466.

Hu, Y.-B., Wang, Z. and Xu, S.-Y. 2008. Treatment of corn bran dietary fiber with xylanase increases its ability to bind bile salts, *in vitro*. Food Chemistry. 106: 113–121.

Jain, A., Sarsaiya, S., Awasthi, M.K., Singh, R., Rajput, R., Mishra, U.C., et al. 2022. Bioenergy and bio-products from bio-waste and its associated modern circular economy: current research trends, challenges, and future outlooks. Fuel. 307: 121859.

Jiang, Y., Zhang, M., Lin, S. and Cheng, S. 2018. Contribution of specific amino acid and secondary structure to the antioxidant property of corn gluten proteins. Food Research International. 105: 836–844.

Kamzon, M.A., Abderafi, S. and Bounahmidi, T. 2015. The efficient process for the conversion of bagasse and beet pulp to bioethanol. 3rd International Renewable and Sustainable Energy Conference (IRSEC). IEEE. 1–6.

Karimi, A., Gavlighi, H.A., Sarteshnizi, R.A. and Udenigwe, C.C. 2021. Effect of maize germ protein hydrolysate addition on digestion, *in vitr* antioxidant activity and quality characteristics of bread. Journal of Cereal Science. 97: 103148.

Kee, S.H., Chiongson, J.B.V., Saludes, J.P., Vigneswari, S., Ramakrishna, S. and Bhubalan, K. 2021. Bioconversion of agro-industry sourced biowaste into biomaterials via microbial factories—a viable domain of circular economy. Environmental Pollution. 271: 116311.

Kee, S.H., Ganeson, K., Rashid, N.F.M., Yatim, A.F.M. and Vigneswari, S. 2022. A review on biorefining of palm oil and sugar cane agro-industrial residues by bacteria into commercially viable bioplastics and biosurfactants. Fuel. 321: 124039.

Kingsly, J.S., Chathalingath, N., Parthiban, S.A. and Sivakumar, D. 2022. Utilization of sugarcane molasses as the main carbon source for the production of polyhydroxyalkanoates from Enterobacter cloacae. Energy Nexus. 6: 100071.

Koo, W.W. and Taylor, R.D. 2014. Outlook of the US and World Sugar Markets. 2013–2023.

Kracher, D., Oros, D., Yao, W., Preims, M., Rezic, I., Haltrich, D., et al. 2014. Fungal secretomes enhance sugar beet pulp hydrolysis. Biotechnology Journal. 9: 483–492.

Lee, J.-K., Patel, S.K.S., Sung, B.H. and Kalia, V.C. 2020. Biomolecules from municipal and food industry wastes: an overview. Bioresource Technology. 298: 122346.

Leeson, S. 2008. Predictions for commercial poultry nutrition. Journal of Applied Poultry Research. 17: 315–322.

Li, J.-T., Zhang, J.-L., He, H., Ma, Z.-L., Nie, Z.-K., Wang, Z.-Z., et al. 2013. Apoptosis in human hepatoma HepG2 cells induced by corn peptides and its anti-tumor efficacy in H22 tumor bearing mice. Food and Chemical Toxicology. 51: 297–305.

Liang, Q., Chalamaiah, M., Ren, X., Ma, H. and Wu, J. 2018. Identification of new anti-inflammatory peptides from zein hydrolysate after simulated gastrointestinal digestion and transport in Caco-2 cells. Journal of Agricultural and Food Chemistry. 66: 1114–1120.

Lim, T.K. 2012. Edible medicinal and non-medicinal plants. Springer.

Lin, C.-Y., Chai, W.S., Lay, C.-H., Chen, C.-C., Lee, C.-Y. and Show, P.L. 2021. Optimization of hydrolysis-acidogenesis phase of swine manure for biogas production using two-stage anaerobic fermentation. Processes. 9: 1324.

Liu, H., Sun, J., Chang, J.-S. and Shukla, P. 2018a. Engineering microbes for direct fermentation of cellulose to bioethanol. Critical reviews in biotechnology. 38: 1089–1105.

Liu, P., Zhao, J., Wang, W., Guo, P., Lu, W., Wang, C., et al. 2018b. Dietary corn bran altered the diversity of microbial communities and cytokine production in weaned pigs. Frontiers in microbiology. 9: 2090.

Liu, Y., Song, M., Almeida, F., Tilton, S., Cecava, M. and Stein, H. 2014. Energy concentration and amino acid digestibility in corn and corn coproducts from the wet-milling industry fed to growing pigs. Journal of Animal Science. 92: 4557–4565.

López-Prieto, A., Rodríguez-López, L., Rincón-Fontán, M., Cruz, J.M. and Moldes, A.B. 2021. Characterization of extracellular and cell-bound biosurfactants produced by Aneurinibacillus aneurinilyticus isolated from commercial corn steep liquor. Microbiological Research. 242: 126614.

Low, S.S., Bong, K.X., Mubashir, M., Cheng, C.K., Lam, M.K., Lim, J.W., et al. 2021. Microalgae cultivation in palm oil mill effluent (POME) treatment and biofuel production. Sustainability. 13: 3247.

Madian, H.R., Hamouda, H.I. and Hosny, M., 2022. Statistical optimization of bioethanol production from giant reed hydrolysate by *Candida tropicalis* using Taguchi design. Journal of Biotechnology. 360: 71–78.

Marcondes, W.F., Milagres, A.M. and Arantes, V. 2020. Co-production of xylo-oligosaccharides, xylose, and cellulose nanofibrils from sugarcane bagasse. Journal of Biotechnology. 321, 35–47.

Marinho, C.M., Lemos, C.O.T., Arvelos, S., de Souza Barrozo, M.A., Hori, C.E. and Watanabe, É.O. 2019. Extraction of corn germ oil with supercritical CO_2 and cosolvents. Journal of Food Science and Technology. 56: 4448–4456.

Modelska, M., Berlowska, J., Kregiel, D., Cieciura, W., Antolak, H., Tomaszewska, J., et al. 2017. Concept for recycling waste biomass from the sugar industry for chemical and biotechnological purposes. Molecules. 22: 1544.

Morão, A. and De Bie, F. 2019. Life cycle impact assessment of polylactic acid (PLA) produced from sugarcane in Thailand. Journal of Polymers and the Environment. 27: 2523–2539.

Mulumpwa, M. and Kang'ombe, J. 2009. Effect of feeding soybean-based diets on the survival, growth, and feed utilization of *Tilapia rendalli* in a semi-intensive pond culture system. Aquaculture Research. 40: 1099–1101.

Mustafa, G., Arshad, M., Bano, I. and Abbas, M. 2020. Biotechnological applications of sugarcane bagasse and sugar beet molasses. Biomass Conversion and Biorefinery. 1–13.

Nagarajan, V., Mohanty, A.K. and Misra, M. 2016. Perspective on polylactic acid (PLA) based sustainable materials for durable applications: focus on toughness and heat resistance. ACS Sustainable Chemistry and Engineering. 4: 2899–2916.

Najari, Z., Khodaiyan, F., Yarmand, M.S. and Hosseini, S.S. 2022. Almond hulls waste valorization towards sustainable agricultural development: production of pectin, phenolics, pullulan, and single cell protein. Waste Management. 141: 208–219.

Nava-Cruz, N.Y., Contreras-Esquivel, J.C., Aguilar-González, M.A., Nuncio, A., Rodríguez-Herrera, R. and Aguilar, C.N. 2016. Agave atrovirens fibers as substrate and support for solid-state fermentation for cellulase production by *Trichoderma asperellum*. 3 Biotech. 6: 1–12.

Niles, M. 2008. Sustainable soils: reducing, mitigating, and adapting to climate change with organic agriculture. Sustainable Development Law and Policy. 19.

Ortiz-Martinez, M., Otero-Pappatheodorou, J.T., Serna-Saldívar, S.O. and García-Lara, S. 2017. Antioxidant activity and characterization of protein fractions and hydrolysates from normal and quality protein maize kernels. Journal of Cereal Science. 76: 85–91.

Ou, S. and Kwok, K.C. 2004. Ferulic acid: pharmaceutical functions, preparation and applications in foods. Journal of the Science of Food and Agriculture. 84: 1261–1269.

Özgür, E., Mars, A.E., Peksel, B., Louwerse, A., Yücel, M., Gündüz, U., et al. Eroğlu, İ. 2010. Biohydrogen production from beet molasses by sequential dark and photofermentation. International Journal of Hydrogen Energy. 35: 511–517.

Padhi, S., Sarkar, P., Abedin, M.M., Sirohi, R., Singh, S.P. and Rai, A.K. 2023. Microbial enzymes and metabolites in health care: an overview. pp. 3–23. *In*: Sirohi, R., Rai, A.K., Vandenberghe, L.P.deS., and Parameswaran, B. (eds). Microbial Enzymes and Metabolites for Health and Well-Being. CRC Press.

Pandit, S., Savla, N., Sonawane, J.M., Sani, A.M.d., Gupta, P.K., Mathuriya, A.S., et al. 2021. Agricultural waste and wastewater as feedstock for bioelectricity generation using microbial fuel cells: recent advances. Fermentation. 7: 169.

Paraskevopoulou, A., Rizou, T. and Kiosseoglou, V. 2019. Biscuits enriched with dietary fibre powder obtained from the water-extraction residue of maize milling by-product. Plant Foods for Human Nutrition. 74: 391–398.

Peighambardoust, S.H., Jafarzadeh-Moghaddam, M., Pateiro, M., Lorenzo, J.M. and Domínguez, R. 2021. Physicochemical, thermal and rheological properties of pectin extracted from sugar beet pulp using subcritical water extraction process. Molecules. 26: 1413.

Philippini, R.R., Martiniano, S.E., Franco Marcelino, P.R., Chandel, A.K., Dos Santos, J.C. and Da Silva, S.S. 2021. Production of β-glucan exopolysaccharide lasiodiplodan by *Lasiodiplodia theobromae* CCT 3966 from corn bran acid hydrolysate. Applied Microbiology and Biotechnology. 105: 2319–2332.

Raina, D., Kumar, V. and Saran, S. 2022. A critical review on exploitation of agro-industrial biomass as substrates for the therapeutic microbial enzymes production and implemented protein purification techniques. Chemosphere. 133712.

Ranum, P., Peña-Rosas, J.P. and Garcia-Casal, M.N. 2014. Global maize production, utilization, and consumption. Annals of the New York Academy of Sciences. 1312: 105–112.

Rasby, R.J., Erickson, G.E., Klopfenstein, T.J. and Adams, D.C. 2003. Value and use of corn milling by-products in the cow herdv. Range Beef Cow Symposium. 62. https://digitalcommons.unl.edu/rangebeefcowsymp/62

Razzaq, A., Shamsi, S., Ali, A., Ali, Q., Sajjad, M., Malik, A. and Ashraf, M. 2019. Microbial proteases applications. Frontiers in Bioengineering and Biotechnology. 7: 110.

Ribeiro Filho, H.M.N., Giacomet, C.D., Dias, K.M., Crestani, S., Setelich, E.A. and Thaler Neto, A. 2009. Corn gluten meal to dairy cows grazing Italian ryegrass. Ciência Rural. 39: 1162–1168.

Robin, F., Schuchmann, H.P. and Palzer, S. 2012. Dietary fiber in extruded cereals: limitations and opportunities. Trends in Food Science and Technology. 28: 23–32.

Rodriguez, A., Rache, L.Y., Brijaldo, M.H., Silva, L.P. and Esteves, L.M. 2020. Common reactions of furfural to scalable process of residual biomass. Ciencia en Desarrollo. 11: 63–80.

Rojan, P.J., Nampoothiri, K.M., Nair, A.S. and Pandey, A. 2005. L (+)– lactic acid production using Lactobacillus casei in solid-state fermentation. Biotechnology Letters. 27: 1685–1688.

Sadh, P.K., Duhan, S. and Duhan, J.S. 2018. Agro-industrial wastes and their utilization using solid-state fermentation: a review. Bioresources and Bioprocessing. 5: 1–15.

Sato, N., Takano, Y., Mizuno, M., Nozaki, K., Umemura, S., Matsuzawa, T., et al. 2013. Production of feruloylated arabino-oligosaccharides (FA–AOs) from beet fiber by hydrothermal treatment. The Journal of Supercritical Fluids. 79: 84–91.

Saulnier, L., Marot, C., Elgorriaga, M., Bonnin, E. and Thibault, J.F. 2001. Thermal and enzymatic treatments for the release of free ferulic acid from maize bran. Carbohydrate Polymers. 45: 269–275.

Schilling, M., Montgomery, S., Titgemeyer, E., Wertz-Lutz, A., Vahl, C., Schilling, A., et al. 2017. Effects of feeding nucleotides in diets containing corn germ meal or dried corn distillers grains and solubles on the performance and health of receiving and growing calves. The Professional Animal Scientist. 33: 440–450.

Serna-Saldivar, S.O. and Carrillo, E.P. 2019. Food uses of whole corn and dry-milled fractions. pp. 435–467. *In*: Serna-Saldivar, S.O. (ed.). Corn. Elsevier.

Sharma, M., Patel, S.N., Lata, K., Singh, U., Krishania, M., Sangwan, R.S., et al. 2016. A novel approach of integrated bioprocessing of cane molasses for production of prebiotic and functional bioproducts. Bioresource Technology. 219: 311–318.

Sharma, M., Sangwan, R.S., Khatkar, B.S. and Singh, S.P. 2021. Development of a prebiotic oligosaccharide-rich functional beverage from sweet sorghum stalk biomass. Waste and Biomass Valorization. 12: 2001–2012.

Shende, D. and Sidhu, G.K. 2016. Response surface methodology to optimize enzyme-assisted aqueous extraction of maize germ oil. Journal of Food Science and Technology. 53: 3282–3295.

Sinha, A.K., Kumar, V., Makkar, H.P., De Boeck, G. and Becker, K. 2011. Non-starch polysaccharides and their role in fish nutrition–a review. Food Chemistry. 127: 1409–1426.

Sun, X.D., Lan, Y., Shi, D., Lu, S.W., Liao, H., Zhang, R.Y., et al. 2015. Determination of molecular driving forces involved in heat-induced corn germ proteins gelation. Journal of Cereal Science. 66: 24–30.

Sutivisedsak, N., Leathers, T.D., Nunnally, M.S., Price, N.P. and Biresaw, G., 2013. Utilization of agricultural biomass in the production of the biopolymer schizophyllan. Journal of Industrial Microbiology and Biotechnology. 40: 105–112.

Tang, W., Yu, G. and Zhou, Y. 2016. Research progress on the characteristics and extraction method of corn germ oil. Cereal Food Industries. 23: 36–38.

Tapia-Hernández, J.A., Del-Toro-Sánchez, C.L., Cinco-Moroyoqui, F.J., Juárez-Onofre, J.E., Ruiz-Cruz, S., Carvajal-Millan, E., et al. 2019. Prolamins from cereal by-products: classification, extraction, characterization and its applications in micro- and nanofabrication. Trends in Food Science and Technology. 90: 111–132.

Tomaszewska, J., Bieliński, D., Binczarski, M., Berlowska, J., Dziugan, P., Piotrowski, J., et al. 2018. Products of sugar beet processing as raw materials for chemicals and biodegradable polymers. RSC Advances. 8: 3161–3177.

Thakur, M., Sharma ,N., Rai ,A.K. and Singh, S.P. 2021. A novel cold-active type I pullulanase from a hot-spring metagenome for effective debranching and production of resistant starch. Bioresource Technology. 320: 124288.

Thakur, M., Rai, A.K. and Singh, S.P. 2023. Structural characteristics, physicochemical properties and digestibility analysis of resistant starch Type-V prepared from debranched corn starch and fatty acid complexation. ACS Omega. 8: 25799−25807.

Tripathi, N.K. and Shrivastava, A. 2019. Recent developments in bioprocessing of recombinant proteins: expression hosts and process development. Frontiers in Bioengineering and Biotechnology. 7: 420.

Urbano, S.A., Ferreira, M.d.A., Madruga, M.S., Azevedo, P.S.d., Bispo, S.V. and Silva, E.C.d. 2014. Corn germ meal as a substitute for corn in the diet of confined Santa Inês sheep: chemical and lipid meat composition. Ciência e Agrotecnologia. 38: 581–588.

Usmani, Z., Sharma, M., Awasthi, A.K., Sivakumar, N., Lukk, T., Pecoraro, L., et al. 2021a. Bioprocessing of waste biomass for sustainable product development and minimizing environmental impact. Bioresource Technology. 322: 124548.

Usmani, Z., Sharma, M., Diwan, D., Tripathi, M., Whale, E., Jayakody, L.N., et al. 2021b. Valorization of sugar beet pulp to value-added products: a review. Bioresource Technology. 126580.

VanBaale, M., Shirley, J., Titgemeyer, E., Park, A., Meyer, M., Lindquist, R. et al. 2001. Evaluation of wet corn gluten feed in diets for lactating dairy cows. Journal of Dairy Science. 84: 2478–2485.

Verbeeck, K., Buelens, L.C., Galvita, V.V., Marin, G.B., Van Geem, K.M. and Rabaey, K. 2018. Upgrading the value of anaerobic digestion via chemical production from grid-injected biomethane. Energy and Environmental Science. 11: 1788–1802.

Vimal, A. and Kumar, A. 2019. Transforming the healthcare system through therapeutic enzymes. pp. 603–625. *In*: Kuddus, Md. (ed.). Enzymes in Food Biotechnology: Production, Applications, and Future Prospects. Elsevier.

Wadhwa, M., Bakshi, M. and Makkar, H. 2016. Wastes to worth: value-added products from fruit and vegetable wastes. CABI Reviews. 1–25.

Wang, Y., Zhou, Y.-L., Cheng, Y.-K., Jiang, Z.-Y., Jin, Y., Zhang, H.-S., et al. 2015. Enzymo-chemical preparation, physico-chemical characterization and hypolipidemic activity of granular corn bran dietary fibre. Journal of Food Science and Technology. 52: 1718–1723.

Wang, B., Dong, F., Chen, M., Zhu, J., Tan, J., Fu, X., et al. 2016. Advances in recycling and utilization of agricultural wastes in China: based on environmental risk, crucial pathways, influencing factors, policy mechanism. Procedia Environmental Sciences. 31: 12–17.

Wang, J., Sun, B. and Tsao, R. 2019. Bioactive Factors and Processing Technology for Cereal Foods. Springer.

Wei, S. 2016. The application of biotechnology on the enhancing of biogas production from lignocellulosic waste. Applied Microbiology and Biotechnology. 100: 9821–9836.

Wrigley, C.W., Corke, H., Seetharaman, K. and Faubion, J. 2015. Encyclopedia of Food Grains. Academic Press.

Xin, H., Sun, F., Sun, K., Fu, Q., Li, Y., Zhang, Y., et al. 2020. Batch-to-batch variation in carbohydrates molecular structures, nutritive value and biodegradation characteristics in corn coproducts. Animal Feed Science and Technology. 263: 114458.

Yadav, S., Malik, A., Pathera, A., Islam, R.U. and Sharma, D. 2016. Development of dietary fibre enriched chicken sausages by incorporating corn bran, dried apple pomace and dried tomato pomace. Nutrition and Food Science.

Yan, K., Wu, G., Lafleur, T. and Jarvis, C. 2014. Production, properties and catalytic hydrogenation of furfural to fuel additives and value-added chemicals. Renewable and Sustainable Energy Reviews. 38: 663–676.

Yang, M., Long, S. and He, T. 2021. Research progress on the application of corn starch wet processing by-products in livestock and poultry production. Chinese Journal of Animal Nutrition. 33: 1–11.

Yatim, A.F.M., Syafiq, I.M., Huong, K.H., Amirul, A.-A., Effendy, A.W.M. and Bhubalan, K. 2017. Bioconversion of novel and renewable agro-industry by-products into a biodegradable poly (3-hydroxybutyrate) by marine *Bacillus megaterium* UMTKB-1 strain. BioTechnologia. Journal of Biotechnology Computational Biology and Bionanotechnology. 98.

Zhang, R., Ma, S., Li, L., Zhang, M., Tian, S., Wang, D., et al. 2021a. Comprehensive utilization of corn starch processing by-products: a review. Grain and Oil Science and Technology. 4: 89–107.

Zhang, S., Wang, J. and Jiang, H. 2021b. Microbial production of value-added bioproducts and enzymes from molasses, a by-product of the sugar industry. Food Chemistry. 346: 128860.

Zhao, S., Yao, S., Ou, S., Lin, J., Wang, Y., Peng, X., et al. 2014. Preparation of ferulic acid from corn bran: Its improved extraction and purification by membrane separation. Food and Bioproducts Processing. 92: 309–313.

Chapter **4**

Management of Waste from Poultry Industry and Slaughterhouses

Rajneesh Thakur[1], Suvarna Bhoj[1], Shwetambri Jamwal[2], Prachurya Biswal[1], Babita Misra[1], Srikanth Vallabhaneni[1] and Ayon Tarafdar*[1]

[1]ICAR-Indian Veterinary Research Institute, Bareilly, India.

[2]ICAR-National Dairy Research Institute, Karnal, India.

INTRODUCTION

Livestock sector is an inspiring agricultural sub-sector providing food security to millions (Hoque et al. 2022). In 2020, the global meat production was recorded to be approximately 252.6 million metric tons (MT), including 99.1 million MT chicken, 95.8 million MT pork, and 57.7 million MT of beef while a rise of 40% in meat consumption has being recorded since the previous decade (Wang et al. 2022). However, livestock and poultry farming in large confined operations to meet the global demands produce nearly unmanageable volumes of dung and slurry in addition to other farming wastes. This has led to inevitable environmental hazards contributing in the production of substantial quantities of solid waste and wastewater. Traditionally, agricultural and livestock production systems are circular in nature involving the three Rs, i.e., residues, reuse and recycling to ensure least waste production. On the other hand, modern day,

* For Correspondence: ayontarafdar@gmail.com

linear agricultural and livestock production systems generate enormous farm wastes resulting in poor waste management, environmental losses and health hazards. Thus, sustainable management of livestock waste generated is very essential to protect the environmental concerns and control the production of obnoxious gases.

Farm-based livestock waste is generally divided into three groups: waste water, soil, and slurry. Major livestock wastes from various farming systems comprise of animal excreta, dung, wash water, feed leftovers, and compost. Thus, apart from this, poultry sector as one of the most emerging livestock industries also generates a considerable amount of waste from bird hatching to slaughtering resulting in challenging situations of waste disposal and management along with environmental and health hazards (Mustafa et al. 2018, Zhang et al. 2023). The solid waste generated from the poultry farming can be majorly categorized in to poultry litter and manure waste, feather waste, mortality waste; hatchery waste and abattoir waste (Zhang et al. 2023, Jana et al. 2023). Solid waste consists of feed, feathers, hatchery waste (empty shells, infertile eggs, dead embryos and late hatchlings), shells, sludge, abattoir waste (offals, blood, feathers and condemned carcasses). Liquid waste generation includes faeces, urine, saw dust, drugs and pesticide residues, disinfection of chicken houses, and abattoirs.

The poultry litter mainly consist of excreta (manure), bedding material, i.e., wood shavings, saw dust wheat straw peanut and rice hulls mixed with waste feed (Chen and Jiang 2014). Poultry litter is mainly composed of water, carbon and nitrogen and phosphorus in minor quantities along with traces of calcium, magnesium iron, copper and zinc etc. This nutrient rich waste is typically applied in fields but an over application of this may lead to eutrophication of water bodies, release of greenhouse gases (GHG) and phytotoxic substances (Billen et al. 2015, Jana et al. 2023). The feather waste from poultry industry includes approximately 91% keratin, 1% lipids, and 8% water. Cystine, glutamine, proline and serine are the chief amino acids present, with Serine being the highest in quantity (16%). Keratin is a highly insoluble protein, which is highly resistant to physical, chemical and biological actions due to the presence of disulfide bonds, hydrogen bonds, salt and cross linkages. Improper disposal of such waste may cause soil fertility, ground water pollution and even human ailments (Muduli et al. 2019). Mortality waste is an unavoidable leftover in poultry production, which includes dead embryos, dead birds, and broken birds. This type of waste is also higher in protein with substantial phosphorus and calcium levels due to mineral rich poultry diet (Gunduz et al. 2019). Abattoir waste or soft waste are inedible by-products of poultry slaughtering in meat processing plants which includes offal, visceral organs, blood, and bones etc. If left untreated, disposing them into the dumpsites may result in high risk of disease and bacterial transmission. However, these by products are rich in nutrients and can be valorised into high protein meat and bone meals for further usage (Alam et al. 2019). In addition, poultry hatchery waste also produces an indispensable solid and liquid waste in form of egg shells, infertile eggs, dead-in-shell and decaying tissue, etc. which should be treated scientifically to be valorised further (Glatz et al. 2011).

Worldwide, rapid growth of poultry industry has contributed into improving the living standards and poverty alleviation of farmers by creating employment opportunities. Along with achievement of development goals, what further needs to be taken care of is the appropriate treatment or disposal of the huge waste generated as the industry outcome to prevent the environmental and health threats to mankind (Dalolio et al. 2017). Apart from traditional practices of dumping, land filling and burial, widely adopted in underdeveloped countries, these poultry wastes can also undergo rational and scientific processes of rendering, composting, anaerobic digestion, chemical hydrolysis and incineration, etc., to combat the nuisance (Muduli et al. 2019). Some economic-environmentally friendly and energy-saving opportunities are to recycle and reuse these by-products as raw materials in wood adhesives, biomaterial and biomedical applications, biocomposites, textiles and leather processing, or as an energy source in biodiesel production (Karuppannan et al. 2021). This chapter entails such eco-friendly alternatives along with the traditional waste management options in a comprehensive manner in further subheads.

Amount of Wastes Generated in Different Sectors

The increasing industrialization and urbanization to cope up with the demands of growing population has also led to high solid waste generation. Global solid waste production is considered to be 11.2 billion tons per year, which is expected to worsen to 19 billion tons per year by 2025. Out of the solid wastes, the production of global municipal solid waste (MSW) was recorded to be about 2.01 billion tons per year (Mozhiarasi et al. 2022). The solid wastes directly contribute to 3% of the GHG emissions (Tahir et al. 2015), out of which livestock wastes are held responsible for major share of GHG release rather than other sectors. The life cycle studies such as carbon footprint, resource consumption, water use, and health impacts suggested that meatless meals contribute 40% less compared to the meat-containing meals with chickens alone emitting 0.6 gigatons of CO_2^- equivalent, representing 8% of the livestock sector's total emissions (Ernstoff et al. 2019, Mozhiarasi et al. 2022).

Slaughterhouse and poultry wastes produce a considerable amount of organic MSW with relatively high levels of suspended solid, liquid, and fat. It is estimated that 68–72% of chicken and 78% of turkey is utilized for meat purpose while the remaining portion contributes to poultry waste. Nearly 32.5–37% of poultry waste is being generated in abattoir shops, with the waste composition consisting of 57.37% feathers and skin; 20.35% intestines; 14.8% legs and < 1% others (Adhikari et al. 2018, Mozhiarasi et al. 2022). A typical abattoir with capacity of 200,000 chickens per day, of 2.3 kg average live weight, will produce 127 metric tons of viscera and offal (Jayathilakan et al. 2012).

The World Bank documented the global waste generation trend from the year 2016–2050, which is shown in Figure 4.1, demonstrating that majority of the global waste is generated from East Asia and the Pacific region.

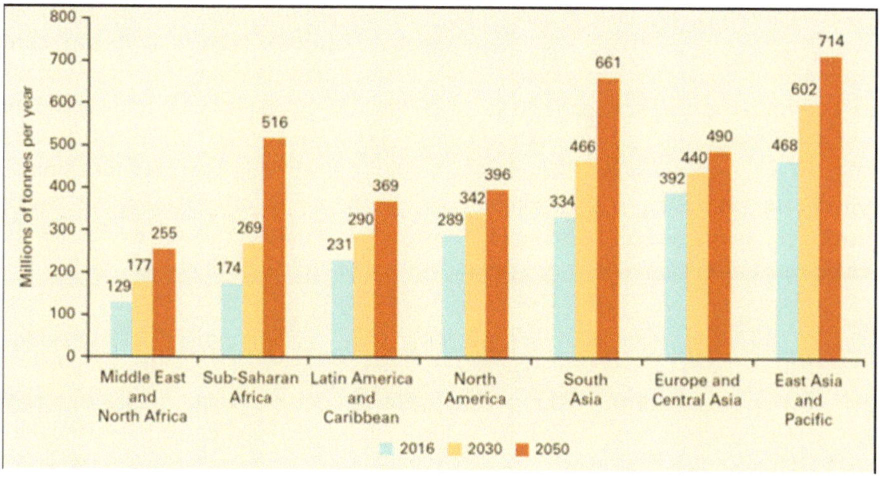

Figure 4.1 Global waste generation trend from the year 2016–2050 (Obtained from Kaza et al. 2018).

The amount of manure production in a poultry farm mainly depends on age, feeding, health status and management practices followed. It is further approximated that a chicken produces 1 kg of fresh manure for each kilogram of feed consumed with variable water content whereas a commercial layer produces about 20 kg waste per year (Mozhiarasi et al. 2022). In his study, Collins et al. (1999) reported that 1000 birds up to market age in commercial farms produce dried manure in range of 1.1 to 2.4 metric tons for chickens, 7.3 to 12.7 metric tons for turkeys and 3.9 metric tons for ducks. However, 90% of excess poultry waste comprising of potential nutrients is spread on land close to the poultry farms for crop production. Similarly, feather waste is commonly processed into feather meal, fertilizers, bedding or for decorative purpose, constituting up to 7–8% of poultry's weight. According to Tesfaye et al. (2017), the poultry meat processing industry generates up to 40 × 109 kg feather by products annually, suggesting 50 g of waste/bird. Another study by Kowalczyk et al. (2018) suggests that 8.5 billion tons of feather waste is produced annually across the globe. India alone produces 350 million tons of feathers from slaughter house and processing industry creating serious disposal issues in developing nations.

Mortality is inevitable in poultry farming, and a mortality rate of 5% is an acceptable practice. Assuming an annual mortality rate of 2%, it is estimated that 85,000 tons of bird carcass is produced annually in the United States alone (Akdeniz 2019, Prabhakaran and Valavan 2021). The documented total solid by-product generation from large slaughter house in Turkey is around one million tons per annum.. In India, 38.33 million metric tons (MT) of poultry manure, 9486 metric MT of hatchery waste and 1.74 million MT of slaughterhouse waste production was recorded in the year 2018–2019.

Poultry slaughterhouses also use huge amount of fresh water for the continuous operations of scalding, feathering, evisceration and washing and continuously discharge considerable volumes of waste water into the environment. This waste water represents 70–75% of the total water consumed in the slaughterhouse (Azadbakht et al. 2021). It is estimated that 7.57 litre of wastewater/poultry are generated in the gutting stage and 4.35 litre of wastewater/poultry in the washing stage. On an average, nearly 20.49–26.5 litre of clean water is consumed to slaughter poultry of 2.3 kg body weight (Ungureanu et al. 2023). Worldwide generation of poultry waste is bound to intensify in coming years owing to rapid growth recorded in poultry industry. Hence, adoption of cost-effective waste processing and recycling technologies is the need of the hour to reduce its impact on the environment and ensure sustainability in poultry production.

Harmful Effects of Non-scientific Waste Handling

Improper management of poultry waste affects the air and water quality severely and acts as a threat to public health due to the presence of several pathogens and pollutants. The solid waste are traditionally burnt in incinerators or disposed in landfills which leads to generation of hazardous GHGs harmful for human health, soil, water and environment. It is estimated that poultry waste contributes 33.7 million MT of CO_2 equivalent/year causing severe threats. Along with GHGs, the irrational and non-sustainable poultry rearing practices includes the release of ammonia, hydrogen sulphide nitrous oxide and volatile organic compounds in to air. The release such of pollutants and particulate matter may cause cold, allergy, cardiovascular and respiratory diseases, and even cancer (Sankoh et al. 2019, Mozhiarasi et al. 2022).

Inappropriate disposal of untreated solid waste greatly contaminates the ground water by leaching of nitrates, phosphates, and pathogens, acting as a potent source of pollution especially for people residing nearby the dump yards. Further, many processing industries and abattoirs illegally discharge blood and animal faeces into streams causing oxygen depletion and nutrient imbalance in water bodies. This also leads to toxin accumulation causing water borne gastro-intestinal diseases such as typhoid fever, cholera, and hepatitis E infections. NO_3 accumulation in water causes blue baby syndrome in humans and abortions in livestock (Mozhiarasi et al. 2022, Bhaskar et al. 2023).

Continuous land disposal of solid waste causes soil contamination through direct contact or leaching, which reduces soil quality and cause nutritional imbalance resulting in low crop production. It is a potential reservoir of bacteria, viruses and even prion proteins. It is reported that about one-third to one-half of the total weight of slaughtered animal remains as unutilized or partly utilized by products of livestock and poultry industry (Adhikari et al. 2018). To meet the demands of poultry market, various metals such as Ca, P, As, Zn, Cu and Fe, etc., and other chemicals such as xanthophylls, antioxidants, antiprotozoals, probiotics and polychlorinated phenols, etc., are added in high doses in poultry diet for higher growth (Marshall and Levy 2011, Jackson et al. 2003). Indiscriminate use of such agents with low standards of biosecurity and waste management causes microbial resistance against *Salmonella* spp., *Campylobacter* spp., *Listeria* spp. and *Cryptosporidium* spp., etc. High levels of pathogenic Salmonella and

Campylobacter jejuni are found in poultry in United States. Farms, slaughterhouse workers and veterinarians are at a direct risk of exposure to such microorganisms. These pathogens have shown high resistance against antibiotics Tetracycline, Trimethoprim-Sulphamethaxazole, Nalidixic acid, Ampicillin, etc. (Bhaskar et al. 2023). A subsequent transport of such pathogenic microbes and chemical agents to the environment through poultry waste calls for an alarming situation in need of thorough management.

Types of Wastes Generated in Poultry Production

The rapidly expanding poultry sector produces a lot of solid and liquid waste. Over 9/10th of the poultry waste is generally applied on the land spread near poultry enterprises on a global scale. Poultry litter materials and excrement, spilled feed, and waste generated from hatcheries such as broken egg shells, infertile eggs, dead-in-shells, dead fledglings, etc., feathers, deceased birds, and abattoir debris make up the majority of these chicken solid wastes. Faeces, urine, sawdust, the drug remains, insecticides, and waste from chicken coops and slaughterhouses are examples of liquid waste (Singh et al. 2018). There are edible and inedible categories for animal by-products. For instance, edible by-products include organs such as the kidneys, liver, and heart, whereas inedible by-products include hairs, horns, and hooves, i.e., around 37% of the carcasses of slaughtered broilers are inedible. One of the biggest animal by-products that are high in protein (approximately 18%) is the blood from slaughterhouses. The types of waste generated by the poultry sector are described in detail under the following headings:

Poultry Excreta and Litter

Depending on the farm's chosen rearing system, poultry enterprises produce two main types of waste, i.e., cage layer waste which is excreta accumulated under the cages, spilled feed, and feathers, and poultry litter that is waste from deep litter systems (Amit 2020). The three wastes that poultry producers should be most concerned about are the bedding materials used to house the birds, the manure produced as a result of that production, and dead birds that are present in all flocks. The management of the litter materials, feed intake, and feed digestibility all affect how much poultry litter is created in a broiler unit. As bedding, a variety of materials are employed, such as wood shavings, cereal straws, husks and hulls, and paper clippings (Swain and Sundaram 2000).

According to Bernhart et al. (2010), for litter control in broiler units, three standard procedures are used, i.e., single-use litter, partial re-use litter, and multi-use litter. In single-use litter after every flock, the entire house must be completely cleaned out, and the bedding must be replaced. In partial reuse, litter from the brooding area of the house is taken out and placed on the growing section, the brooding area is then covered with fresh bedding material. It is common practice to partially compost the used litter for a few days in order to eradicate microorganisms by raising its temperature. The used litter is replaced after each batch, followed by completely cleaning out the house after two to five batches. In the multi-use litter, the house is disinfected and only congealed litter material is taken out as found by Sistani et al. (2003). As per Kelley et al. (1996), the multi-flock litter creates a spent litter that has a considerably greater intensity of nutrients, which may increase the occurrence of parasites and harmful microorganisms.

The quantity of total overall solids on dry matter basis that birds defecate, can be predicted using the diet's dry matter digestibility. Up until 35-day of age, chickens eat about 2.5–3.0 kg, and up to 49-days of age, they eat about 5–6 kg on a dry matter basis. Based on the dry matter digestibility of 87.5% of the diet (FSA 2007), a 35- and 49-day-old bird is thought to excrete 0.34 kg and 0.63 kg of solids, respectively, and the total manure production with a moisture content of 90% is roughly 4 kg and 6 kg, respectively. According to estimates, broilers excrete about 55% of the entire N, 70% of the total P, and 80% of the total K. The overall amount of solids and nutrients still present in the litter can be considerably impacted by the amount of feed spilled during feeding (Leytem et al. 2007). Iron, zinc, copper, and arsenic are the most common elements found in poultry litter and manure, along with calcium, chlorine, sodium, magnesium, manganese, and phosphorus.

According to ASABE (2005), the day-to-day fresh excrement generation rate for broiler chicken is around 43 kg per 1,000 kg of live weight. This translates to 6.9 kg per 1,000 kg of live weight per day for broilers when the amount of dry manure is collected. The poultry litter deposition varies between 0.7 and 2 tons per 1,000 broilers per flock on a dry weight basis.

Based on average daily live weights throughout the birds' production cycle, approximations of the amount of excrement by 1,000 birds per day are as follows: 120 kg for layer chickens, 80 kg for broilers, 200–350 kg for turkeys (grower jennies and grower heavy toms, respectively), and 150 kg for ducks (Charles 2013). Depending on the breed (grower females or grower heavy males, respectively), for 1,000 poultry raised for the market, approximations of the litter formed fluctuate between 1 to 2.5 tons for the chickens, 7.4 to 12.8 tons for turkeys, and 4 tons for ducks (Charles 2013).

The kind, age, nutrition, and health of the birds, as well as farming practices, have an impact on the quantity and composition of manure. Additionally, the amount and value of manure mixed with litter produced by different kinds of poultry units are impacted by handling and storage conditions (Malone 1992, Maguire et al. 2006). These include the type of bedding used, how often the crust is removed, how thoroughly the house is cleaned out, how many flocks are housed in a house before the bedding needs to be replaced, the ultimate live body weight of the chickens, and managing techniques.

Feathers, manure, bedding material, and spilled feed are the main components of chicken litter (Kelley et al. 1996). Nutrients for plants like nitrogen, phosphorus, and potassium, trace metals like copper, zinc, and arsenic, pesticide remains, medications like coccidiostats, endocrine disruptors, and microbes can all be found in the litter. Poultry kinds, diets, nutritional supplements, litter types, handling practices, and storage procedures have all been proven to significantly affect the moisture percentage, pH, and fundamental composition of poultry excreta and litter. Also, the age of the bird, kind of diet, style of shelter provision, and waste managing technique, affect the concentration of minerals in excreta significantly. The quantity, regularity, holding capacity, and constancy of the chemical, as well as the composting of the litter, all affect the intensity of chemical residue observed in chicken waste.

Endocrine-disrupting substances (EDS) are a class of substances that have been synthesized or are naturally occurring in the environment and are thought to have negative effects on both humans and animals. These EDC include a number of the animal's endogenous hormones as well as bug killers, weedkillers, plant estrogens, and additional substances that interfere with the endocrine system. The EDCs are mainly excreted in the form of steroid hormones by broilers. According to studies, broiler litter includes measurable quantities of 17-oestradiol, oestrone, oestriol, and testosterone (Bolan et al. 2010).

A sizable and varied micro-population of bacteria, viruses, protozoa, and fungi can be found in poultry manure. In poultry litter, microbial concentrations might reach 10^{10} cells/g (Bolan et al. 2010); gram-positive bacteria, such as Lactobacilli in the bacilli group, Clostridia in Eubacteria, Actinomycetes, make up nearly 9/10th of the microbial range. While bacteria carry out a range of diverse enzymatic and metabolic procedures in the litter, the microbes of particular importance to the poultry commercial are nitrogen-mineralizing microbes (Bacillus species and Arthrobacter species), and pathogens. Poultry litter is a reservoir for a number of zoonotic pathogens, such as Salmonella species, *E. coli*, Campylobacter species, Clostridium perfringens, and Listeria species (Bolan et al. 2010).

Hatcheries Waste

Empty eggshells, blighted eggs, dead-in-shell embryos, tardy hatchlings, deceased birds, and a sticky substance made of eggs and decomposing flesh make up solid hatchery waste. Water needed to clean incubators, hatcheries, and spaces where chicks are handled becomes waste.

According to Das et al. (2002), only in the United States, hatcheries produce roughly 140,000 tons of trash each year. The GOP (2008) projected that 9974 tons of hatchery waste (HW) were produced in Pakistan from the eggs that were planned for incubation between 2008 and 2009. Ghana data are not easily accessible; however, according to Djang-Fordjour (2016), the sector produces about a thousand tons of hatchery waste annually.

According to the specific exercise carried out to ascertain the amounts of HW produced, 1,676.64 kg of hatchery waste, consisting of dead-in-shells, egg shells, and rejected chicks, was produced on average each month, along with 5,000 eggs. However, the 1.68 tons of hatchery waste per year estimated in this study was less than the 10.4 tons of HW that the biggest chicken meat hatcheries in Australia produced in one province (Glatz and Miao 2009).

The quality of the egg laying and generation of incubated egg residues is impacted by the age of the birds. The eggs laid at the beginning of a hen's productive life are smaller and less fertile. On the other side, as birds age, their eggshells grow less robust, more porous, and finer cuticles, which decreases their ability to hatch (Orrico et al. 2020). This shows that there is an increasing tendency of the younger and older hens to have greater incubated egg residues than the medium-aged hens.

From receiving the eggs to transporting them internally to sorting and moving them to the incubator, residue creation occurs at several stages of embryonic development, producing varying waste compositions depending on the phase.

According to Asiedu et al. (2022), the first location where culling starts and hatchery trash is collected is where the egg is received in the egg reception area. The eggs are then candled to evaluate their state and jumbo eggs (75 g or more) or pee-wee eggs (46 g or less), eggs that have dual yolks, are broken, or have mottling, or are unclean are thrown away. The minute the eggs are relocated manually or mechanically to the hatcher after 19 days in the incubator, again culling steps are carried out. To preserve only the eggs with intact visual elements, manual removal is performed on the eggs with obvious cracks, broken eggs, or, eggs with visible insides, and hatched premature chicks. The third phase involves candling, which is used to mechanically remove any eggs that are thought to be infertile or deformed. After hatching, the chicks are sorted in the sex sorting room, where stillbirths and those with malformations are segregated. The eggs that are unable to hatch, combined along with the egg shells, stillborn, deceased, and discarded chicks, frame the hatcher dregs, which is disposed of in the last phase.

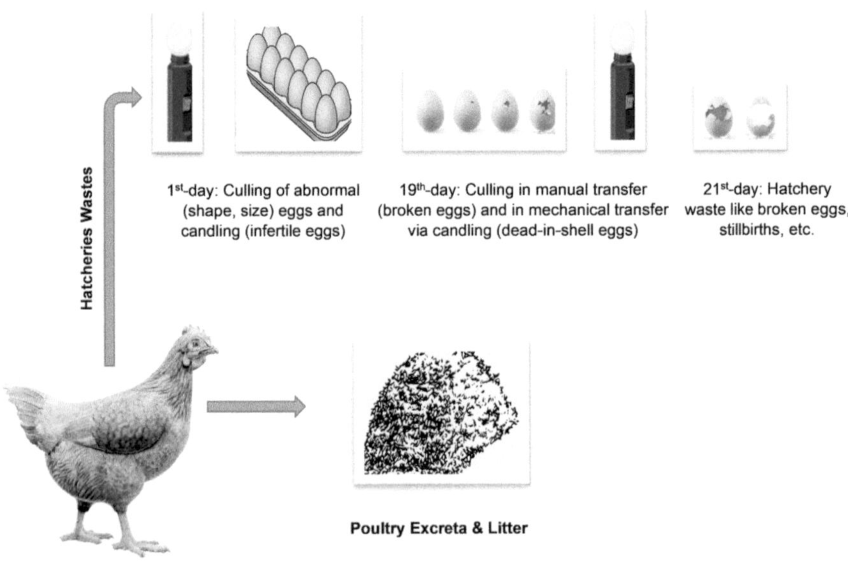

Figure 4.2 Illustration of poultry hatchery wastes and excreta and litter materials.

According to Orrico et al. (2020), a minimum of 79% of the overall trash in the hatchery happens during the hatcher stage. Only 1.8% of the waste produced overall is attributable to discarding during egg receiving and the manually performed relocation process, which makes them insignificant contributors to the makeup of the total waste. However, as much as 19.0% of the entire dregs produced by the hens of a medium laying age may be attributable to the mechanized transfer procedure. Late-laying hens produced material that had larger concentrations of lipid, volatile substances, or, nitrogen, and carbon. For every kg of day-old chicks, an average of 0.16 kg of hatchery dregs in terms of dry matter is generated (Figure 4.2). The quantities of debris produced depended on the entire numeral of eggs set in the hatcher and the number that hatched, which fluctuated between 5 to 10% of all eggs laid (Asiedu et al. 2022).

Slaughterhouse Waste

Slaughterhouse wastes are wastes produced during broiler slaughter and include offal, soft visceral organs, blood, and bones, which are typically regarded as inedible by-products in many nations as mentioned by Ferreira et al. (2018). Approximately 23% of the waste from chickens is classified as soft waste and includes the head, feet, organs, and intestines, which are primarily made of chicken protein and fat (Figure 4.3).

Blood, feet, heads, bones, trimmings, and organs are among the organic solid wastes and byproducts that are created during the raising and slaughter of broilers. According to Salminen and Rintala (2002), offal contains 0.6 to 0.9% of the ability to produce methane, 32% proteins, 54% lipids, and 5.3% of the total Kjeldahl nitrogen content.

According to Jayathilakan et al. (2012), an average abattoir with a daily capacity of 2 lakh chickens with an average live weight of around 2 kg would create 127 metric tons of viscera and offal. After removing the saleable carcass (70 and 75% of the weight of the live bird), the quantity of inedible offal determined on the basis of the live weight is equal to 30% of the weight of the live fowl.

In addition to viruses, bacteria, and residues, chicken wastes such as feathers, feet, trimmings, and intestinal contents are said to include several hundred different species of microorganisms (Muduli et al. 2019) including dangerous pathogens including Salmonella species, Clostridium species, and Staphylococcus species.

Carcass Waste

In the production of poultry, carcass waste includes dead-in-shell embryos, deceased birds, and severed bird parts. In the 42-day production cycle, a mortality rate of 5% is regarded as acceptable (Delgado et al. 2003). Mortality is unavoidable in the chicken industry. According to a report, the average monthly death rate for poultry farms in Nigeria is 7.4%, which represents 83.34% of all farms (Ka et al. 2014). Assuming an annual mortality rate of 2%, the United States would produce 85,000 tons of bird carcasses each year (Akdeniz 2019). Due to the high mineral content of the bird's diet, mortality waste has a high protein content and significant amounts of phosphorus and calcium (Gunduz et al. 2019, Nebiyu et al. 2016).

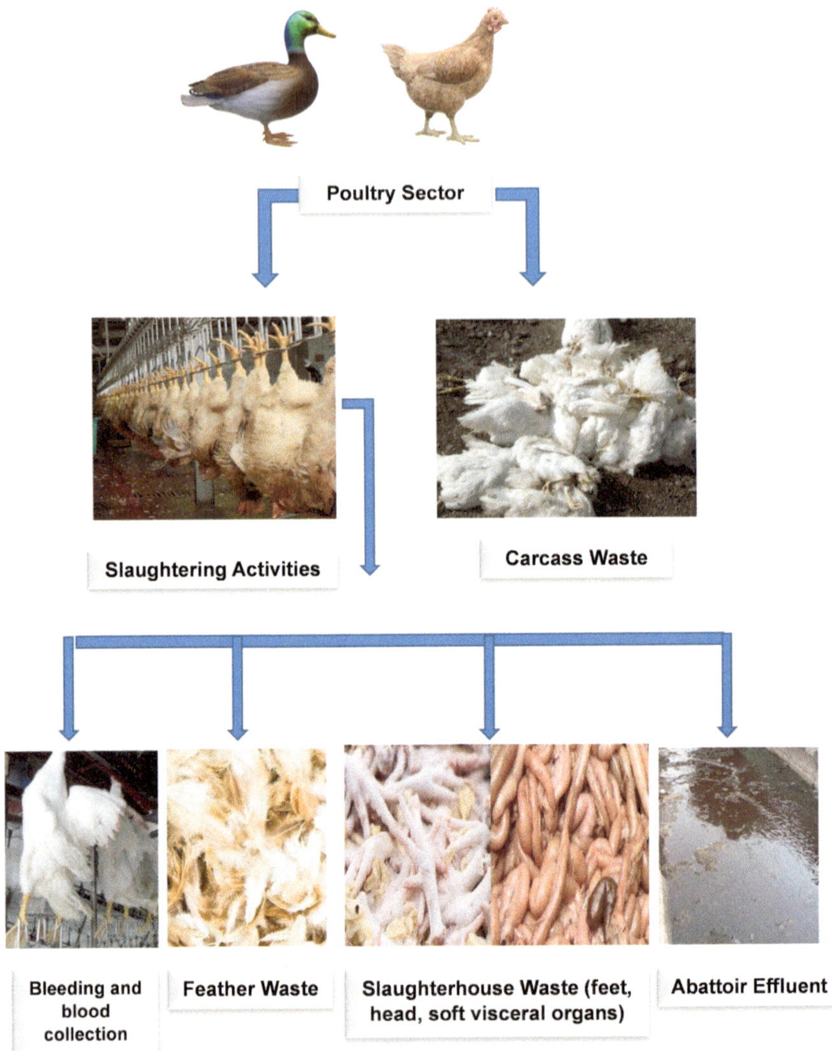

Figure 4.3 Illustration of the slaughterhouse, carcass, and miscellaneous waste.

Feathers

Another by-product that is produced in large quantities by the chicken industry worldwide is feather waste (Sinkiewicz et al. 2017). Poultry feathers are a protein rich and keratinous waste that can make up to 5–10% (Zhang et al. 2023) of the weight of chicken, or more than 50 g of waste per bird. Up to 40 to 109 kg of feather by products are generated yearly from the processing of poultry bird's meat worldwide (Tesfaye et al. 2017).

Every year, 8.5 billion tons of waste poultry feathers are produced worldwide. India produced 4.6 million tons of broiler chicken in 2018, while 350 million tons of feather debris are produced per year and discarded carelessly into the environment (Thiyagarajan et al. 2023). Chlorosis, chicken cholera, mycoplasmosis, and H5N1 virus outbreaks are just a few of the human diseases caused by feather build-up in the environment (Thiyagarajan et al. 2023). The main protein found in feathers, which makes up more than 91% of them, is keratin followed by 8% moisture and 1% fat (Muduli et al. 2019). Keratin is stiff and insoluble in water due to extensive disulphide bonds and cross-linkages. Hard keratins, such as those found in feathers, hairs, hooves, and nails, and soft keratins, such as those found in skin and calluses, are distinguished by their sulphur concentration (Thiyagarajan et al. 2023).

Miscellaneous

The majority (70–75%) of the water used in the abattoir comes from wastewater (Azadbakht et al. 2021). Due to the constant processes of cutting, cleaning, scalding, singeing, de-feathering, eviscerating, and washing and packing of the meat, poultry abattoirs produce enormous amounts of wastewater/effluents that are released into the environment. For instance, the gutting stage generates 7.6 L whereas the washing stage generates 4.4 L of effluent. A 2.3 kg of poultry needs, on average, between 20.49 and 26.5 L of clean water to slaughter, according to other studies.

A significant amount of organic matter, as evaluated by chemical oxygen demand (COD) and biochemical oxygen demand (BOD), is present in the wastewater produced in poultry slaughterhouses. Blood, lipids, oils, fats, and proteins are also present in significant amounts, as well as nitrogen and phosphate. Thus, its discharge from the poultry abattoir carries a high risk of contaminating freshwater sources with serious environmental and health hazards if it is not properly treated.

Methods of Poultry Waste Management

Poultry waste methods are further categorized as traditional methods and recent advances. Both are discussed in detail in the upcoming sections.

Traditional Methods

Anaerobic Digestion

Anaerobic digestion is an age-old practice which is utilized to manage different types of waste along with production of biogas. This technique gives environmental as well as economical benefits and is hence considered as a sustainable method of handling waste. Anaerobic digestion can be used to treat poultry and slaughterhouse waste, converting it into biogas and nutrient-rich digestate. The biogas can be used as a renewable energy source, while the digestate can be used as a fertilizer in agricultural fields. The biogas produced through anaerobic digestion can further be utilized as a source of electricity, either for on-site use, such as at a specific farm, or for sale to a broader distribution network. Another application of this biogas is as Compressed Natural Gas (CNG), offering an eco-friendly option for vehicle fuel or heat generation (Holm-Nielsen et al. 2009).

Anaerobic digestion is defined as a biological process that involves breaking down of complex organic materials, such as poultry waste, into simpler ones in the absence of oxygen. Different steps of anaerobic digestion includes hydrolysis, acidogenesis and acetogenesis, and methanogenesis. Poultry waste can also be treated with anaerobic digestion with certain modifications such as pretreatments, co-digestion, ammonia stripping, etc. Poultry waste can not be used directly for anaerobic digestion because of certain factors which are present in it and are responsible for inhibition of anaerobic digestion. These include higher ammonia concentration, basic pH level, lower carbon-nitrogen ratio (C:N), presence of non-degradable bedding materials, and lack of methanogens. Modifications were required for the elimination of such factors. Different steps of anaerobic digestion are explained in Figure 4.4.

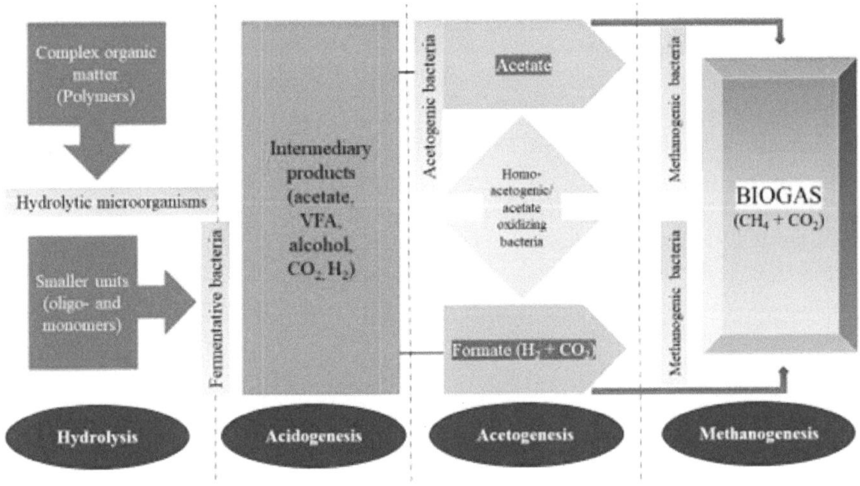

Figure 4.4 Different stages of anaerobic digestion process.

Techniques Used for Successful Anaerobic Digestion of Poultry Manure

(a) Conditioning technique: To optimize anaerobic digestion process, various conditioning techniques can be employed with the objective of improving the efficiency and effectiveness of the process. Some of the key factors involved for optimization of anaerobic digestion process for poultry manure are as under:

Temperature control: Anaerobic digestion is influenced by temperature. There are three common temperature ranges for digestion: Mesophilic: 30–40°C (86–104°F); Thermophilic: 50–60°C (122–140°F) and Psychrophilic: <20°C (68°F). Maintaining the appropriate temperature range for the specific type of anaerobic digestion is crucial for the activity and growth of methane-producing microorganisms.

pH regulation: The pH level affects the microbial activity in the digester. Typically, the optimal pH range for anaerobic digestion is between 6.5 and 7.5.

Regular monitoring of the pH level and use of buffering agents is needed to maintain a stable pH, which can help in enhancing efficiency of anaerobic digestion of poultry waste.

Mixing/Agitation: Efficient mixing helps to disperse the microbes uniformly, ensures equal distribution of nutrients, and prevents the formation of stratified layers that can hinder anaerobic digestion and biogas production.

Retention time: The retention time is the duration for which feedstock remains in the digester. Longer retention times allow microbes more time to break down complex organic matter, resulting in higher biogas yields. However, it's essential to strike a balance between retention time and digester capacity.

Inoculation: Inoculating the digester with an active and diverse microbial population can jumpstart the biogas production process and reduce the start-up time. Commercial microbial inoculants or using digestate from a functioning digester can be used for this purpose. As poultry waste is having lower number of methanogens, addition of inoculum becomes more important in case of poultry waste.

Inhibition management: During anaerobic digestion, some substances or intermediary metabolites have the tendency to inhibit the microbial activity in the digester. For instance, in case of poultry waste, high levels of ammonia or sulfide negatively impact biogas production. Monitoring and managing inhibitory substances are vital for optimizing anaerobic digestion process.

Digester design and optimization: Proper design and sizing of the digester based on the feedstock characteristics, process conditions, and intended use of biogas is essential for efficient biogas production.

(b) Codigestion of poultry waste: Codigestion is a technique which involves combination of different substrates in order to make the properties of whole substrate suitable for anaerobic digestion. Biogas production from poultry waste was demonstrated successfully by mixing it with carbonaceous co-substrates such as wheat straw and meadow grass (Rahman et al. 2017). Additionally, it was found that higher amount of biogas was produced through codigestion of poultry droppings with lignocellulosic co-substrates in comparison to their monodigestion. Codigestion of chicken manure from cage layers with food waste was also reported and significantly higher amount of biogas yield along with lower hydraulic retention time was found through codigestion in comparison to poultry waste alone (Wang et al. 2014).

Codigestion also depends on the amount and type of substrates used. In case of poultry litter substrates like corn stover, cow dung, municipality waste and many others were studied and found to be effective in producing biogas. Mohd Johari et al. (2020) investigated the production of biogas using poultry litter in combination with corn stover through codigestion. Different ratios of poultry litter and corn stover (1:1, 1:2, and 2:1) were mixed, resulting in a consistent total solids concentration of 8% across all ratios. The experiment revealed that the substrate with a 2:1 ratio of poultry litter to corn stover produced the highest amount of biogas. Also, codigestion with cow substrate was attempted, and positive results were demonstrated in a study by Miah et al. (2016). To evaluate the impact of codigestion with cow dung, the researchers conducted trials in four laboratory-scale reactors: R1 (solely poultry litter), R2 (75% poultry litter and 25% cow dung), R3 (50% poultry litter and 50% cow dung), and R4 (70% poultry litter and 30% poultry droppings), all with a total solids concentration of 6%. After a digestion period of 50 days at room temperature (around 32°C), it was observed that R2 showed the highest degradation of volatile solids (51.99%) and gas production (0.469 liters per gram of volatile solid feed). Another successful codigestion of poultry litter with municipal solid waste was reported by Matheri et al. (2017). This codigestion process demonstrated effective biogas production, leading to the conclusion that organic waste from municipal landfills could be codigested with cow manure to produce methane.

(c) Ammonia stripping: Poultry waste differs from other kind of animal waste in terms of higher ammonia production. This ammonia affects the anaerobic digestion process and biogas production. In order to eliminate this ammonia for successful anaerobic digestion, ammonia stripping technique was employed. It was demonstrated that methane production was possible from excreta of layer birds using technique of ammonia stripping (Abouelenien et al. 2010). This technique involves digestion of poultry manure in anaerobic conditions for a duration of 4 days at 55°C with an initial pH of 8–9; ammonia was effectively extracted by recycling biogas followed by gas washing in sulfuric acid to collect ammonia. This approach converted 80% of total nitrogen in poultry manure to ammonia and eliminated 82% of the produced ammonia along with production of higher amount of methane.

(d) Pretreatment: The application of pretreatment to poultry excreta offers a potential solution to address the existing limitations for undergoing successful anaerobic digestion. The main objective of pretreatment is to remove lignin and hemicelluloses, reduce cellulose crystallinity, and enhance porosity in the lignocellulosic materials found in the excreta (Kumar et al. 2009). The content of cellulose, hemicellulose, and lignin in poultry manure has been reported by various researchers to range from 20 to 44%, 11 to 25%, and 1.5 to 8.5%, respectively (Li et al. 2013, 2015, Wei et al. 2020).

Various types of pretreatments can be employed to prepare poultry waste containing lignocellulosic materials for digestion, including physical, chemical, thermal (heat) or thermochemical, and biological methods (Muhammad Nasir and Mohd Ghazi 2015, Orlando and Borja 2020). For physical treatments, mechanical degradation and the application of heat in different forms can be utilized. Chemical pretreatment involves the use of alkaline hydrolysis, acid hydrolysis, organosolv process, wet oxidation, and alkaline peroxide. On the other hand, biological pretreatment involves the use of enzymes, microbial consortia, and fungi. Each of these pretreatment methods can play a crucial role in optimizing the efficiency of biogas and vermicompost production from poultry excreta containing lignocellulosic materials. In case of poultry waste, successful anaerobic digestion was demonstrated by application of heat pretreatment (Dornelas et al. 2017), chemical pretreatment (Assefa et al. 2014) and combination of physical and chemical pretreatment (Esendir 2012).

Benefits of anaerobic digestion for poultry and slaughterhouse waste management includes energy generation, waste reduction, pathogen reduction, nutrient recovery and production of bio-fertilizers. Overall, anaerobic digestion offers a sustainable and environmentally friendly solution for managing poultry and slaughterhouse waste by converting it into valuable resources, reducing greenhouse gas emissions, and minimizing the environmental impact.

Incineration

Incineration is a waste management method that involves the combustion of waste materials at high temperatures, i.e., 750–1100°C (Ghosh et al. 2020). In other words, it is the firing of waste at high temperatures in an enclosed structure. This process reduces the large amount of waste in to a mass of ash that is less than 20% of its original volume. With the reduction in land availabilities and urbanization, incineration acts as one of the most viable options to handle waste from animal husbandries (poultry) and slaughterhouses.

For incineration of waste material, incineration systems need to be installed. Incineration systems generally consist of an incinerator (furnace) and one or more air pollution control devices. The two most commonly used incineration systems are multiple hearth furnaces (MHFs) and fluidized-bed reactors.

a) Multiple-hearth furnace: The multiple-hearth furnace operates using a vertical arrangement of stacked chambers, also known as hearths. Air is introduced from the lower part of the furnace, while waste is introduced from the top. Within each chamber, the waste is moved to designated "drop holes," causing it to fall into the hearths situated below. The primary combustion occurs in the central hearths, while the lower levels serve for ash cooling.

b) Fluidized-bed reactors: The fluidized bed reactor is a specialized furnace used for the controlled combustion of various substances, such as wastewater sludge, bio-solids, industrial sludge, and fine solids. This reactor incorporates heat recovery, ensuring efficient utilization of thermal energy, while maintaining a stable process of fluidization and combustion. Additionally, integrated flue gas cleaning enhances the environmental performance of the system. In this reactor, a sand bed is utilized, and air is blown from the bottom upwards at a carefully controlled velocity, allowing the sand particles to remain suspended in the air. The reactor operates as a "bubbling bed," ensuring that the particles are not entrained with the air flow. The sand bed behaves like a boiling liquid due to the continuous fluidization. To initiate the combustion process, the material to be incinerated is introduced just above the fluidized bed. The air required for fluidization is introduced into the sand bed through a specially designed distributor plate. This plate ensures an even distribution of air throughout the bed and prevents sand from flowing back into the air ducts.

c) Air pollution control equipment: Typically, air pollution control systems serve the purpose of either eliminating fine particles and metals from the exhaust gas or facilitating the decomposition of organic compounds. Methods like wet scrubbers, dry, and wet electrostatic precipitators are employed to eliminate metals. On the other hand, afterburners are used to thoroughly combust organic substances present in the exhaust gases. In modern times, regenerative thermal oxidizers (RTOs) are increasingly integrated into incinerators' air pollution control systems.

While incineration can be used for waste management from poultry and slaughter houses, it is important to consider both the advantages and disadvantages of this approach before being employed for an operation. The advantages of incineration include volume reduction, pathogen destruction, energy generation whereas the disadvantages include air emissions, cost and infrastructure, and ash disposal. It's important to note that the appropriateness of incineration as a waste management method for poultry and slaughterhouse waste depends on various factors, including local regulations, environmental considerations, availability of alternative technologies, and specific waste characteristics. Environmental impact assessments and careful planning are necessary to ensure that incineration is implemented in a manner that minimizes its potential negative effects on air quality and the environment.

Land Application

Land application, also known as land spreading or land disposal, is a practice to handle waste and is employed by most of the small farmers. In case of poultry waste, it involves applying poultry and slaughterhouse waste directly to land as a form of fertilizer or soil amendment. This method utilizes the nutrient content of the waste to enhance soil fertility and improve crop growth. This approach provides a practical means of recycling the essential nutrients Nitrogen (N), Phosphorus (P), and Potassium (K) found in manure (Kaiser et al. 2009). Manure serves as an organic nutrient-rich resource for crop cultivation. The poultry excreta typically contain significant amounts of nitrogen along with small quantities of ammonia and urea. When manure is stored in open-air conditions, there is a rapid loss of nitrogen due to its high proteolytic activity. As a result, swift processing is essential to preserve the nutritional properties of the manure. Incorporating poultry manure into the soil not only addresses waste management concerns but also improves soil fertility (McGrath et al. 2010). Land application can be taken as a method of waste disposal up to certain level with considerations to ensure its effectiveness and minimize potential negative impacts. Some of these considerations are as follows:

- **Waste characterization and testing:** Before land application, it's important to analyze the waste to determine its nutrient content, moisture content, pH, and potential presence of contaminants. This information helps determine appropriate application rates and any necessary treatment or processing steps.
- **Nutrient management planning:** A nutrient management plan should be developed to guide the land application process. This plan considers factors such as soil nutrient requirements, crop nutrient demands, waste nutrient content, and appropriate application rates to prevent nutrient imbalances or overloading.
- **Application methods:** There are different application methods, including surface spreading, injection, or incorporation into the soil. The choice of

method depends on the waste characteristics, soil conditions, and local regulations. Proper equipment and techniques should be used to ensure even distribution and minimize odors and runoff.

- **Application timing and weather conditions:** The timing of waste application should align with crop nutrient requirements and weather conditions. Avoiding application during heavy rainfall or on frozen or saturated soils can help prevent nutrient runoff and minimize the risk of water contamination.
- **Setback distances and buffer zones:** To protect water bodies, wells, and sensitive areas, setback distances and buffer zones should be established based on local regulations and environmental considerations. These distances ensure that waste is applied at a sufficient distance from water sources to prevent contamination.
- **Monitoring and record-keeping:** Regular monitoring of soil and water quality is important to assess the effectiveness of the land application and detect any potential issues. Keeping records of application rates, dates, and locations can help with compliance, planning, and future management decisions.
- **Compliance with regulations:** It's crucial to comply with local regulations, permits, and guidelines related to land application. These regulations may include restrictions on application rates, setback distances, and storage requirements.

When compared to other animal manures, poultry manure has a higher nutritional content, faster mineralization rates, and hence higher plant availability of these nutrients when distributed in the fields (Eghball et al. 2002, Preusch et al. 2002). However, this practice can lead to environmental pollution if soil and climatic conditions are not suitable for agronomic utilization of poultry manure. Moreover, this practice can also lead to obnoxious odor problems and attract insects, which can act as vectors for different diseases and ultimately affectthe human and animal health (Kyakuwaire et al. 2019). Depletion of water sources, nutrient imbalances in the soil and emission of greenhouse gases like methane (CH_4), carbon dioxide (CO_2), ammonia (NH_3) and nitrous oxide (N_2O) are among the other concerns related to improper and non-scientific poultry waste management (Moore et al. 2011). Land application of poultry and slaughterhouse waste can be an effective waste management practice when implemented with proper planning, nutrient management, and adherence to regulatory requirements. However, local conditions, waste characteristics, and environmental considerations should be carefully assessed to determine the suitability and feasibility of this method.

Composting

Composting is another traditional method of organic waste disposal which utilises oxygen (aerobic process) to convert organic waste materials into a valuable manure. In this biological process of decomposition, the degradation is sped up by the activity of microbes in a raised temperature condition which produces CO_2, water, minerals and stabilised organic product, i.e., compost from the organic wastes (Moreki and Chiripasi 2011). The objective of this process is to enhance the degradation process by microbes in presence of thermophilic bacteria as poultry waste acts as a hub of offensive odor, breeding place for flies and rodents and a potent source of disease transmission; composting can act as a boon to reduce and reutilise the poultry litter (Thyagarajan et al. 2013). In this method of decomposition, the waste is converted into a more stabilized and useful product which can act as soil conditioner and organic fertiliser which is more environmentally acceptable as compared to the chemical fertilizer. This has a role in enhancing soil structure as well as improving the water holding capacity, and by reducing the pathogens, it maintains environmental balance. The whole process of composting is presented in Figure 4.5.

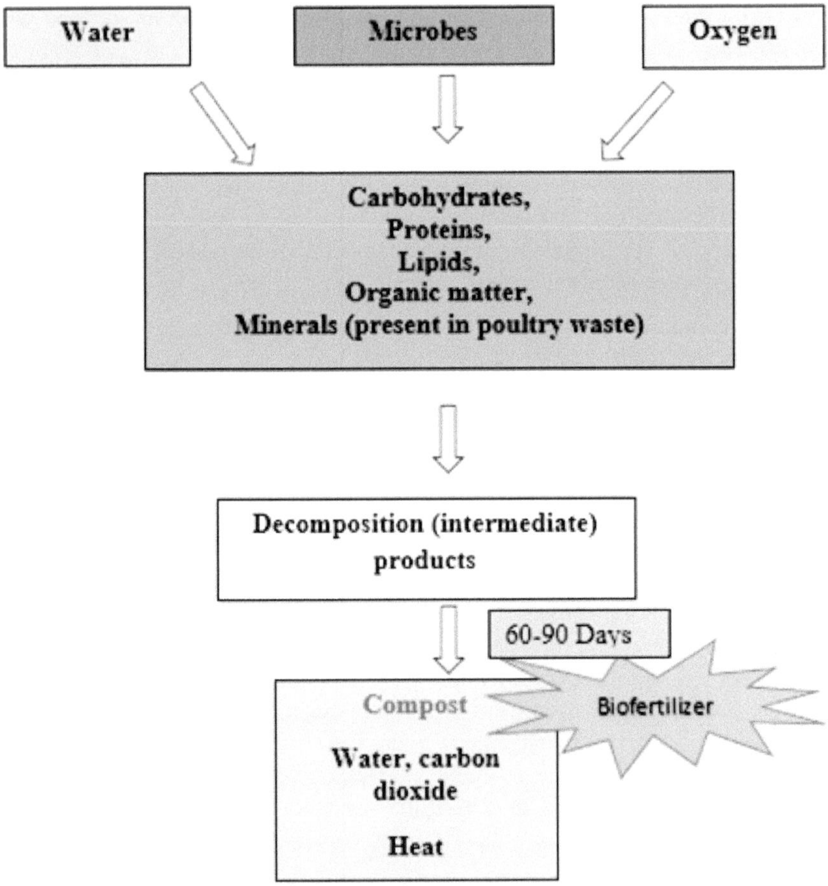

Figure 4.5 Different stages of composting process.

Composting is a biological process carried out by microorganisms that breakdown the organic matter into a stable and nutrient rich product known as compost (Epstein 2017). Certain chemical transformations occur throughout the process, which are as follows:

(a) **Decomposition of organic matter:** Organic matter contains complex compounds such as carbohydrates, proteins, lipids and lignin. During composting, the microorganisms secrete enzymes, which degrade complex organic matters to simpler ones.

(b) **Oxidation:** Composting is an aerobic process, which requires oxygen, which is utilised by the microorganisms for their metabolism. During this process, organic molecules combine with oxygen to produce heat, carbon dioxide and water and by products.

(c) **Humification and stabilization:** With proper C:N ratio, temperature and pH composting progresses and produces a simpler and more stable compound by humification. The compound produced is known as humus—a dark, stable and nutrient rich organic matter that makes up the bulk of compost.

(d) **Nutrient release:** The composting process breaks down organic molecules into their elemental forms, releasing essential plant nutrients such as nitrogen (N), phosphorus (P), potassium (K), sulphur (S), and various micronutrients. These nutrients become available to plants when the compost is applied to the soil, promoting plant growth.

Also, composting is influenced by a variety of factors that can impact the speed and quality of the composting process. Understanding these factors is essential for successful composting. Some of the key factors affecting composting are discussed as under:

(a) **Carbon-to-nitrogen (C/N) ratio:** The carbon-to-nitrogen ratio (C/N ratio) is a critical factor in composting. Microorganisms require carbon for energy and nitrogen for protein synthesis. The ideal C/N ratio for efficient composting is generally considered to be around 25:1 to 30:1. Too much carbon (high C/N ratio) can slow down the composting process, while too much nitrogen (low C/N ratio) can lead to the release of ammonia and odors.

(b) **Moisture content:** Proper moisture is essential for composting. Microorganisms need humidity to thrive, but excessive humidity can lead to anaerobic conditions and unpleasant odors. The compost pile should ideally have a humidity content of about 50–60%. Regularly monitoring and adjusting moisture levels is essential for successful composting.

(c) **Oxygen (Aeration):** Composting is an aerobic process. Adequate aeration ensures that enough oxygen is available to support the aerobic microorganisms responsible for efficient decomposition. Turning or mixing the compost pile regularly helps introduce oxygen and maintain aerobic conditions.

(d) **Temperature:** Temperature plays a crucial role in composting. Microorganisms are most active within specific temperature ranges.

In well-managed compost piles, the microbial activity generates heat, and the internal temperature can rise significantly. Higher temperatures between 40°C and 71°C are favorable for accelerating the composting process and killing pathogens and weed seeds.

(e) **Particle size:** Chopping or shredding compost materials into smaller pieces increases the surface area available for microorganisms to work on. Smaller particle sizes accelerate the composting process, as it allows for quicker decomposition.

(f) **Composting method:** The chosen composting method can affect the speed and ease of composting. For example, traditional backyard composting may take longer compared to hot composting methods that rely on higher temperatures to speed up the process.

(g) **Climate and weather:** Climate and weather conditions can influence composting. In cold climates, composting may slow down during the winter months due to lower temperatures. Proper insulation or choosing a suitable composting method can help address this issue.

Composting can be considered as an option to handle poultry and slaughterhouse waste considering its merits and demerits. Merits of composting include reduction in volume of waste generated from poultry enterprises, reduction in pathogen count and hence disease outbreak, production of valuable organic manure, less inputs required and hence an economical method. On the other hand, demerits include loss of nutrients, time taking process, space requirement and attracting pests.

In poultry waste sector, some studies have been conducted with some interventions to enhance the composting process. Mixing of biochar in poultry waste increased the temperature and shortened the thermophilic phase of the composting process, which also resulted in an increase in carbon dioxide emissions (Czekała et al. 2016). Biochar has been reported to reduce ammonia emission, nitrogen loss and also speed up the decomposition process (Czekała et al. 2016). More initiatives are required to reduce the duration of composting without affecting the quality of compost.

Vermicomposting

Large amounts of nutrients found in poultry waste contribute to environmental issues including eutrophication and phytotoxicity. By utilising earthworms to stabilise its nutrients, these difficulties can be lessened (Yuvaraj et al. 2019). Vermicomposting is also known as worm composting which is carried out by the presence of earthworms in a non-thermophilic environment. The end product of this process is known as vermicompost which is dark in color, more stable, has earthy smell and numerus micronutrients, along with worm castings which enhances the plant growth (Alshehrei and Ameen 2021). For the process of vermicomposting, mostly *Eisenia fetida* species of earthworms are employed. It is a European species of earthworm having higher rate of reproduction, which was introduced into other developing countries. They ingest the complex organic materials and excrete simpler substances as worm castings generally known as vermicompost. It also reduces the bioavailability of some heavy metals which are harmful to plants and subsequently to animals (Alshehrei and Ameen 2021). By colonising and consuming, digesting, and absorption of large amounts of organic waste, earthworms played a significant part in the organic waste system. In addition, they have high reproduction rates and can survive a variety of environmental challenges. With the assistance of aerobic and anaerobic microflora found in earthworms' guts, organic waste is decomposed. Both soil and organic waste are consumed by earthworms and move through their guts, where they combine with digestive enzymes and are broken down by the grinding motion. Following digestion, the worms excrete nutrient-rich material known as "castings." Moisture-rich, well-aerated soils with a low acidity value are best for getting good quality vermicompost (Ali et al. 2015). Different factors affecting the vermicomposting process are presented in Table 4.1.

Table 4.1 Factors affecting vermicompost production.

Sl. No.	Parameters	Optimum range
1.	Stocking density or worm load	27–53 earthworms/kg feed
2.	Temperature	25–35 °C
3.	Feeding rate	1.25 kg feed/kg worm/day
4.	Humidity	65–70%
5.	Carbon: Nitrogen	25–30
6.	pH	4.2–7

The advantages of vermicomposting include production of a soil conditioner, which improves soil structure and water holding capacity, neutralises soil pH by addition of nutrients and enhances soil health and plant growth. Other advantage include the reduction in pathogens and pests and providing an economical and environment friendly process. On the other hand, the disadvantages of this process include that it is highly temperature and moisture sensitive due to death of earthworms in high temperature and low moisture.

Even though many studies have been done on the subject of vermicomposting using poultry waste, there are still unanswered questions that require further research in order to fully realise vermicompost's potential using poultry waste. Setting up a suitable composting system, ensuring correct maintenance, and frequently checking the progress are necessary when using vermicomposting for poultry waste. By doing this, we may turn poultry waste into useful vermicompost, minimising waste and producing an organic fertiliser for crops and plants. In case of poultry waste, large amount of ammonia production, lower C/N ratio and presence of non-degradable materials hinders the process of vermicomposting. These limiting factors need to overcome in order to produce vermicompost from poultry waste. Different techniques used for this involve conditioning, codigestion and composting process followed by vermicomposting process.

Rendering

Wastes from poultry slaughterhouses, such as offal and trimmings are quickly perishable and may degrade if kept at room temperature for a long period of time.

Rendering is the most promising method to obtain and preserve the nutritional quality of the end product, which quickly converts the animal by-products into high-grade hygienic products. Because of the high content of biodegradable organic matter in poultry abattoir waste, the availability of resource-based product is rare (Ozdemir and Yetilmezsoy 2019). Traditional rendering techniques transform organic wastes generated from poultry processing plants into items that can be sold. These are poultry meal, edible and inedible fats that can be used as feed ingredients, and like fuel blending agents. Each rendering product is a great resource for its intended application and often offers an affordable source of animal protein. Due to the complete sterilisation of the waste product by high temperature and pressure, the generated products are environment friendly (Leiva et al. 2018). The high lipid and protein content of the rendering products makes them an easy source for animal feed, fuel for biogas plants, or as an alternative renewable biofuel (Farmanesh et al. 2019). Rendering utilises the poultry slaughter wastes and by grinding, autoclaving and pressing produces blood meal, feather meal, fat and poultry meal (Sari et al. 2016). Some of the products obtained through rendering are discussed as follows:

a) Feather meal: Poultry feather has a high potential for a variety of uses due to its nitrogen- and protein-rich composition. Keratin, on the other hand, is an insoluble structural protein that has excellent mechanical stability and is resistant to being broken down by typical proteolytic enzymes. Feather meal is typically produced by the rendering process with high-pressure autoclaving at a temperature more than 133°C, 3 bar pressure after the hydrolysed feathers are dried and ground into powder for large-scale manufacturing. However, steam hydrolysed feather has a lower rate of digestion than regular feather meal (Leiva et al. 2018). An inadequate quantity of lysine and methionine may result in a nutritional imbalance in the diet, necessitating the usage of unique feed formulations for various animal species. Depending on the autoclaving conditions employed and the animal species utilised, the digestion rate of the nutrients in feather meal may vary to some extent (Ozdemir and Yetilmezsoy 2019).

b) Fats from rendering: Poultry rendering facilities create less priced edible or inedible fat and grease that may be used as blending ingredients in fuel, non-livestock animal feed, and chemical sector products. The raw ingredients are heated during the rendering process, and the fat is then separated from the raw materials by boiling them in water. Following the dewatering procedure, the rendered product is pushed through a screw press to remove some of the leftover fat and moisture before a large portion of the melted fat is removed from the batch by draining. Depending on the source material and the rendering conditions used, the quantity of extractable fat from the chicken wastes ranges from 2 to 30% (Jayathilakan et al. 2012).

c) Poultry meal: Poultry meal is made by combining soft solid waste and byproducts such as feet, head, bone, trimmings, and digestive organs produced from the slaughter of broilers. These byproducts are abundant in fat and protein, making them highly degradable in short time because of enzymatic and chemical reactions. The solid elements of the waste materials are swiftly treated by rendering to create a product of protein-rich chicken meal and fat, which is then used to further convert the waste materials into high-grade new products. The raw poultry meal has a high capacity for digestion and contains 32% protein and 54% fat (Ozdemir and Yetilmezsoy 2019).

If not effectively handled, the extremely degrading nature of such organic wastes poses significant threats to the environment, human health, and economic value. Prior to regulatory limits, solid wastes from poultry slaughterhouses were processed by traditional rendering, which produces valuable byproducts and hygienic feed components for ruminants and birds. Rendering is still considered as the most preferred method for handling poultry waste, which produces a high valued microbiologically safe product, which can be used as animal feed (Ozdemir and Yetilmezsoy 2019).

Recent Advances

Apart from the traditional methods of waste management, there are some recent technologies being implemented in several countries so as to convert the waste to wealth and also to reduce the dreadful impacts of waste on environment. In addition to environmental and safety concerns, meat, poultry, and fish processing wastes frequently have the potential to be recycled as raw materials or transformed into beneficial goods of greater worth (Kulkarni and Devatkal 2015). By adopting low-cost technology for processing and utilising the resulting waste, the poultry industry needs to take steps to reduce its adverse effects on the planet and improve sustainability in chicken production, ensuring more profit for the farmers in the course of the operation. Some of the advances emerged in efficient utilization of poultry wastes are discussed in the following heads.

Biodiesel

Biodiesel is "the mono alkyl ester of long chain fatty acids generated from renewable lipid feed materials such as vegetable oil or animal fats, for use in compression ignition (diesel) engines" (National Biodiesel Board 1996). Biodiesel is appealing because it is a renewable fuel that can help lessen our reliance on imported fossil fuels. It has a smaller carbon footprint than petro-diesel and hence can help decrease greenhouse gas emissions, thereby creating less air pollution making it more environmental friendly.

Different technologies are being exploited for the production of biodiesel viz. transesterification, esterification/hydro-esterification (Brito et al. 2008), hydrocracking (Gusmão et al. 1989) and pyrolysis (Arvanitoyannis et al. 2007, Demirbas and Dincer 2008, Huber and Corma 2007, Maher et al. 2008, Tian et al. 2008). Transesterification of vegetable or animal fats is the most commonly employed technique for producing biodiesel, essentially because it is simple and very affordable.

Biodiesel can be generated from high-quality chicken fat obtained as slaughter waste with low free fatty acids using acid/alkaline transesterification (Wyatt et al. 2005). This process involves the acid catalysed esterification of the free fatty acid (FFA) portion of chicken fat followed by the base catalysed transesterification of the triglyceride portion (Hoque et al. 2011).

The process of blending a triglyceride molecule with an excess of alcohol in the presence of a catalyst (KOH, NaOH, $NaOCH_3$, etc.) to form glycerin and fatty esters is known as transesterification. Figure 4.6 depicts the chemical reaction with methanol.

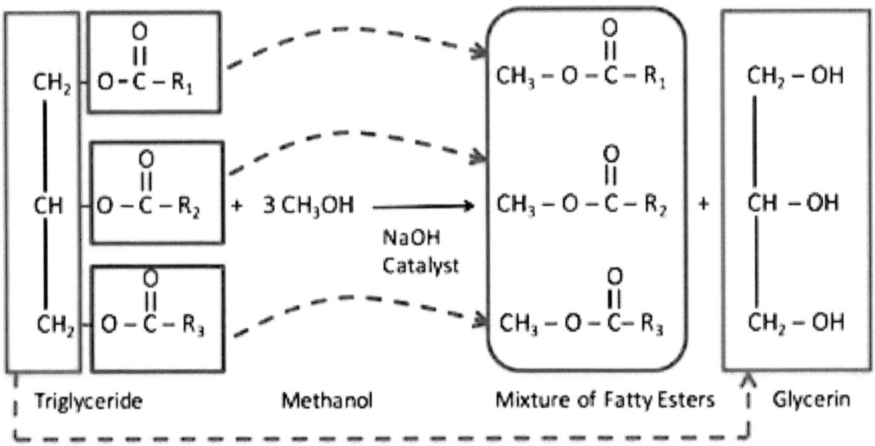

Figure 4.6 Triglyceride transesterification reaction. *R_1, R_2 and R_3 are long hydrocarbon chains sometimes called fatty chain.

At each stage of the transesterification reaction, one mole of ester is created as the triglycerides are successively transformed into diglycerides, monoglycerides, and glycerol. One mole of triglyceride interacts with three moles of methanol to generate three moles of fatty esters (biodiesel) and one mole of glycerol in this transesterification process. Each mole of triglyceride weighs approximately 884 grams, the three moles of methanol weigh approximately 96 grams, the resultant three moles of biodiesel weigh approximately 888 grams, and the mole of glycerol weighs approximately 92 grams. To guarantee that the reaction is driven to completion, most biodiesel producers will inject 1.6 to 2.0 times the chemically adequate quantity of methanol.

Since chicken products are not considered forbidden by any religion, their consumption has been growing as well as their production has increased substantially in recent years due to the relative low price of poultry meat. Chicken fat was found to be highly suitable to produce biodiesel with recommended fuel properties (Nagaraja et al. 2014). The best fat comes from the three main areas of skin: the bottom, body, and wings, which is easier to extract with less contamination (Figure 4.7). Hoque et al. (2011) found that potassium hydroxide was suitable for ethyl ester biodiesel production. Lin and Tsai (2015) revealed that using ethanol or a methanol/ethanol mixture (3:3) for transesterification of chicken oil can be a good alternative method for biodiesel production. Abid et al. (2019) produced biodiesel from the skin of chickens by homogeneous transesterification.

Purandaradasa et al. (2018) found that layer chicken and rooster feather fat have better potential for processing into biodiesel than broiler chicken feather fat because of high energy, fatty acid composition values.

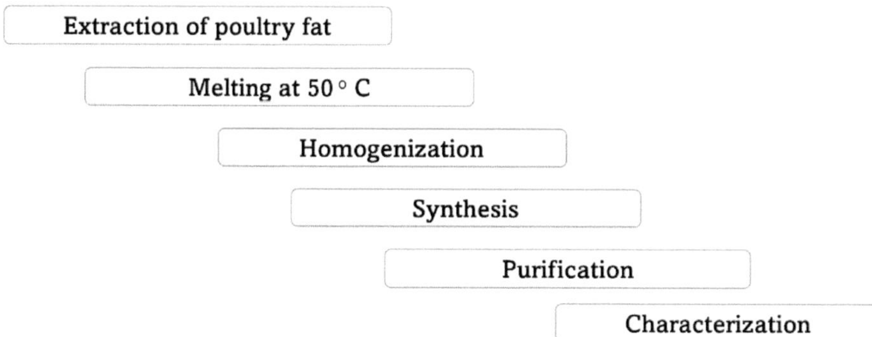

Figure 4.7 Production of biodiesel from poultry fat.

Initiatives have been taken by Indian Railways, Indian Oil Corporation, and Haryana Roadways to use biodiesel for commercial reasons. Poultry waste can be used as an important renewable asset for the cultivation of microalgae *Chlorella* sp., for biomass, biodiesel and lipid production (Agwa and Abu 2014).

The biodiesel thus produced can be blended with other fossil fuels so as to reduce the negative impact on the environment. According to Abraham (2012), adding 20% biodiesel to commercial diesel can significantly cut down on smoke in CRDI engines (by 47.14%). Table 4.2 summarizes different studies related to the production of biodiesel from poultry wastes in different corners of the world.

Table 4.2 Production of biodiesel from poultry waste in different parts of world.

Raw material	Process	Catalyst	Temp (°C)	Time (hrs)	Yield (%)	References
Chicken fat	Transesterification	NaOH	30	1.5	78.9–96.3	Moreira et al. (2009)
Poultry fat	Transesterification	H_2SO_4	50	24.0	99.01	Bhatti et al. (2008)
Chicken fat	Transesterification	NaOH	62	2.0	89.0	Guru et al. (2009)
Duck fat	Transesterification	KOH	65	3.0	97.1	Chung et al. (2009)
Poultry fat	Pyrolysis	–	350–400	1.3	40.0	Santos et al. (2010)
Feather meal	Transesterification	KOH	70	1.0	7–11	Kondamudi et al. (2009)

Dry rendering is used to turn waste from broiler slaughterhouses into Rendered Chicken Oil (RCO). In standard diesel engines, chicken oil methyl ester might be utilised as an alternate fuel without requiring significant adjustments. It raises mechanical efficiency, boosts brake thermal efficiency, and cuts down on smoke emissions by 47.14 percent, lowering pollution and averting climatic changes (Abraham et al. 2015).

Biochar

While the Amazonian terra preta soil suggested that biochar may be a viable remedy for the worldwide soil degradation and climate change, interest in the pyrolytic conversion of animal dung to biochar first emerged in the early 2000s (Guo 2016).

A black, porous solid residue known as "biochar" is produced when animal manure is thermochemically processed using the pyrolysis and gasification procedures (Ro 2012). Biochar is a fine-grained or granular charcoal generated by heating vegetative biomass, bones, manure solids, or other plant-derived organic residuals in an oxygen-free or oxygen-limited environment. It is utilised as a soil supplement for agricultural and environmental applications as it is environmentally resilient and retains water and nutrients. Biochar is a potential soil supplement that can steadily improve soil quality and promote plant development (Guo et al. 2016). Although biochar is greatly desired by crop growers, its high production cost and limited availability currently restricts it from being used on an extensive scale (Guo et al. 2016).

Given this, the viability of producing affordable biochar from readily accessible agricultural byproducts such as animal manure, food processing waste, and crop residues has been researched. It functions better at delivering nutrients than raw manure, which is easily biodegradable and releases nutrients in a flush, and plant-derived biochar, which is often deficient in nutrients (Li et al. 2018).

The thermochemical methods that are currently employed to transform sewage sludge, bovine manure, swine manure, and poultry litter into useful bioenergy and biochar products include pyrolysis, gasification, hydrothermal liquefaction, and combustion.

Poultry litter may easily be transformed into biochar by slow agriculture scale pyrolysis because of its high lignocellulosic content and low humidity. Pyrolysis, the thermal breakdown of biomass in an environment with a low oxygen content and a wide temperature range that typically ranges between 300 and 1000 °C, produces biochar (Zhang et al. 2015). The use of different biomass as poultry litter (bark, straw, seeds, bagasse, and sawdust, among others), its permanence in the poultry house (which can be reused in more than one batch), the feeding of the birds and their wastes, and other factors all affect the chemical and physical properties of poultry litter biochar, making it one of the challenges associated with its use.

The pyrolysis process is extremely reliant on the heating rate, pressure, residence duration, peak temperature, heat transfer rate, and vapor-solid interactions (Libra 2011). There are three main stages (Demirbas 2004) involved in pyrolysis process:

(i) **Hydration:** Dry organic matter plus water equals biomass.

(ii) **Initial charring:** Primary char plus tarry vapors equals dry organic matter.

(iii) **Final charring:** Primary char is created by adding secondary char to tarry vapors.

Heavy hydrocarbons, water vapors, and non-condensable gases including CO_2, CO, H_2, CH_4, C_2H_2 and C_3H_6 are all mixed together to form the tarry vapor.

Pyrolysis can be done in muffle furnaces, bed reactors, kiln furnaces, and tube furnaces with N_2 flow to generate an oxygen-free atmosphere. There are also two forms of pyrolysis: slow and quick. High heating rates and brief vapour residence durations are the hallmarks of fast pyrolysis, which typically occurs at a moderate (in pyrolysis standards) temperature of roughly 500°C (Brownsort 2009). Slow pyrolysis is distinguished from fast pyrolysis by slower heating rates, prolonged solid and vapour residence durations, and typically lower temperatures, about 400°C.

Biochar made from poultry litter may be used to improve soil, feed, litter, and plants' nutritional value. It makes chicken litter disposal less expensive, facilitates simple storage, lessens farm odours, prevents eutrophication, makes transportation simpler (Amita et al. 2017).

Value-added Products

In addition to management techniques and the generation of renewable energy, wastes from slaughterhouses and poultry enterprises may be used to create products with added value (such as protein and protein hydrolysate), which might be beneficial substitutes for their commercial equivalents. This next section goes over in brief about value-added goods from chicken by-products and wastes.

Blood

Blood, constituting about 7% of chicken liveweight, can be used for the production of both non-edible and edible products. Non-edible products include binding agents, fertilizers, feedstuffs, emulsifiers, stabilizers, and color additives. Some edible goods made from blood include blood meal, sausages, blood cake, biscuits, and bread (Karuppannan et al. 2021). It is a component in blood curd, blood cake and blood pudding in Asia. Blood sausages, blood pudding, biscuits and bread have all been made with animal blood in Europe for a very long time (Ghost 2001). It can also be utilized for laboratory media preparation viz. blood agar, animal tissue culture and peptone (Kurbanoglu and Kurbanoglu 2004).

Feathers

Bacillus licheniformis, a bacteria that can degrade feathers, has the ability to ferment and transform feathers into feather lysate, a feed-grade protein source. This bacteria secretes a powerful proteinase enzyme called keratinase, which hydrolyzes collagen, elastin, and feather keratin (Jayathilakan et al. 2012, Padhi et al. 2023). Feather keratin can be used in wood adhesives, biomaterial development, biofertilizers, biomedicals, animal feeds, environmental remediation, textile and leather industry, flame retardants, biocomposites, etc. (Karuppannan et al, 2021).

Egg Shell Waste

The eggshell wastes are converted to valuable products such as bone substitute, preparation of calcium phosphate, growth and mineralization of bone, catalysts, biofertilizers and used in medical and dental low-cost adsorbent. In addition to these, they may be utilised for mosaic and 3D artwork, chicken feed, reinforcing cement, and slug and snail management. Additionally, eggshell membranes are used in a variety of medical procedures and treatments, including the filling of corneal wounds with collagen glue, skin grafts, angioplasty sleeves, dental implants, osteoporosis therapy, food casings, film emulsions, reconstructive surgery, drug development, bioplastics, effluent treatment, and heavy metal removal from waste waters (Karuppannan et al. 2021).

Spent Hens

From the spent chickens, mechanically deboned meat (MDM) is made. MDM makes up roughly 20% of the weight of live birds, with the rest of it being processed to provide animal food supplements. Wood adhesives were made possible by chemically processing recovered proteins from dead chickens using proteinase inhibitors like urea and sodium dodecyl sulphate. The acceptability is increased by using emulsion-based mutton nuggets in conjunction with by-products from spent hens such as skin, gizzards, and hearts (Kim and Patterson 2000).

Textiles

The cost of making non-woven textiles from chicken feathers is relatively inexpensive. The use of chicken feathers to make textiles has a lot of potential (Chinta et al. 2013).

Biodegradable Plastics in the Environment

By using a method called polymerization, chicken feathers can also be transformed into biodegradable polymers. To create thermoplastics, feather keratin is ground into a fine dust and then chemically polymerized at 170°C (Thyagarajan et al. 2013). Several goods, such as plastic cups and sheets for furniture manufacturing, may be produced using this.

Gelatin

Chickens employing the skin, bone, etc., that are left behind after additional processing are a new prospective or replacement source for gelatin. This area is yet to be explored as there are various challenges during the process (Seidavi et al. 2019). Having a substantially greater real value (0.55) than bovine gelatin (0.22), chicken gelatin gel is stronger than bovine gelatin (Norizah et al. 2013).

Fibreboard Production

Recent studies have focused on creating natural insulating composite fiberboard prototypes using a mixture of leftover chicken feathers and wood scraps. Fiberboard samples were made by combining feathers in various ratios with either wood shavings (for a coarse structure) or mixed wood residues (for a finer and denser structure) (Safari et al. 2020).

Bioplastic Sheet Fabrication

One of the sustainable sources of protein for the production of bioplastic films is abattoir and poultry waste. Bioplastics may be made from discarded feathers (Ramakrishnan et al. 2018), meat and bone meal (Lukubira and Ogale 2013) and blood meals (Bier et al. 2013).

CONCLUSIONS AND FUTURE PERSPECTIVES

Poultry industry is blooming at a rapid rate in terms of production but along with this, there is rapid generation of waste both from poultry houses and slaughter houses. Generation of huge amount of waste requires scientific and sustainable management and hence leads to development of new avenues. These measures not only reduce the volume of waste but also promote resource conservation and energy efficiency. Traditional methods like anaerobic digestion, incineration, land application, composting, vermicomposting and rendering were already studied at large scale for poultry waste management and their utilization depends on different factors. The new avenues of poultry waste management comprise of the recent advances such as production of biodiesel, biochar and different value added products. These advances need to be explored at large scales and can be adopted as sustainable methods of waste utilization along with generation of economy. Effective management of waste from the poultry industry and slaughterhouses is essential to mitigate environmental pollution, protect public health, and ensure the industry's long-term viability.

REFERENCES

Abid, M., Touzani, A. and Benhima, R. 2019. Synthesis of biodiesel from chicken's skin waste by homogeneous transesterification. International Journal of Sustainable Engineering. 12(4): 272–280.

Abouelenien, F., Fujiwara, W., Namba, Y., Kosseva, M., Nishio, N. and Nakashimada, Y. 2010. Improved methane fermentation of chicken manure via ammonia removal by biogas recycle. Bioresource Technology. 101(16): 6368–6373.

Abraham, J. 2012. Utilization of dead poultry birds for biodiesel production. Ph.D Thesis Submitted to the Tamil Nadu Veterinary and Animal Science University. Chennai.

Abraham, J., Kumar, R.S., Xavier, F. and Mathew, D. 2015. Biodiesel Production from broiler chicken waste. International Journal of Biological, Bio-molecular, Agricultural. Food and Biotechnological Engineering. 9: 1181–1185.

Adhikari, B.B., Chae, M. and Bressler, D.C. 2018. Utilization of slaughterhouse waste in value-added applications: recent advances in the development of wood adhesives. Polymers. 10(2): 176.

Agwa, O.K. and Abu, G.O. 2014. Utilization of poultry waste for the cultivation of Chlorella sp. for biomass and lipid production. International Journal of Current Microbiology and Applied Sciences. 3(8): 1036–1047.

Akdeniz, N. 2019. A systematic review of biochar use in animal waste composting. Waste Management. 88: 291–300.

Alam, M.U., Rahman, M., Islam, M.A., Asaduzzaman, M., Sarker, S., Rousham, E., et al. 2019. Human exposure to antimicrobial resistance from poultry production: assessing hygieneand waste-disposal practices in Bangladesh. International Journal of Hygiene and Environmental Health. 222(8): 1068–1076.

Ali, U., Sajid, N., Khalid, A., Riaz, L., Rabbani, M.M., Syed, J.H., et al. 2015. A review on vermicomposting of organic wastes. Environmental Progress and Sustainable Energy. 34(4): 1050–1062.

Alshehrei, F. and Ameen, F. 2021. Vermicomposting: a management tool to mitigate solid waste. Saudi Journal of Biological Sciences. 28(6): 3284–3293.

Arvanitoyannis, I.S., Kassaveti, A. and Stefanatos, S. 2007. Current and potential uses of thermally treated olive oil waste. International Journal of Food Science Technology. 42: 852–867.

ASABE. 2005. Manure production characteristics. ASABE standard D384.2. American Society of Agricultural and Biological Engineers. St. Joseph, MI.

Asiedu, P., Odoi, F.N.A., Adu, E.K., Wallace, P., Prah, A.A., Osei, D.Y., et al. 2022. Quantities of hatchery waste generated, handling, and disposal practices of selected hatcheries in the greater ACCRA region of Ghana. Ghanaian Journal of Animal Science. 13(1).

Assefa, A., Egigu, M.C. and Kebede, A. 2014. Thermal and chemical pre-treatments of cow dung and poultry litter enhance biogas production in batch fermentation. International Journal of Scientific and Technology Research. 3: 165–170.

Azadbakht, M., Safieddin, A.S. and Rahmani, M. 2021. Potential for the production of biofuels from agricultural waste, livestock, and slaughterhouse waste in Golestan province, Iran. Biomass Conversion and Biorefinery. 1–11.

Bernhart, M., Fasina, O., Fulton, J. and Wood, C. 2010. Compaction of poultry litter. Bioresource Technology. 101: 234–238.

Bhaskar, T., Varjani, S., Krishna, B.B. and Pandey, A. (eds). 2023. Biotic Resources: Circular Bioeconomy Perspective. CRC Press.

Bhatti, H., Hanif, M., Qasim, M. and Rehman, A. 2008. Biodisel production from waste tallow. Fuel. 87: 2961–2966.

Bier, J.M., Verbeek, C.J.R. and Lay, M.C. 2013. Thermal and mechanical properties of Bloodmealblood meal-based thermoplastics plasticized with tri (ethylene glycol). Macromolecular Materials and Engineering. 299: 85–95.

Billen, P., Costa, J., Van der Aa, L., Van Caneghem, J. and Vandecasteele, C. 2015. Electricity from poultry manure: a cleaner alternative to direct land application. Journal of Cleaner Production. 96: 467–475.

Bolan, N., Szogi, A.A., Chuasavathi, T., Seshadri, B., JR, M.J. and Panneerselvam, P. 2010. Uses and management of poultry litter. World's Poultry Science Journal. 66: 673–698.

Brito, Y.C., Mello, V.M., Macedo, C.S., Meneghetti, M.R., Suarez, P.A.Z. and Meneghetti, S.M.P. 2008. Fatty acid methyl esters preparation in the presence of maltolate and nbutoxide Ti(IV) and Zr(IV) complexes. Applied Catalysis A. 351: 24–28.

Brownsort, P.A. 2009. Biomass pyrolysis processes: performance parameters and their influence on biochar system benefits. A dissertation presented for the degree of Master of Science, University of Edinburgh.

Charles, M.W. 2013. North Carolina State University, Department of Poultry Science, Raleigh NC, United States of America. FAO. ISBN 978–92-5-108067-2. https://www.fao.org/3/i3531e/i3531e05.pdf..

Chen, Z. and Jiang, X. 2014. Microbiological safety of chicken litter or chicken litter-based organic fertilizers: a review. Agriculture. 4(1): 1–29.

Chinta, S.K., Landage, S.M. and Yadav, K. 2013. Application of chicken feathers in technical textiles. International Journal of Innovative Research in Science, Engineering and Technology. 2(4): 1158–1165.

Chung, K., Kim, J. and Lee, K. 2009. Biodiesel production by transesterification of duck tallow with methanol on alkali catalysts. Biomass and Bioenergy. 33: 155–158.

Collins, E., Barker, J.C., Carr, L.E., Brodie, H.L. and Martin, J.H. 1999. Poultry Waste Management Handbook (NRAES 132-FRONT MATTER ONLY).

Czekała, W., Malińska, K., Cáceres, R., Janczak, D., Dach, J. and Lewicki, A. 2016. Co-composting of poultry manure mixtures amended with biochar–The effect of biochar on temperature and C-CO_2 emission. Bioresource Technology. 200: 921–927.

Dalolio, F.S., da Silva, J.N., de Oliveira, A.C.C., Tinôco, I.D.F.F., Barbosa, R.C., de Oliveira Resende, M., et al. 2017. Poultry litter as biomass energy: a review and future perspectives. Renewable and Sustainable Energy Reviews. 76: 941–949.

Das, K.C., Minkara, M.Y., Melear, N.D. and Tollner, E.W. 2002. Effect of poultry litter amendment on hatchery waste composting. Journal of Applied Poultry Research. 11: 282–290.

Delgado, C., Narrod, C. and Tiongco, M. 2003. Policy, technical and environmental determinants and implications of the scaling-up of livestock production in four fast-growing developing countries: a synthesis. International Food Policy Research Institute. 188.

Demirbas, A. 2004. Effects of temperature and particle size on bio-char yield from pyrolysis of agricultural residues. Journal of Analytical and Applied Pyrolysis. 72(2): 243–248.

Demirbas, A. and Dincer, K. 2008. Sustainable green diesel: a futuristic view. Energy Sources Part A—Recovery Utilization Environmental Effects. 30: 1233–1241.

Djang-Fordjour, H. 2016. Identifying broiler meat production challenges in Ghana: focusing on hatchery performance. A Dissertation Submitted to the School of Graduated Studies, Kwame Nkrumah University of Science and Technology, Kumasi, Ghana, in Partial Fulfilment of the Requirements for the Award of the Degree of Master of Philosophy Animal Breeding and Genetics.

Dornelas, K.C., Schneider, R.M. and Do Amaral, A.G. 2017. Biogas from poultry waste—production and energy potential. Environmental Monitoring and Assessment. 189(8): 1–11.

Eghball, B., Wienhold, B.J., Gilley, J.E. and Eigenberg, R.A. 2002. Mineralization of manure nutrients. Journal of Soil and Water Conservation. 57(6): 470–473.

Epstein, E. 2017. The Science of Composting. CRC Press.

Ernstoff, A., Tu, Q., Faist, M., Del Duce, A., Mandlebaum, S. and Dettling, J. 2019. Comparing the environmental impacts of meatless and meat-containing meals in the United States. Sustainability. 11(22): 6235.

Esendir, E. 2012. Effect of heat-, alkali-, and ultrasonic-pretreatment on anaerobic digestion of chicken manure. Environmental Science and Technology. 608–614.

Farmanesh, A., Mohtasebi, S.S. and Omid, M. 2019. Optimization of rendering process of poultry by-products with batch cooker model monitored by electronic nose. Journal of Environmental Management. 235: 194–201.

Ferreira, A., Kunh, S.S., Cremonez, P.A., Dieter, J., Teleken, J.G., Sampaio, S.C., et al. 2018. Brazilian poultry activity waste: destinations and energetic potential. Renewable and Sustainable Energy Reviews. 81: 3081–3089.

FSA. 2007. Feedlot Services Australia Pty Ltd. Available at: http://www.fsaconsulting.net/.

Ghosh, P., Sengupta, S., Singh, L. and Sahay, A. 2020. Life cycle assessment of waste-to-bioenergy processes: a review. Bioreactors. 105–122.

Ghost, R. 2001. Fractionating of biological macromolecules using carrier phase ultrafiltration. Biotechnology and Bioengineering. 74: 1–11.

Glatz, P. and Miao, Z. 2009. High-value Products from Hatchery Waste. © 2009 Rural Industries Research and Development Corporation. Electronically published by RIRDC in April 2009. Canberra at www.rirdc.gov.au.

Glatz, P., Miao, Z. and Rodda, B. 2011. Handling and treatment of poultry hatchery waste: a review. Sustainability. 3(1): 216–237.

GOP. 2008. Economic survey. Economic Advisor's Wing, Finance Division, Government of Pakistan, Islamabad.

Gunduz, S., Aslanova, F. and Abdullah, K.S. 2019. Poultry waste management techniques in urban agriculture and its implications: a case study of Tripoli, Libya. Ekoloji. 28(107): 4077–4084.

Guo, M. 2016. Pyrogenic carbon in Terra Preta soils. pp. 15–27. *In*: M. Guo, Z. He and S.M. Uchimiya (ed.). Agricultural and Environmental Applications of Biochar: Advances and Barriers. SSSA Special Publications, 63. Soil Science Society of America, Madison.

Guo, M., He, Z. and Uchimiya, S.M. 2016. Introduction to biochar as an agricultural and environmental amendment. *In*: M. Guo, Z. He and S.M. Uchimiya (eds). Agricultural and Environmental Applications of Biochar: Advances and Barriers. SSSA Special Publications, 63. SSSA, Madison, WI. 1–14.

Guru, M., Artukoglu, B.D., Keskin, A. and Koca, A. 2009. Biodiesel production from waste animal fat and improvement of its characteristics by synthesized nickel and magnesium additive. Energy Conversion and Management. 50: 498–502.

Gusmão, J., Brozki, D., Djéga-Mariadassou, G. and Frety, R. 1989. Utilization of vegetable oils as an alternative source for diesel-type fuel: hydrocracking on reduced Ni/SiO_2 and sulphided Ni-Mo/c-Al_2O_3. Catalysis Today. 5: 533.

Holm-Nielsen, J.B., Al Seadi, T. and Oleskowicz-Popiel, P. 2009. The future of anaerobic digestion and biogas utilization. Bioresource Technology. 100(22): 5478–5484.

Hoque, M.E., Singh, A. and Chuan, Y.L. 2011. Biodiesel from low cost feed stocks: the effects of process parameters on the biodiesel yield. Biomass and Bioenergy. 35: 1582–1587.

Hoque, M., Mondal, S. and Adusumilli, S. 2022. Sustainable livestock production and food security. *In*: Emerging Issues in Climate Smart Livestock Production. Academic Press. 71–90.

Huber, G.W. and Corma, A. 2007. Synergies between bio- and oil refineries for the production of fuels from biomass. Angewandte Chemie International Edition. 46: 7184–7201.

Jackson, B.P., Bertsch, P.M., Cabrera, M.L., Camberato, J.J., Seaman, J.C. and Wood, C.W. 2003. Trace element speciation in poultry litter. Journal of Environmental Quality. 32(2): 535–540.

Jana, A., Dasgupta, D., Bhaskar, T. and Ghosh, D. 2023. Poultry waste biorefinery: opportunities for sustainable management. Biotic Resources: Circular Bioeconomy Perspective. 85.

Jayathilakan, K., Sultana, K., Radhakrishna, K. and Bawa, A.S. 2012. Utilization of byproducts and waste materials from meat, poultry, and fish processing industries: a review. Journal of Food Science and Technology. 49: 278–293.

Ka, A. and Benson, O. 2014. Poultry wastes management strategies and environmental implications on human health in Ogun state of Nigeria. Advances in Economics and Business. 2: 164–171.

Kaiser, D.E., Mallarino, A.P., Haq, M.U. and Allen, B.L. 2009. Runoff phosphorus loss immediately after poultry manure application as influenced by the application rate and tillage. Journal of Environmental Quality. 38(1): 299–308.

Karuppannan, S.K., Dowlath, M.J.H., Raiyaan, G.D., Rajadesingu, S. and Arunachalam, K.D. 2021. Application of poultry industry waste in producing value-added products—a review. Concepts of Advanced Zero Waste Tools. 91–121.

Kaza, S., Yao, L., Bhada-Tata, P. and Woerden, F.V. 2018. What a waste 2.0 a global snapshot of solid waste management to 2050. *In*: Urban Development. The World Bank Group, Washington.

Kelley, T.R., Pancorbo, O.C., Merka, W.C., Thompson, S.A., Cabrera, M.L. and Barnhart, H.M. 1996. Elemental concentrations of stored whole and fractionated broiler litter. The Journal of Applied Poultry Research. 5: 276–281.

Kim, W.K. and Patterson, P.H. 2000. Recycling dead hens by enzyme or sodium hydroxide pretreatment and fermentation. Poultry Science. 79(6): 879–885.

Kondamudi, N., Strull, J., Misra, M. and Mohapatra, S.K. 2009. A green process for producing biodiesel from feather meal. Journal of Agricultural and Food Chemistry. 57(14): 6163–6166.

Kowalczyk, P., Mahdi-Oraibi, S., Misiewicz, A., Gabzdyl, N., Miskiewicz, A. and Szparecki, G. 2018. Feather-degrading bacteria: their biochemical and genetic characteristics. Arabian Journal for Science and Engineering. 43: 33–41.

Kulkarni, V.V. and Devatkal, S.K. 2015. Utilization of byproducts and waste materials from meat and poultry processing industry: a review. Journal of Meat Science. 11(1): 1–10.

Kumar, P., Barrett, D.M., Delwiche, M.J. and Stroeve, P. 2009. Methods for pretreatment of lignocellulosic biomass for efficient hydrolysis and biofuel production. Industrial and Engineering Chemistry Research. 48(8): 3713–3729.

Kurbanoglu, E.B. and Kurbanoglu, N.I. 2004. Utilization as peptone for glycerol production of ram horn waste with a new process. Energy Conversion and Management. 45(2): 225–234.

Kyakuwaire, M., Olupot, G., Amoding, A., Nkedi-Kizza, P. and Ateenyi Basamba, T. 2019. How safe is chicken litter for land application as an organic fertilizer?: a review. International Journal of Environmental Research and Public Health. 16(19): 3521.

Leiva, A., Granados-Chinchilla, F., Redondo-Solano, M., Arrieta-González, M., Pineda-Salazar, E. and Molina, A. 2018. Characterization of the animal by-product meal industry in Costa Rica: manufacturing practices through the production chain and food safety. Poultry Science. 97(6): 2159–2169.

Leytem, A.B., Plumstead, P.W., Maguire, R.O., Kwanyuen, P. and Brake, J. 2007. What aspect of dietary modification in broilers controls litter water-soluble phosphorus: dietary phosphorus, phytase, or calcium?. Journal of Environmental Quality. 36(2): 453–463.

Li, Y., Zhang, R., Chen, C., Liu, G., He, Y. and Liu, X. 2013. Biogas production from co-digestion of corn stover and chicken manure under anaerobic wet, hemi-solid, and solid state conditions. Bioresource Technology. 149: 406–412.

Li, K., Liu, R. and Sun, C. 2015. Comparison of anaerobic digestion characteristics and kinetics of four livestock manures with different substrate concentrations. Bioresource Technology. 198: 133–140.

Li, W., Feng, X., Song, W. and Guo, M. 2018. Transformation of phosphorus in speciation and bioavailability during converting poultry litter to biochar. Frontiers in Sustainable Food Systems. 2: 1–10.

Libra, J.A., Ro, K.S., Kammann, C., Funke, A., Berge, N.D., Neubauer, Y., et al. 2011. Hydrothermal carbonization of biomass residuals: a comparative review of the chemistry, processes, and applications of wet and dry pyrolysis. Biofuels. 2(1): 89–124.

Lin, C.W. and Tsai, S.W. 2015. Production of biodiesel from chicken wastes by various alcohol-catalyst combinations. Journal of Energy in Southern Africa. 26: 36–45.

Lukubira S. and Ogale, A.A. 2013. Thermal processing and properties of bioplastic sheets derived from meat and bone meal. Journal of Applied Polymer Science. 130: 256–263.

Maguire, R.O., Plumstead, P.W. and Brake, J. 2006. Impact of diet, moisture, location, and storage on soluble phosphorus in broiler litter breeder manure. Journal of Environmental Quality. 35: 858–865.

Maher, K.D., Kirkwood, K.M., Gray, M.R. and Bressler, D.C. 2008. Pyrolytic decarboxylation and cracking of stearic acid. Industrial and Engineering Chemistry Research. 47: 5328–5336.

Malone, G.W. 1992. Nutrient enrichment in integrated broiler production systems. Poultry Science. 71: 1117–1122.

Marshall, B.M. and Levy, S.B. 2011. Food animals and antimicrobials: impacts on human health. Clinical Microbiology Reviews. 24(4): 718–733.

Matheri, A.N., Ndiweni, S.N., Belaid, M., Muzenda, E. and Hubert, R. 2017. Optimising biogas production from anaerobic co-digestion of chicken manure and organic fraction of municipal solid waste. Renewable and Sustainable Energy Reviews. 80: 756–764.

McGrath, S., Maguire, R.O., Tracy, B.F. and Fike, J.H. 2010. Improving soil nutrition with poultry litter application in low-input forage systems. Agronomy Journal. 102(1): 48–54.

Miah, M.R., Rahman, A.K.M.L., Akanda, M.R., Pulak, A. and Rouf, M.A. 2016. Production of biogas from poultry litter mixed with the co-substrate cow dung. Journal of Taibah University for Science. 10(4): 497–504.

Mingxin Guo, Y.S. and He, Z. 2012. Poultry litter-based biochar: preparation, characterization, and utilization. Applied Research of Animal Manure.

Mohd Johari, S.A., Aqsha, A., Osman, N.B., Shamsudin, M.R., Ameen, M. and Dol, S.S. 2020. Enhancing biogas production in anaerobic co-digestion of fresh chicken manure with corn stover at laboratory scale. SN Applied Sciences. 2(7): 1–6.

Moore Jr, P.A., Miles, D., Burns, R., Pote, D., Berg, K. and Choi, I.H. 2011. Ammonia emission factors from broiler litter in barns, in storage, and after land application. Journal of Environmental Quality. 40(5): 1395–1404.

Moreira, A.L. 2009. Produção de Biodiesel a Partir de Gordura de Frango, Mestrado Integrado em Engenharia Química, Faculdade de Engenharia, Universidade do Porto, Porto.

Moreki, J.C. and Chiripasi, S.C. 2011. Poultry waste management in Botswana: a review. Online Journal of Animal and Feed Research. 285: 292.

Mozhiarasi, V. and Natarajan, T.S. 2022. Slaughterhouse and poultry wastes: management practices, feedstocks for renewable energy production, and recovery of value-added products. Biomass Conversion and Biorefinery. 1–24.

Muduli, S., Champati, A., Popalghat, H.K., Patel, P. and Sneha, K.R. 2019. Poultry waste management: an approach for sustainable development. International Journal of Advanced Science Research. 4: 08–14.

Muhammad Nasir, I. and Mohd Ghazi, T.I. 2015. Pretreatment of lignocellulosic biomass from animal manure as a means of enhancing biogas production. Engineering in Life Sciences. 15(7): 733–742.

Mustafa, E.A., Hamad, E.M., Elhassan, M.M.O., Salman, A.M., Elsiddig, M.M.E. and Lamyia, M.A. 2018. Disposal of dead birds and manure in poultry farms under different production and management systems in Khartoum State, Sudan. World Journal of Pharmacy and Pharmaceutical. 7(11): 61–70.

Nagaraja, Y.P., Venkatesh, H.S. and Vinay, M.S. 2014. Production of biodiesel from poultry waste. International Journal of Pharma and Bio Sciences. 5: 791–799.

Nebiyu, Y., Berhan, T. and Ashenafi, M. 2016. Poultry waste management practices under small scale intensive urban poultry production in Addis Ababa, Ethiopia. Academia Journal of Agricultural Research. 4(4): 212–217.

Norizah, M.S., Badii, F. and Howell, N.K. 2013. Preparation and characterization of chicken skin gelatin as an alternative to mammalian gelatin. Food Hydrocolloids. 30: 143–151.

Orlando, M.Q. and Borja, V.M. 2020. Pretreatment of animal manure biomass to improve biogas production: a review. Energies. 13(14): 3573.

Orrico, A., Schwingel, A., Costa, M., Orrico, M., Aspilcueta-Borquis, R., Alves, G., et al. 2020. Characterization and valuing of hatchery waste from the broiler chicken productive chain. Waste Management. 105: 520–530.

Ozdemir, S. and Yetilmezsoy, K. 2019. A mini literature review on sustainable management of poultry abattoir wastes. Journal of Material Cycles and Waste Management. 1–11.

Padhi, S., Sarkar, P., Abedin, M.M., Sirohi, R., Singh, S.P. and Rai, A.K. 2023. Microbial enzymes and metabolites in health care: an overview. *In*: Microbial Enzymes and Metabolites for Health and Well-Being. CRC Press. 3–23.

Prabakaran, R. and Valavan, S.E. 2021. Wealth from poultry waste: an overview. World's Poultry Science Journal. 77(2): 389–401.

Preusch, P.L., Adler, P.R., Sikora, L.J. and Tworkoski, T.J. 2002. Nitrogen and phosphorus availability in composted and uncomposted poultry litter. Journal of Environmental Quality. 31(6): 2051–2057.

Purandaradasa, A., Silambarasan, B.T., Murugan, K., Babujanarthanam, A., Arumugam, D.G., Dhandapani, K.V., et al. 2018. Development and quantification of biodiesel production from chicken feather meal as a cost-effective feedstock by using green technology. Biochemistry and Biophysics Reports. 14: 133–139.

Ragasri, S. and Sabumon, P.C. 2023. A critical review on slaughterhouse waste management and framing sustainable practices in managing slaughterhouse waste in India. Journal of Environmental Management. 327: 116823.

Rahman, M.A., Møller, H.B., Saha, C.K., Alam, M.M., Wahid, R. and Feng, L. 2017. Optimal ratio for anaerobic co-digestion of poultry droppings and lignocellulosic-rich substrates for enhanced biogas production. Energy for Sustainable Development. 39: 59–66.

Ramakrishnan, N., Sharma, S., Gupta, A. and Alashwal, B.Y. 2018. Keratin-based bioplastic film from chicken feathers and its characterization. International Journal of Biological Macromolecules. 111: 352–358.

Ro, K.S. 2012. Thermochemical conversion technologies for production of renewable energy and value-added char from animal manures. *In*: Z. He (Ed.), Applied Research of Animal Manure: Challenges and Opportunities Beyond the Adverse Environmental Concerns. Nova Science Publishers. 63–82.

Safari, C.R., Zemljiˇc, L.F., Novak, M., Dugonik, B., Bratina, B., Gubeljak, N., et al. 2020. Preparation and characterization of waste poultry feathers composite fibreboard. Materials. 13: 4964.

Salminen, J.E. and Rintala, J. 2002. Anaerobic digestion of organic slaughterhouse waste—a review. Bioresource Technology. 83(1): 13–26.

Sankoh, F.P., Yan, X. and Tran, Q. 2013. Environmental and health impact of solid waste disposal in developing cities: a case study of Granville Brook dumpsite, Freetown, Sierra Leone. Journal of Environmental Protection. 4: 665–670.

Santos, A.L.F., Martins, D.U., Iha, O.K., Ribeiro, R.A., Quirino, R.L. and Suarez, P.A. 2010. Agro-industrial residues as low-price feedstock for diesel-like fuel production by thermal cracking. Bioresource Technology. 101: 6157–6162.

Sari, O.F., Ozdemir, S. and Celebi, A. 2016. Utilization and management of poultry slaughterhouse wastes with new methods. In Eurasia 2016 Waste Management Symposium. 02–04.

Seidavi, A., Zaker-Esteghamati, H. and Scanes, C.G. 2019. Poultry byproducts. In Byproducts from Agriculture and Fisheries: Adding Value for Food, Feed, Pharma, and Fuels. 123–146.

Shakya, A. and Agarwal, T. 2017. Poultry litter biochar: an approach towards poultry litter management—a review. International Journal of Current Microbiology and Applied Sciences. 6(10): 2657–2668.

Singh, P., Mondal, T., Sharma, R., Mahalakshmi, N. and Gupta, M. 2018. Poultry waste management. International Journal of Current Microbiology and Applied Sciences. 7(8): 701–712.

Sinkiewicz, I., Śliwińska, A., Staroszczyk, H. and Kołodziejska, I. 2017. Alternative methods of preparation of soluble keratin from chicken feathers. Waste and Biomass Valorization. 8: 1043–1048.

Sistani, K.R., Brink, G.E., McGowen, S.L., Rowe, D.E. and Oldham, J.L. 2003. Characterization of broiler cake and broiler litter, the by-products of two management practices. Bioresource Technology. 90(1): 27–32.

Swain, B.K. and Sundaram, R.N. 2000. Effect of different types of litter material for rearing broilers. British Poultry Science. 41(3): 261–262.

Tahir, M., Hussain, T., Behaylu, A. and Tilahun, A. 2015. Scenario of present and future of solid waste generation in metro cities of India. Journal of Environmental Earth Science. 5: 164–170.

Tesfaye, T., Sithole, B. and Ramjugernath, D. 2017. Valorisation of chicken feathers: a review on recycling and recovery route–current status and future prospects. Clean Technologies and Environmental Policy. 19: 2363–2378.

Thyagarajan, D., Barathi, M. and Sakthivadivu, R. 2013. Scope of poultry waste utilization. Journal of Agriculture and Veterinary Science. 6(5): 29–35.

Thiyagarajan, A. and Cyril, R. 2023. Isolation, characterization, and optimization of keratinolytic bacteria from chicken feather waste dumping site. Journal of Applied Biology and Biotechnology. 11(3): 144–152.

Tian, H., Li, C.Y., Yang, C.H. and Shan, H.H. 2008. Alternative processing technology for converting vegetable oils and animal fats to clean fuels and light olefins. Chinese Journal of Chemical Engineering. 16: 394–400.

Ungureanu, N., Vladuţ, V., Biriş, S.Ş., Dinca, M. and Gheorghiţa, N.E. 2023. Management of by-products and waste from poultry meat industry. Annals of the Faculty of Engineering Hunedoara-International Journal of Engineering, (1).

Wang, M., Sun, X., Li, P., Yin, L., Liu, D., Zhang, Y. et al. 2014. A novel alternate feeding mode for semi-continuous anaerobic co-digestion of food waste with chicken manure. Bioresource Technology. 164: 309–314.

Wang, S., Jena, U. and Das, K.C. 2022. Long term performance of pilot methanogenic digester filled with seashell wastes treating slaughterhouse wastes: biogas production and environmental impact. Biochemical Engineering Journal. 187: 108651.

Wei, Y., Wachemo, A.C., Yuan, H. and Li, X. 2020. Enhanced hydrolysis and acidification strategy for efficient co-digestion of pretreated corn stover with chicken manure: digestion performance and microbial community structure. Science of the Total Environment. 720: 137401.

Wyatt, V.T., Hess, M.A., Dunn, R.O., Foglia, T.A., Haas, M.J. and Marmer, W.N. 2005. Fuel properties and nitrogen oxide emission levels of biodiesel produced from animal fats. Journal of the American Oil Chemists' Society. 82(8): 585–591.

Yuvaraj, A., Thangaraj, R. and Maheswaran, R. 2019. Decomposition of poultry litter through vermicomposting using earthworm Drawida sulcata and its effect on plant growth. International Journal of Environmental Science and Technology. 16: 7241–7254.

Zhang, J., Liu, J. and Liu, R. 2015. Effects of pyrolysis temperature and heating time on biochar obtained from the pyrolysis of straw and lignosulfonate. Bioresource Technology. 176: 288–291.

Zhang, L., Ren, J. and Bai, W. 2023. A review of poultry waste-to-wealth: technological progress, modeling and simulation studies, and economic-environmental and social sustainability. Sustainability. 15(7): 5620.

Chapter **5**

Utilization of Agriculture Waste for Development of Value-added Products

Shilpi Bansal[1], Saumya Shukla[2], Pradeep Kumar Yadav[3], Sudhir Pratap Singh[4] and Jyoti Singh Jadaun*[2]

[1]Department of Science and Humanities, SRM Institute of Science and Technology, Delhi NCR Campus, Modinagar, India.

[2]Department of Botany, Dayanand Girls PG College, Civil Lines, Kanpur Nagar, UP, India.

[3]Department of Botany, University of Allahabad, Prayagraj, UP, India.

[4]Department of Industrial Biotechnology, Gujarat Biotechnology University, GIFT City, Gandhinagar-382355, Gujarat, India.

INTRODUCTION

Agricultural trash is considered as a low-value, underutilized sustainable commodity. The volume and complexity of garbage generated have increased due to economic expansion and rising living standards which is turning out to be a serious issue. Inadequate management and disposal strategies, which include burning, dumping and land-fillers, leave a negative impact on the environment. Accumulation of agricultural waste results in the emission of harmful gases (greenhouse gases), releases unwanted chemicals, foul odour, and contaminate soil,

* For Correspondence: jyotisingh.jadaun@gmail.com

air, water which ultimately harms the health of living beings and environment. Agriculture industry produces approximately 1300 million tonnes of garbage every year and due to the rising demand for agricultural products, this figure is anticipated to increase only (Gustavsson et al. 2011). Additionally, it has been calculated that in India, around 75% of all agricultural field wastes produced are used for thatch, mulching, composting, firewood, and other domestic and agricultural tasks in addition to producing feed. Finding suitable disposal methods for the remaining 25% of the trash, however, is a considerable issue (Cardoen et al. 2015). It is quite concerning for the environment when crop yield and the removal of agricultural wastes differ so greatly. Therefore, it is crucial to investigate sustainable and cutting-edge technologies that can efficiently use the feedstocks that are already accessible in bio-based businesses, benefitting humanity while minimizing negative consequences and the production of harmful by-products. Agricultural waste problem can be managed by converting them into beneficial products simultaneously lowering the concentration of organic matter present in the environment. Even though they pose threat to the atmosphere, organic compound wastes offer a potential source for the production of meals, nutraceuticals, and other bio-based goods including bio-energy and biofertilizers. Agro-wastes contain significant amounts of complex carbohydrates, proteins, fibre, polyphenolic components, bioactive substances and other nutrients. Because of their high nutritional content, these residues are not regarded "wastes" but rather raw materials for the synthesis and growth of other products. Effective agro-waste biomass pretreatment methods, biochemical properties, and improved conversion procedures can improve the cost-effectiveness of conversion processes for bioproduct development. Biomass is constituted of plant materials, agricultural products, and animal manure which can act as renewable resources. It is a desirable substitute for petroleum in the production of chemicals due to its long-term viability and overall low cost. Additionally, the chemical composition of biomass and fossil feedstocks remains the same, i.e., carbon, hydrogen, nitrogen and oxygen making it a suitable choice for the production of biofuel and other sustainable energy sources. Agricultural leftovers, with their availability and potential, present a viable and sustainable carbon source for the manufacture of value-added goods through bioprocessing. The nutrient-rich biomass creates favourable circumstances and moisture for the development of microorganisms, boosting the possibilities for numerous industrial processes. Fruit and vegetable wastes are the best sources of substances with antioxidative, anti-inflammatory, anti-microbial, and anti-allergenic qualities, including polysaccharides, protein, phytochemicals, dietary fibre, and flavour. Also, the use of cost-effective procedures is crucial since the economic factor is so important in the manufacture of diverse chemicals. This is especially important for emerging nations. Many chemicals may be produced utilizing cheap feedstocks and waste products, including acetic acid, starch, ethanol, lactic acid, and formic acid (Varanasi et al. 2013).

In this chapter, various methods have been discussed for recovering valuable resources from agricultural waste. Some light has also been shed upon some common agri-waste. Recent technologies are gaining popularity in the development of strategies for extracting valuable compounds from crop wastes.

TYPE OF AGRICULTURE WASTES

Agriculture sector generates multiple types of waste materials along with its major contribution in food production. Field and process residues are further subcategories of agricultural waste. Field residues such as leaves, stems, stalks, seed pods etc. are the remnants which are left over post harvesting of the crops (Figure 5.1). Processed residues generated during the extraction of valuable resources are also categorized as remnants. Molasses, bagasse, husks, seeds, stems, leaves, straw, shells, stalks, pulp, stubble, roots, peel, etc., are some examples of it. These leftovers are subsequently used in manufacturing, soil improvement, fertilizer production, animal feeding, and several other processes. Quite often a significant portion of the field waste remains unutilized. If these agricultural byproducts are utilised and managed properly, soil erosion management and irrigation will be far easier to implement (Mitchell et al. 2020). Farmers must abide by the Control of Substances Hazardous to Health Regulations (COSHH) because the quantity is typically enormous and ordinarily hazardous. If the garbage is not managed, it can impair farming operations and pollute the air and water. Waste is either reduced, recycled, or repurposed during waste management. The two main cereal crops farmed are wheat and barley, primarily in the Middle East region of the world. In most parts of the world, other crops like rice, lentils, maize, chickpeas, fruits, and vegetables are grown. Agricultural leftovers are sorted according to their availability and other characteristics that set them apart from other solid fuels like charcoal, wood, and char briquettes (Mohamed et al. 2021). Figure 5.1 gives a brief overview on the type of waste products and its management techniques.

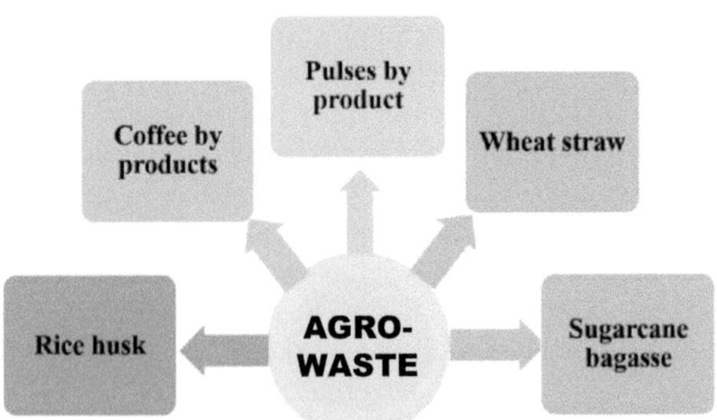

Figure 5.1 Different types of agricultural waste.

POTENTIAL EXAMPLES OF AGRICULTURAL WASTE

Rice Husk

There is a substantial amount of agricultural waste, specifically rice husk, which is produced in large quantities and is expected to grow in the coming years.

Each year, the country has access to approximately 0.83 million tonnes of rice husk. Currently, rice husk serves as animal feed and is utilized in brick kilns. However, if this waste is mishandled, it can lead to environmental issues. Rice husk, the outer covering of white rice grains, contains a significant amount of silica, typically exceeding 80–88% in concentration. When rice husk is burnt, it turns into rice husk ash, which makes up 20% of the original weight of the material. Rice husk ash is acknowledged as a highly pozzolanic substance. Rice husk is utilized for the production of rice bran oil (Lakshmi et al. 2017). People are becoming more interested in using mixed cement for construction and development work due to their advantageous effects on the manufacture of concrete, such as improving concrete qualities and minimizing detrimental environmental consequences; thus, industrial and agricultural wastes that act as pozzolans are gaining more attention (Ferraro et al. 2012). Rice husk ash with a cellular microstructure is created when rice husk is burnt at temperatures lower than 700°C. This ash is useful as an additional cementitious component due to its high silica concentration in the form of non-crystalline or amorphous silica.

An innovative alternative use for rice husk has also emerged, involving its transformation into insulating material. Insulating materials play a crucial role in conserving energy by reducing heat gain or loss. Several researchers have successfully created insulating materials using various combinations such as durian peel and coconut fibre, fly ash and asphalt, coal fly ash and scrap tire fibres, as well as rice husk ashes (Koller et al. 2013). In 1986, Mishra and his colleagues conducted a study using the transient heat flow method to examine the thermal properties of rice husk. Their findings indicated that rice husk could be effectively employed as a heat-resistant material.

Coffee by Products

The yearly output of coffee is over 105 million tonnes worldwide. The shell and mucilaginous components of the cherries must be removed during the industrial preparation of coffee. Coffee generates different by-products at different stages of harvesting. Coffee can be processed by two methods, namely, dry method which yields coffee husk (CH), and wet method which gives a pool of pulp, parchment husk, silver skin, and mucilage. Since coffee pulp and mucilage is obtained by wet processing, it contains high moisture content (78–84%) and is composed of cellulose (8–36%), pectic polysaccharides (21%), hemicellulose (9–18%), proteins and sugar. Coffee husk, coffee parchment and silver skin are low in moisture content (4–9%). Similar to CP and CM, these coffee by-products are rich in polysaccharides, proteins, lipids, phenolic compounds (lignin, xylose, cellulose) at varying concentration. After pre-treating these byproducts with viscozyme, phenolic extraction revealed that silver skin had the greatest phenolic content (25%) followed by wasted waste (19%) and cherry husk (17%). Bioactive preserves derived from coffee by-products displayed antioxidant activity ranging from 65% to 70% in terms of chlorogenic acid. The antioxidant activity of these by-products was measured to be 1.5 to 2.0 mmol trolox per 100 g (Murthy et al. 2012). Coffee husk facilitates adhesion of microbes on its surface and promotes biodegradation as was evident in the study conducted by researchers Borghesis et al. (2016). The researchers synthesized a composite film containing poly-caprolactones (PCL) and coffee husk and checked its biodegradation after 30 days upto 120 days. They observed that the degradation rate of composite film was twice that of PCL film alone, which is also an effective way to control environmental pollution also. Chlorogenic acids, compounds with immense therapeutic potential in the human system, can be produced from coffee pulp. *S. cerevisiae* mediated fermentation of coffee pulp with 0.5% wt. sulfite as a stabilizing agent for phenols upregulated chlorogenic acid level by 400% (da Silveira et al. 2019).

Pulses by Product

Agro-wastes consist mainly of cellulose and lignin, along with various other components that contain functional groups. These waste materials have the ability to form complexes with heavy metal ions, facilitating their removal through adsorption. To enhance their metal adsorption capabilities, agro-wastes need to undergo processing and be transformed into a carbonized form, which increases their porosity and surface area. This carbonized form is more effective in the metal adsorption process (Onipe et al. 2015). Pigeon pea husk, which accounts for approximately 8% of the pulse grain, is a low-cost byproduct of the pulse processing industry. It has been found to contain around 5.6% crude protein, 0.3% fat, 31.9% crude fibre, 3.5% ash, and 58.7% carbohydrates. Pigeon pea husk shows potential for use as a metal adsorbent. The husk is its carbon form with adsorption properties. In a study, green gram and Bengal gram husks were used washed, dried, and carbonized in a sealed container at a temperature of 500°C for 1 hour to obtain as substrates for the production of single-cell protein in a maltose-enriched medium (7.7 mg). The maximum yield of 9.2 mg was achieved within 6–10 days of fermentation (Anbuselvi et al. 2014).

Wheat Straw

Wheat straw constitutes 60% of the wheat crop. On average, more than 4.8 tonnes of straw is produced per hectare of wheat. In the United States alone, it is estimated that over 90 million metric tonnes of wheat straw are generated annually. Currently, wheat straw is primarily utilized as a livestock comforter or as low-grade animal feed, resulting in minimal economic returns. Merely 3.2% of the economic value derived from wheat comes from straw.

Wheat straw is a flexible and environmentally beneficial material due to its numerous benefits. It can be easily formed, is renewable, and widely accessible in the area. Wheat straw is non-abrasive, shows anisotropic behaviour, and both absorbs and releases moisture. It is recyclable or combustible, viscoelastic, and porous.

It is a valuable source of bioactive compounds, including policosanols, phytosterols, phenolic compounds, and triterpenoids, which have beneficial properties. The recovery of these bioactive compounds during or before the conversion of wheat straw to ethanol enhances the feasibility of the conversion process (Dunford and Edwards 2010). Moreover, wheat straw can be utilized to develop composite products such as sorbents, geotextiles, structural and nonstructural composites, filters, molded products, packaging, and combinations with other materials (Rogers et al. 2007). Overall, wheat straw offers a range of possibilities for various applications and the extraction of valuable compounds.

Sugarcane Bagasse

Sugarcane (*Saccharum officinarum*), or noble cane, has high sugar and low fibre. This composition makes sugarcane the most important industrial crop in the world. During sugarcane processing, the cane is peeled using abrasive tools, and the extracted sucrose juice is obtained by squeezing. By-products such as bagasse are also generated. Studies have indicated that around 100 tonnes of sugarcane can yield approximately 14.3 tonnes of raw sugar, 5.2 tonnes of filter cake, 27.2 tonnes of bagasse, 2.6 tonnes of molasses, and 50.7 tonnes of wastewater (Luo et al. 2018).

Sugarcane is cultivated primarily for the production of sugar, jaggery, ethanol, syrups, and various other products. It consists of a range of secondary metabolites viz. alkaloids, phenolics, flavonoids, etc. The phytochemicals present in sugarcane, particularly flavonoids and anthocyanins, exhibit robust antioxidant properties, providing hepatoprotection against oxidative liver injury (Khan et al. 2015, Sarkar et al. 2022). Traditional medicine has utilized sugarcane juice for generations as an external and internal treatment for conditions like renal disease, enlarged prostate, gonorrhoea, and cystitis (Khuimphukhieo et al. 2023). It also contains bioactive compounds in its bark that help prevent microbial attacks, particularly due to its high sucrose content, which can be susceptible to such attacks (Abbas et al. 2014). Furthermore, scientists have discovered antioxidant and free radical scavenging activities in sugarcane components such as luteolin.

A process for pre-treating sugarcane bagasse utilizing the recyclable organic acid furoic acid was described in research. The goal was to extract the xylo-oligosaccharides component from the hemicellulose and improve the subsequent cellulose saccharification technique. Furoic acid pretreatment recovered 120 g of xylo-oligosaccharides and 335 g of glucose from 1000 g of sugarcane bagasse. Additionally, cooling crystallisation could be employed to recover the furoic acid quickly. The study reported a novel process for converting sugarcane bagasse into xylo-oligosaccharides and glucose, which could help cut production costs in the sugar and nutraceutical industries. Bioactive chemicals were extracted from sugarcane bagasse employing orbital shaking and 90% methanol at a constant temperature of 50°C. The overall phenolic content of this extraction technique was 5.91 mg GAE/g and included flavonoids, alkaloids, and lignan (–) Podophyllotoxin. The extract has potent antibacterial and antioxidant effects that were sustained for 30 days of storage and were effective against *E. coli*, *B. cereus*, *S. aureus*, and the mutant yeast SGS1.

MD and GA were used to create microencapsulated sugarcane bagasse extracts, and neither time nor storage temperature had an impact on their

TPC ($p > 0.05$). These results imply that sugarcane bagasse extracts might be an important source of bioactive chemicals for nutraceuticals or as a food additive (Velazquez et al. 2021).

Coconut Fibre

Coconut (*Cocos nucifera*), a fruit originating from Southeast Asia, has gained widespread popularity worldwide. The cultivation of coconut generates agricultural waste, which is commonly found in tropical regions. Improper disposal or burning of these waste materials can have adverse environmental effects. So, there is a pressing need to find productive uses for this waste and mitigate improper solid waste disposal practices. Coconut waste exhibits several favourable characteristics, including low cost, high lignin content, low density, availability, high elongation at break, and low modulus of elasticity (Faria et al. 2020).

Extensive study has been undertaken on the application of coconut fibre, which can be processed and integrated as particles or fibres prior to use. It can be used as a filler material within the matrix or as reinforcement, depending on the intended use. When compared to other natural fibres, coconut fibre is often characterised as a short fibre with relatively low flexibility and higher hardness. The enhanced hardness is due to the presence of lignin, which gives it its characteristic colour. According to a comprehensive study, coconut fibre possesses 5% water solubility, 3% pectin, 0.3% hemicellulose, 46% lignin, and roughly 2% ash, defining it as a natural lignocellulosic fibre. According to Loaiza et al. (2018), coconut fibre has already made a name for itself in the market and among scientists, finding use in cords, carpets, brooms, and various types of boards.

Since the coconut must be hauled across lengthy, twisting, and uneven up-and-down hills owing to stream and river routes, it is bulky, heavy, and difficult to carry due to the coco-husks. The primary factor is the cost of collecting coco husks for bulk transportation along major roadways. The same reasoning is true for coconut fronds. Coconut husks are used as fuel for biomass (wood) stoves when coconuts with the husks are dehusked in the villages where people live (and, consequently, prepare their meals).

Banana Fibres

Bananas are a popular tropical fruit known for their significant nutritional value. These can be consumed raw or processed into a variety of products, such as snacks, insecticides, or food colouring agent, paper and other items (Majaliwa et al. 2019). Scientists highlighted the qualities and traits of bananas that make them appropriate for such uses.

As reported by Vu et al. (2017), the peels of bananas make up around 35% of the total weight of the fruit. This indicates that over 39.9 million tonnes of banana peels are produced each year. These wastes are unfortunately generally disposed of in landfills as a result of inadequate utilization practices. Banana peels, despite being considered as waste, possess significant antioxidant capacity and antimicrobial properties (Ibrahim 2015). These peels have been used historically to treat a variety of conditions, including burns, diarrhoea, ulcers, and inflammation (Amini et al. 2019). Banana peels are therefore an important raw resource with enormous promise in the nutraceutical and pharmaceutical sectors. Banana peels are frequently used as animal fodder, notably for cows and buffaloes. However, little study has been done on the useful uses of banana peels in the field of health and wellbeing.

The importance of using different non-wood plant fibres instead of wood pulp to make pulp, paper, and paperboard has been well established on a worldwide scale. The banana stem serves no purpose once the fruit has been harvested. Pulping is the process of removing the fibres from the stem using either chemical, mechanical, or a mix of the two processes. Chemical pulping is the process of separating the lignin from the cellulose in lignocellulosic materials using chemicals. The length of the fibres is not significantly affected by this technique. According to Dejene (2020), five pulping methods have been investigated: the kraft process, the sulphite process, the soda process, the formic acid/acetic acid solvent, and the urea/NaOH solvent system. These methods present several pulp manufacturing philosophies.

In Costa Rica, industrial paper is being produced using pseudostems from bananas. A vast range of goods, including paper, cardboard, notepads, letter paper, envelopes, postcards, packing materials, and notebooks, have been released under the brand name Earth (Vigneswaran et al. 2015). Banana paper is prized for its extraordinary tensile strength and is even used to make tea bags and banknotes. Ongoing work is being done in Germany to create banana fibre inside automobile linings.

Banana fibre extraction is now done on a big scale. Banana fibres have been used to make handbags and other fashionable products by being spun using jute spinning equipment. Banana fibres are an example of an agro-based biofiber whose composition, structure, and characteristics make them suitable for food, chemical, fuel, pulp, paper, textiles, and composites industry (Sharma et al. 2022). The structure, characteristics, and appropriateness of banana fibres for many industrial uses are the subject of much research.

VALUE-ADDED PRODUCTS FROM AGRICULTURAL WASTE

Different types of value added products have been generated by using various types of agricultural waste materials (Figure 5.2). Some examples of them are discussed below:

Prebiotics from Agricultural Waste

Prebiotics can be broadly defined as a non-digestible dietary component that ameliorates the growth and activity of gut microbiota, thereby positively affecting the host's well-being. Due to number of advantages which include expansion in the population of probiotic bacteria, improvement in the immune system by upregulating the better absorption of minerals, vitamins, and production of health beneficiary metabolites, prebiotics is in high demand (Al-Sheraji et al. 2013, Chourasia et al. 2021, Abedin et al. 2023). Owing to the benefits provided by prebiotics, from 2022 to 2030, the prebiotics market is projected to increase at a compound annual growth rate (CAGR) of 14.9%, from a market size of USD 6.05 billion in 2021. (https://www.grandviewresearch.com/industry-analysis/prebiotics-market). The prebiotic carbohydrate compounds like inulin, fructooligosaccharides (FOS), oligofructose, fructans, and galactan have properties to stimulate the growth of gut microflora. Recent researches have focused exploration of diverse ecological habitats to identify and report enzymes that can hydrolyse biomass derived macromolecules, e.g., cellulose, xylan, pectin, into cello-oligosaccharides, xylo-oligosaccharides, and pectic-oligosaccharides (Joshi et al. 2020, 2021, Kumari et al. 2021, Phukon et al. 2020). Chemo-enzymatic processes have been developed to transform the biomass derived digestible starch into non-digestible crystals with prebiotic property, known as resistant starch (Thakur et al. 2021, 2023).

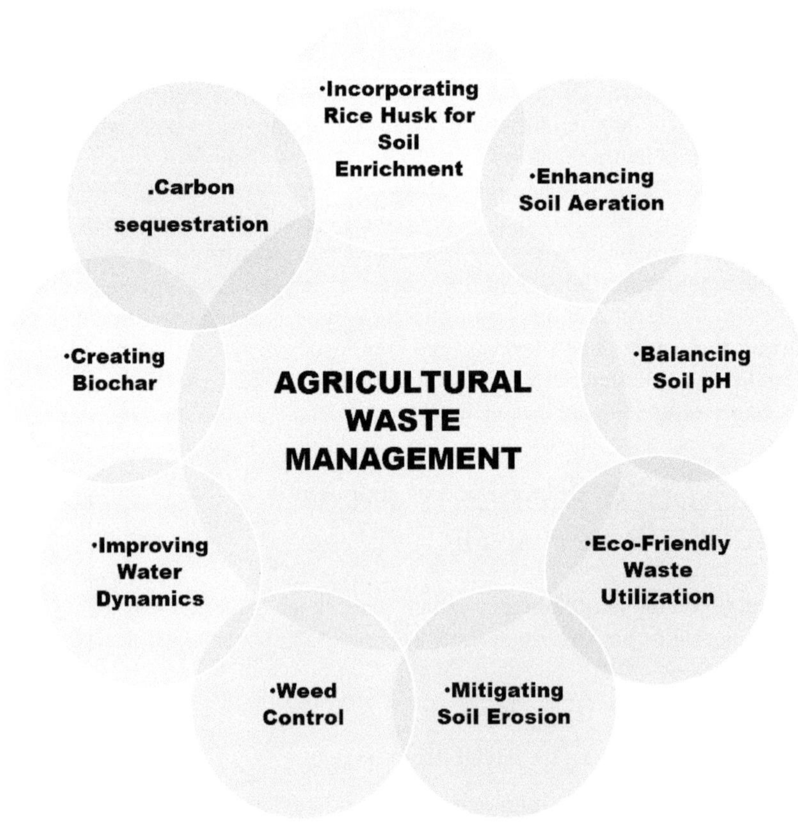

Figure 5.2 Different strategies applicable for agriculture waste management.

Using solid-state fermentation techniques, bran, straw, peels, and pomace obtained from a variety of agro-wastes were tested for their efficiency to produce FOS. Among the wide host range, *Aspergillus flavus* NFCCI 2364 FTase acted best upon sugarcane bagasse with a yield of 35.95% and 73.42% under unoptimized and optimized conditions, respectively (Ganaie et al. 2017). From sugarcane bagasse, xylan was extracted using an alkali process, and then enzymatic hydrolysis was employed to produce xylooligosaccharides (XOS). This XOS was then tested for its prebiotic potential which includes fulfilling the criterion of not getting digested in the stomach and increasing the population of probiotic bacteria such as *Bifidobacterium* and *Lactobacillus* in the lower bowel (Kathiresan et al. 2021). Similarly, *Miscanthus* lignocellulosic biomass has also been identified as a source of XOS which induces *Lactobacillus brevis* (Hong et al. 2019). The production of the type of prebiotics and subsequent proliferation of probiotic microorganisms is dependent on the raw material used. Switchgrass, hybrid poplar and southern pine rich in hemicellulose were fractioned using hot water technique which resulted in production of different concentrations of galactan, glucan, xylan and arabinan oligosaccharides. The elicitation responses for different cultures varied according to the lignocellulosic biomass initially used such as *Lactobacillus casei* was more responsive to switch grass while *B. fragilis* was more responsive towards southern pine (Rajan et al. 2021). Mixture of XOS compounds were produced from wheat husk using *A. fumigatus* R1 strain as a catalyst which further induced the generation of probiotic micro-organisms and harbored anti-oxidant activity. Prebiotics not only improves the production of probiotic microbes but also are helpful in combating pathogens. Action of Mannanase from *Penicillium oxalicum* KUB-SN2-1 on copra meal, coffee residue yielded mannotetraose, mannobiose and mannose. These mannooligosaccharides stimulated the multiplication of lactic acid bacteria isolates and simultaneously suppressed the growth of *Staphylococcus aureus* TISTR 029 and *E. coli* (Chantorn et al. 2018). To be converted into a final edible form, the crop undergoes a series of processing events, which removes some important components from the crop which are actually beneficial as prebiotics. By-products obtained from cereal crops such as wheat, rice, barley and finger millet have been used to extract xylo-oligosaccharides (XOS) and arabinoxylo-oligosaccharides (AXOS) which are categorized as prebiotics. Among the several wastes, seed waste is the most overlooked, despite the fact that it contains a pool of prebiotics. Seeds of mango, date, grapes, sesame, buckwheat are some of the examples, which harbours an array of prebiotic compounds and supports the proliferation of probiotic microflora (Bamigbade et al. 2022). Tamarind seeds were shown to function as prebiotics for the growth of lactic acid bacteria when incubated at 70°C for 180 minutes. Coffee silverskin contains a modest amount of free phenol compounds, but it has significant antioxidative activity, which can be linked to melanoidins. Static batch culture fermentation tests revealed that it promotes the development of bifidobacteria over *Clostridia* and *Bacteroides* spp. In light of the significant antioxidant activity, high concentration of soluble dietary fibre, and probable prebiotic action, Coffee silverskin can be suggested as a potential candidate to be used as prebiotics.

Agricultural Waste as Bio-fertilizers

Nitrogen, phosphorous, and potassium, the key constituent of chemical fertilizers, have been extensively used all over the world throughout the year since long which has had a huge negative influence on the ecosystem. Manure, compost, home waste and sewage sludge are the several forms of organic waste that may be used as agricultural fertilizers (Yong et al. 2021). Agricultural biomass is one of the most common sources of organic fertilizer; however, due to probable presence of pathogens and harmful substances, its direct application to soil is restricted. Similarly, animal manures and excretions are the rich source of nitrogen and phosphorus but their continuous aggregation on soil surfaces may lead to infections, bad odour, and emission of greenhouse gas (GHG). Also, rainfall or surface runoff can carry accumulated nutrients from the soil surface into water streams, causing eutrophication and the development of algae (Lee and Oa, 2013). This necessitates the recycling of diverse biomass wastes, including food waste, sewage sludge waste, and animal manure, into organic fertilizers. These wastes are treated to increase their nutritional value and boost their agricultural potential in order to contribute to a more environmentally and economically sustainable society. The term biofertilizer refers to fertilizer that includes living microorganisms. These organisms are anticipated to affect the soil ecosystem and provide additional plant nutrients.

The ashes made from plant biomass include large amounts of macronutrients including calcium (Ca), potassium (K), phosphorous (P), and sulphur (S), while displaying modest concentrations of hazardous substances like arsenic (As) and lead (Pb). Thus, these ashes might be viewed as important components in goods made for agricultural use. Wood ash rich in nitrogen and phosphorus has proved to be an effective fertilizer in peatland forests (Väätäinen et al. 2011). Anaerobic digestion can be used to process various organic matter to create digestate and biogas (CH_4 and CO_2). Among both, digestate is believed to be a valuable fertiliser for crop nutrition and productivity as well as from the standpoint of climate mitigation because it is rich in micro and macronutrients. In one of the study, rice husk was subjected to anerobic digestion, which resulted in the production of biogas, solid and liquid digestates. The ability of these digestates to be used as fertilizers were determined according to which of these products were rich in nitrogen, phosphorus and potassium compounds and lacked any heavy metals, making it safe to be utilized as biofertilizer (Olugbemide and Likozar 2022). The use of biofertilizers alone or in concomitance with organic fertilizers can have a good impact on nutrient availability and organic matter dynamics. In one study, manures from different animal sources and NPK were tested to determine their effect on physio-chemical and biological properties of the soil and crop yield. As was expected, the use of manures in different combinations enriched the properties of the soil thereby making them effective biofertilizer (Al-Suhaibani et al. 2020). Compost made from olive milling waste is an excellent fertilizer because it contains less phenolic and lipidic components and water-soluble organic materials. Furthermore, when compared to the starting material, this finished product exhibits less phytotoxicity. This advancement makes it possible to obtain compost that is high-quality, has excellent agronomic qualities, and has little to no environmental impact. Comparative analysis of organic fertilization and mineral fertilization on durum wheat revealed similar yield of 3.63 t/ha. It was seen that use of industrial orange waste increased wheat growth upto 400%. Addition of greenhouse tomato plant wastes, mushroom composts and used cocopeat wastes in the soil significantly upregulated the yield and growth of *Lactuca sativa* L. plant along with its nutrient content (Sönmez et al. 2017). Liquid fertilizers obtained through solvent extraction method or microwave assisted extraction have been found to contain higher levels of micronutrients than organic fertilizers (Leno et al. 2021). These liquid fertilizers generated from municipal solid waste facilitates better water retention ability of the soil and also are a perfect solution to the landfill problems. Using technology in a way that single action can resolve multiple issues is always appreciated. Orange peel and egg shell was calcinated at 700°C to produce a unique low cost bio-composite which had the ability to absorb phosphorus from waste water to form brushite and apatite (calcium phosphate compounds). These compounds could serve as a good substitute for phosphate fertilizer (Pérez et al. 2022). Compost obtained from sewage sludge, fish waste, manure from chicken or beef mixed with plant biomass; biochar from *E. coli*, banana peduncle, sugarcane waste straw; digestate from slaughterhouse waste, distillers waste, pharmaceutical organic waste are some of the examples of effective biofertilizers which could enrich plants with nutrients and simultaneously fight with environmental problems of pollution and degradation.

Biofuel from Agriculture Waste

Biofuels, which are derived from renewable organic biomass, have the potential to reduce dependency on fossil fuels, reduce greenhouse gas (GHG) emissions, notably in the transportation sector, and increase fuel supply reliability. The manufacture of biofuels from lignocellulosic materials and other waste sources has been studied by a number of researchers (Pathak et al. 2022). A variety of agricultural raw resources may be used to make ethanol, a crucial molecule with many uses in the food and petroleum industries. Following physio-chemical preparation, enzymes are essential for separating monosaccharides from polysaccharides. To maximise the release of sugar monomers, enzymes that can degrade cellulose, hemicellulose, and lignin are required. However, it has been shown that the majority of the fermentation organisms participating in the process are unable to digest the oligosaccharides produced by thermo-chemical pre-treatment procedures.

The yeast *Saccharomyces cerevisiae* is frequently used to produce ethanol from glucose and other hexoses efficiently. Yeasts and bacteria with industrial uses in the generation of ethanol include *Brettanomyces bruxellensis* and *Zymomonas mobilis*. For instance, oat straw hydrolysate has been effectively fermented into ethanol using *B. bruxellensis* (Cheng and Brewer 2021). Anaerobic digestion of methane (biogas) can be used as an alternative to using straw as a source of biofuel. This switch from ethanol to biogas production increases biomass conversion, lowers greenhouse gas emissions, and improves energy efficiency (Yong et al. 2022). Because it has a greater energy density, is less corrosive, and is more compatible with current engines than ethanol, butanol—including both n-butanol and iso-butanol—offers superior fuel qualities. Acetone, butanol, and ethanol may now be produced on a wide scale from starchy materials thanks to the utilization of solventogenic *Clostridium* (Karthick and Nanthagopal 2021). Straw or agricultural waste may be used as raw materials to make biofuels, which has the potential to be environmentally sustainable. Wheat straw has proven to be effective in the synthesis of bioethanol, biogas, and biohydrogen in biorefineries, aiming to improve overall biomass utilization efficiency. Wheat straw is first heated hydrothermally, which separates it into two fractions: one with cellulose-rich fibre and one with hemicellulose-rich liquid. When cellulose is hydrolyzed by enzymes and then fermented, 0.41 g of ethanol/g of glucose is produced, whereas fermentation of hemicellulose results in the production of 178.0 mL of hydrogen/g of sugar (Chu and Lee 2007). Furthermore, when evaluating the potential of wheat straw for biofuel production, it has been observed that utilizing wheat straw for biogas production or multi-fuel production demonstrates significant advancements compared to the production of a single fuel, such as bioethanol. Consequently, the production of multiple biofuels from wheat straw enhances material and energy efficiency, establishing a more effective biomass consumption process.

A study was carried out by Meenakshi and Kumaresan (2014) to produce ethyl alcohol using corn and potato peel waste as raw materials. They concentrated on pH, substrate concentration, and particle size optimisation, among other factors. The scientists discovered that submerged fermentation was a very efficient method for the fermentation process. They discovered that the maximum yield of ethyl alcohol occurred at a substrate concentration of 10%. The research also showed that a pH of 5.5 was ideal for maximising production. Additionally, the researchers discovered that yields of ethyl alcohol increased with an intermediate particle size (0.157 mm). Research was done by Bhatt and his colleagues on the creation of ethanol from used groundnut shells (Bhatt 2014). They used a procedure known as simultaneous saccharification and fermentation. The trash was subjected to a number of chemical pre-treatments, with hydrochloric acid being determined to be the most successful. A 2% (v/v) culture of *Saccharomyces cerevisiae* and *Bacillus stearothermophilus* was added to the fermentation media and allowed to develop for the duration of the next day. The potassium dichromate technique was used to gauge ethanol output. The groundnut shell waste was analysed by the researchers, who found that it included 35% cellulose, 32.10% lignin, 4.3% miscellaneous materials, 27.7% organic carbon content, and 23.4% nitrogen content.

Comparative investigations on the generation of bioethanol from diverse agricultural organic waste residues was conducted by Singh and Singh 2015. They noticed that 41–46% alcohol was produced following fermentation and various treatments. They investigated conversion using potato peel, banana peel, rice straw and corn cob. In potato peel and corn cob, the highest potential sugar to alcohol conversion was achieved.

Saccharomyces cerevisiae with sorghum stover served as the substrate for the research conducted by Udhayaraja and Narayanan (2012). They looked at the effects of temperature, inoculum level, and pH on ethanol production. The researchers discovered that a 10% inoculum level produced the maximum production of ethanol, followed by 8% and 6%. They found that the temperature at which the largest yield was produced was 35°C. In addition, they found that the maximum ethanol output occurred at a pH of 6.5.

Organic Acid from Agriculture Waste

Organic acids rank third in terms of the volume of biological products, according to Ali and Zulkali (2011). In the presence of water, these compounds only partially dissociate and have mild acidic characteristics. Given the importance of organic acids as chemical building blocks, microbial processing is often used to create them. Citric acid, lactic acid, succinic acid and acetic acid are some of the many organic acids that are used in many different sectors, including food preservatives, flavoring, medicines, polymers, cosmetics, etc.

In order to get crucial chemical building blocks from sustainable carbon sources, organic acids present a possible method. Further, the transition from conventional chemical production techniques towards microbial-based organic acid synthesis is eco-friendly. Integration of microbes with agricultural waste for the production of organic acids is highly economical and environmentally friendly. Some of the waste products have the ability to produce different types of organic acid such as kimchi cabbage waste, which alone can generate three different organic acids. Using simultaneous saccharification and fermentation processes, two *Lactobacillus* strains, namely *L. sakei* WiKim31 and *L. curvatus* WiKim38, catalysed the conversion of kimchi cabbage waste into a significant quantity of lactic acid (~12.0 g/L), fumaric acid (~7.0 g/L), and acetic acid (~4.0 g/L) (Kim et al. 2018).

Citric Acid

Citric acid, an intermediary compound of the tricarboxylic acid (TCA) cycle used widely as a natural preservative, was produced in an ample amount of 82 g/kg of dry weight by Kareem and Rahman (2013) in their exploration to use banana peel as a substrate for *Aspergillus niger*. The efficiency of the fermentation medium, which mostly comprised dried banana peel, was further improved on the addition of nutrients and trace elements.

Based on the consumption of sugar, the output of citric acid from bananas was 90%. Fruit pulp waste is also an excellent source of citric acid which was confirmed by utilizing Indian jujube, beetroot, guava, papaya, wood apple, and pineapple as a substrate by *A. niger* under two distinct circumstances (with and without methanol). Pineapple pulp waste turned out to be the best candidate with maximal citric acid production capabilities of 5.25 g/kg with methanol and 3.25 g/kg without methanol, respectively (Pratik et al. 2013).

Prabha and Rangaiah (2014) in their study discovered that with a combination of *A. niger* and methanol (1%), pineapple peel could produce citric acid more easily.

Citric acid was created with a yield of 14.71 g/L when methanol was present, compared to the control sample's yield of just 8.58 g/L. Another investigation showed that apple pomace was a superior substrate for *A. niger's* synthesis of citric acid. 5N hydrochloric acid treated wheat straw has been used for the production of citric acid (25.60 g/kg) using *A. niger* as a catalyst. Under ideal temperature and pH conditions of 30°C and 6.0, respectively, the resultant yield was 0.65 g/100 ml (Egbe et al. 2022). *A. niger* alone has the capability to induce citric acid production; however, under optimized conditions of orange peel as a substrate (25 gm), incubation period (6 days), pH (6.0) and temperature (50°C)m co-cultivation with *A. fumigatus* further increased upto (114.68 142 $\pm$ 0.73 mg/mL), which was higher than individual inoculation (Zafar et al. 2021).

Acetic Acid

Acetic acid or vinegar, as it is sometimes referred to as, has wide-scale applications across the globe and can be produced by both aerobic and anaerobic fermentation techniques. While *Clostridium* and *Acetobacterium* may directly convert glucose into acetic acid by anaerobic fermentation, oxidative fermentation using Acetobacter, a kind of acetic acid bacteria, is the favoured technique for producing food-grade acetic acid. Alcohol dehydrogenase and aldehyde dehydrogenase play key roles in *Acetobacter aceti* mediated oxidation of ethanol to acetate through a series of processes.

Pineapple peels were used as the substrate in a study on the synthesis of acetic acid by Raji et al. (2012). *S. cerevisiae* was used to ferment the peels for 48 hours to generate ethanol from carbohydrates, which was followed by a 9-day incubation period with *A. aceti* to produce acetic acid. Under ideal conditions, acetic acid production reached a maximum of 4.77%. Apple processing industry produces large apart of apple pomace which is rich in metabolites and various enzymes. This pomace, when acted upon by yeast followed by another strain of acetobacter, i.e., *A. pasteurianus* SKYAA25, yields 52.4 g of acetic acid/100 g of DM. Similar to it, vegetable waste is yet another source for acetic acid production as was seen in the case of onion peel. 6% of onion waste alone produces 19.3 g/L acetic acid via simultaneous saccharification and two-step fermentation method and 14.9g/L via simultaneous saccharification and co-fermentation method (Kim et al. 2019).

Lactic Acid

It can be produced by both methods viz. bacterial fermentation and chemical synthesis with a major contribution by bacterial fermentation. According to Vijayakumar et al. (2008), homo-fermentative lactic acid bacteria may create lactic acid with up to 90% purity. Apart from its regular use in various industries, lactic acid is essential for the synthesis of high molar mass polymers. Polylactic acid, a type of eco-friendly biodegradable plastic, is in current times a beneficial substitute to chemical based plastic products (Bangar et al. 2022). Several research focused on the synthesis of lactic acid from fruit, vegetable and agricultural leftovers. In one research, lactic acid was produced by fermenting five different forms of fruit and vegetable waste using the bacteria *Lactobacillus casei* and *L. delbruckii*, including mango peel, potato peel, sweet corn waste, orange peel and green pea waste. Mango peel had the maximum lactic acid output, measuring 63.33 g/L when *L. casei* was used. Other than the standardization of normal parameters, continuous efforts are made to include new approaches which could elevate the levels of target compound. In an attempt to produce lactic acid from molasses of corn steep liquor using *L. casei,* one such new approach was introduced. Immobilization of *L. casei* on sodium alginate matrix (2.5 mm bead diameter) resulted in high yield 123.67 ± 2.1 g/L lactic acid, improved stability and decreased contamination (Thakur et al. 2019).

Optimization of parameters such as temperature, pH of the medium, incubation time, and fermentation time is a pre-requisite for efficient recovery of the organic acids as was observed in the case of mango peel. Here, maximum lactic acid production of 17.484 g/L was achieved under the following conditions: initial pH of the medium at 10, the incubation period of 6 days, and temperature of 35°C (Jawad et al. 2013).

Succinic Acid

Succinic acid (SA) is a tricarboxylic acid (TCA) cycle intermediate that may also be generated as an end product by different bacteria through mixed acid fermentation. Both oxidative and reductive mechanisms may be used for its synthesis (Raab and Lang 2011). There are two suggested mechanisms in fungi for succinate accumulation. The first approach uses the TCA cycle's oxidative pathway, where citrate is changed into -ketoglutarate by the action of isocitrate and succinate is created by the formation of succinyl CoA from succinyl CoA ligase. However, since two CO_2 molecules are produced from the six-carbon substance (glucose), this pathway is regarded as ineffective for the buildup of organic acids.

In one of a kind of study, engineered strain of *Yarrowia lipolytica* was employed for the synthesis of succinic acid using fruit and vegetable waste. Initially hydrolysis parameters were standardized to produce 56.7 g/L glucose, which was subsequently utilized at a concentration of 100 g/L along with 4% corn steep liquor to yield 43.1 g/L succinic acid. Further replacing free cell batch fermentation technique with *in situ* fibrous bed bioreactor and fed batch fermentation resulted in enhanced SA production upto 140.6 g/L (Luo et al. 2018). Similarly, in other study hydrolysis of sugarcane bagasse generated glucose and xylose. This glucose (40 g/L) and xylose (8.4 g/L) together in combination was acted upon by *Y. lipolytica* PSA02004 to provide 33.2 ± 0.3 g/L SA (Ong et al. 2019).

Agrowaste in Construction Industry

Increase in population is directly associated to increase in demand for residence, which is further linked to elevated production of construction materials. Concrete, bricks, aluminium, wood, aggregates, portioning materials, etc., constitute the construction materials, which are mostly derived from natural sources. The stock of these natural resources is depleting due to its excess usage and also the production of these materials required for construction are not nature friendly, thereby generating a need to develop for alternate sources. Usage of agro waste is a satisfactory solution to the problem and there are several examples associated with it. Silicon oxide has been categorized as one of the important inorganic compounds due to multiple usage in different sectors such as cement industry, as a filler for rubber, catalysis, as an adsorbent, etc. Extraction of silica from different agro waste involving rice and oats waste, wheat straw, corn cob, etc., has already been established. Pretreatment of corn and coconut husk, yam and cassava peelings with citric acid followed by combustion at high temperature generated significant amount of amorphous silica with reduced concentration of metallic impurities (Prempeh et al. 2021). Another important component of construction industry is brick and various studies have been conducted where agro-waste especially rice waste has been utilized for its production. Sintering mixtures of dried sludge obtained from drinking water treatment plant and rice husk has resulted in the creation of new lightweight bricks. The bricks obtained were of high compressive strength and high porosity which contributed to its light weight. Concentration of rice husk (0–20%) and sintering temperature were the regulatory factors controlling the quality of bricks. Taurino et al. (2019) determined the effect of wine waste in the manufacturing of bricks and studied its thermo-physical mechanical properties. The best results in terms of characteristics were obtained when 5% of wine waste was added to the combination at various concentrations. Characteristic features of rice hush ash (RHA) and its potential as a component of concrete was explored. They replaced up to 20% of the cement with RHA and determined its effect on compressive strength and workability of concrete mixtures, which happened to be directly proportional to the increase in RHA concentration. A sand-cement block developed from RHA and commercially available clay brick was compared with each other. The rice husk ash-based sand-cement block was found to minimise solar heat transmission by 46 W. The results of Ganesan et al. (2008) showed that concrete with a 15% rice husk ash (RHA) concentration had the best compressive strength, and that strength decreased when the RHA percentage was increased above 15%. The impact of using RHA in high strength concrete was examined in separate research by using two different percentages: 5% and 7.5% by weight of cement. Magnani et al. (2014) explored how rice husk ash (RHA) was used to create high strength concrete. Their research centered on evaluating the results of substitution or addition of silica fume on the mechanical characteristics of fresh and cured concrete.

Acoustic panels made of different types of cork, rice husk, residues from vine pruning could serve the function of ceiling tiles. These agrowaste bounded by acrylic resin at different concentrations showed good thermal conductivity, and better sound absorption qualities. Measurement of sound absorption coefficients showed value near to 0.8 with an appropriate flexural strength of the bio composites suggesting its usage as ceiling tiles. Introduction of an air gap behind the sample further reduced the amount of material required for building up of the panel making it more cost-effective as well as light weighted (Maderuelo et al. 2022). A number of scientists are continuously working to determine the physical and chemicals properties of huge spectra of available agro-industrial waste, which can be utilized for construction purposes. Comparative study of the sand, crushed granite and agricultural waste showed that their physio-chemical properties are nearly similar and follow the standard set up by various agencies for construction. Sada et al. (2013) investigated the effect of partially replacing groundnut shells by the weight of aggregates from river sand in concrete production in ratios of 0%, 5%, 15%, 25%, 50% and 75%. It was seen that increase in replacement percentage after 5% resulted in decrease in compressive strength and density. Similar kind of observations were made by Kimeng et al. (2015) thereby suggesting that groundnut shell is a good replacement for aggregates but only for lightweight or non-load bearing applications. Concrete from oil palm shell, yet another agro-waste, satisfied the criterion of being used as a structural concrete. In comparison to stone aggregate, the bulk density of oil palm shell (1850 kg/m^3) is lower making it lightweight. Also, the 28-day compressive strength ranged from 20–24 N/mm^2 that was achieved using 480 kg/m^3 cement with free water binding ratio as 0.41. Hybrid particleboards prepared using sawdust and coconut fibre were assessed for its properties under laboratory conditions. These boards were sprayed and hot pressed (180°C) after being combined with 8% wt. resin. The design density was kept at 650 kg/m^3. Different percent/ratio of both the materials were used with three different board thickness, i.e., 16 mm, 20 mm and 32 mm. After 7 days of production, the optimal conditions found were 0% sawdust to 100% coconut fibre and 20 mm thickness. At this condition, the properties of the particle board were swelling thickness 11.98%, density 761.99 kg/m^3, and water absorption at 37.64% (Tawasil et al. 2021). Coconut fibre possesses structural attributes that make it suitable for construction industry, particularly in relation to cementitious materials (Khieng et al. 2021).

Particleboards, bricks, thermal insulator, cement or binder material, aggregate, and fibre reinforcement are some of the construction materials which have been attempted to be synthesized from agri-waste by employing various techniques and understanding the physical, chemical, and mechanical properties of these compounds. It was also established that these alternates are more economical, and lightweight with lower thermal conductivity.

CONCLUSION

Value addition can be considered as alteration of one substance in to other forms, which have higher value than the previous one. Nearly all of the agricultural crop by-products that are now being commercialized have the potential to be recovered for achieving zero waste target. It is also feasible for the recovery process to be environmentally friendly and to guarantee its sustainability in the industrial process by using the right extraction method. Customers became more conscious about their health, nutrition, and convenience. Technological advancements enable producers to supply the products which can meet to the expectation of customers.

Innovation and collaboration can be used to increase the value of the waste items. We have to develop new processes, procedures, goods, and services or improve the existing ones for conversion of raw (waste) material in to profitable products. We should keep in mind that value addition cannot compensate for the production efficiency as well as ultimate production cost. Value-added product development offers fantastic chances to boost economic growth in the rural sector, where the number of farmers and jobs has been continuously declining. Moreover, additional research work is still required to achieve the target of sustainable development goals through utilization of agriculture waste.

REFERENCES

Abbas, S.R., Sabir, S.M., Ahmad, S.D., Boligon, A.A. and Athayde, M.L. 2014. Phenolic profile, antioxidant potential and DNA damage protecting activity of sugarcane (*Saccharum officinarum*). Food Chemistry. 147: 10–16.

Abedin, M.M., Chourasia, R., Phukon, L.C., Sarkar, P., Ray, R.C., Singh, S.P., et al. 2023. Lactic acid bacteria in the functional food industry: biotechnological properties and potential applications. Critical Reviews in Food Science and Nutrition. 1–19. DOI: 10.1080/10408398.2023.2227896.

Ali, H.K.Q. and Zulkali, M.M.D. 2011. Utilization of agro-residual ligno-cellulosic substances by using solid-state fermentation: a review. Croatian Journal of Food Technology, Biotechnology and Nutrition. 6(1–2): 5–12.

Al-Sheraji, S.H., Ismail, A., Manap, M.Y., Mustafa, S., Yusof, R.M. and Hassan, F.A. 2013. Prebiotics as functional foods: a review. Journal of Functional Foods. 5(4): 1542–1553.

Al-Suhaibani, N., Selim, M., Alderfasi, A. and El-Hendawy, S. 2020. Comparative performance of integrated nutrient management between composted agricultural wastes, chemical fertilizers, and biofertilizers in improving soil quantitative and qualitative properties and crop yields under arid conditions. Agronomy. 10(10): 1503.

Amini Khoozani, A., Birch, J. and Bekhit, A.E.D.A. 2019. Production, application and health effects of banana pulp and peel flour in the food industry. Journal of Food Science and Technology. 56: 548–559.

Anbuselvi, S., Mahalanobis, S. and Jha, M. 2014. Optimizing the production of single cell proteins using Bengal gram husk and pigeon pea husk using yeast. International Journal of Pharmaceutical Sciences Review and Research. 28(1): 188–190.

Bamigbade, G.B., Subhash, A.J., Kamal-Eldin, A., Nyström, L. and Ayyash, M. 2022. An updated review on prebiotics: insights on potentials of food seeds waste as a source of potential prebiotics. Molecules. 27(18): 5947.

Bangar, S.P., Suri, S., Trif, M. and Ozogul, F. 2022. Organic acids production from lactic acid bacteria: a preservation approach. Food Bioscience. 46: 101615.

Bhatt, S.M. 2014. Bioethanol production from economical agro waste (groundnut shell) in SSF mode. Research Journal of Pharmaceutical, Biological and Chemical Sciences. 5(6): 1210–1218.

Borghesi, D.C., Molina, M.F., Guerra, M.A. and Campos, M.G.N. 2016. Biodegradation study of a novel poly-caprolactone-coffee husk composite film. Materials Research. 19: 752–758.

Cardoen, D., Joshi, P., Diels, L., Sarma, P.M. and Pant, D. 2015. Agriculture biomass in India: Part 1. Estimation and characterization. Resources, Conservation and Recycling. 102: 39–48.

Chantorn, S., Piyapittayanun, C. and Dangpram, P. 2018. Bioconversion of agricultural wastes to mannooligosaccharides and their prebiotic potential. Chiang Mai Journal of Science. 45(1): 60–67.

Cheng, F. and Brewer, C.E. 2021. Conversion of protein-rich lignocellulosic wastes to bio-energy: review and recommendations for hydrolysis+ fermentation and anaerobic digestion. Renewable and Sustainable Energy Reviews. 146: 111167.

Chourasia, R., Abedin, M.M., Phukon, L.C., Sahoo, D., Singh, S.P. and Rai, A.K. 2021. Biotechnological approaches for the production of designer cheese with improved functionality. Comprehensive Reviews in Food Science and Food Safety. 20(1), 960–979.

Chu, B.C. and Lee, H. 2007. Genetic improvement of *Saccharomyces cerevisiae* for xylose fermentation. Biotechnology Advances. 25(5): 425–441.

da Silveira, J.S., Durand, N., Lacour, S., Belleville, M.P., Perez, A., Loiseau, G., et al. 2019. Solid-state fermentation as a sustainable method for coffee pulp treatment and production of an extract rich in chlorogenic acids. Food and Bioproducts Processing. 115: 175–184.

Dejene, W. 2020. Physicochemical properties, antioxidant and antimicrobial activities of oil from unripe and ripe banana (*Musa* sp.) peels and pulp (Doctoral dissertation, Haramaya University).

Dunford, N.T. and Edwards, J. 2010. Nutritional bioactive components of wheat straw as affected by genotype and environment. Bioresource Technology. 101(1): 422–425.

Egbe, N.E., Ihediwa, L., Abdulsalami, M.S. and Adebayo, A. 2022. Citric acid production from agricultural wastes using Aspergillus niger isolated from some locations within Kaduna Metropolis, Nigeria. Journal of Applied Sciences and Environmental Management. 26(9): 1607–1614.

Faria, D.L., Júnior, L.M., Resende, A.A., Lopes, D.E., Mendes, L.M., Martins, M.A., et al. 2020. Physical and mechanical properties of polyurethane thermoset matrices reinforced with green coconut fibers. Journal of Composites Materials. 54(30): 4841–4852.

Ferraro, R.M. and Nanni, A. 2012. Effect of off-white rice husk ash on strength, porosity, conductivity and corrosion resistance of white concrete. Construction and Building Materials. 31: 220–225.

Ganaie, M.A., Soni, H., Naikoo, G.A., Oliveira, L.T., Rawat, H.K., Mehta, P.K., et al. 2017. Screening of low-cost agricultural wastes to maximize the fructosyltransferase production and its applicability in the generation of fructooligosaccharides by solid-state fermentation. International Biodeterioration and Biodegradation. 118: 19–26.

Ganesan, K., Rajagopal, K. and Thangavel, K. 2008. Rice husk ash blended cement: assessment of optimal level of replacement for strength and permeability properties of concrete. Construction and Building Materials. 22(8): 1675–1683.

Gustavsson, J., Cederberg, C., Sonesson, U., Van Otterdijk, R. and Meybeck, A. 2011. Global food losses and food waste. Report by Food and Agricultural Organization of the United Nations. Accessed on: 25/09/2023.

Hong, C., Corbett, D., Venditti, R., Jameel, H. and Park, S. 2019. Xylooligosaccharides as prebiotics from biomass autohydrolyzate. LWT—Food Science and Technology. 111: 703–710.

Ibrahim, H.M.M. 2015. Green synthesis and characterization of silver nanoparticles using banana peel extract and their antimicrobial activity against representative microorganisms. Journal of Radiation Research and Applied Sciences. 8(3): 265–275.

Jawad, A.H., Alkarkhi, A.F., Jason, O.C., Easa, A.M. and Norulaini, N.N. 2013. Production of lactic acid from mango peel waste–factorial experiment. Journal of King Saud University—Science. 25(1): 39–45.

Joshi, N., Sharma, M. and Singh, S.P. 2020. Characterization of a novel xylanase from an extreme temperature hot spring metagenome for xylooligosaccharide production. Applied Microbiology and Biotechnology. 104: 4889–4901.

Joshi, N., Kaushal, G. and Singh, S.P. 2021. Biochemical characterization of a novel thermo-halo-tolerant GH5 endoglucanase from a thermal spring metagenome. Biotechnology and Bioengineering. 118: 1531–1544.

Kareem, S.O. and Rahman, R.A. 2013. Utilization of banana peels for citric acid production by *Aspergillus niger*. Agriculture and Biology Journal of North America. 4(4): 384–387.

Karthick, C. and Nanthagopal, K. 2021. A comprehensive review on ecological approaches of waste to wealth strategies for production of sustainable biobutanol and its suitability in automotive applications. Energy Conversion and Management. 239: 114219.

Kathiresan, N., Gopal, L., Karuppiah, V., Naveenethan, R., Abraham, D.R. and Thangavel, K. 2021. Utilization of agricultural waste for the production of xylooligosaccharides using response surface methodology and their *in vitro* prebiotic efficacy. Preprint.

Khan, S.W., Tahir, M., Lone, K.P., Munir, B. and Latif, W. 2015. Protective effect of *Saccharum officinarum* L. (sugar cane) juice on isoniazid induced hepatotoxicity in male albino mice. Journal of Ayub Medical College. 27(2): 346–350.

Khieng, T.K., Debnath, S., Liang, E.T.C., Anwar, M., Pramanik, A. and Basak, A.K. 2021. A review on mechanical properties of natural fibre reinforced polymer composites under various strain rates. Journal of Composites Science. 5(5): 130.

Khuimphukhieo, I., Marconi, T., Enciso, J. and da Silva, J.A. 2023. The use of UAS-based high throughput phenotyping (HTP) to assess sugarcane yield. Journal of Agricultural and Food Research. 11: 100501.

Kim, H.M., Park, J.H., Choi, I.S., Wi, S.G., Ha, S., Chun, H.H., et al. 2018. Effective approach to organic acid production from agricultural kimchi cabbage waste and its potential application. PloS One. 13(11): e0207801.

Kim, H.M., Choi, I.S., Lee, S., Yang, J.E., Jeong, S.G., Park, J.H., et al. 2019. Biorefining process of carbohydrate feedstock (agricultural onion waste) to acetic acid. ACS Omega. 4(27): 22438–22444.

Kimeng, H.T., Ekundayo, O.O., Sani, M. and Frederick, K. 2015. Feasibility study of the use of groundnut shells as fine aggregates in lightweight concrete construction. International Journal of Applied Engineering Research. 1(1): 13.

Koller, M., Sandholzer, D., Salerno, A., Braunegg, G. and Narodoslawsky, M. 2013. Biopolymer from industrial residues: life cycle assessment of poly (hydroxyalkanoates) from whey. Resources, Conservation and Recycling. 73: 64–71.

Kumari, M., Padhi, S., Sharmas, S., Phukon, L.C., Singh, S.P. and Rai, A.K. 2021. Biotechnological potential of psychrophilic microorganisms as the source of cold-active enzymes in food processing applications. 3 Biotech. 11(11): 479.

Lakshmi, N.B., Jessie, S.W., Maheswari, K.U., Kumari, B.A. and Prabhakar, B.N. 2017. Antioxidant potential of rice bran and vegetable waste powders incorporated extrudates. The Pharma Innovation. 6: 12.

Lee, Y. and Oa, S.W. 2013. Nutrient transport characteristics of livestock manure in a farmland. International Journal of Recycling of Organic Waste in Agriculture. 2: 1–6.

Leno, N., Sudharmaidevi, C.R., Byju, G., Thampatti, K.C., Krishnaprasad, P.U., Jacob, G., et al. 2021. Thermochemical digestate fertilizer from solid waste: characterization, labile carbon dynamics, dehydrogenase activity, water holding capacity and biomass allocation in banana. Waste Management. 123: 1–4.

Loaiza, A., Garcia, E. and Colorado, H.A. 2018. Evaluation of asphalt binder blended with coconut coir dust and residual coconut fibers for structural applications. Journal of Construction Engineering. 17(3): 542–554.

Luo, Y., Li, S. and Ho, C.T. 2018. Sugarcane rind: applications and health benefits: a review. Journal of Food Bioactives. 3: 1–7.

Maderuelo-Sanz, R., García-Cobos, F.J., Sánchez-Delgado, F.J., Mota-López, M.I., Meneses-Rodríguez, J.M., Romero-Casado, A., et al. 2022. Thermal and acoustical evaluation of biocomposites made of agricultural waste for ceiling tiles. Applied Acoustics. 191: 108689.

Magnani, C.D., Shah, H.S., Lad, J.J., Mali, D. and Patel, V.N. 2014. A review on compressive and tensile strength of concrete containing rice husk ash and coir fiber. Scholarly Journal of Engineering and Technology. 2(5B): 750–754.

Majaliwa, N., Kibazohi, O. and Alminger, M. 2019. Optimization of process parameters for mechanical extraction of banana juice using response surface methodology. Journal of Food Science and Technology. 56: 4068–4075.

Meenakshi, A. and Kumaresan, R. 2014. Ethanol production from corn, potato peel waste and its process development. International Journal of ChemTech Research. 6(5): 2843–2853.

Mishra, P., Chakraverty, A. and Banerjee, H.D. 1986. Studies on physical and thermal properties of rice husk related to its industrial application. Journal of Materials Science. 21: 2129–2132.

Mitchell, E.J.S., Gudka, B., Whittaker, C., Shield, I., Price-Allison, A., Maxwell, D., et al. 2020. The use of agricultural residues, wood briquettes and logs for small-scale domestic heating. Fuel Processing Technology. 21: 106552.

Mohamed, H., Shah, A.M. and Song, Y. 2021. Conversion of agro-industrial wastes into value-added products. pp. 197–217. *In*: Sarker, M.I., Liu, L.S., Yadav, M.P., Yosief, H.O. and Hussain, S.A. (eds). Conversion of Renewable Biomass into Bioproducts. ACS Symposium Series. American Chemical Society., Washington, DC.

Murthy, P.S. and Naidu, M.M. 2012. Recovery of phenolic antioxidants and functional compounds from coffee industry by-products. Food and Bioprocess Technology. 5: 897–903.

Olugbemide, A.D. and Likozar, B. 2022. Assessment of liquid and solid digestates from anaerobic digestion of rice husk as potential biofertilizer and nutrient source for microalgae cultivation. Processes. 10(5): 1007.

Ong, K.L., Li, C., Li, X., Zhang, Y., Xu, J. and Lin, C.S. 2019. Co-fermentation of glucose and xylose from sugarcane bagasse into succinic acid by *Yarrowia lipolytica*. Biochemical Engineering Journal. 148: 108–115.

Onipe, O.O., Jideani, A.I. and Beswa, D. 2015. Composition and functionality of wheat bran and its application in some cereal food products. International Journal of Food Science and Technology. 50(12): 2509–2518.

Pathak, N., Singh, S., Singh, P., Singh, P.K., Singh, R., Bala, S., et al. 2022. Valorization of jackfruit waste into value-added products and their potential applications. Frontiers in Nutrition. 9: 1061098.

Pérez, S., Giraldo, S., Forgionny, A., Flórez, E. and Acelas, N. 2022. Eco-friendly reuse of agricultural wastes to produce biocomposites with high potential in water treatment and fertilizers. Biomass Conversion and Biorefinery. 14: 8537–8547.

Phukon, L.C., Abedin, M.M., Chourasia, R., Sahoo, D., Parameswaran, B. and Rai, A.K. 2020. Production of xylanase under submerged fermentation from *Bacillus firmus* HS11 isolated from Sikkim Himalayan region. Indian Journal of Experimental Biology (IJEB). 58(08): 563–570.

Prabha, M.S. and Rangaiah, G.S. 2014. Citric acid production using *Ananas comosus* and its waste with the effect of alcohols. International Journal of Current Microbiology and Applied Sciences. 3(5): 747–754.

Pratik, B., Gomashe, A.V., Sanap, H.M. and Gulhane, P.A. 2013. Production and optimization of citric acid by *Aspergillus niger* using fruit pulp waste. International Journal of Current Microbiology and Applied Sciences. 2(10): 347–352.

Prempeh, C.O., Formann, S., Schliermann, T., Dizaji, H.B. and Nelles, M. 2021. Extraction and characterization of biogenic silica obtained from selected agro-waste in Africa. Applied Sciences. 11(21): 10363.

Raab, A.M. and Lang, C. 2011. Oxidative versus reductive succinic acid production in the yeast *Saccharomyces cerevisiae*. Bioengineered Bugs. 2(2): 120–123.

Rajan, K., D'Souza, D.H., Kim, K., Choi, J.M., Elder, T., Carrier, D.J., et al. 2021. Production and characterization of high-value prebiotics from biorefinery-relevant feedstocks. Frontiers in Microbiology. 12: 675314.

Raji, Y.O., Jibril, M., Misau, I.M. and Danjuma, B.Y. 2012. Production of vinegar from pineapple peel. International Journal of Advanced Scientific Research. 3(2): 656–666.

Rogers, P.L., Jeon, Y.J., Lee, K.J. and Lawford, H.G. 2007. *Zymomonas mobilis* for fuel ethanol and higher value products. pp. 263–288. *In*: Olsson, L. (ed.). Biofuels. Advances in Biochemical Engineering/Biotechnology, vol 108. Springer, Berlin, Heidelberg. https://doi.org/10.1007/10_2007_060.

Sada, B.H., Amartey, Y.D. and Bakoc, S. 2013. An investigation into the use of groundnut as fine aggregate replacement. Nigerian Journal of Technology. 32(1): 54–60.

Sarkar, P., Abedin, M.M., Singh, S.P., Pandey, A. and Rai, A.K. 2022. Microbial production and transformation of polyphenols. pp. 189–208. *In*: Rai, A.K., Singh, S.P., Pandey, A., Larroche, C. and Soccol, C.R. (eds). Current Developments in Biotechnology and Bioengineering: Technologies for Production of Nutraceuticals and Functional Food Products. Elsevier.

Sharma, N., Sahoo, D., Rai, A.K. and Singh, S.P. 2022. A highly alkaline pectate lyase from the Himalayan hot spring metagenome and its bioscouring applications. Process Biochemistry. 115: 100–109.

Singh, A. and Singh, A. 2015. A comparative overview of bioethanol production from organic residues of agro-waste materials. European Journal of Biotechnology and Bioscience. 3(3): 11–14.

Sönmez, İ., Kalkan, H., Demir, H., Külcü, R., Yaldiz, O. and Kaplan, M. 2017. Mineral composition and quality parameters of greenhouse-grown lettuce (*Lactuca sativa* L.) depending on fertilization with agricultural waste composts. Acta Scientiarum Polonorum Hortorum Cultus. 16(3): 85–95.

Taurino, R., Ferretti, D., Cattani, L., Bozzoli, F. and Bondioli, F. 2019. Lightweight clay bricks manufactured by using locally available wine industry waste. Journal of Building Engineering. 26: 100892.

Tawasil, D.N., Aminudin, E., Abdul Shukor Lim, N.H., Nik Soh, N.M., Leng, P.C., Ling, G.H., et al. 2021. Coconut fibre and sawdust as green building materials: a laboratory assessment on physical and mechanical properties of particleboards. Buildings. 11(6): 256. https://doi.org/10.3390/buildings11060256.

Thakur, A., Panesar, P.S. and Saini, M.S. 2019. L (+)–Lactic acid production by immobilized *Lactobacillus casei* using low cost agro-industrial waste as carbon and nitrogen sources. Waste Biomass Valorization. 10: 1119–1129.

Thakur, M., Sharma, N., Rai, A.K. and Singh, S.P. 2021. A novel cold-active type I pullulanase from a hot-spring metagenome for effective debranching and production of resistant starch. Bioresource Technology. 320: 124288.

Thakur, M., Rai, A.K. and Singh, S.P. 2023. Structural characteristics, physicochemical properties and digestibility analysis of resistant starch type-V prepared from debranched corn starch and fatty acid complexation. ACS Omega. 8: 25799–25807.

Udhayaraja, P. and Narayanan, J.S. 2012. Optimization for production of bioethanol using *Sorghum* stovar by *Saccharomyces cerevisiae*. Journal of Pure and Applied Microbiology. 2(4): 64–67.

Väätäinen, K., Sirparanta, E., Räisänen, M. and Tahvanainen, T. 2011. The costs and profitability of using granulated wood ash as a forest fertilizer in drained peatland forests. Biomass and Bioenergy. 35(8): 3335–3341.

Varanasi, P., Singh, P., Auer, M., Adams, P.D., Simmons, B.A. and Singh, S. 2013. Survey of renewable chemicals produced from lignocellulosic biomass during ionic liquid pre-treatment. Biotechnology for Biofuels. 6(1): 1–9.

Velazquez-Martinez, V., Valles-Rosales, D., Rodriguez-Uribe, L., Holguin, O., Quintero-Quiroz, J., Reyes-Jaquez, D., et al. 2021. Antimicrobial, shelf-life stability and effect of maltodextrin and gum arabic on the encapsulation efficiency of sugarcane bagasse bioactive compounds. Foods. 10(1): 116.

Vigneswaran, C., Pavithra, V., Gayathri, V. and Mythili, K. 2015. Banana fiber: scope and value-added product development. Journal of Textile and Apparel, Technology and Management. 9(2): 1–7.

Vijayakumar, J., Aravindan, R. and Viruthagiri, T. 2008. Recent trends in the production, purification and application of lactic acid. Chemical Engineering and Processing: Process Intensification. 22(2): 245–264.

Vu, H.T., Scarlett, C.J. and Vuong, Q.V. 2017. Optimization of ultrasound-assisted extraction conditions for recovery of phenolic compounds and antioxidant capacity from banana (*Musa cavendish*) peel. Journal of Food Processing and Preservation. 41(5): 13148.

Yong, K.J. and Wu, T.Y. 2022. Second-generation bioenergy from oilseed crop residues: Recent technologies, techno-economic assessments and policies. Energy Conversion and Management. 267: 115869.

Yong, Z.J., Bashir, M.J. and Hassan, M.S. 2021. Biogas and biofertilizer production from organic fraction municipal solid waste for sustainable circular economy and environmental protection in Malaysia. Science of the Total Environment. 776: 145961.

Zafar, M., Bano, H.S. and Anwar, Z. 2021. Orange peels valorization for citric acid production through single and co-culture fermentation. Jordan Journal of Biological Sciences. 14(2): 261–266.

Chapter **6**

Strategies for the Transformation of Waste Vegetable Oils into Biofuels

Anisha GS[1], Bessy Raj BN[2], Anil Kumar Patel[3] and Reeta Rani Singhania*[3]

[1]Post-Graduate and Research Department of Zoology, Government College for Women, Thiruvananthapuram 695014, Kerala, India.

[2]Post-Graduate and Research Department of Chemistry, University College, Thiruvananthapuram 695034, Kerala, India.

[3]Institute of Aquatic Science and Technology, College of Hydrosphere, National Kaohsiung University of Science and Technology, Kaohsiung City 81157, Taiwan.

INTRODUCTION

The past decades have witnessed a tremendous increase in the world population and their fuel needs, which were defied by a dwindling of fossil fuel reserves. This has urged scientists to find out alternative sources of sustainable energy resources that would be environmentally friendly with lower carbon emissions and lesser contribution to global warming. Biofuels originating from biomass resources have emerged as an environmentally friendly, cost-effective and sustainable substitute for fossil fuels (Altalhi et al. 2022). Waste vegetable oils form an exceptional raw

* For Correspondence: reetasinghania@nkust.edu.tw

material for the generation of biofuels such as biodiesel, bio-oil, bio-kerosene, syngas and biohydrogen (Nascimento et al. 2022).

The non-edible oils have the advantage of providing a renewable and sustainable supply of feedstock without compromising the oil crops grown for edible purposes. Moreover, non-edible oleaginous crops can be grown on wastelands. Along with non-edible vegetable oils, waste cooking oils also can be transformed into biodiesel and other fuels. The disposal of waste vegetable oils from households, restaurants, hotels and food manufacturing units jeopardizes environmental safety and poses threats to humans and animals (Awogbemi et al. 2021). The usual practice of disposing of waste cooking oil is to drain them in cesspits, open spaces, or water bodies which hamper the natural ecosystem. Disposing of waste vegetable oils in aquatic bodies hinders oxygen dissolution and thereby increases the chemical and biological oxygen demand of water leading to a decline of aquatic flora and fauna. Instead of disposing of waste cooking oil, valorization can convert and transform them into value-added products such as biofuels, bio-lubircants, bisabolene, plasticizers, biosurfactants, microbial lipases and lipids, etc. (Lopes et al. 2020, Phukon et al. 2020, Awogbemi et al. 2021, Kumari et al. 2021).

Generating bioenergy from waste vegetable oil is an interesting and state-of-the-art method that can not only mitigate the problem of waste disposal but also meet the rising energy demand and contribute to the circular bioeconomy. Moreover, biodiesel is environmentally amicable because of its lower carbon emission in comparison to conventionally used fossil fuels. It is reported that the combustion of biodiesel emits about 48% less carbon monoxide, 47% less particulate matter and 67% less hydrocarbon as opposed to petroleum-based diesel (Nascimento et al. 2022).

SOURCES OF WASTE VEGETABLE OILS USED FOR BIOFUEL PRODUCTION

The feedstock for biofuel production must be cheap and abundantly available to ensure sustainable supply, or else there will be a rise in the cost of production and the market price of biofuels. Waste cooking oil and non-edible oils are excellent feedstocks for biofuel production. The non-edible vegetable oils such as those from *Jatropha curcas* (Ong et al. 2014a), *Ceiba pentandra* L. Gaertn or kekabu (Ong et al. 2014a), *Calophyllum inophyllum* L. or polanga (Ong et al. 2014a, Arumugam and Ponnusami 2019), *Pongamia pinnata* (Jaichandar and Annamalai 2012), *Thevetia peruviana* or yellow oleander (Basumatary et al. 2020), castor (Valente et al. 2011), *Azadirachta indica* or neem (Brahma et al. 2022), *Hevea brasiliensis* or rubber (Brahma et al. 2022) and *Cerbera manghas* (Ong et al. 2014b) can provide sufficient feedstock for biodiesel production without compromising edible vegetable oils. The used cooking oils which have lost the quality standards for edible purposes can also be used for the production of biofuels.

TECHNOLOGIES FOR THE PRODUCTION OF BIOFUELS FROM WASTE VEGETABLE OIL

The waste cooking oils collected from households, cafes, restaurants and hotels have to be pre-treated before being used for the synthesis of biofuels. The waste cooking oil may contain water droplets that may infiltrate the oil from the food items while frying. The water droplets have to be removed from the oil by heating. The waste cooking oil may also contain suspended food particles and other solid residues that are to be removed by filtration (El-Araby et al. 2020). The technology adopted for the generation of biofuels from waste cooking oil differs for different types of biofuels (Table 6.1).

Pyrolysis for Biodiesel

Pyrolysis is an upgrading process involving the heating of vegetable oil at substantially high temperatures ranging from 300–1300 °C under anoxic conditions leading to the alteration of its structure by the rupturing of bonds of long-chain fatty acids into small combustible molecules. The process can be done with or without the help of suitable catalysts. Pyrolysis involves not only the depolymerisation and monomer formation of larger molecules, but also the condensation and reformation of the molecules to value-added and energy-rich compounds. Thermal cracking of long-chain fatty acids in vegetable oil yields alkanes, alkadienes, aromatics, alkenes, carboxylic acids and volatile compounds which are combustible (Brahma et al. 2022).

The thermal activation energy required for the cracking of long-chain fatty acids in crude vegetable oil can be lowered via catalytic cracking (Anisha and Shyni Raphel 2023). The fatty acid or triglyceride content of the vegetable oil, process temperature, rate of heating, duration of heating and type and load of catalysts are the influential factors that determine the yield of pyrolysis product (Altalhi et al. 2022). As the process temperature increases, the thermal cracking of macromolecules proceeds faster and the yield of pyrolysis increases correspondingly with a minimum amount of undesirable by-products. Pyrolysis, when carried at a faster heating rate in a shorter duration of time, is called fast pyrolysis and yields more amount of biofuel product than the conventional method. The pyrolysis oil or bio-oil obtained by fast pyrolysis is called fast pyrolysis oil (Zhang et al. 2019). The low viscosity, high cetane number, low sulphur and water content are beneficial fuel characteristics of bio-oil (Altalhi et al. 2022, Brahma et al. 2022). However, the high energy input required for the heating process is a major disadvantage that increases the cost of operation.

The catalysts used for the cracking-upgrading of crude vegetable oil include natural minerals, alkali metals, alkaline earth metals and transition metals (Anisha and Shyni Raphel 2023). Silica alumina and zeolites are examples of natural mineral catalysts used for the catalytic upgrading of pyrolysis oil. As catalysis proceeds at high temperatures, carbon deposition occurs on the surface of catalysts clogging the pores and active sites and reducing the catalytic efficiency. Chemical

Table 6.1 Technologies for the transformation of waste vegetable oil and cooking oil into biofuels.

Source of oil	Production technique	Operation condition	Biofuel and yield	Fuel characteristics	References
Waste cooking oil	Catalytic cracking	Single step batch reactor, 450°C, (Co/Zn) Al_2O_4 nano catalyst (2%), 15 min	Bio-kerosene	A convenient freezing point (–55°C) of a binary sample of 5% bio jet distillate product with 95% jet A-1 was achieved	(El-Araby et al. 2022)
Cerbera manghas	Catalytic transesterification	Two-step acid-alkaline transesterification, methanol to oil ratio 9:1, 50°C, 1 hour	Biodiesel (98.5%)	869.7 kg/m^3 fuel density at 15°C 4.86 mm^2/s viscosity at 40°C 159.5°C flash point 6.0°C pour point 8.0°C cloud point 39.925 MJkg calorific value 8.21 oxidation number 0.17 mg KOH/g Acid value	(Ong et al. 2014b)
Plum waste seed oil	Catalytic transesterification	Calcinized bentonite–potassium ferricyanide composite catalyst (0.3%); 60°C, 4 h, and methanol to oil ratio of 10:1	Biodiesel (>90%)	Acid value 0.7846 mg KOH/g Iodine value 156.3 g I_2/100 g Cetane number 67.8	(Saeed et al. 2021)
Waste cooking oil	Catalytic transesterification	NaOH catalyst and methanol	Biodiesel	Household waste cooking oil is better than restaurant waste cooking oil	(Li and Yu 2015)
Waste cooking oil	Catalytic transfer hydrogenation	Continuous flow reactor with a fixed bed of granular activated carbon; pressure 2 bar; temperature 300–400 °C; weight hourly space velocity 6.7 h^{-1}; H-donor 2-propanol (99.9 wt%)	Jet/diesel-fuel liquid hydrocarbon 72% (32% alkanes, 16% aromatics, 37% alkenes)	Reduction of volume ratio of hydrogen source to oil Catalyst maintained its activity despite the inherent coke formation	(Asiedu et al. 2021)

(Contd.)

Table 6.1 Technologies for the transformation of waste vegetable oil and cooking oil into biofuels (*Contd.*).

Source of oil	Production technique	Operation condition	Biofuel and yield	Fuel characteristics	References
Palm kernel oil, palm olein and waste cooking oil (produced from palm olein)	Single-step catalytic deoxygenation	Deoxygenation catalysts (CaO, Zeolite, V_2O_5, Pd/C, TiO_2); Catalyst loading (0 wt%, 5 wt%, 6 wt%, 7 wt%, 8 wt%, 9 wt% and 10 wt%); 400 °C; 2 h	96% liquid product (n-paraffins, isoparaffins, olefins, naphthenes, aromatic) and 73% of jet paraffins	Low temperature fluidity, combustion characteristics and fuel volatility are compliant with the ASTM standards	(Why et al. 2022)
Waste cooking oil	Single-step catalytic hydrocracking followed by fractional distillation	Stainless steel reactor (450°C), pressure of (120 bars), $ZnAl_2O_4$ (2.5% w/v), hydrogen pressure 4 atmosphere, 60 min	Bio-kerosene (70 to 85%)	Freezing point –55°C Compliant with the international standard specifications of aviation turbine fuel	(El-Araby et al. 2020)
Waste edible oil	Single-step catalytic pyrolysis	Sulphonated chloromethyl dimethyl phenyl silane surface grafted γ-alumina catalyst (0.8% , wt/wt) in a closed stainless steel cylindrical reactor, 300°C, 40 min	Biodiesel (88%)	Acid value 3.15 mg KOH/g Viscosity 6.11 cSt Flash point 45°C; Cloud point –9°C Pour point –12°C Cetane number 51.2	(Altalhi et al. 2022)
Madhuca indica oil	Lipase-catalyzed transesterification	Cross-linked enzyme aggregates; 6 h; 50 mg of lipase Protein-coated microcrystals	Biodiesel 92% in 2.5 h 99% in 2.5 h	n.m.	(Kumari et al. 2007)
Jatropha oil	Lipase-catalyzed transesterification	Chromobacterium viscosum lipase immobilized in Celite-545; 8 h; 40°C	Biodiesel 92%	n.m.	(Shah et al. 2004)

modification of natural mineral catalysts can prevent carbon deposition on active sites, provide more number of acidic active sites and enhance the porosity of catalysts, all of which contribute to increasing the catalytic efficiency and yield of the biofuel product (Altalhi et al. 2022). Chemical modification of alumina efficiently converts larger triglycerides into short-chain hydrocarbons (Altalhi et al. 2022). The catalytic cracking can be carried out either in a single-step or a two-step process. In a single-step process, the catalysts are mixed with the crude vegetable oil and subjected to cracking-upgrading at high temperatures. On the other hand, the two-step process involves the pyrolysis of vegetable oil in the first step followed by the catalytic upgrading of pyrolysis products to energy-rich volatile compounds (Altalhi et al. 2022, Anisha and Shyni Raphel 2023). Several researchers have reported the production of bio-oil from waste cooking oil by pyrolysis with the assistance of thermal technologies such as the microwave technique (Anis et al. 2021, Wu et al. 2020). Gad et al. (2021) and Shitao et al. (2020) have reported the production of bio-oil by catalytic pyrolysis.

Transesterification for Biodiesel

Transesterification, also called alcoholysis, is the chemical method for the synthesis of biodiesel from oils, which can be used in compression ignition engines in automobiles or as fuel for industrial purposes (Valente et al. 2011). The non-edible vegetable oils and waste cooking oil contain enormous amounts of triglycerides. Triglycerides are esters of glycerol and three fatty acids. They can undergo transesterification reaction with another alcohol, in the presence of a catalyst, either acid or base, to produce long-chain fatty acid alkyl esters, such as methyl esters (FAME) constituting biodiesel (Figure 6.1). Transesterification is simply the modification of one ester to another by changing its alcohol part.

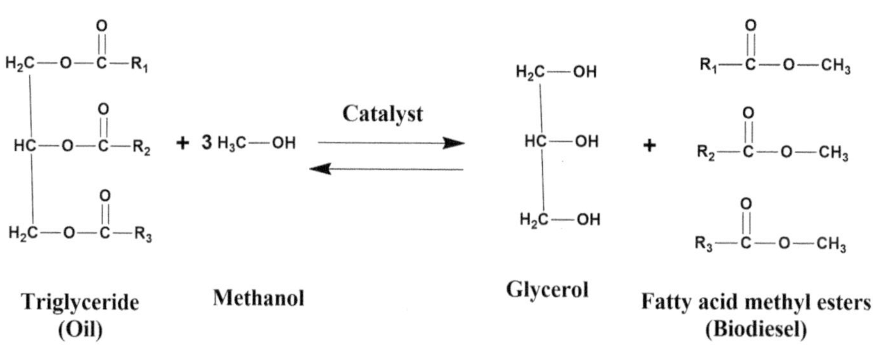

Figure 6.1 Transesterification of triglycerides to fatty acid methyl esters (biodiesel).

Among the several methods for the synthesis of biodiesel from waste vegetable oils, transesterification is the most chosen owing to the higher cetane numbers, lower emission and higher combustion efficiency of the resultant fuel (Brahma et al. 2022). The important factors influencing the transesterification reaction include the alcohol-to-oil ratio, reaction temperature, type of catalyst and the amount of catalyst used. For preventing the reversal of transesterification, a surplus amount of alcohol is needed to facilitate the reaction to proceed in the forward direction forming FAMEs. An alcohol-to-oil ratio of 3:1 is considered to be ideal for the transesterification reaction to proceed optimally favouring the formation of FAMEs (Brahma et al. 2022). The alcohols generally used for the production of biodiesel from vegetable oils via transesterification include methanol, ethanol, propanol, butanol and amyl alcohol (Brahma et al. 2022). Owing to the simplest chemical structure and polar nature, methanol is generally used for the transesterification of vegetable oils. Although transesterification is the most preferred method of biodiesel synthesis from waste vegetable oils, the surplus amounts of methanol required for the reaction is one of the main drawbacks.

The transesterification of vegetable oils to biodiesel can be mediated by chemical catalysts or enzymatic catalysts.

Chemical Catalysts

Conventionally, transesterification is carried out using chemical catalysts such as sodium hydroxide (NaOH) or potassium hydroxide (KOH). Table 6.2 enlists the chemical catalysts used for the synthesis of biodiesel via transesterification. The chemical catalysts used for transesterification are of two types—homogenous and heterogeneous catalysts. The homogenous and heterogeneous catalysts for transesterification can be either acids or bases, of which the latter are reported to accelerate the reaction considerably higher than the former (Brahma et al. 2022).

Homogenous Catalysts

Homogenous catalysts used for the transesterification process are soluble in methanol and the most commonly used catalysts are NaOH and KOH, of which NaOH is economic because of its availability, low cost and improved yield. The methanol reacts with the alkaline metal hydroxide and form methoxide ions as intermediates. Methoxide ions are strong nucleophiles which can react with the carbonyl groups of the glycerides forming alkyl esters and accomplishing the hydrolysis at a faster rate (Salaheldeen et al. 2021) (Figure 6.2). The methoxide salts used as catalysts can also dissociate and form methoxide ions (Figure 6.2). In comparison to the hydroxide catalysts, metal methoxide bases in methanol combination are recommended for a continuous production of biodiesel.

$$NaOH + CH_3OH \longrightarrow CH_3O^- + Na^+ + H_2O \quad \text{(Equation 1)}$$

$$CH_3ONa \longrightarrow CH_3O^- + Na^+ \quad \text{(Equation 2)}$$

Figure 6.2 Formation of methoxide ions by the reaction of NaOH with methanol (Equation 1) and dissociation of sodium methoxide (Equation 2).

However, homogeneous base catalysts are not recommended for feedstocks having high carboxylic acid content, as they form metal carboxylate salts (soap) and do not facilitate esterification. Also, the presence of soap makes it difficult to

Table 6.2 Chemical catalysts used for transesterification of triglycerides and fatty acids in waste vegetable oils into FAMEs in biodiesel.

Oil source	Catalyst	Catalyst load (wt% to oil)	Methanol to oil molar ratio	Temperature (°C)	Time	Biodiesel yield (%)	References
Carthamus oxyacantha (Wild safflower)	Nano-magnetic Fe/SnO catalyst	1	5:1	50	120 min	98.1 ± 0.6	(Hanif et al. 2022)
Citrullus colocynthis (bitter apple)	Nano-magnetic Fe/SnO catalyst	1	10:1	50	120 min	98.5 ± 0.5	(Hanif et al. 2022)
Pongamia pinnata (Karanja)	Nano-magnetic Fe/SnO catalyst	1.5	10:1	50	120 min	97.4 ± 0.6	(Hanif et al. 2022)
Rapeseed oil	*Peganum harmala* spice seed extract modified GO-$CuFe_2O_4$ (SSE@GO-$CuFe_2O_4$) nano-composite	8	13:1	40	7 h	92.81	(Tamoradi et al. 2022)
Ricinus communis (Castor oil)	Nano-magnetic Fe/SnO catalyst	1.5	15:1	60	120 min	97.0 ± 0.6	(Hanif et al. 2022)
Sinapis arvensis (Wild mustard)	Nano-magnetic Fe/SnO catalyst	1	5:1	50	90 min	97.6 ± 0.8	(Hanif et al. 2022)
Waste corn oil	*Peganum harmala* spice seed extract modified GO-CuFe2O4 (SSE@GO-$CuFe_2O_4$) nano-composite	8	13:1	40	7 h	87.26	(Tamoradi et al. 2022)
Waste cooking oil	CaO nano-catalyst	1	8:1	50	90 min	96	(Degfie et al. 2019)
Waste cooking oil	Egg shell-derived CaO nano-catalyst	2.5	12:1	60	120 min	94	(Erchamo et al. 2021)

separate the transesterification products (glycerol and the biofuel). In the case of feedstocks with high carboxylic acid content, homogeneous acid catalysis using sulphuric acid and sulphonic acid can be employed. But it is not economical because of low reaction rate and high alcohol consumption. Moreover, the reaction condition should be kept anhydrous to prevent the hydrolysis of triglycerides instead of transesterification.

The vegetable oils with a high amount of free fatty acids (FFAs) are converted into FAMEs by a two-step acid-alkaline transesterification process. This technique involves an acid-esterification process followed by a base-catalyzed transesterification reaction. In the first step, the FFAs in the crude oil are completely esterified using H_2SO_4 in the presence of methanol at methanol-to-oil ratio of 9:1. In the subsequent step, a solution of methanol and NaOH (Ong et al. 2014a) or KOH (Ong et al. 2014b) is added to allow transesterification of triglycerides in the oil and separating the solution into two layers, the top layer with FAMEs and bottom layer with glycerine. The FAMEs are then washed using distilled water and dried using anhydrous $CaCl_2$. Further purification of FAMEs is then achieved using a rotary evaporator. Ong et al. (2014b) synthesized biodiesel from the crude *Cerbera manghas* oil via the two-step process with a yield of 98.5%.

Heterogeneous Catalysts

Heterogeneous catalysts are solid catalysts which are insoluble in methanol and can be easily separated from the solution after the transesterification process (Hájek et al. 2021). Heterogeneous catalysts include natural mineral catalysts such as zeolites, alkaline metals, transition metals, metal oxides and carbonates. Heterogeneous catalysts are industrially advantageous since they can be easily recovered from the reaction mixture and reused for another round of transesterification reaction (Hájek et al. 2021).

Calcium oxide (CaO) is a heterogeneous catalyst used to mediate transesterification reactions in the production of biodiesel. The unreacted methanol can be removed from the reaction mixture by heating to temperatures higher than the boiling point of methanol (64.7°C). The catalysts can be separated by allowing the reaction mixture to stand until the solid catalysts settle down at the bottom. Degfie et al. (2019) synthesized biodiesel from waste cooking oil using methanol and calcium oxide as catalysts. The catalyst calcium oxide was of nanoparticle size which could increase the surface area of the catalysts thus increasing the efficiency of the transesterification reaction. Degfie et al. (2019) obtained a 96% yield of biodiesel with transesterification carried out for 90 minutes at 50°C using waste cooking oil and methanol in the molar ratio of 1:8 (v/v) and CaO catalyst at 1% concentration (w/w). The final biodiesel obtained was also reported to demonstrate viscosity, specific gravity, water and sediment, total acidity, ash content and sulphur content corresponding with that of the standard values (Degfie et al. 2019).

Enzyme Catalysts

In enzyme-catalyzed transesterification, lipase enzymes are used as catalysts to facilitate the reaction (Almeida et al. 2021, Arumugam and Ponnusami 2019,

Toldrá-Reig et al. 2020). Lipases or triacylglycerol ester hydrolases (EC 3.1.1.3) are hydrolytic enzymes with a serine, histidine and aspartate triad in the active site (Toldrá-Reig et al. 2020). Lipases can break down fats and oils into their individual fatty acid and alcohol components. They can also facilitate the synthesis of esters, which are the main components of biodiesel by esterification or transesterification reaction. The vegetable oil mixed with an alcohol, typically methanol or ethanol, when subjected to the action of lipase enzyme catalyst, produces FAMEs and glycerol as a by-product. Lipases hydrolyze the triacylglycerides and facilitate the esterification of the fatty acids with the alcohol. The reaction takes place at a moderate temperature, usually around 40–60°C which is most favourable for the enzyme action. The glycerol is separated from the biodiesel, and the resulting biodiesel can be further purified and refined to meet the required quality standards.

Lipases commonly used for the transesterification process are of microbial origin. The common microbial sources of lipases include *Pseudomonas, Aspergillus, Rhizopus* sp., *Penicillium* sp., *Geotrichum* sp., *Mucor* sp. and *Thermomyces* sp. (Almeida et al. 2021, Phukon et al. 2022). Akin to any enzyme-catalyzed reaction, the rate of lipase-catalyzed transformation of lipids into biodiesel is susceptible to reaction conditions such as temperature, pH and type and concentration of solvent used. Lipases obtained from extremophilic microorganisms such as those dwelling in hot springs, polar ice belts, salt lakes, acid lakes, etc., have the ability to maintain the enzyme activity even under stringent reaction conditions (Vivek et al. 2022). Table 6.3 enlists the microbial lipases and their application in the synthesis of biodiesel. Lipase obtained from the thermophilic fungus *Thermomyces lanuginosus* is commercially available with the trade name Callera™ Trans, Eversa® Transform or Eversa® Transform 2.0 and is used for the synthesis of biodiesel (Vivek et al. 2022). Christopher et al. (2015) reported biodiesel production from waste cooking oil using thermo-alkaline lipase isolated from the thermophilic bacterium *Geobacillus thermodenitrificans* AV-5 with a transformation rate of 76% at 65°C.

The lipases can be used either as free enzymes or as immobilized enzymes for the transesterification process (Vivek et al. 2022, Phukon et al. 2022). The immobilized enzymes have the advantage of being recovered from the reaction mixture and reused for several cycles of the transesterification process. In addition, immobilization also confers improved thermal stability, pH stability and solvent tolerance to the lipases which enable them to retain their catalytic performance under repeated use. Hence, immobilized enzymes can also be employed in bioreactors designed for the transesterification of vegetable oils. Enzyme-catalyzed transesterification offers several advantages over the chemical process, including milder reaction conditions, reduced energy requirements, and the ability to use a wider range of feedstocks.

Hydrocracking for Biokerosene

Hydrocracking is a technique of removing sulphur, nitrogen, oxygen and metals from the glycerides in waste vegetable oils using high saturations of hydrogen at

Table 6.3 Microbial lipases employed for the transesterification of vegetable oils to biodiesel.

Microbial source of lipase	Oil source	Significant findings on biodiesel production	Conversion rate	References
Thermomyces lanuginosus	Soybean oil	Lipase hydrolysis of triacylglyceride involves hydroesterification reaction with 50% (v/v) soybean oil and 2.3% (v/v) lipase (25 U/mL of reaction medium) in distilled water and at 80°C; 48 hours of reaction	89%	(Cavalcanti-Oliveira et al. 2011)
Burkholderia ubonensis SL-4	Soybean oil	Thermostable and solvent-tolerant extracellular lipase	92.4%	(Yang et al. 2016)
Geobacillus thermodenitrificans AV-5	Waste cooking oil	Thermo-alkali stable: at 70°C, it retained 81% activity and 45% stability; at pH 10 it lost only 15% and 2.6% of its maximum activity and stability, respectively	76%	(Christopher et al. 2015)
Thermomyces lanuginosus lipase (Eversa® Transform 2.0)	Sunflower oil	Lipase-catalyzed semi-continuous hydroesterification with lipase immobilized on octyl agarose beads; stable in the presence of 90% methanol or dioxane at 25°C and pH 7	n.m.	(Martínez-Sanchez et al. 2020)
Bacillus thermocatenulatus lipase 2 (BTL2) overexpressed in *E. coli*	Rapeseed oil	Resistant to neutral pH, high temperature, and organic solvents	86.04%	(Zhang et al. 2021)
Pseudomonas fluorescens B68	—	Transesterification with psychrophilic lipase at 20°C; highly competitive energy-saving biocatalyst	92%	(Luo et al. 2006)
Idiomarina sp. W33	Jatropha oil	Organic solvent-tolerant lipase in free and immobilized form resulted in high stability and activity and yielded biodiesel from *Jatropha* oil	84% (free enzyme); 91% (immobilized enzyme)	(Li et al. 2014)
Ralstonia RL74	Palm oil	Resistant to oxidizing and reducing agents; highest activity at 50–55°C and pH 8.0–9.5 and stable at pH 7.0–12.0 and up to 45°C.	n.m.	(Yoo et al. 2011)
LipK107®*:* lipase gene from *Proteus sp.* expressed in *E. coli*	Olive oil	Whole cell biocatalyst with 100% yield of biodiesel, the alkaline lipase LipK107® had optimum activity at pH 9 and 35°C.	100%	(Gao et al. 2009)

a high temperature and pressure of 52–300°C and 0.1–20 MPa, respectively, to produce pure and higher-quality straight chain alkane hydrocarbons in biodiesel and biokerosene which can be used as aviation fuel or bio-jet fuel (El-Araby et al. 2020, Nascimento et al. 2022). Several oil companies such as Neste Oil, Petrobras, UOP/ENI, etc., have employed hydrocracking treatment of oil feedstocks for the production of bio-jet fuel of higher quality (El-Araby et al. 2020). Liquid fuels with short-chain saturated hydrocarbons have low freezing points and are suitable as aviation fuel or bio-jet fuel. Unlike biodiesel, bio-kerosene obtained by catalytic hydrocracking contains more short-chain saturated hydrocarbons, branched alkanes and cycloalkanes which lowers its freezing point (El-Araby et al. 2020). The short-chain saturates obtained by hydrocracking can then be separated by fractional distillation and used as bio-kerosene or bio-jet fuel directly or by blending with petroleum-based jet fuel (El-Araby et al. 2020). The fuels produced by the hydrocracking method have low oxygen content, better atomization, viscosity and lubrication, which are favourable for use as transportation fuels in aviation. Catalysts such as Ni-Mo/Al_2O_3, Co-Mo/Al_2O_3, zeolites and bentonites are used along with saturated hydrogen to increase the yield of hydrocarbons and for the better removal of undesired atoms. Zeolites catalyse the deoxygenation and decarbonylation of heptadecane ($C_{17}H_{36}$) and pentadecane ($C_{15}H_{30}$) chains in the waste vegetable oils and break them into alkanes of eight to sixteen carbon chain lengths. The fuel composition resulting from the hydrocracking of waste vegetable oil depends on the temperature of the operation. At temperatures below 350°C, the resulting fuel contains more amount of kerosene that can be used as jet fuel, whereas at around 350°C together with high liquid hourly space velocity in the reactor, the production of biodiesel is favoured. Increasing the temperature to 370°C supports the formation of bio-oils. Analogous to gasoline and diesel from fossil reserves, the biofuels from waste cooking oil also contain 2-propanone, dodecane, octanoic acid, decanoic acid, dodecanoic acid and 1,2-benzene dicarboxylic acid.

Anaerobic Digestion for Biomethane

Anaerobic digestion is a biological process in which anaerobic microorganisms are employed to ferment and break down biomass in the absence of oxygen to generate biomethane or biogas which can be used for bioenergy purpose. Fats are good sources for the production of biomethane when subjected to anaerobic treatment. The waste cooking oils and non-edible vegetable oils have higher biogas generation potential than the feedstocks rich in protein and carbohydrate content and are promising inputs for anaerobic digestion (Thakur et al. 2022). The biogas yield varies depending on the fatty acid composition of the feedstock oil and the microorganism employed for anaerobic fermentation.

The complete degradation of large lipids into biomethane requires the concerted action of several microbial communities forming a consortium. The microorganisms used for anaerobic fermentation of waste cooking oils and non-edible oils include the acetogenic and methanogenic bacteria such as *Clostridium,*

Methanosphaera, Methanobrevibacter, etc. (Thakur et al. 2022). The richness of bacterial genera such as *Syntrophomonas* and *Methanosarcina* establishes a syntrophic relationship which facilitates faster breakdown of long-chain fatty acids and generation of biomethane (Kurade et al. 2020). The acetogenic bacteria produce enzymes such as acetyl-Co A C-acetyltransferase and enoyl-Co A hydratase involved in the β-oxidation of fatty acids and which act on the long-chain fatty acids to produce the precursors of biomethane. The biochemical steps involved in the transformation of vegetable oils to biomethane involves hydrolysis, acidogenesis, acetogenesis and methanogenesis. The long-chain fatty acids subjected to the first two steps are converted to short-chain volatile fatty acids. The short-chain volatile fatty acids are then converted to acetate, carbon dioxide and hydrogen by the acetogenic bacteria. The products of acetogenesis are then converted into biomethane by the action of methanogenic bacteria.

The accumulation of long-chain fatty acids in the anaerobic digester inhibits the growth and action of bacteria and poses limitations in the generation of biomethane. Several strategies are practised for relieving this inhibition by long-chain fatty acids. Co-digestion of waste oils with other feedstocks is one of the strategies to overcome the inhibition of long-chain fatty acids. Intermittent feeding of waste oils into the anaerobic digester thus allowing the bacteria to regain their growth and metabolism has proved effective in circumventing the diffusional limitations posed by large lipids. Employing suitable adsorbents also can curtail the inhibitory effects of long-chain fatty acids (Kurade et al. 2020).

Waste cooking oil can be used along with sewage sludge and subjected to anaerobic digestion for the production of biomethane. The addition of waste cooking oil to the sewage sludge and slurry from farmhouses improves the methane yield and renders the process cost-effective. Marchetti et al. (2019) reported an increase in the biomethane yield by 15.5% on co-digestion of waste cooking oil with pig slurry in an anaerobic bioreactor. Similarly, Oliveira et al. (2018) also reported a boost in the biomethane yield on co-digestion of waste frying oil along with sewage sludge.

Acclimatizing the microbial communities prior to introducing into anaerobic digester can enhance the biomethane yield several times when compared to the unacclimatized cultures. This is because the acclimatized microbial cultures can fine tune the metabolic processes that facilitate the breakdown of large lipids and generate biomethane. Kurade et al. (2020) reported that the biomethane yield obtained from fats, oils and grease employing acclimatized cultures was 65–76% higher than the yield obtained with unacclimatized cultures. The increase in the biomethane generation with acclimatization of the microbial consortium can be attributed to the selective increase in the population of lipid-degrading microbes such as Firmicutes, Bacteroidetes, Synergistetes and Euryarchaeota (Kurade et al. 2020).

PHYSICOCHEMICAL PROPERTIES OF BIODIESEL FROM WASTE VEGETABLE OILS

The physicochemical properties of the biofuels produced from vegetable oils such as acid value, fuel density, kinematic viscosity, flash point, pour point, cloud point, cetane number, etc., have to comply with the standards specified by the American Society for Test and Materials (ASTM) (Table 6.4) (Altalhi et al. 2022). The physicochemical properties and fatty acid profile of the waste vegetable oils play a decisive role in adopting the techniques for their value-addition (Awogbemi et al. 2021). The vegetable oils used for cooking are subjected to heating at temperatures as high as 150 to 200°C, which can alter their physical properties such as colour, viscosity and density. The high-temperature heating also affects the chemical properties such as acid value, iodine value, peroxide value, saponification value and polar nature due to the alteration of their structure, degree of saturation/unsaturation and fatty acid composition (Awogbemi et al. 2021). During the frying of food items, the moisture content may also infiltrate the oil and contaminate it.

Table 6.4 ASTM D-6751 specifications of fuel characteristics for biodiesel fuel (B100) blend stock

Fuel characteristic	ASTM method	Specified limit	Permissible value
Flash point	D93	Minimum	130.0°C
Water and sediment	D2709	Maximum	0.050 Vol%
Kinematic viscosity at 40°C	D445	Range	1.9–6.0 mm^2s^{-1}
Sulphated ash	D874	Maximum	0.020 mass%
Sulphur emission	D5453	Maximum	0.05 mass%
Cetane number	D613	Minimum	47 min
Cloud point	D2500	Range	–3 to 12°C
Carbon residue	D4530	Maximum	0.050 mass%
Acid number	D664	Maximum	0.80 mg KOH/g
Oxidation stability	—	Minimum	3 h
Distillation temperature	D1160	Maximum	360°C

Source: https://afdc.energy.gov/fuels/biodiesel_specifications.html

The biodiesel has to be compatible with the fuel specifications of the engine to safeguard its efficient performance (Valente et al. 2011). Biodiesel is considered a good substitute for petroleum-based diesel because of its several beneficial properties as a fuel for automotive and machinery. Ideally, any fuel should have sufficient fluidity at the operating temperature to facilitate an efficient engine performance which is achievable when it is wax-free. In addition, effective combustion with comparatively fewer emissions is the desirable characteristic of biodiesel (Valente et al. 2011). In addition to lower carbon emission, the biodiesel from vegetable oils has reduced viscosity, increased flash point and cetane number (Brahma et al. 2022). Moreover, faster ignition and higher combustion efficiency of biodiesel also prolong the life of engines when used either alone or in combination with petroleum-based diesel. The quickness of fuel combustion in a diesel engine is a measure of its ignition quality and is indicated by the cetane number and cetane index of biodiesel. The cetane index and cetane number of biodiesel are dependent on its fuel density and volatility or vaporization. The biofuels with a lower degree of unsaturation have higher cetane numbers. Correspondingly, the biodiesel from waste coconut oil will have a higher cetane number than the biodiesel from castor oil or soybean oil which have higher amounts of unsaturated hydrocarbons (Altalhi et al. 2022). The minimum cetane number recommended by UNE-EN 14214 and ASTM D6751-02 are 51 and 47, respectively (Saeed et al. 2021). The source of cooking oil and the process conditions to which the oil is exposed while cooking impact the molecular structure, fuel density, vaporization and cetane number of biodiesel.

The saturated fatty acid content of vegetable oil influences the kinematic viscosity and combustion efficiency of biodiesel. In advance of combustion, the fuel should be sufficiently vaporized and atomized, for which adequate lubrication of the fuel injection system is obligatory. As viscosity increases, fuel vaporization and atomization decrease and reduce its combustion efficiency. For instance, the biodiesel generated from crude palm oil contains more saturated bonds which cause the quick solidification of oil at high altitudes where the temperature is low causing the filters to clog and damaging the engines of vehicles. Hence, palm oil biodiesel is preferred for use along with petroleum-based diesel with a blended rate not higher than B5 (up to 5% biodiesel). In Brazil and USA, the permissible blending of waste cooking oil biodiesel with diesel oil is up to 25% (Valente et al. 2011). The viscosity of fuel blends increases with an increase in the concentration of biodiesel. The maximum recommended viscosity of biodiesel blends according to European standards is $5.0 \times 10^{-6}\ m^2s^{-1}$ (Valente et al. 2011). The presence of long-chain hydrocarbons and oxygenated hydrocarbons increases the viscosity of biofuels while increasing the unsaturation of hydrocarbons can reduce its viscosity (Altalhi et al. 2022).

Flash point is an important characteristic of any fuel and denotes the minimum ignition temperature on exposure to a source of ignition (Arumugam and Ponnusami 2019). Highly volatile fuels have a lower flash point and a higher risk of fire peril. The biofuels with higher flash points characteristically have longer chain hydrocarbons that require higher temperatures to undergo vaporization and flashing (Altalhi et al. 2022) which is an important safety attribute during transportation and storage of biofuels. The flash point of biodiesel synthesized by catalytic pyrolysis of waste edible oil was reported to be 45°C (Altalhi et al. 2022), whereas the flash point of algal biofuel reported was 98°C (Vijayaraghavan and Hemanathan 2009). The pour point denotes the temperature at which a fuel ceases to flow when the temperature drops. The biofuels with a high degree of unsaturated hydrocarbons have low point values. On the other hand, the biodiesel made from palm oil and Jatropha oil have high pour points which is attributable to their low degree of unsaturation. The cloud point denotes the lowest temperature at which the biodiesel starts gel formation or crystal formation. The biofuels with highly unsaturated hydrocarbons have higher cloud points and *vice versa* (Altalhi et al. 2022). The acid value is an indication of the amount of free fatty acids in the oil and is measured as the amount of KOH in milligrams needed to neutralize the free fatty acids in the oil. The iodine value is a measure of the amount of unsaturated bonds in the FAMEs.

Another important chore for oil companies is the reduction of sulphur emissions while fuel combustion in engines and automobiles. According to the endorsed stipulations in Brazil, Europe and USA, the permissible sulphur content in fuel is 50 ppm, 10 ppm and 15 ppm, respectively (Valente et al. 2011). Diesel combustion results in higher emissions of sulphur which has several environmental impacts such as acid rain. On the contrary, blending diesel oil with biodiesel can reduce sulphur emissions considerably. The sulphur emission reduces in proportion to the increase in biodiesel concentration in the blended fuel, whereas the presence of sulphur is insignificant in pure biodiesel. Valente et al. (2011) suggested that blending waste cooking oil biodiesel with diesel fuel at a concentration of 83% reduced the sulphur content to less than 50 ppm. High oxygen content in biodiesel can reduce particulate emissions while combustion in compression ignition engines.

CONCLUSION

Generating biofuels such as biodiesel and bio-kerosene from waste vegetable oils is a sustainable strategy to meet the increasing fuel demand of the world population and to mitigate the menace of disposing of waste cooking oils. Biofuels which are compliant with the ASTM standards can be used as transportation fuels and industrial fuels. Given the declining global reliance on fossil fuels, biofuels from waste vegetable oils are promising candidates to replace petroleum-based fuels.

REFERENCES

Almeida, F.L.C., Travália, B.M., Gonçalves, I.S. and Forte, M.B.S. 2021. Biodiesel production by lipase-catalyzed reactions: bibliometric analysis and study of trends. Biofuels, Bioproducts and Biorefining. 15: 1141–1159. https://doi.org/10.1002/bbb.2183.

Altalhi, A.A., Morsy, S.M., Abou Kana, M.T.H., Negm, N.A. and Mohamed, E.A. 2022. Pyrolytic conversion of waste edible oil into biofuel using sulphonated modified alumina. Alexandria Engineering Journal. 61: 4847–4861. https://doi.org/10.1016/j.aej.2021.10.038.

Anis, S., Alhakim, R., Wahyudi, Khoiron, A.M. and Kusumastuti, A. 2021. Microwave-assisted pyrolysis and distillation of cooking oils for liquid bio-fuel production. Journal of Analytical and Applied Pyrolysis. 154: 105014. https://doi.org/10.1016/j.jaap.2020.105014.

Anisha, G.S. and Shyni Raphel, M. 2023. Cracking upgrading process of biomass. pp. 1–10. *In*: Rahimpour, Md. Reza (ed.). Encyclopedia of Renewable Energy, Sustainability, and the Environment, Vol. 1. Elsevier. https://doi.org/10.1016/B978-0-323-93940-9.00057-8.

Arumugam, A. and Ponnusami, V. 2019. Biodiesel production from *Calophyllum inophyllum* oil a potential non-edible feedstock: an overview. Renewable Energy. 131: 459–471. https://doi.org/10.1016/j.renene.2018.07.059.

Asiedu, A., Barbera, E., Naurzaliyev, R., Bertucco, A. and Kumar, S. 2021. Waste cooking oil to jet-diesel fuel range using 2-propanol via catalytic transfer hydrogenation reactions. Biofuels. 12: 723–736. https://doi.org/10.1080/17597269.2018.1532754.

Awogbemi, O., Kallon, D.V. Von, Aigbodion, V.S. and Panda, S. 2021. Advances in biotechnological applications of waste cooking oil. Case Studies in Chemical and Environmental Engineering. 4: 100158. https://doi.org/10.1016/j.cscee.2021.100158.

Basumatary, B., Nath, B., Kalita, P., Das, B. and Basumatary, S. 2020. Yellow oleander (*Thevetia peruviana*) seed as a potential bioresource for industrial applications. Mini-Reviews in Organic Chemistry. http://dx.doi.org/10.2174/1570193X17666191230122142.

Brahma, S., Nath, B., Basumatary, B., Das, B., Saikia, P., Patir, K., et al. 2022. Biodiesel production from mixed oils: a sustainable approach towards industrial biofuel production. Chemical Engineering Journal Advances. 10: 100284. https://doi.org/10.1016/j.ceja.2022.100284.

Cavalcanti-Oliveira, E.A., da Silva, P.R., Ramos, A.P., Aranda, D.A.G. and Freire, D.M.G. 2011. Study of soybean oil hydrolysis catalyzed by *Thermomyces lanuginosus* lipase and its application to biodiesel production *via* hydroesterification. Enzyme Research. 8. https://doi.org/10.4061/2011/618692.

Christopher, L.P., Zambare, V.P., Zambare, A., Kumar, H. and Malek, L. 2015. A thermo-alkaline lipase from a new thermophile *Geobacillus thermodenitrificans* AV-5 with potential application in biodiesel production. Journal of Chemical Technology and Biotechnology. 90: 2007–2016.

Degfie, T.A., Mamo, T.T. and Mekonnen, Y.S. 2019. Optimized biodiesel production from Waste Cooking Oil (WCO) using Calcium Oxide (CaO) nano-catalyst. Scientific Reports. 9: 1–8. https://doi.org/10.1038/s41598-019-55403-4.

El-Araby, R., Abdelkader, E., El Diwani, G. and Hawash, S.I. 2020. Bio-aviation fuel *via* catalytic hydrocracking of waste cooking oils. Bulletin of the National Research Centre. 44. https://doi.org/10.1186/s42269-020-00425-6.

El-Araby, R., Ibrahim, M.A., Abdelkader, E. and Ismail, E.H. 2022. (Co/Zn) Al_2O_4 nano catalyst for waste cooking oil catalytic cracking. Scientific Reports. 12: 1–13. https://doi.org/10.1038/s41598-022-10596-z.

Erchamo, Y.S., Mamo, T.T., Workneh, G.A. and Mekonnen, Y.S. 2021. Improved biodiesel production from waste cooking oil with mixed methanol–ethanol using enhanced eggshell-derived CaO nano-catalyst. Scientific Reports. 11: 1–12. https://doi.org/10.1038/s41598-021-86062-z.

Gad, M.S., Abu-Elyazeed, O.S., Mohamed, M.A. and Hashim, A.M. 2021. Effect of oil blends derived from catalytic pyrolysis of waste cooking oil on diesel engine performance, emissions and combustion characteristics. Energy. 223: 120019. https://doi.org/10.1016/j.energy.2021.120019.

Gao, B., Su, E., Lin, J., Jiang, Z.B., Ma, Y. and Wei, D. 2009. Development of recombinant Escherichia coli whole-cell biocatalyst expressing a novel alkaline lipase-coding gene from *Proteus* sp. for biodiesel production. Journal of Biotechnology. 139: 169–175. https://doi.org/10.1016/j.jbiotec.2008.10.004.

Hájek, M., Vávra, A., de Paz Carmona, H. and Kocík, J. 2021. The catalysed transformation of vegetable oils or animal fats to biofuels and bio-lubricants: a review. Catalysts. 11: 1–30. https://doi.org/10.3390/catal11091118.

Hanif, M., Bhatti, I.A., Zahid, M. and Shahid, M. 2022. Production of biodiesel from non-edible feedstocks using environment friendly nano-magnetic Fe/SnO catalyst. Scientific Reports. 12: 1–14. https://doi.org/10.1038/s41598-022-20856-7.

Jaichandar, S. and Annamalai, K. 2012. Effects of open combustion chamber geometries on the performance of pongamia biodiesel in a DI diesel engine. Fuel. 98: 272–279. https://doi.org/10.1016/j.fuel.2012.04.004.

Kumari, V., Shah, S. and Gupta, M.N. 2007. Preparation of biodiesel by lipase-catalyzed transesterification of high free fatty acid containing oil from *madhuca indica*. Energy and Fuels. 21: 368–372. https://doi.org/10.1021/ef0602168.

Kumari, M., Padhi, S., Sharmas S., Phukon, L.C., Singh, S.P. and Rai, A.K. 2021. Biotechnological potential of psychrophilic microorganisms as the source of cold-active enzymes in food processing applications. 3 Biotech. 11(11): 479.

Kurade, M.B., Saha, S., Kim, J.R., Roh, H.S. and Jeon, B.H. 2020. Microbial community acclimatization for enhancement in the methane productivity of anaerobic co-digestion of fats, oil, and grease. Bioresource Technology. 296: 122294. https://doi.org/10.1016/j.biortech.2019.122294.

Li, X., Qian, P., Wu, S.G. and Yu, H.Y. 2014. Characterization of an organic solvent tolerant lipase from *Idiomarina* sp. W33 and its application for biodiesel production using *Jatropha* oil. Extremophiles. 18: 171–178. https://doi.org/10.1007/s00792-013- 0610-0.

Li, H.L. and Yu, P.H. 2015. Conversion of waste cooking oils into environmentally friendly biodiesel. Springerplus. 4: 1–2. https://doi.org/10.1186/2193-1801-4-s2-p7.

Lopes, M., Miranda, S.M. and Belo, I. 2020. Microbial valorization of waste cooking oils for valuable compounds production – a review. Critical Reviews in Environmental Science and Technology. 50: 2583–2616. https://doi.org/10.1080/10643389.2019.1704602.

Luo, Y., Zheng, Y., Jiang, Z., Ma, Y. and Wei, D. 2006. A novel psychrophilic lipase from *Pseudomonas fluorescens* with unique property in chiral resolution and biodiesel production *via* transesterification. Applied Microbiology and Biotechnology. 73: 349–355. https://doi.org/10.1007/s00253-006-0478-3.

Marchetti, R., Vasmara, C. and Fiume, F. 2019. Pig slurry improves the anaerobic digestion of waste cooking oil. Applied Microbiology and Biotechnology. 103: 8267–8279. https://doi.org/10.1007/s00253-019-10087-8.

Martínez-Sanchez, J.A., Arana-Peña, S., Carballares, D., Yates, M., Otero, C. and Fernandez-Lafuente, R. 2020. Immobilized biocatalysts of Eversa® Transform 2.0 and lipase from *Thermomyces lanuginosus*: comparison of some properties and performance in biodiesel production. Catalysts. 10: 1–19. https://doi.org/10.3390/catal10070738.

Nascimento, L., Ribeiro, A., Ferreira, A., Valério, N., Pinheiro, V., Araújo, J., et al. 2022. Turning waste cooking oils into biofuels—valorization technologies: a review. Energies. 15. https://doi.org/10.3390/en15010116.

Oliveira, J.V, Duarte, T., Costa, J.C., Cavaleiro, A.J., Pereira, M.A. and Alves, M.M. 2018. Improvement of biomethane production from sewage sludge in co-digestion with glycerol and waste frying oil, using a design of experiments. BioEnergy Research. 11: 763–771. https://doi.org/10.1007/s12155-018-9933-0.

Ong, H.C., Masjuki, H.H., Mahlia, T.M.I., Silitonga, A.S., Chong, W.T. and Yusaf, T. 2014a. Engine performance and emissions using *Jatropha curcas*, *Ceiba pentandra* and *Calophyllum inophyllum* biodiesel in a CI diesel engine. Energy. 69: 427–445. https://doi.org/10.1016/j.energy.2014.03.035.

Ong, H.C., Silitonga, A.S., Mahlia, T.M.I., Masjuki, H.H. and Chong, W.T. 2014b. Investigation of biodiesel production from *Cerbera manghas* biofuel sources. Energy Procedia. 61: 436–439. https://doi.org/10.1016/j.egypro.2014.11.1143.

Phukon, L.C., Chourasia, R., Kumari, M., Godan, T.K., Sahoo, D., Parameswaran, B., et al. 2020. Production and characterisation of lipase for application in detergent industry from a novel *Pseudomonas helmanticensis* HS6. Bioresource Technology. 309: 123352.

Phukon, L.C., Chourasia, R., Padhi, S., Abedin, M.M., Godan, T.K., Parameswaran, B., et al. 2022. Cold-adaptive traits identified by comparative genomic analysis of a lipase-producing *Pseudomonas* sp. HS6 isolated from snow-covered soil of Sikkim Himalaya and molecular simulation of lipase for wide substrate specificity. Current Genetics. 68(3–4): 375–391.

Saeed, A., Hanif, M.A., Nawaz, H. and Qadri, R.W.K. 2021. The production of biodiesel from plum waste oil using nano-structured catalyst loaded into supports. Scientific Reports. 11: 1–18. https://doi.org/10.1038/s41598-021-03633-w.

Salaheldeen, M., Mariod, A.A., Aroua, M.K., Rahman, S.M.A., Soudagar, M.E.M. and Fattah, I.M.R. 2021. Current state and perspectives on transesterification of triglycerides for biodiesel production. Catalysts. 11: 1121. https://doi.org/10.3390/catal11091121.

Shah, S., Sharma, S. and Gupta, M.N. 2004. Biodiesel preparation by lipase-catalyzed transesterification of *Jatropha* oil. Energy and Fuels. 18: 154–159. https://doi.org/10.1021/ef030075z.

Shitao, Y., Cao, X., Wu, S., Chen, Q., Li, L. and Li, H. 2020. Effective pyrolysis of waste cooking oils into hydrocarbon rich biofuel on novel mesoporous catalyst with acid and alkali coexisting. Industrial Crops and Products. 150: 112362. https://doi.org/10.1016/j.indcrop.2020.112362.

Tamoradi, T., Kiasat, A.R., Veisi, H., Nobakht, V. and Karmakar, B. 2022. RSM process optimization of biodiesel production from rapeseed oil and waste corn oil in the presence of green and novel catalyst. Scientific Reports. 12: 1–15. https://doi.org/10.1038/s41598-022-20538-4.

Thakur, N., Jalalah, M., Alsareii, S.A., Harraz, F.A., Salama, E.S., Sharma, M., et al. 2022. Waste cooking oils (WCOs) to biogas nexus: kinetics, active microbes, and functional enzymes. Fuel. 330: 125525. https://doi.org/10.1016/j.fuel.2022.125525.

Toldrá-Reig, F., Mora, L. and Toldrá, F. 2020. Developments in the use of lipase transesterification for biodiesel production from animal fat waste. Applied Sciences. 10(15): 5085. https://doi.org/10.3390/app10155085.

Valente, O.S., Pasa, V.M.D., Belchior, C.R.P. and Sodré, J.R. 2011. Physical-chemical properties of waste cooking oil biodiesel and castor oil biodiesel blends. Fuel. 90: 1700–1702. https://doi.org/10.1016/j.fuel.2010.10.045.

Vijayaraghavan, K. and Hemanathan, K. 2009. Biodiesel production from freshwater algae. Energy and Fuels. 23: 5448–5453.

Vivek, K., Sandhia, G.S. and Subramaniyan, S. 2022. Extremophilic lipases for industrial applications: a general review. Biotechnology Advances. 60: 108002. https://doi.org/10.1016/j.biotechadv.2022.108002.

Why, E.S.K., Ong, H.C., Lee, H.V., Chen, W.H., Asikin-Mijan, N., Varman, M., et al. 2022. Single-step catalytic deoxygenation of palm feedstocks for the production of sustainable bio-jet fuel. Energy. 239: 122017. https://doi.org/10.1016/j.energy.2021.122017.

Wu, Q., Wang, Y., Peng, Y., Ke, L., Yang, Q. and Jiang, L. 2020. Microwave-assisted pyrolysis of waste cooking oil for hydrocarbon bio-oil over metal oxides and HZSM-5 catalysts. Energy Conversion and Management. 220: 113124. https://doi.org/10.1016/j.enconman.2020.113124.

Yang, W., He, Y., Xu, L., Zhang, H. and Yan, Y. 2016. A new extracellular thermo-solvent-stable lipase from *Burkholderia ubonensis* SL-4: Identification, characterization and application for biodiesel production. Journal of Molecular Catalysis B: Enzymatic. 126: 76–89. https://doi.org/10.1016/j.molcatb.2016.02.005.

Yoo, H.Y., Simkhada, J.R., Cho, S.S., Park, D.H., Kim, S.W., Seong, C.N., et al. 2011. A novel alkaline lipase from *Ralstonia* with potential application in biodiesel production. Bioresource Technology. 102: 6104–6111. https://doi.org/10.1016/j.biortech.2011.02.046.

Zhang, S., Yang, X., Zhang, H., Chu, C., Zheng, K., Ju, M., et al. 2019. Liquefaction of biomass and upgrading of bio-oil: a review. Molecules. 24: 1–30. https://doi.org/10.3390/molecules24122250.

Zhang, J., Chen, H., Wang, Z., Xu, H., Luo, W., Xu, J., et al. 2021. Heat-induced overexpression of the thermophilic lipase from *Bacillus thermocatenulatus* in *Escherichia coli* by fermentation and its application in preparation of biodiesel using rapeseed oil. Biotechnology and Applied Biochemistry. 69: 1812–1820. https://doi.org/10.1002/bab.2247.

Chapter 7

Recent Advancements for Management of Biomedical Wastes

Sonia Yesenia Silva Belmares[1], Luis Cobos Puc[1], Hilda Aguayo Morales[1], Crystel Aleyvick Sierra Rivera[1] and Maria del Carmen Rodriguez Salazar*[1]

[1]School of Chemistry, Autonomous University of Coahuila, Blvd. Venustiano Carranza, Col. Republica Oriente, 25280, Saltillo, Coahuila, México.

INTRODUCTION

According to the World Health Organization (WHO), medical waste encompasses waste produced by healthcare-related activities, encompassing a broad spectrum from used needles and syringes to soiled dressings, body parts, diagnostic samples, blood, chemicals, pharmaceuticals, medical devices, and radioactive materials (Chartier et al. 2014). In accordance with the guidelines provided by the Environmental Protection Agency (EPA), medical waste pertains to waste generated within healthcare facilities, which falls under the category of healthcare waste contaminated with bodily fluids, blood, or other substances that have the potential to transmit infections. This form of waste is frequently denoted as regulated medical waste (EPA 2016).

Out of the overall waste produced by medical practices, 15% is categorized as hazardous waste, which may encompass infectious, toxic, or radioactive materials.

* For Correspondence: carmen_rodriguez@uadec.edu.mx

The remaining 85% constitutes general waste and is classified as non-hazardous waste. Various healthcare institutions, such as hospitals, healthcare facilities, research centers, mortuaries, autopsy centers, animal testing laboratories, blood banks, collection services, and elderly nursing homes, contribute significantly to the generation of healthcare waste. In developed nations, hospitals typically generate around 0.5 kilograms of hazardous waste per bed on a daily basis. In contrast, in underdeveloped countries, the estimated amount of hazardous waste produced per hospital bed is approximately 0.2 kilograms (Chartier et al. 2014).

Unfortunately, some countries do not adequately manage the collection of biomedical waste, which can put people's health at risk by exposing them to this waste. Shockingly, inadequate hospital waste management leads to 5.2 million deaths worldwide yearly (Rahman et al. 2020).

REGULATION OF BIOMEDICAL WASTE

The regulation of waste stemming from healthcare activities is of utmost importance, as it determines the proper handling of waste to prevent harm to both health and the environment during its final disposal. Nonetheless, various factors contribute to inadequate or subpar waste management. The primary factors include a lack of awareness regarding the risks posed by hazardous medical waste to both health and the environment, insufficient or deficient information on medical waste management, the absence of waste management and disposal systems, limited financial resources, a shortage of personnel, and the insufficient attention given to waste management issues. Additionally, in many countries, either appropriate regulations are lacking, or they are not effectively enforced.

In 1999, the World Health Organization (WHO) and its European Center for Environment and Health collaborated to produce an all-encompassing guidance document on healthcare waste management. The document was developed in Nancy, France, with a specific focus on addressing the challenges of biomedical waste management in developing countries. The primary objective of this guide was to provide governments and healthcare establishments across the globe with a framework to create their own guidelines. The guide aims to reduce health problems and eliminate potential risks for health personnel by establishing programs to manage medical waste properly. In addition to affecting public health, it also greatly impacts the environment; this guide is crucial for protecting the environment from the harmful effects of biomedical waste. Subsequently, in 2014, the second edition of this guide was published to expand and update the knowledge that had been collected since the first edition came out on the requirements of healthcare waste generators and new methodologies (Chartier et al. 2014, Singh et al. 2022).

WASTE CLASSIFICATION

Medical waste is typically classified into two main categories: hazardous and non-hazardous waste (Table 7.1). Hazardous waste poses significant risks to both human health and the environment, whereas non-hazardous waste is considered harmless. The World Health Organization (WHO) further categorizes medical waste into eight distinct categories, which encompass infectious waste, pathological waste, sharps waste, chemical waste, pharmaceutical waste, cytotoxic waste, radioactive waste, and general waste. This categorization helps to address the diverse nature of medical waste and tailor appropriate disposal methods for each type (Chartier et al. 2014, Singh et al. 2022).

Table 7.1 Classification of healthcare refuse

Types of waste	Definitions and samples
Hazardous healthcare waste	
Puncture waste	New or used sharp items
Infectious waste	Waste with potential pathogen presence and disease transmission risk
Pathological waste	Human tissue, organ, or fluid samples; body segments; fetal tissues; and surplus blood products.
Pharmaceutical waste	Pharmaceuticals that have expired or are no longer required; objects polluted by or housing pharmaceutical compounds.
Cytotoxic waste	Waste with cytotoxic substances that have genotoxic properties
Chemical waste	Waste with chemical components
Radioactive waste	Waste with radioactively active agents
Non-hazardous or general healthcare waste	Waste lacking any discernible biological, chemical, radioactive, or physical hazards.

Source: Data from Y. Chartier et al. 2014. "Safe management of wastes from healthcare activities", World Health Organization, 2nd ed., 4.

BIOMEDICAL WASTE MANAGEMENT

Waste management comprises a set of principles that aid organizations in systematically and cost-effectively overseeing the entire waste cycle, encompassing collection through disposal, all the while minimizing the environmental footprint. This is achieved by continuously reviewing, evaluating, and enhancing their environmental performance through an Environmental Management System (EMS). The EMS handles various aspects of environmental management and is built on the principles of the 3Rs—reduce, reuse, and recycle—which promote the sustainable utilization of resources (Table 7.2).

At present, the effective environmental management of biomedical waste represents a significant challenge, particularly for developing nations. Due to the increase in the world population, it is increasingly difficult to face this problem since the services that exist for treating this biomedical waste are insufficient. Managing biomedical waste is a significant challenge, particularly for developing nations, due to the rise in global population. The services for treating this waste must be improved to address this problem effectively. Institutions responsible for

Table 7.2 Key points and benefits of establishing an EMS

The key points that an EMS must contain are:	The benefits of establishing an EMS in an organization are:
Analyze environmental impacts and compliance obligations	Improved environmental performance
Review the environmental goals of the organization	Enhanced compliance
Establish requirements and goals to reduce environmental impact	Pollution prevention
Establish programs to meet the established requirements and goals	Resource conservation
Supervise the achievement of the objectives set	New customers/markets
Ensure the environmental awareness of employees	Increased efficiency/reduced costs
Review EMS progress and make improvements	Improved employee morale
	Enhanced reputation with the public, regulators, lenders, and investors.
	Employee consciousness of environmental concerns and obligations.

Source: Data from United States Environmental Protection Agency. 2022. https://www.epa.gov/ems

treating and managing patients often encounter challenges where their staff lacks knowledge about biomedical waste management. This issue can result in hazardous waste negatively affecting the environment and health. A recent qualitative study conducted in two hospitals in Pakistan found that only 35.4% of the hospital staff knew the regulations surrounding biomedical waste. Additionally, the study pointed out that there is a shortage of personnel training in waste management, and the proper procedures for its treatment and disposal are not being followed (Chartier et al. 2014, Hooshmand et al. 2020, Lee and Lee 2022, Shabani and Jerie 2023).

HEALTH RISKS OF INADEQUATE WASTE MANAGEMENT

The working personnel of places that provide medical services for health sometimes need to be trained in managing and disposing of medical waste, which increases the health risks due to bad practices.

The people with the highest risk of exposure to waste are:

- Doctors, nurses, healthcare aides, and maintenance staff
- Patients within healthcare facilities or receiving home care
- Visitors to healthcare institutions
- Support service employees, including janitors, laundry workers, and porters
- Personnel involved in waste transportation to treatment or disposal facilities
- Workers at waste management facilities (Chartier et al. 2014)

The waste may contain potentially infectious pathogenic microorganisms that can infect and harm the health of hospital patients, health personnel, and visitors. Other common risks to which one is exposed due to the inadequate management of biomedical waste are sharp injuries contaminated with potentially infectious microorganisms, exposure, and contact with chemical products that are harmful to health. These risks can manifest during waste handling, collection, and disposal. The chosen method for disposing of the residues also poses risks to the individuals involved in disposal, the operators of incineration equipment, and the general public. It can cause damage to health. For example, during the incineration of waste, particles are produced that are released into the ambient air and are breathed by living beings. The operation of incineration machines by untrained personnel increases the risk of thermal and radiation burns. Finally, there is a risk of chemical burns that can occur during the sterilization, disinfection, or treatment processes of biomedical waste (Chartier et al. 2014).

Risks of Inadequate Waste Management for the Environment

Improper handling and disposal of medical waste poses significant environmental hazards. It can lead to soil and water contamination and harm or even kill crucial ecological microorganisms (Singh et al. 2022, Kanwar et al. 2023). For example, burying or dumping biomedical waste in sites not designated for these activities causes direct soil and water contamination. Poor wastewater and sludge management can contaminate water and soil with harmful chemicals or pathogenic microorganisms.

The production of biomedical waste contributes to pollution and damage to the environment. Therefore, looking for strategies that minimize waste production as much as possible is crucial. For example, purchase material that does not have so much packaging or purchase easily recyclable plastic material, such as polyethylene, polyethylene terephthalate, and polypropylene, minimizing the use of difficult-to-recycle plastics such as polyvinyl chloride (PVC) and polycarbonate, which is made from bisphenol A, which has harmful health effects. Therefore, latex and silicone are the materials that should be used as a replacement for PVC in medical materials (Lee and Lee 2022, Shabani and Jerie 2023).

Recycling is a widely used practice in various organizations and is an effective way to minimize the production of biomedical waste. It's important to note that medical materials can be safely reused after disinfection processes that significantly diminish the risk of infection transmission.

- Disposable material (reutilize avoided)
- Medical supplies that do not present a risk of cross-infection
- Medical material designed for reuse.

Material recycling is associated with a series of steps: cleaning, decontamination, reconditioning, disinfection, and sterilization. Hazardous hospital waste should not reutilized, but in some cases, the material from which it is made can be recycled after sterilization—for example, plastic from syringes and catheters.

There are two methods for processing collected waste. The first involves the institution organizing the material to reuse it as fuel or energy for generating heat or electricity. The second consists of transforming the recovered waste into new products, which requires sending the material to recycling plants (Manzoor and Sharma 2019).

BIOMEDICAL WASTE TREATMENT

Currently, there are different techniques for the treatment of medical waste. However, selecting an appropriate method depends on economic resources, easy and safe handling, care for the environment, and the safety of living beings (Liu et al. 2015).

When selecting a method for waste disposal, it is essential to consider the physicochemical characteristics of the waste. These characteristics help establish recycling plans for non-hazardous waste and set equipment specifications or operation parameters for treating hazardous waste material. Efficient management of biomedical waste requires various techniques and treatments to eliminate hazardous materials without harming living beings or the environment. Finding affordable and sustainable methods for treating and disposing of biomedical waste is crucial due to the large volume produced while minimizing environmental risks (Hooshmand et al. 2020).

Mechanical Treatment

Various mechanical treatments are employed in biomedical waste management to reduce waste volume by 60% or more. This facilitates the subsequent treatment of the waste with other techniques aimed at eliminating pathogenic microorganisms. Using this technique on solid biomedical waste additionally increases the surface area before its treatment with heat or chemicals. These techniques include granulation, spraying, shredding, grinding, mixing, crushing, and agitation, using equipment such as crushers, millers, shattering machines, and splinters (Chartier et al. 2014, Hooshmand et al. 2020).

Treatment with Sterilization

The process of sterilization involves the elimination of all microorganisms. In hospitals, pressure steam sterilization is used to treat waste. This method involves subjecting the waste to temperatures ranging from 121 to 149°C and 100 kPa for 20 minutes. Through this method, microorganisms are eliminated by causing protein coagulation and denaturation. Subsequently, the biomedical waste is either incinerated or disposed of in a sanitary landfill after disinfection.

It is also frequently used to disinfect contaminated clothing or sterilize culture media to isolate microorganisms. It is unsuitable for organic pathological biomedical waste and formaldehyde, mercury, phenol, and other substances. One of the downsides of this method is that the amount of waste remains the same, and it may release harmful gases or unpleasant odors that can harm the environment (Chartier et al. 2014, Xu et al. 2020, Kanwar et al. 2023).

Chemical Disinfection

The sterilization process entails the complete eradication of all microorganisms. In hospitals, pressure steam sterilization is used to treat waste. This method involves subjecting the waste to temperatures ranging from 121 to 149°C and 100 kPa for 20 minutes. Through this method, microorganisms are eliminated by causing protein coagulation and denaturation. Subsequently, the biomedical waste is either incinerated or disposed of in a sanitary landfill after disinfection. It is also frequently used to disinfect contaminated clothing or sterilize culture media to isolate microorganisms. It is unsuitable for organic pathological biomedical waste and formaldehyde, mercury, phenol, and other substances. One of the downsides of this method is that the amount of waste remains the same, and it may release harmful gases or unpleasant odors that can harm the environment.

The process of disinfection aims to reduce or eliminate the presence of microorganisms to a safe level. In biomedical waste management, disinfection with chemicals is a common practice. This method is used after the waste is treated using techniques that ensure maximum grinding, ensuring that the chemicals are in contact with every surface of the waste. The most common chemicals used for disinfection are sodium hypochlorite, chlorine compounds, peracetic acid, glutaraldehyde, ethylene oxide (EtO), ozone gas, and lime solution. Additionally, dry inorganic compounds like calcium oxide powder are used. EtO is widely used for material disinfection, medical waste treatment, and frequently used equipment that cannot be sterilized by methods that use heat or moisture and is effective for disinfection of liquid medical waste and pathological waste.

Encapsulating compounds can solidify materials like sharps, blood, or other bodily fluids into a solid matrix prior to disposal. They have the advantage of being a fast, easy-to-use, and low-cost technique. The disadvantages are that it is not a cost-effective technique for use in equipment and should not be used to treat biomedical waste disposed of in landfills. Not for use on chemotherapy residues, volatile wastes driven from organic compounds, or radioactive wastes. The disinfectant selection depends on the microorganism to be eliminated. On the other hand, some disinfectants are for a wide variety of microorganisms, and others for particular microorganisms (Chartier et al. 2014, Hooshmand et al. 2020, Xu et al. 2020).

Microwave

Microwave radiation is a technique that uses electromagnetic waves at high frequencies of 2450 MHz for disinfection of medical waste by generating heat due to the effect of microwaves on the cell membrane of microorganisms when absorbed by water molecules, proteins, and lipids of the structure of the microorganism, causing rapid vibrations of the molecules, causing death due to the increase in temperature.

Maximizing the treatment's effectiveness on the waste requires the initial use of mechanical methods to reduce its volume. In order to guarantee effective disinfection, it is necessary to moisten the dry residues by employing humidifiers in the microwave chambers, allowing sufficient time for exposure to the microwaves, and stirring the residues throughout the treatment process. The advantages are that it has a minimal environmental impact, the equipment is easy to operate and is automated, and it significantly reduces the volume of medical waste. The disadvantages are that the construction and operation of the equipment can be expensive. This substance cannot be utilized for treating hazardous materials or blood, and it may emit unpleasant smells (Hooshmand et al. 2020, Xu et al. 2020).

Gamma Radiation

It is a sterilization technique using gamma rays generated by radioactive isotopes such as cobalt to destroy bacterial DNA. The effectiveness of the technique depends on the energy released. The disadvantage is that the shadow effect can occur, which consists of a more significant sterilization of the side of the object that is in direct contact with the gamma rays (Hooshmand et al. 2020).

Photocatalysis

The use of photocatalysts in sterilization activities is highly effective in killing microorganisms. This technique relies on metal oxide nanostructure films, which can photo-oxidize organic substances like surfactants, alkanes, and alkenes. When it comes to photocatalysts, the metal oxides that are most frequently utilized are TiO_2, ZnO, alpha-Fe_2O_3, MoO_3, WO_3, ZeO_2, and SnO_2. Unfortunately, these compounds have proven ineffective in treating biomedical waste. However, researchers are currently exploring new compounds that may be more effective in eliminating this type of waste (Hooshmand et al. 2020).

Incineration

It is a thermal method carried out in incinerators that uses high and stable temperatures in conjunction with other specific conditions. Considered the best to destroy pathogenic microorganisms and dangerous organic waste, it significantly reduces the volume of waste. The disadvantages are that the incineration of materials not suitable for disposal by this method can produce and release polluting gases into the air, such as polychlorodibenzofurans (PCDD/Fs), polycyclic aromatic hydrocarbons (PAHs), CO_2, nitrogen oxides, gases from inorganic acids and metals. Aside from releasing ash into the atmosphere, it is also hazardous to the environment and living organisms. The associated expenses are significant, thus requiring the use of a gas purification system to prevent air pollution (Cai and Du 2020, Hooshmand et al. 2020, Xu et al. 2020, Kanwar et al. 2023).

Plasma Pyrolysis

A technique that uses very high temperatures with the plasma arc without oxygen produces complete waste decomposition. The residues are pyrolyzed to CO, H_2, and hydrocarbons when they come in contact with the plasma arc. In this technique, the gases are incinerated, raising the temperature above 1200°C. The key distinction lies in the fact that the hot gases are subsequently quenched, bringing the temperature down to a range between 500°C and 70°C to prevent the formation of dioxins and furans. The advantage is reducing the volume of waste, especially organic; however, it can be used with all kinds of waste, from hazardous to non-hazardous, and potential thermal energy can be recycled. The disadvantages are the high operation and construction cost, the system's stability can be easily affected, and the reliability of the technique must be improved and verified (Nema and Ganeshprasad 2002, Hooshmand et al. 2020, Xu et al. 2020, Guangcan et al. 2021).

Pyrolysis Incineration

This process involves subjecting materials to high temperatures ranging between 600°C and 900°C in an environment with minimal or no oxygen. Organic waste is converted into liquid and gaseous fuels. The gases produced by pyrolysis contain methane, hydrogen, CO, CO_2, and volatile organic compounds. For the treatment to be effective, the waste must meet several conditions, for example, the degree of moisture of the waste, the size of the waste, and the chemical structure of the waste. The advantages are that fewer harmful substances are produced, mainly dioxins. Medical waste must not be pre-sorted or pre-treated before the pyrolysis incineration process (Nema and Ganeshprasad 2002, Hooshmand et al. 2020, Xu et al. 2020, Guangcan et al. 2021).

GENERATION OF HEALTHCARE WASTE AMID THE PANDEMIC

The COVID-19 pandemic represented a challenge not only in health but also in caring for the environment and in the management of biomedical waste. During the pandemic, biomedical waste generation increased abruptly, mainly masks, protective equipment, gloves, blood collection bags, and plastic waste from clinical laboratories.

The increase was seen alarmingly in healthcare centers, clinical laboratories, and houses where there were sick and isolated people, which caused challenges in the disposal, storage, transport, and disposal of waste. These changes originated the approach of new strategies in managing biomedical waste by the world's countries. Moreover, new protocols were established for the proper disposal of waste during the pandemic, aiming to prevent contaminated waste from entering municipal waste streams, which could otherwise lead to environmental contamination and an elevated risk of contagion among personnel without adequate protection

(Rahman et al. 2020, Gowda et al. 2022). Various international institutions spearheaded these modifications in hospital waste management amid the COVID-19 pandemic. As the regulations recommend, biomedical waste contaminated with COVID-19 must be decontaminated by sterilization techniques at temperatures between 121 to 149°C or by spraying with 0.1% sodium hypochlorite and disposed of by incineration or shredding techniques (Kanwar et al. 2023).

It also increased the production of protective equipment, mostly made of non-biodegradable plastic, which caused the amount of medical waste and led to the generation of new strategies for its correct disposal or recycling. Developed countries primarily consume these protection products, but they have the advantage of policies for properly treating these residues, unlike other countries that have deficiencies in the management of biomedical waste, where waste is not disposed of correctly, which has consequences for health and the environment (Dehal et al. 2022).

CONCLUSION

The proper management of biomedical waste is a matter of great significance, as the consequences have a profound impact on the environment and the living organisms that inhabit it. Globally, there are situations where regulatory authorities do not accord the necessary importance to enforcing established standards for appropriate waste management. Surprisingly, one in every three healthcare facilities worldwide does not implement proper waste management, a problem that more frequently affects developing countries.

The absence of an established and proper management of medical waste in healthcare facilities and waste collection and disposal companies, following the guidelines of organizations like the WHO, poses a significant risk to health. When not handled by adequately trained personnel with appropriate protection, these wastes can lead to health hazards. Moreover, they can result in severe and irreversible environmental contamination, often due to a lack of knowledge and training in waste disposal. This can lead to the mixing of hazardous and non-hazardous waste, with subsequent detrimental effects on both public health and the environment.

The COVID-19 pandemic has dramatically reshaped the management of waste collection, storage, transportation, and disposal. The sudden surge in the production of medical waste worldwide has necessitated swift action to develop new strategies. This includes the implementation of infrastructure and reorganizing waste classification, storage, and disposal in healthcare facilities globally. All of this has been crucial to preventing the virus's spread while ensuring the complete destruction of viral particles.

The primary objective of the treatment and final disposal of biomedical waste is to reduce risks to both the environment and health. The selection of specific processes for waste treatment is determined by factors such as the characteristics of the material to be disposed of, the quantity of waste, volume, treatment efficiency, the facility's capacity to handle specific waste quantities, and the available technology for treatment.

The final disposal of biomedical waste occurs after treatment, and various methods are used depending on each country's resources. It is essential, however, that these methods meet the necessary criteria to prevent environmental contamination and adverse health effects.

For a long time, waste treatment and disposal processes have been employed for their efficiency and speed in waste disposal. However, they have also displayed drawbacks that often negatively impact environmental integrity. Therefore, it is crucial in today's context to seek waste management strategies that do not harm the environment or the health of living beings. Some of these strategies include:

- Raising awareness among countries worldwide about the importance of proper management and final disposal of biomedical waste to prevent negative impacts on the environment and health.
- Reducing improper waste management practices according to each country's capabilities.
- Implementing strategies based on waste prevention and minimization.
- Promoting recycling and environmental sustainability practices.

These measures are essential to ensure responsible waste management and to protect our environment and health. Hence, it is crucial for the responsible authorities in each country to take seriously the importance of establishing proper management of biological waste and implementing the various strategies that have been developed. The goal is to prevent irreversible damage to the environment and protect public health.

REFERENCES

Cai, X. and Du, C. 2021. Thermal plasma treatment of medical waste. Plasma Chemistry and Plasma Processing. 41(1): 1–46.

Chartier, Y., Emmanuel, J., Pieper, U., Prüss, A., Rushbrook, P., Stringer, R., et al. 2014. Safe management of wastes from health-care activities. WHO Press, Switzerland.

Dehal, A., Vaidya, A.N. and Kumar, A.R. 2022. Biomedical waste generation and management during COVID-19 pandemic in India: challenges and possible management strategies. Environmental Science and Pollution Research. 29(10): 14830–14845.

Gowda, N.R., Siddharth, V., Inquillabi, K. and Sharma, D.K. 2022. War on waste: challenges and experiences in COVID-19 waste management. Disaster Medicine and Public Health Preparedness. 16(6): 2358–2362.

Hooshmand, S., Kargozar, S., Ghorbani, A., Darroudi, M., Keshavarz, M., Baino, F., et al. 2020. Biomedical waste management by using nanophotocatalysts: the need for new options. Materials. 13(16): 3511.

Kanwar, V.S., Sharma, A., Rinku, Kanwar, M., Srivastav, A.L. and Soni, D.K. 2023. An overview for biomedical waste management during pandemic like COVID-19. International Journal of Environmental Science and Technology. 20(7): 8025–8040.

Lee, S.M. and Lee, D. 2022. Effective medical waste management for sustainable green healthcare. International Journal of Environmental Research and Public Health. 19(22): 14820.

Liu, H.C., You, J.X., Lu, C. and Chen, Y.Z. 2015. Evaluating health-care waste treatment technologies using a hybrid multi-criteria decision-making model. Renewable and Sustainable Energy Reviews. 41: 932–942.

Manzoor, J. and Sharma, M. 2019. Impact of biomedical waste on environment and human health. Environmental Claims Journal. 31(4): 311–334.

Nema, S.K. and Ganeshprasad, K.S. 2002. Plasma pyrolysis of medical waste. Current Science. 83(3): 271–278.

Rahman, M.M., Bodrud-Doza, M., Griffiths, M.D. and Mamun, M.A., 2020. Biomedical waste amid COVID-19: perspectives from Bangladesh. The Lancet. Global Health. 8(10): e1262.

Shabani, T. and Jerie, S. 2023. A review of the applicability of environmental management systems in waste management in the medical sector of Zimbabwe. Environmental Monitoring and Assessment. 195(6): 789.

Singh, N., Ogunseitan, O.A. and Tang, Y. 2022. Medical waste: current challenges and future opportunities for sustainable management. Critical Reviews in Environmental Science and Technology. 52(11): 2000–2022.

Su, G., Ong, H.C., Ibrahim, S., Fattah, I.M.R., Mofijur, M. and Chong, C.T. 2021. Valorisation of medical waste through pyrolysis for a cleaner environment: progress and challenges. Environmental Pollution. 279: 116934.

US EPA. 2016. Model guidelines for state medical waste management [Other Policies and Guidance]. https://www.epa.gov/rcra/model-guidelines-state-medical-waste-management.

Xu, L., Dong, K., Zhang, Y. and Li, H. 2020. Comparison and analysis of several medical waste treatment technologies. IOP Conference Series: Earth and Environmental Science. 615(1): 012031.

Chapter **8**

Innovative Technologies for Minimization of Plastic Wastes

Ranju Kumari Rathour[1], Ravi Kant Bhatia[2], Amit K. Rai[3], Jyoti S. Jadaun[4] and Sudhir Pratap Singh*[5]

[1]Department of Biotechnology, Chandigarh College of Technology, Chandigarh group of Colleges, Landran, Mohali, Punjab 140307, India.

[2]Department of Biotechnology, Himachal Pradesh University, Shimla-05, India.

[3] National Agri-Food and Biomanufacturing Institute (NABI), Sector 81, S.A.S Nagar, Mohali, Punjab, India.

[4]Department of Botany, Dayanand Girls' PG College, Civil Lines, Kanpur Nagar-208001, UP, India.

[5]Department of Industrial Biotechnology, Gujarat Biotechnology University, GIFT City, Gandhinagar, India.

INTRODUCTION

An omnipresent man-made substance built on organic substrates having high molecular weight and a rigid conformation is popularly known as "plastic". Notwithstanding its terminological *Greek* lineage, the word "plastic" has allegedly been derived from "pliable" also meaning easy to shape as it is somewhat true for the plastics which undergo different shaping processes before being embodied into the final product. Most plastics are made up of non-biodegradable synthetic polymers or long-chain hydrocarbons derived from petrochemicals (Shin et al. 2020). Historically, John Wesley Hyatt is considered as the inventor of the process

* For Correspondence: sudhir.singh@gbu.edu.in

of synthetic polymer synthesis, and its subsequent conversion into plastic after the treatment with cellulose (cotton fiber polymer) and camphor. The addition of cellulose enhances the properties of plastic in terms of versatility, durability, and robustness (Gilbert 2017). Plastic is widely used for different purposes including manufacturing goods, construction, packaging, toys, satellites, automobiles, and a myriad other disposable product, etc. (Idumah and Nwuzor 2019). Plastic has been an unavoidable part of daily chores due to its high thermal and electrical insulation power, resistance to chemicals, durability, and comparatively low cost. However, the indiscriminate use of plastic products has also resulted into various environmental hazards, which apparently need urgent statutory as well as scientific rectification efforts.

Since the onset of industrial revolution, millions of tons of plastics are generated every year to fulfill an ever-increasing demand. The majority of synthesized plastic (more than 40%) is used for packaging, storage, and transportation purposes. In 2018, the plastic production was 0.36 billion tons, and was expected to reach approximately 0.50 billion tons by the end of year 2025 (Facts 2019). According to the recent reports, about 390.7 metric ton production of plastic (with a 4% annual increment) was reported in the year 2021 (Statista 2023). In a somewhat similar trend, the worldwide plastic market observed a 3.7% compound annual growth rate (CAGR) in the year 2022, which was further anticipated to reach an approximate fiscal of USD885 billion by the end of financial year 2023 (Plastic Industry 2023). Notwithstanding the gigantic share of plastic in the total solid waste generated, only 8% of it gets recycled and 12% is incinerated for energy. However, about 80% gets unsustainably segregated into the landfills, riverine, ponds, etc., causing severe environmental pollution (Rafey and Siddiqui 2021).

The most severely impacted environmental habitats due to plastic wastes are the water bodies. Plastic litter is known to damage several ecological habitats. Being a synthetic or semi-synthetic polymer and having lipophilic stable configuration, plastic structures are generally resistant to water assisted degradation under natural conditions. Therefore, the plastic can sustain severe environmental conditions, and remains intact for several years. Moreover, the culmination of plastic into micro and nano plastics having a miniscule particle size in micro and nano meter through the course of slow natural coercion is an urgent emergency to tackle. The accumulation of micro/nano plastics has been reported in fresh water bodies, ponds, lakes, rivers, beaches, and oceans, etc. (Chatterjee and Sharma 2019, Kung et al. 2022). The accumulation of plastic has also been reported in human body (Yee et al. 2021). Previously, it was reported that from a single beverage cup, millions of nano-plastic particles were consumed by humans (Lai et al. 2022). The seepage of micro/nano plastics into water bodies is one of the emerging threats to the aquatic as well as terrestrial flora and fauna including humans (Chatterjee and Sharma 2019). A group of scientists critically studied the effects of virgin plastic especially micro and nano plastic on fishes and revealed that 32% of total area significantly showed alteration in intestinal normal flora, behavior, metabolism, and neurological functioning of fishes when exposed to virgin plastic (Jacob et al. 2020). The plastic can enter into food web in the form of micro or nano plastics causing severe physiological ailments (stress, apoptosis, inflammation, necrosis etc.) (Haegerbaeumer et al. 2019, Yee et al. 2021). It has been estimated that if the use of plastic continues with the similar pace without any robust technologies or scientific interventions for its post use disposal, approximately 12 billion tons of waste will accumulate in the nature by the end of year 2050 (Faraca and Astrup 2019). The management of non-recyclable plastic waste is the major concern due to their one time use (Idumah and Nwuzor 2019). The management of this waste in sustainable way is of greater concern; therefore, this chapter details the different innovative strategies (biodegradation, controlled thermal decomposition, and conversion of plastic into bioplastics) for recyclability and less perilous disposal of plastic waste.

COMPOSITION AND TYPES OF PLASTIC WASTE

According to the Resin Identification Code (RIC) designed by the Society of Plastic Industry, to differentiate between recycling, recovery, and reusability of materials, plastic waste can be divided into seven different types polyethylene (low and high-density): Polyethylene terephthalate (PET), Polypropylene (PP), Polystyrene, Polyvinyl chloride (PVC) and others as shown in Figure 8.1. The concentration of these seven type of wastes may vary depending on the source of generated waste (Areeprasert et al. 2017). The highest amount of waste is generated by the packaging sector which mainly uses polyethylene followed by polyethylene terephthalate and polypropylene (Kurniawati et al. 2021). As the content of polyethylene is highest (40–45%) in plastic waste (especially in municipal solid waste), it is further divided into Low-density (LDPE) and High-density polyethylene (HDPE). Besides this, excessive use of disposable items (polystyrene) is the second prime source of plastic wastes (Huang et al. 2022).

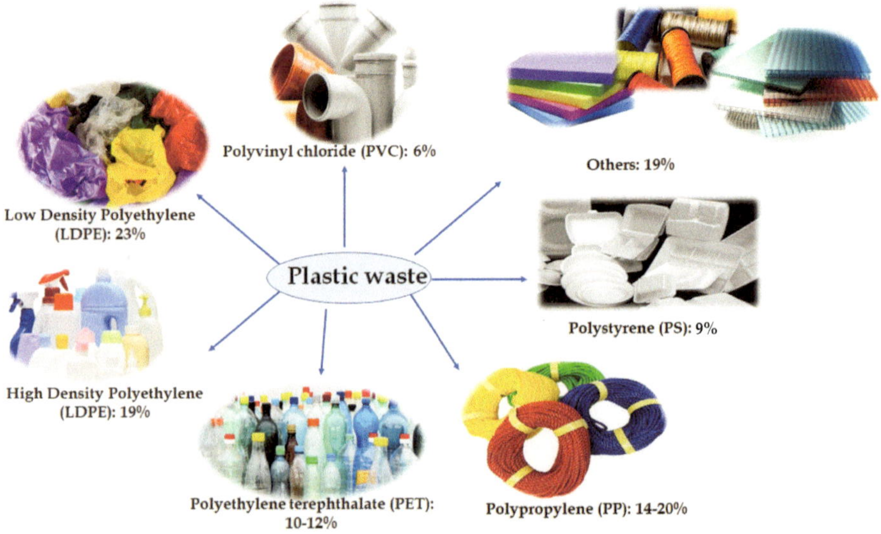

Figure 8.1 Different types of plastic commonly used in day-to-day life.

Low density polyethylene: LDPE is the first-grade polyethylene thermoplastic, made up of monomeric units of ethylene. The free radical polymerization technique was used for the first time to synthesize LDPE by Imperial Chemical Industry in 1933 (Malpass 2010). Plastic waste contains 23% of LDPE. It is mainly used for the synthesis of synthetic bags, wraps, trays, bottles, containers, packaging materials, battery cases, and playground slides, etc. However, in comparison to the waste generated, a very small amount of this type of plastic is usually recycled (Jordan et al. 2016).

High density polyethylene: HDPE is considered as the safest among all types of plastic. This thermo tolerant petroleum polymer is one of the most versatile among different plastics. HDPE is reported to withhold elevated levels of heating, cooling, and other weather conditions due to its high density to strength ratio. Plastic waste mainly contains 19% of HDPE. It is used for manufacturing bottles (plastic, shampoo, bleach), pipes, cutting boards, toys, jugs (milk, water, juice), stools, and chairs, etc. (Kurniawati et al. 2021).

Polyethylene terephthalate (PET): PET or PETE is an adherent of the polyester family. It is strong, stiff, and durable synthetic fiber. Moreover, it could either be turned into fibers or molded into bottles. The composition of PET in plastic waste is nearly 10–12% (Raheem et al. 2019).

Polypropylene (PP): 14–20% of total plastic waste is PP. It is a synthetic resin synthesized by polymerizing the propylene. PP is a lightweight, tough, flexible, and heat-resistant polymer generally used to manufacture diapers, ropes, kitchenwares, boxes, etc. (Canopoli et al. 2020).

Polystyrene: This type of plastic is not frequently recycled, and usually gets trashed after single-time use. Polystyrene is mainly used for the synthesis of disposable materials and forms about 9–10% of total plastic waste (Maafa 2021).

Polyvinyl chloride (PVC): Plastic waste contains 6% of PVC. Due to the stringent standards for recycled material as well as constituted by several materials with varied origins, waste containing PVC (curtains, pipes, windows, door frames, etc.) is difficult to recycle (Miliute-Plepiene et al. 2021).

Others: There are a large number of miscellaneous plastic components such as polycarbonates, nylon, and bisphenols, etc. These aforesaid plastics are used in car parts, medical equipment, dental caress accessories, etc. (Aurpa 2021). A large fraction (19%) of used or leftover plastic forms this portion of generated wastes.

CONVENTIONAL METHODS OF PLASTIC WASTE MANAGEMENT

Plastic is useful, and indeed has tremendous applications in multitude. But the use of hazardous chemical additives and other contaminants for its manufacturing possess certain risks to health as well as environmental concerns. Accumulation of plastic in the environment significantly hampers the ecosystem by altering habitat chemistry, diversity, natural processes, and climate as a whole (Pan et al. 2020). Plastic are reported to cause cancer, alter the activity of human hormones especially endocrine disruption, reduce growth, and reproductive capabilities. Besides this, many other social and economic depreciations have been reported to be caused by plastic (Kumar et al. 2021). Therefore, the management of this catastrophic waste is necessarily important in both ethical as well as in sustainable ways. The conventional ways of plastic waste management include recycle, landfilling, and incineration (Figurr 8.2). Further, the four R's approach "Reduce, Reuse, Recover, and Recycle" could be the key to solve this issue.

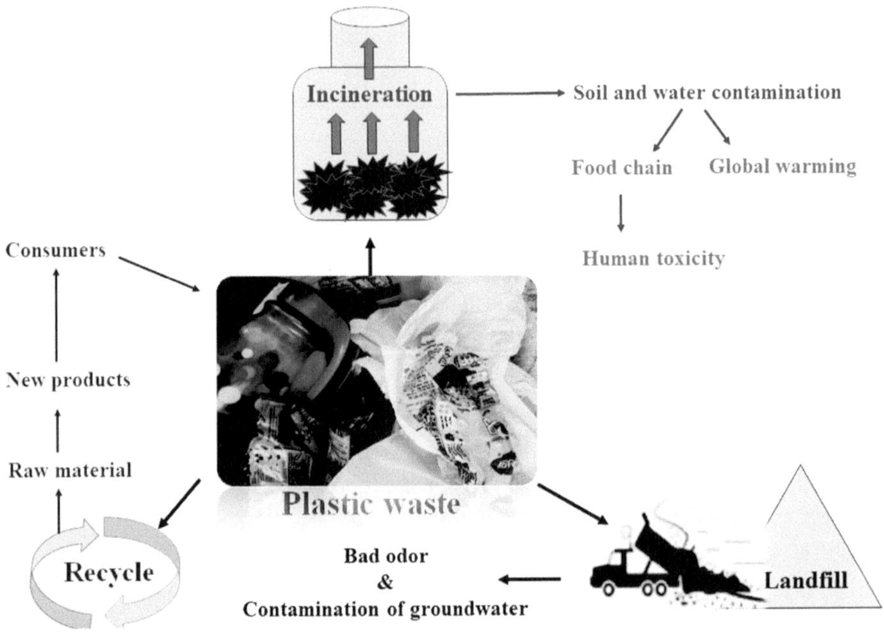

Figure 8.2 An overview of different traditional methods applicable for plastic waste management.

Recycling: One of the most crucial measures to lessen the harmful effects of plastic waste is recycling. It is a method of creating a new product out of recovered, discarded, or waste materials, and regarded as a dynamic segment of the plastic business. Recycling offers a chance to cut down on the amount of waste segregated, its disposal, reduced carbon dioxide emissions, and oil use for subsequent neutralization (Hopewell et al. 2009). Different commercial technologies are there for recycling plastic waste including mechanical, chemical, and physicochemical (thermochemical) recycling (Beghetto et al. 2021). The first step in recycling is the collection of waste, followed by sorting of plastics, washing to remove adhesives, grinding, and shredding to fine raw materials. After grinding, the raw powder is melted and molded into a new one (Lange 2021). Four different strategies of recycling are as follows (Beghetto et al. 2021, Schwarz et al. 2021):

- ***Primary or closed-loop recycling:*** The loop of polymer in this process remains the same as the waste product, which means both recycled and new products might have similar attributes after recycling.
- ***Secondary or open-loop recycling:*** The open loop recycling generates the products of low quality from recovered wastes. However, most of the plastic is recycled in this manner till nowadays.
- ***Tertiary recycling:*** The polymer is broken down into its monomeric units, and these monomeric units are then used as feedstock for the synthesis of value-added products.
- ***Quaternary recycling:*** The polymer is incinerated to generate energy, preferably heat and electricity.

Landfilling: A landfill is a natural phenomenon in which solid waste and water are used as inputs to form leachate and gas. Particularly, landfills designate a place/area/site decided to dispose of plastic waste in the form of segregate waste after utilization. The formation and processing of landfills is a combination of physical, chemical, and biological processes, which converts plastic waste into gases. Theoretically, it is an end of life for plastic. However, the empirical evidences suggest otherwise, and plastic remains in the environment for long time (Wojnowska-Baryła et al. 2022). Among all methods, landfilling is considered as most preferred technique for management of plastic due to its low processing cost, convenience, easier disposal and construction of infrastructure. It was observed that the dumping of plastic in landfills is increasing faster than its decomposition. According to an estimate, approx. 300 million tons of waste is landfilled annually, which is almost equal to the weight of global pollution (Pandey et al. 2023). Conversely, the degradation of waste using the landfill strategy is a cumbersome five-step process that starts with aerobic biodegradation and ends with the stabilization of solids. The five stages of landfilling (pH 4.5–9.0, temperature ranges 60–90°C, and time may vary depending upon the type of plastic) are (Kazour et al. 2019, Wojnowska-Baryła et al. 2022):

***Stage 1*:** Aerobic biodegradation, also called as initial adjustment, in which plastic waste is dumped at a particular site where it undergoes aerobic degradation.

***Stage 2*:** The aerobic condition changes into anaerobic conditions.

***Stage 3*:** Formation of acid from the polymer (hydrolysis).

Stage 4: Methanogenesis or formation of methane.

***Stage 5*:** Stabilization and maturation of solid residues.

Incineration: It is an alternative process to the landfilling and is also known as the complete combustion of waste under high temperature. In incineration, the waste is burned in the presence of oxygen and water resulting in the release of carbon dioxide and other gases into the atmosphere. After complete incineration of the plastic waste, the leftover constitutes volatile organic chemicals (VOCs), toxic heavy metals, polycyclic organic matters (POMs) and ash as solid waste, etc. (Evode et al. 2021). Incineration of plastic is a highly unlikely used technique to dispose this waste as compared to landfilling and recycling. The only anticipated advantage of incineration could be the generation of energy and that impact on the cost of several environmental hazards (Adeniran and Shakantu 2022).

CONVENTIONAL PRACTICES AS BENEFITS AND LIABILITIES

Plastic offers a number of advantages, including adaptability, longevity, portability due to lightweight, and more importantly, the cost effectiveness. Furthermore, the water resistive attributes of the plastic allow it to be modified into a variety of products satisfying the diverging requirements of consumer market. Notwithstanding the non-biodegradability and the accumulation of trash in the environment, plastic is used rampageously throughout the world. Moreover, the traditional practices of plastic waste management act as double-edged swords, e.g. they assist plastic degradation on one hand, and release several toxic byproducts on the other. More pros and cons associated with different traditional plastic management practices have been listed in Table 8.1 in detail.

Recycling offers a refurbished use of plastic by transforming it into new products. Therefore, recycling not only reduces the environmental burden but also provides cheap raw materials for plastic manufacturing industries (Vollmer et al. 2020). Besides being cost-effective and an economically sustainable management process, recycling also has some disadvantages associated with it. Processing of waste into new products releases some toxic chemicals, trashes, volatile gases, and plastic monomers into the environment. These toxic components alter the climatic conditions of the ecosystem, and make it unfavorable for the survival of animals, plants, and other living beings. Similarly, when processed raw material is melted at a very high temperature to synthesize new products, it emits toxic byproducts like dioxins, sulfur, furans, bisphenols, carbon, phthalates, microplastics, and many other pollutants into the atmosphere (Evode et al. 2021). These resultant chemicals, gases, etc. cause acid rain, greenhouse effects, and other global warming incidents (Shen and Worrell 2024).

Landfilling is the most preferred among all other ways of plastic management but it is a very slow process and needs many preventive steps to avoid secondary adverse effects such as groundwater pollution, and soil damage that might come from inadequate processing (Polzer and Persson 2015). Among all, incineration is a fast, season independent process, offers no or very less transportation cost if installed near dumping site, and eliminates hazardous germs and chemicals (Evode et al. 2021). But this is the least used process due to its number of disadvantages w.r.t. release of toxic heavy metals, expensive installation, greenhouse gas emission, ash, and many more (Kristo 2020).

The traditional methodologies as summarized in Table 8.1 used for the plastic waste disposal seems more as liabilities to manage than to exert any substantial benefits. Therefore, there is an inevitable need to develop some innovative technologies for sustainable utilization of plastic waste.

Table 8.1 Pros and cons associated with traditional practices of waste management

Process	Recycling	Landfilling	Incineration
Principle	Generation of new product from old one	Complete degradation of waste	Burning of waste at higher temperature
Outcome	New product with low quality	Production of methane and acid	Energy production
Physiological conditions	Varies with demand of product	pH 4.5–9.0, temperature ranges 60–90°C	590–650°C
Carbon emission	Moderate	High	High
Pros	• Recovery and reusability of waste • Reduces emissions of greenhouse (CO_2, CH_4, CFCs, furans) etc. • Conserves energy • Provides livelihood opportunities • Generation of wealth from waste • Economic • Reduced demand of raw material	• Excellent source of energy • Cheaper than rest of the process • Hazardous waste gets segregated • Easy to construct • Easy disposal of waste • Eco-friendly	• Generation of energy • Production of high value chemicals
Cons	• Sometimes it can lead to pollution • Requires energy • Costly processing	• Partially responsible for alteration in weather • CH_4 produced at the end is highly inflammable • Bad odor • Contaminates soil and groundwater • Infiltration of toxic heavy metals to environment • Negative impact on wild life of particular area • Susceptible to collapses due to accumulation of gases • In humans, it causes respiratory disorders, reduction in growth, cancer, and anemia etc.	• Higher installation cost • Major source of air pollution as performed on high temperature • Production of harmful byproducts • Release of solid volatile compounds • In humans, increases the rate of heart related ailment, respiratory diseases (asthma and emphysema), causes nausea, rashes, headaches, and damages the CNS • Skilled labor is required
References	(Beghetto et al. 2021, Schwarz et al. 2021, Lange 2021)	(Wojnowska-Baryla et al. 2022, Pandey et al. 2023)	(Evode et al. 2021, Wojnowska-Baryła et al. 2022)

INNOVATIVE STRATEGIES FOR PLASTIC WASTE MANAGEMENT

The demand for a 'zero-plastic' environment is extremely overdue, and in the present scenario, a shift is required to protect the ecosystem from the plastic litter menace. Unmanaged plastic waste is a severe threat. Though it can be managed using the aforementioned conventional techniques, these have certain limitations of their own as discussed in the previous section. However, the research and development (R&D) efforts are perpetually coming up with innovative strategies to manage plastic in a sustainable manner, e.g., biodegradation, controlled thermal decomposition, conversion of plastic into bioplastics, etc. (Figure 8.3). Today, all these interventions are relatively new but are continuously evolving from their nascent stages to achieve a target of zero waste alternative to plastic management. Some of the emerging innovative technologies for efficient plastic waste management have been summarized in the following sections discussed below.

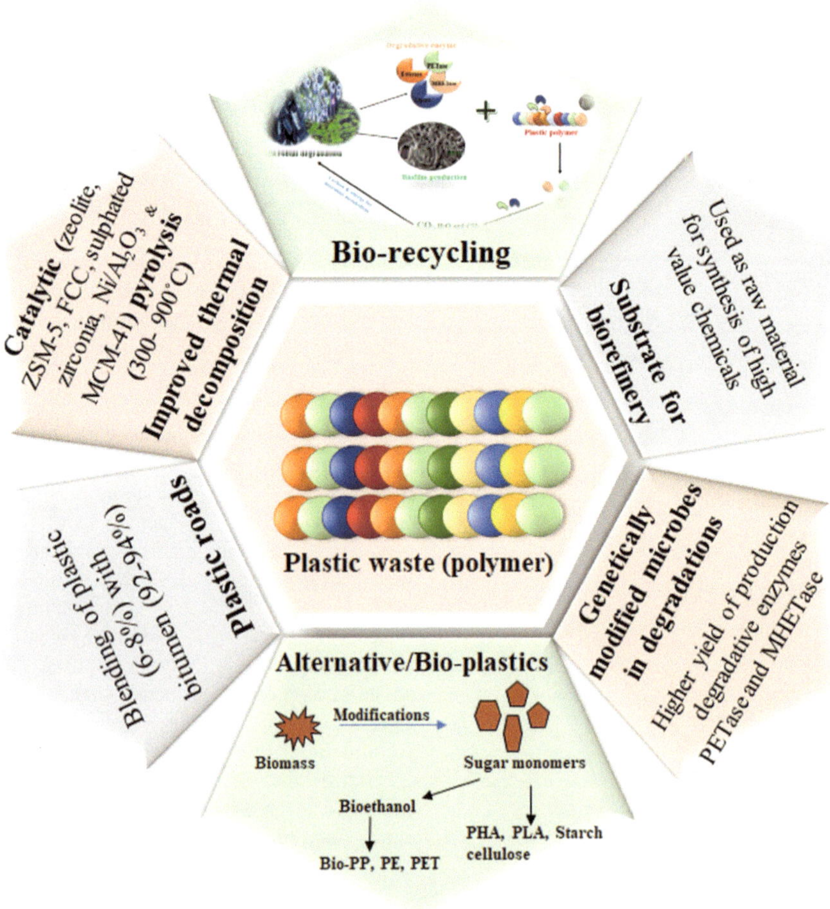

Figure 8.3 Innovative strategies for sustainable management of plastic waste.

Biological recycling and biodegradation: Recycling and reuse are the two most preferred ways to limit the amount of plastic waste. The conventional methods of recycling (chemical and physicochemical recycling) have certain limitations associated with them, and biological recycling is definitely a solution to it. Though biological recycling is a slow process (Jiang et al. 2022, Ardolino et al. 2023), plastic waste can be utilized by microbial entities with minimum release of hazardous byproducts. Biological recycling, is an emerging method in which plastic is broken down into its monomeric constituents by using microorganisms particularly bacteria and fungi (Figure 8.4). The end product of this biological degradation thus formed is reused to synthesize a new value-added product (Kumar et al. 2023). The microbiota of soil and marine habitats are diverse and capable of plastic degradation. In fact, the microbes have developed this ability to degrade plastic waste by selectively modifying their genetic weaponry through evolution. Microbes generally degrade and metabolize the plastic as source of energy using a variety of degradative and substrate transforming biocatalysts (Lee and Liew 2021).

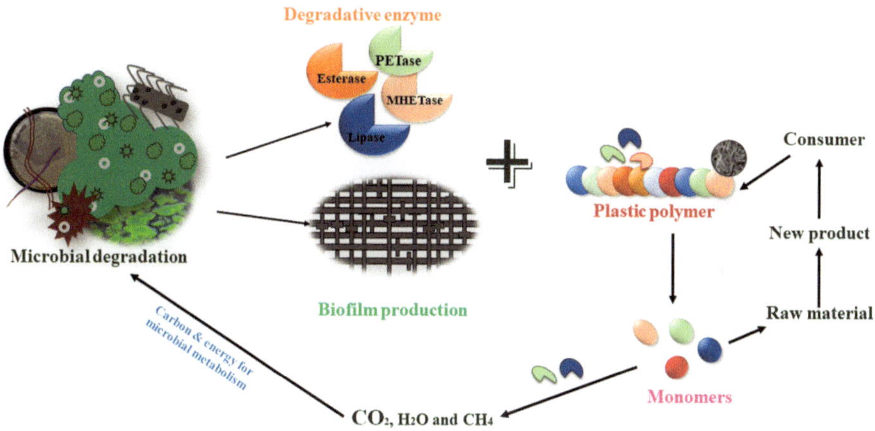

Figure 8.4 Schematic representation of biological recycling of plastic.

A variety of bacterial and fungal genera have been reported to degrade plastic and synthesize several polymers. Bacteria of genera *Bacillus*, *Pseudomonas*, *Azotobacter*, *Erythrobacter*, and *Ideonella* (Bhardwaj et al. 2013, Bollinger et al. 2020, Cassone et al. 2020, Puspitasari et al. 2021, Khandare et al. 2022), and fungi *Ralstonia eutropha*, *Halomonas* sp., *Fusarium solani* and some endophytic fungi (Kale et al. 2015, Ekanayaka et al. 2022, Mazariego 2023) have been reported in recent times as efficient degraders of plastic. For the degradation of plastic, microbes either form biofilm on polymer surfaces or synthesize degradative enzymes (Koshti et al. 2018). Microbial degradation of polymer is a four stage process and it is highly influenced by environmental and cultural conditions (Emadian et al. 2017, Koshti et al. 2018, Lee and Liew 2021). The microbial strains previously reported for plastic degradation have been summarized in Table 8.2.

Table 8.2 List of microorganisms having potential for plastic degradation

Microorganism	Plastic material	Mechanism	Results	References
Bacillus fexus	Polypropylene (PP) flux	—	Decrease in tensile strength 51.8%; Crystallinity increases by 28%	Arkatkar et al. 2009
Nocardia nova SH22a	Gutta-percha and Natural rubber	Enzymatic oxidation	*Nocardia nova* strains were mutated and degradative gene was analysed	Luo et al. 2013
Pseudomonas aestusnigri	Polyethylene terephthalate (PET)	Hydrolyses using PET hydrolytic enzymes	PET hydrolytic enzymes genes were cloned and expressed in *E. coli* and degradation of PET was reported	Bollinger et al. 2020
Endophytic fungi	PET	PETase hydrolyses	—	Mazariego 2023
Ideonella sakaiensis 201-F6	PET	PETase hydrolyses gene was cloned in *E. coli*	Monomeric units of PET (MHET and TPA) were reported after 24 h	Puspitasari et al. 2021
Bacterial strain NA26, NB6, NB26	Polystyrene	—	Degradation of polystyrene	Naima et al. 2010
Brevibacillus borstelensis	Plastic waste	—	Degradation of polymer	Bhardwaj et al. 2013
*Fusarium solan*i	Polyurethane	—	Degradation of Polyurethane	Caruso 2015
Genetically engineered *E. coli*	PET	Enzymatic hydrolyses (PETase)	Depolymerization of PET with 9.1% efficiency	Zhu et al. 2022
Galleria melonelaa	LDPE	—	Degradation of PE	Cassone et al. 2020
Marinobacter sp. H-246	LDPE	Esterase mediated degradation	1.68% reduction in plastic weight	Khandare et al. 2022
Ideonella sakaiensis	PET	PETase mediated degradation	Degrades PET polymer	Yoshida et al. 2016
Shewanella CT12; *Psychrobacter* JT05; *Pseudomonas* JT08	Polycaprolactone (PCL)	—	Degradation of PCL aliphatic polyesters	Sekiguchi et al. 2011
*Erythrobacte*r	Polycyclic aromatic hydrocarbons (PAHs)	Formation of biofilm	Degradation of PHAs	Urbanek et al. 2018
Vibrio alginolyticus and *V. parahaemolyticus* consortia	Polyvinyl alcohol (PVA) and LDPE	—	Visible cracks and grooves were observed on PVA-LDPE surface	Raghul et al. 2014

Different environmental factors such as moisture content, temperature, pH, salinity, concentration of oxygen, sunlight, water, stress, and many other physiochemical conditions positively or negatively boost the growth, metabolism, and degradative enzyme synthesis in microbes. To summarize a number of examples, lignolytic enzyme production in fungi is efficient under acidic pH while in case of bacteria, neutral pH is required for the same (Kale et al. 2015). Similarly, anaerobic bacteria work efficiently in absence of oxygen. Besides this, strength, crystallinity, and molecular weight of polymer also play a significant role in its effective degradation. In the recent past, it has been reported that microbes effectively degrade low molecular weight polymer compared to high molecular weight ones (Ru et al. 2020). It was also reported that the rate of degradation is negatively influenced by the addition of additives and polymer stabilizers. However, the additives are generally added to strengthen the plastic polymer during its processing, and some might render the antimicrobial activity to the plastic polymer (Kale et al. 2015, Maity et al. 2021a). The four stages of degradation have been summarized in the sections below.

Stage 1 Adherence: Microorganisms are attached to the surface of polymer for acclimatization, growth, and metabolism sustenance.

Stage 2 Biofilm formation: Microorganisms colonize the surface of plastic polymer, e.g., polyethene through biofilm formation and utilizing/degrading/metabolizing the polymer for carbon and energy (Mohanan et al. 2020). The hydrophilic nature of cell surface makes the attachment easier, and the degradation of polymer can be confirmed either by morphological changes occurring on plastic film or by evaluating CO_2 evolution (Koshti et al. 2018).

Stage 3 Enzymatic degradation: Different hydrolytic enzymes, e.g., esterase, cutinase, lipase, **polyethylene terephthalate hydrolases** (PETase) and monohydroxyethyl terephthalate hydrolase (MHETase) are known to be synthesized by microorganisms for the degradation of high molecular weight hydrocarbons. These microbial enzymes help the secreting host to hydrolyze the polymer into short chain molecules for their complete degradation (Lee and Liew 2021).

Stage 4 Complete degradation: The monomeric units formed after enzymatic degradation are further derivatized into low molecular weight constituents (oligos and dimers), and subsequently metabolized into CO_2, H_2O, and CH_4 (Emadian et al. 2017).

Improved thermal decomposition: The traditional method of thermal decomposition has certain disadvantages due to their functioning under supercritical conditions as mentioned under section 'Conventional practices as benefits and liabilities'. Incineration and pyrolysis are the two most commonly used methods of thermal decomposition of plastic. Among these two, pyrolysis is the thermal decomposition/combustion of organic matter in the absence of oxygen, and the primary product at the end of pyrolysis is a blend of either liquid or wax (a blend of C6 to C25 hydrocarbons) depending on the process parameters (Qureshi et al. 2020). Pyrolysis is generally performed at higher temperatures ranging from 300–900°C (Rehan et al. 2017). Moreover, it has been considered to be an advanced thermal decomposition technology which is used to overcome the disadvantages of the traditional method.

To improve the process and enhance the efficiency of the system, catalysts are used during pyrolysis. The type, concentration, and the ability of catalysts to perform solitary or in consortium could vary depending on the need and composition of the product to be pyrolyzed. Isomerization, oligomerization, cracking, and aromatization are the catalytic reactions that commonly occur during plastic pyrolysis, whereas zeolites, ZSM-5 (Zeolite Socony Mobil-5), FCC (Fluid Catalytic Cracking), sulphated zirconia, Ni/Al_2O_3, and MCM-41 (Mobil Composition of Matter No. 41) are most preferred catalysts for pyrolysis (Miandad et al. 2019). In a previous study, plastic pyrolysis has been catalyzed with sulfate zirconia, cold plasma, HZMS-5, and Ni/Al_2O_3. It was also reported that the cold plasma-catalyzed pyrolysis of polyethylene and polypropylene produces high-value chemicals (hydrogen, carbon nanotubes, and ethylene) as end products. The addition of zeolites to the pyrolysis reduces the temperature of processes, and also improves the quality of the hydrocarbon blend (wax) (Diaz Silvarrey 2019). In a study Dia et al. (2022) used pillared clays as catalyst and reported high yield of oil and other high-value chemicals with reduced wax formation. In a similar study, 22–25% ethylene monomer was recovered from high-density polyethylene by cold plasma catalyzed pyrolysis which is approx. 55 times more than that of ethylene recovered using the conventional combustion method (Diaz-Silvarrey et al. 2018). Further, to make catalytic pyrolysis more effective, and to prevent the catalyst from deactivation caused by impurities present in waste plastic, the process is divided into two steps (two-reactor pyrolysis). In the first step, plastic waste is decomposed at a high temperature (500°C) to volatile matter, and in the second step, these volatile constituents are passed through a catalytic bed (Dia et al. 2022). Therefore, the advanced pyrolysis provides an efficient way of plastic waste management with a high yield of value added products at low temperature compared to the conventional combustion methods. Furthermore, the end product formed as well as the integrity of catalysts used remain contamination free, and both can be easily recovered as they don't have any direct contact with each other as well as wastes during the entire process.

Plastic roads: The concept of 'plastic road' is relatively new, and was first proposed by Prof. Rajagopalan Vasudevan two decades back around circa 2001. For this innovative idea, Prof. Vasudevan got the prestigious 'Padma Shri' award in 2018 (Thiagarajan 2018). The plastic roads are made up by blending plastic with other important constituents, e.g. bitumen. In India, the first plastic road was constructed by infusing plastic with tar, and a law supporting the incorporation of this invention was made in 2015. Since then, approx. 33,000 km of roads have been built using plastic in the Asian continent, and this is also gaining popularity worldwide (Tsui 2022). Plastic waste is incorporated into the roads through blending with bitumen (6–8% plastic and 92–94% bitumen) (Trimbakwala 2017). The addition of plastic increases the melting temperature of bitumen, and roads thus formed retain their flexibility in winter. For the construction of roads, plastic waste is firstly shredded (to make it an effective binder), then melted at 170°C, and mixed with bitumen at 45°C temperature (Trimbakwala 2017, Enfrin and Giustozzi 2022). Packaging plastics (PET, PP, HDPE, and LDPE) are the most commonly used for this purpose. It is reported that the utilization of plastic waste for the construction of roads not only reduces the environmental burden but it also attributes a smoother texture, durability, economic efficiency, sustainability, less probability of cracks and pot holes, and an ability to withstand extreme environmental conditions compared to the conventional concrete roads (Conlon 2022). However, the major drawbacks associated with this approach are the efficiency of this technology, and also its long term effects on infrastructure, ecology and environment as it is a comparatively young technology. Moreover, the scientific community is also contemplating the eventual break down of plastic into more ecologically perilous micro/nano plastics during any stage of disposal including incorporation as "plastic roads" (Saraswathi et al. 2022).

Alternative plastic/bioplastic: The best solution developed till date to manage plastic accumulation in environment is either recycling or replacing it with biodegradable plastics also known as bioplastic (Singh and Parma 2011). These biodegradable materials can be a prudent alternative to the traditional petro-chemical based polymers as these are sustainable, easily decomposable, environment friendly, and produce less CO_2 during their synthesis (Narancic et al. 2020). Based on the route of decomposition, these bio-based polymers are classified as biodegradable and non-biodegradable. The biodegradable ones include polyhydroxyalkanoates (PHA), polyhydroxybutyrate (PHB), polylactic acid (PLA), starch, cellulose, etc. The biodegradable polymers could be degraded by microbial action, composting, anaerobic digestion, etc. (Narancic et al. 2018). On the other hand, bio-PET, bio-PE, polyol-PUs are some of the non-biodegradable bioplastics (Narancic et al. 2020). In the recent years, the genesis of bioplastics using petrochemicals as the main ingredients of plastics have gained perpetual impetus through development of several technologies insinuating microbiological machinery as well as the approach of genetic engineering (Obruča et al. 2022, Rosenboom et al. 2022).

The types and beneficial attributes of biodegradable plastics are discussed in further sections.

Polyhydroxyalkanoates **(PHA):** It is a bio-degradable polymer synthesized by bacteria under the stressed conditions, i.e. excess carbon, less nitrogen, and other nutrients or *vice versa* (Sasidhara 2014). PHA has properties similar to available plastic (low density polyethene, polypropylene, and polystyrene properties). PHA is usually bio-compatible, and can be completely degraded under natural geo-climatic conditions (Bhubalan et al. 2011). PHA is an intracellular polymer synthesized and stored in cytoplasm of microorganisms. It mainly has two main groups' short chain length (SCL-3-5 carbon) and medium chain length (MCL, i.e. more than 6 carbon). The most commonly synthesized PHA is Poly (R)-3-hydroxybutyrate (PHB) (Tan et al. 2014, Shantini et al. 2015). Structurally, PHB is a homopolymer of PHA which is stiff, brittle, crystalline, thermally unstable having properties almost similar to the synthetic polyethylene and polypropylene (McAdam et al. 2020). Notwithstanding their phenomenal features and ecological advantages, broader utilization of PHAs is drained by their high production costs. One of the primary components for this is the utilization of costly carbon sources, which represent around 40% of the absolute costs. However, this can be reduced by using inexpensive biomass, wastes residues and byproducts of agriculture and food processing (whey, food waste, maple sap, oil palm, molasses, waste cooking oil, wheat straw etc.) (Aslan 2016, Nielsen et al. 2017, Amaro et al. 2019, McAdam et al. 2020). The use of waste biomass as a carbon source for the production of PHA will not only reduce the production cost but will simultaneously help in the eco-friendly disposal of generated wastes (Koller et al. 2017, Nielsen et al. 2017, Amaro et al. 2019).

***Polylactic acid* (PLA):** It is an aliphatic polyester obtained after polymerization of the lactic acid. PLA is a thermoplastic, mainly synthesized from renewable biomass, and used for the synthesis of disposable items such as cups, trays, plates, and containers, etc. (Atiwesh et al. 2021). After their synthesis, PLA formed is quite fragile, and additives (Joncryl ADR 4368 and 4370, acryl, bending with LDPE (1:1), 3D printing of natural fibers, etc.) are used to make it durable (Wittbrodt and Pearce 2015, Borrega et al. 2021, Tripathi et al. 2021). PLA is a commercially available bioplastic, formed by molding (extrusion or injection) and thermoforming (Bernat et al. 2021). Commercially, these bio-based polymers are used for the synthesis of disposable garments, diapers, and hygiene products that can easily be degraded by hydrolyses (Atiwesh et al. 2021, Bernat et al. 2021).

***Polysaccharide based bioplastics*:** These are mainly constituted by starch and cellulose as their inherent biodegradable polymers. Lignocellulosic biomass (LCB) contains a significant amount of starch and cellulose. Therefore, they are the favorable as well as renewable polymer that are used to synthesize biodegradable films and matrix. Structurally, starch is a linear polymer consisting of 20–30% amylose (α-1, 4-D-glucopyranoside)$_n$ and 70–80% amylopectin (poly-α-1, 4-D-glucopyranoside)$_n$ and (α-1, 6-D-glucopyranoside)$_n$ (Rendón-Villalobos et al. 2022). On the other hand, cellulose is also a renewable, crystalline, and biodegradable polymer. However, the cellulose-based polymers are less permeable to the water vapors compared to the starch-based plastics (Zhou et al. 2021). However, a combination of both starch and cellulose-based bioplastics is preferred due to their lower tensile strength (Abe et al. 2021a). For the synthesis of polysaccharide-based bioplastics, natural polymers are extracted from the biomass and modified into the bioplastics via acetylation, enhancing their hydrophobicity (Abe et al. 2021b). These modifications reduce the polarity as well as solubility of the polysaccharides, and enhance their enzymatic degradation (Reshmy et al. 2021).

***Protein based bioplastics*:** These are created by cross-linking the amino acid residues of proteins using different enzymes (Jones et al. 2015). A wide range of proteins of both animal (keratin, casein, collagen, and fibroin, etc.) as well as plant (algae and glutens, etc.) origin act as raw material for the synthesis of such bioplastics. Glycerol is the most commonly and preferably used plasticizer in the synthesis of protein-based bioplastics (Álvarez-Castillo et al. 2021). These natural polymers have plastic-like properties, and can be used as a replacement for plastics, especially in the food packaging enterprises (Jones et al. 2015). Besides this, in a recent study, it has been also reported that protein-based bioplastics possess antibacterial properties, and could help in enhancing the shelf life of packed food by preventing any bacterial contaminations (Perez-Puyana et al. 2022).

***Non-biodegradable bioplastics*:** These are plastic polymers mainly produced using biological sources. In such case of bioplastics, the biomass (LCB) is mainly converted into sustainable plastic alternatives such as bio-PE, bio-PP, bio-PET by a series of modifications and reactions including doping of 5-HMF with water or halide salt, catalysis at higher temperature, pyrolysis, etc. (Goel et al. 2021, Lee et al. 2022). For that purpose, the biomass is firstly pretreated and hydrolyzed into the simple sugars, and these simple sugars are then fermented into ethanol (bio-ethanol) by microbial action. The bioethanol thus obtained is converted into bioplastics (PET, PP, and PE, etc.) (Rahman and Bhoi 2021). However, bioplastics synthesized in this way are usually non-biodegradable, and have properties similar to the conventional plastics (Rahman and Bhoi 2021).

Engineered organism and biocatalyst assisted biodegradation: During past few decades, a number of studies have summarized the degradation of recalcitrant plastics as well as other petrochemical products by using different microbial strains or by the products of their genetic machinery, especially enzymes (Emadian et al. 2017, Koshti et al. 2018, Lee and Liew 2021). Bio-degradation of plastic is highly influenced by a number of its configurational and conformational characteristics such as shape, size, crystallinity, molecular weight, chemical bonds, hydrophobicity, and the presence of a number of benzene rings in the plastic polymer, etc. (Lee and Liew 2021). These factors sometime also reduce the rate of plastic waste neutralization using wild strains through hindrance (Perera and Hemamali 2022). Though the genetic modification of organisms or engineering of efficient biocatalysts enhances the propensity of plastic degradation by increasing stability, specificity, and the fidelity of enzymes as well as the ability of microbes to hydrolyze the versatile synthetic resin polyethylene (Rathour et al. 2022). The diversifying applications of biotechnology, and the advancements in the technology of genetic engineering has allowed the scientific community to disclose the role of complex genes in microbial metabolism and their uses, especially for plastic degradation (Huang et al. 2018, Jadaun et al. 2022). In the present scenario, powerful biological approaches such as genetic manipulation, CRISPR-cas, and synthetic biology as a whole have boosted the scopes of bioremediation beyond its traditional limits (Kumari and Chaudhary 2020). In a previous study, Maity et al. (2021b) discerned the role of genetically modified polyethylene terephthalate degrading hydrolase, i.e., PETases and monohydroxyethyl terephthalate hydrolases, i.e. MHETases in the degradation of polyethylene terephthalate (PET). Both these enzymes have been the products of a versatile aerobically thriving, gram negative, rod shaped, non-spore forming, and most importantly PET films degrading microorganism *Ideonella sakaiensis* 201-F6. In another study, the genetically modified *Bacillus subtilis* observed 3.8 folds' increase in PETase activity with respect to the PET degradation (Huang et al. 2018). Similarly, the PETase of *Ideonella sakaiensis* engineered using the chassis of a photosynthetic algae *Phaeodactylum tricornutum* was reported to degrade PET in saltwater (Moog et al. 2019).

Genetic modifications of wild microbial strains using recombinant DNA technology, protein engineering, and gene insertion/deletion/mutation approaches provide profound opportunities to manage the plastic waste on needed parameters such as varied temperature and pH ranges. In a recent study, an engineered cutinase from *Thermomonospora curvata* with an improvised anchor peptide enhanced the polyurethane degradation by 6–7 fold (Anjana et al. 2020). Moreover, a microbial cell factory created by genetically engineering *Chlamydomonas reinhardtii* was used for the production of enzyme PETase, and the cell lysate was incubated with PET films to evaluate its degradative effects. After incubation, visible holes were observed on PET films (Kim et al. 2021). Notwithstanding the benefits of genetic modifications as technology, it is indeed a costly and time-consuming intervention as thousands of mutant strains are constructed and screened to get the best one. Besides this, there are some issues related to the stability, alteration under varied environmental conditions and risk of contamination through genetically altered strains (Chia et al. 2020). However, the major issue is ethical acceptance as the release of any foreign gene into the environment (for the degradation of plastic) may cause an allergic reaction as plastic is very common among the human population (Chia et al. 2020).

Conversion of plastic waste into high value products: Plastic is a synthetic polymer of long carbon chains; therefore, the plastic waste can be used as a carbon source, substrate, or raw material for the synthesis of high-value products (Johnston et al. 2022). Valorization of plastic waste on the one hand solves the problem associated with plastic accumulation, and on the other hand provides different products (polyalphaolefin, lubricants, feedstock for refinery, monomer products, fuels and chemical additives). Energy generation by thermochemical transformation (e.g., pyrolysis and combustion) of waste is an attractive approach (Mahari et al. 2018). Co-pyrolysis has also gained much attention in the past few years due to its improvements over traditional technologies. In this process, one substrate improves the quality and amount of liquid fuel, and the plastic counterpart acts as a hydrogen donor (Zheng et al. 2023). The propensity of converting plastic waste into fuels (liquid fuel) has increased over the past few decades, and the researchers are continuously working on this aspect to find out new possibilities (Zheng et al. 2023). In a previous study, HDPE and waste oil were catalytically co-pyrolyzed using a polycrystalline ceramic ZrO_2 foam catalyst and transformed into short-chain hydrocarbon fuels with 63.1 weight (wt.) % yield (Wang et al. 2017). In a similar study, 81 wt.% yield of liquid oil was obtained by Mahari et al. (2018) on microwave assisted co-pyrolysis of frying cooking oil with plastic. Moreover, in order to study the comparative effect of single substrate pyrolysis and co-pyrolysis, the sugarcane bagasse, low-density polyethylene, and a combination of both (in 1:1 ratio) were pyrolyzed at 350 to 600°C temperature and 47.15, 74.40 and 52.75 wt.% yields of liquid fuel were obtained, respectively (Dewangan et al. 2016).

Plastic valorization into bioplastics through microbial fermentation using polyethylene, polystyrene, and polypropylene as substrates is the most sought biotechnological intervention of present times (Johnston et al. 2018, 2019, 2022).

The metabolic pathway of bioplastic synthesis is well documented in *Cupriavidus necator* having three cycles (I, II, and III). Cycle I involves the utilization of sugar monosaccharides (glucose and fructose) to generate PHB, and its subsequent metabolization in II and III pathways (Emadian et al. 2017). Also, the plastic polymer in microbial metabolism enters the β-oxidation and forms acetyl-CoA, which on further metabolism forms PHA (Palmeiro-Sánchez et al. 2022, Johnston et al. 2022). In a recent study, biosynthesis of PHA from plastic waste has been observed in a bacterial strain *C. necator* H16. The *C. necstor* strain was incubated in a culture media supplemented with polyethylene, and about 40% (w/w) accumulation of PHA in cellular biomass was observed after 72 h incubation at 30°C temperature (Ekere et al. 2022). Furthermore, plastic waste was utilized as substrate to convert it into lipids. The waste was supplemented along with other minimal media components to the microbial strain *Yarrowia lipolytica*, and under optimized conditions, about 1.9 g/L fatty acids were observed, i.e., 41% (w/w) of the total cellular content generated (Mihreteab et al. 2021). In another study, polypropylene is thermally depolymerized into fatty acids (Mihreteab et al. 2019). Further, reports are there on the catalytic conversion of plastic into chemicals such as benzenes, toluene, alkanes, gasoline, and many other heavier liquids or aromatics (Costa et al. 2021, Munir and Usman 2022, Zheng et al. 2023). Exposure of plastics to the higher temperature leads to the breaking of C–C and C–H bonds of long-chain polymers and generation of abundant hydrocarbon radicals. These radicals are further transformed into short chain fuels and chemicals (Zheng et al. 2023). Hence, the humongous amount of plastic waste generated throughout the post-industrial revolution era till date and beyond could be utilized sustainably by coupling the traditional wisdom with modern day cutting edge technologies.

CONCLUSIONS

Plastic has broadened our way of life through bestowing numerous advantages including life-saving inventions. However, it is important to assess the effects of our expansionist instincts in terms of contaminating mother Earth by perpetually adding plastic waste. Probably, the current time is pretty much rife to reconsider the adverse effects of plastic waste on the cost of modern conveniences. The polymer economy is advancing linearly, and the perpetual addition of waste into the environment is also unprecedented. The need to switch from our existing plastic paradigm to a circular economic based model is concomitant. Therefore, the traditional methods of plastic management like incineration and landfills which generate large amounts of heat, particulate matter, and greenhouse gas emissions need to be replaced with sustainable modern technologies. There is a need to develop innovative methods to significantly increase the economic value of plastic waste through its valorization into value-added chemicals or other products. The aforementioned concepts could be further utilized to develop methods for the full recyclability of plastic. Apart from that, multi-omics approaches undoubtedly play an effective role in revealing microorganism assisted bioconversions which can address a number of open questions related to plastic waste utilization. Recent innovation compendiums place a strong emphasis on providing a thorough understanding of the methods available for turning plastic trash into the wealth. The development of biodegradable materials from plastic waste provides an affordable and secure alternative, and significantly rely on evolving biotechnological research.

REFERENCES

Abe, M.M., Martins, J.R., Sanvezzo, P.B., Macedo, J.V., Branciforti, M.C., Halley, P., et al. 2021a. Advantages and disadvantages of bioplastics production from starch and lignocellulosic components. Polymer. 13(15): 2484.

Abe, M.M., Branciforti, M.C. and Brienzo, M. 2021b. Biodegradation of hemicellulose-cellulose-starch-based bioplastics and microbial polyesters. Recycling. 6(1): 22.

Adeniran, A.A. and Shakantu, W. 2022. The health and environmental impact of plastic waste disposal in South African Townships: a review. International Journal of Environmental Research and Public Health. 19(2): 779.

Álvarez-Castillo, E., Bengoechea, C., Felix, M. and Guerrero, A. 2021. Protein-based bioplastics from biowastes: sources, processing, properties and applications. Bioplastics for Sustainable Development. 137: 176.

Amaro, T.M., Rosa, D., Comi, G. and Iacumin, L. 2019. Prospects for the use of whey for polyhydroxyalkanoate (PHA) production. Frontiers in Microbiology. 10: 992.

Anjana, K., Hinduja, M., Sujitha, K. and Dharani, G. 2020. Review on plastic wastes in marine environment: biodegradation and biotechnological solutions. Marine Pollution Bulletin. 150: 110733.

Ardolino, F., Palladini, A. and Arena, U. 2023. Social life cycle assessment of innovative management schemes for challenging plastics waste. Sustainable Production and Consumption. 37: 344–355.

Areeprasert, C., Asingsamanunt, J., Srisawat, S., Kaharn, J., Inseemeesak, B., Phasee, P., et al. 2017. Municipal plastic waste composition study at transfer station of Bangkok and possibility of its energy recovery by pyrolysis. Energy Procedia. 107: 222–226.

Arkatkar, A., Arutchelvi, J., Bhaduri, S., Uppara, P.V. and Doble, M. 2009. Degradation of unpretreated and thermally pretreated polypropylene by soil consortia. International Biodeterioration & Biodegradation. 63(1): 106–111.

Aslan, A.N., Ali, M.M., Morad, N.A. and Tamunaidu, P. 2016. Polyhydroxyalkanoates production from waste biomass. *In*: IOP Conference Series: Earth and Environmental Science. 36: 012040.

Atiwesh, G., Mikhael, A., Parrish, C.C., Banoub, J. and Le, T.A.T. 2021. Environmental impact of bioplastic use: a review. Heliyon. 7(9): e07918.

Aurpa, S.S. 2021. Characterization of MSW and plastic waste volume estimation during COVID-19 pandemic. The University of Texas at Arlington. 2021.

Beghetto, V., Sole, R., Buranello, C., Al-Abkal, M. and Facchin, M. 2021. Recent advancements in plastic packaging recycling: a mini-review. Materials. 14(17): 4782.

Bernat, K., Kulikowska, D., Wojnowska-Baryła, I., Zaborowska, M. and Pasieczna-Patkowska, S. 2021. Thermophilic and mesophilic biogas production from PLA-based materials: possibilities and limitations. Waste Management. 119: 295–305.

Bhardwaj, H., Gupta, R. and Tiwari, A. 2013. Communities of microbial enzymes associated with biodegradation of plastics. Journal of Polymers and the Environment. 21: 575–579.

Bhubalan, K., Chuah, J.A. and Shozui, F. 2011. Characterization of the highly active polyhydroxyalkanoate synthase of *Chromobacterium* sp. strain USM2. Applied and Environmental Microbiology. 77: 2926–2933.

Bollinger, A., Thies, S., Knieps-Grünhagen, E., Gertzen, C., Kobus, S., Höppner, A., et al. 2020. A novel polyester hydrolase from the marine bacterium Pseudomonas aestusnigri–structural and functional insights. Frontiers in Microbiology. 11: 114.

Borrega, M., Pihlajaniemi, V., Liitiä, T., Wikström, L. and Tamminen, T. 2021. Evaluation of chemical additives in hydrothermal pre-treatment of wood for the integrated production of monosugars and hydrolysis lignins for PLA-based biocomposites. Biomass Conversion and Biorefinery. 1–13.

Canopoli, L., Coulon, F. and Wagland, S.T. 2020. Degradation of excavated polyethylene and polypropylene waste from landfill. Science of the Total Environment. 698: 134125.

Caruso, G. 2015. Plastic degrading microorganisms as a tool for bioremediation of plastic contamination in aquatic environments. Journal of Pollution Effects and Control. 3(3): 1–2.

Cassone, B.J., Grove, H.C., Elebute, O., Villanueva, S.M. and LeMoine, C.M. 2020. Role of the intestinal microbiome in low-density polyethylene degradation by caterpillar larvae of the greater wax moth, Galleria mellonella. Proceedings of the Royal Society B. 287(1922): 20200112.

Chatterjee, S. and Sharma, S. 2019. Microplastics in our oceans and marine health. Field Actions Science Reports. 19: 54–61.

Chia, W.Y., Tang, D.Y., Khoo, K.S., Lup, A.N. and Chew, K.W. 2020. Nature's fight against plastic pollution: algae for plastic biodegradation and bioplastics production. Environmental Science and Ecotechnology. 4: 100065.

Conlon, K. 2022. Plastic roads: not all they're paved up to be. International Journal of Sustainable Development and World Ecology. 29(1): 80–83.

Costa, C.S., Muñoz, M. Ribeiro, M.R. and Silva, J.M. 2021. H-USY and H-ZSM-5 zeolites as catalysts for HDPE conversion under a hydrogen reductive atmosphere. Sustainable Energy and Fuels. 5(4): 1134–1147.

Dai, L., Zhou, N., Lv, Y., Cheng, Y., Wang, Y., Liu, Y., et al. 2022. Pyrolysis technology for plastic waste recycling: a state-of-the-art review. Progress in Energy and Combustion Science. 93: 101021.

Dewangan, A., Pradhan, D. and Singh, R.K. 2016. Co-pyrolysis of sugarcane bagasse and low-density polyethylene: influence of plastic on pyrolysis product yield. Fuel. 185: 508–516.

Diaz-Silvarrey, L.S., Zhang, K. and Phan, A.N. 2018. Monomer recovery through advanced pyrolysis of waste high-density polyethylene (HDPE). Green Chemistry. 20(8): 1813–1823.

Diaz Silvarrey, L.S. 2019. Advanced pyrolysis of plastic waste for chemicals, fuel and materials (Doctoral dissertation, Newcastle University).

Ekanayaka, A.H., Tibpromma, S., Dai, D., Xu, R., Suwannarach, N., Stephenson, S.L., et al. 2022. A review of the fungi that degrade plastic. Journal of Fungi. 8(8): 772.

Ekere, I., Johnston, B., Tchuenbou-Magaia, F., Townrow, D., Wojciechowski, S., Marek, A., et al. 2022. Bioconversion process of polyethylene from waste Tetra Pak® packaging to polyhydroxyalkanoates. Polymers. 14(14): 2840.

Emadian, S.M., Onay, T.T. and Demirel, B. 2017. Biodegradation of bioplastics in natural environments. Waste Management. 59: 526–536.

Enfrin, M. and Giustozzi, F. 2022. Recent advances in the construction of sustainable asphalt roads with recycled plastic. Polymer International. 71(12): 1376–1383.

Evode, N., Qamar, S.A., Bilal, M., Barceló, D. and Iqbal, H.M. 2021. Plastic waste and its management strategies for environmental sustainability. Case Studies in Chemical and Environmental Engineering. 4: 100142.

Facts, P.E. 2019. An analysis of European plastics production, demand, and waste data. https://plasticseurope.org/wp-content/uploads/2021/10/2019-Plastics-the-facts.pdf.

Faraca, G. and Astrup, T. 2019. Plastic waste from recycling centres: characterization and evaluation of plastic recyclability. Waste Management. 95: 388–398.

Gilbert, M. 2017. Plastics materials: introduction and historical development. *In*: Brydson's Plastics Materials. Butterworth-Heinemann. 1–18.

Goel, V., Luthra, P., Kapur, G.S. and Ramakumar, S.S.V. 2021. Biodegradable/bio-plastics: myths and realities. Journal of Polymers and the Environment. 29: 3079–3104.

Haegerbaeumer, A., Mueller, M.T., Fueser, H. and Traunspurger, W. 2019. Impacts of micro- and nano-sized plastic particles on benthic invertebrates: a literature review and gap analysis. Frontiers in Environmental Science. 7: 17.

Hopewell, J., Dvorak, R. and Kosior, E. 2009. Plastics recycling: challenges and opportunities. Philosophical Transactions of the Royal Society B: Biological Sciences. 364(1526): 2115–2126.

Huang, S., Wang, H., Ahmad, W., Ahmad, N., Ivanovich Vatin, A.M., Mohamed, A., et al. 2022. Plastic waste management strategies and their environmental aspects: a scientometric analysis and comprehensive review. International Journal of Environmental Research and Public Health. 19(8): 4556.

Huang, X., Cao, L., Qin, Z., Li, S., Kong, W. and Liu, Y. 2018. Tat-independent secretion of polyethylene terephthalate hydrolase PETase in *Bacillus* subtilis 168 mediated by its native signal peptide. Journal of Agricultural and Food Chemistry. 66(50): 13217–13227.

Idumah, C.I. and Nwuzor, I.C. 2019. Novel trends in plastic waste management. SN Applied Sciences. 1: 1–14.

Jacob, H., Besson, M., Swarzenski, P.W., Lecchini, D. and Metian, M. 2020. Effects of virgin micro- and nanoplastics on fish: trends, meta-analysis, and perspectives. Environmental Science and Technology. 54(8): 4733–4745.

Jadaun, J.S., Bansal, S., Sonthalia, A., Rai, A.K. and Singh, S.P. 2022. Biodegradation of plastics for a sustainable environment. Bioresource Technology. 347: 126697.

Jiang, J., Shi, K., Zhang, X., Yu, H., Zhang, J. and Liu, J. 2022. From plastic waste to wealth using chemical recycling: a review. Journal of Environmental Chemical Engineering. 10(1): 106867.

Johnston, B., Radecka, I., Hill, D., Chiellini, E., Ilieva, V.I., Sikorska, W., et al. 2018. The microbial production of polyhydroxyalkanoates from waste polystyrene fragments attained using oxidative degradation. Polymers. 10(9): 957.

Johnston, B., Radecka, I., Chiellini, E., Barsi, D., Ilieva, V.I., Sikorska, W., et al. 2019. Mass spectrometry reveals molecular structure of polyhydroxyalkanoates attained by bioconversion of oxidized polypropylene waste fragments. Polymers. 11(10): 1580.

Johnston, B., Adamus, G., Ekere, A.I., Kowalczuk, M., Tchuenbou, F.M. and Radecka, I. 2022. Bioconversion of plastic waste based on mass full carbon backbone polymeric materials to value-added polyhydroxyalkanoates (PHAs). Bioengineering. 9(9): 432.

Jones, A., Mandal, A.A. and Sharma, S. 2015. Protein-based bioplastics and their antibacterial potential. Journal of Applied Polymer Science. 132: 18.

Kale, S.K., Deshmukh, A.G., Dudhare, M.S. and Patil, V.B. 2015. Microbial degradation of plastic: a review. Journal of Biochemical Technology. 6(2): 952–961.

Kazour, M., Terki, S., Rabhi, K., Jemaa, S., Khalaf, G. and Amara, R. 2019. Sources of microplastics pollution in the marine environment: importance of wastewater treatment plant and coastal landfill. Marine Pollution Bulletin. 146: 608–618.

Khandare, S.D., Agrawal, D., Mehru, N. and Chaudhary, D.R. 2022. Marine bacterial based enzymatic degradation of low-density polyethylene (LDPE) plastic. Journal of Environmental Chemical Engineering. 10(3): 107437.

Kim, J.W., Park, S.B., Tran, Q.G., Cho, D.H., Choi, Y.J., Lee, Y.J., et al. 2020. Functional expression of polyethylene terephthalate-degrading enzyme (PETase) in green microalgae. Microbial Cell Factories. 19(1): 1–9.

Koller, M., Maršálek, L., de Sousa Dias, M.M. and Braunegg, G. 2017. Producing microbial polyhydroxyalkanoate (PHA) biopolyesters in a sustainable manner. New Biotechnology. 37: 24–38.

Koshti, R., Mehta, L. and Samarth, N. 2018. Biological recycling of polyethylene terephthalate: a mini-review. Journal of Polymers and the Environment. 26: 3520–3529.

Kristo, K. 2020. Pros and Cons for the Incineration of Solid Waste. International Conference.

Kumar, R., Sadeghi, K., Jang, J. and Seo, J. 2023. Mechanical, chemical, and bio-recycling of biodegradable plastic: a review. Science of the Total Environment. 163446.

Kumar, S., Singh, E., Mishra, R., Kumar, A. and Caucci, S. 2021. Utilization of plastic wastes for sustainable environmental management: a review. ChemSusChem. 14(19): 3985–4006.

Kumari, A. and Chaudhary, D.R. 2020. Engineered microbes and evolving plastic bioremediation technology. *In*: Bioremediation of Pollutants. 417–443. Elsevier.

Kung, H.C., Wu, C.H., Cheruiyot, N.K., Mutuku, J.K., Huang, B.W. and Chang-Chien, G.P. 2022. The current status of atmospheric micro/nanoplastics research: characterization, analytical methods, fate, and human health risk. Aerosol and Air Quality Research. 23: 220362.

Kurniawati, D., Nandy, P., Abdullah, N., Hakim, I.I. and Nurrokhmat, A. 2021. An experimental analysis of diesel fuel produced from HDPE (high-density polyethylene) waste using thermal and catalytic pyrolysis with passive heat pipe cooling system. Thermal Science and Engineering Progress. 23: 100917.

Lai, H., Liu, X. and Qu, M. 2022. Nanoplastics and human health: hazard identification and biointerface. Nanomaterials. 12(8): 1298.

Lange, J.P. 2021. Managing plastic waste-sorting, recycling, disposal, and product redesign. ACS Sustainable Chemistry & Engineering. 9(47): 15722–15738.

Lee, A. and Liew, M.S. 2021. Tertiary recycling of plastics waste: an analysis of feedstock, chemical and biological degradation methods. Journal of Material Cycles and Waste Management. 23(1): 32–43.

Lee, K., Jing, Y., Wang, Y. and Yan, N. 2022. A unified view on catalytic conversion of biomass and waste plastics. Nature Reviews Chemistry. 6(9): 635–652.

Luo, Q., Hiessl, S., Poehlein, A. and Steinbüchel, A. 2013. Microbial gutta-percha degradation shares common steps with rubber degradation by Nocardia nova SH22a. Applied and Environmental Microbiology. 79(4): 1140–1149.

Maafa, I.M. 2021. Pyrolysis of polystyrene waste: a review. Polymers. 13(2): 225.

Mahari, W.W., Chong, C.T., Cheng, C.L., Lee, C.L., Hendrata, K., Yek, P.Y., et al. 2018. Production of value-added liquid fuel via microwave co-pyrolysis of used frying oil and plastic waste. Energy. 162: 309–317.

Maity, S., Banerjee, S., Biswas, C., Guchhait, R., Chatterjee, A. and Pramanick, K. 2021a. Functional interplay between plastic polymers and microbes: a comprehensive review. Biodegradation. 32(5): 487–510.

Maity, W., Maity, S., Bera, S. and Roy, A. 2021b. Emerging roles of PETase and MHETase in the biodegradation of plastic wastes. Applied Biochemistry and Biotechnology. 193: 2699–2716.

Malpass, D.B. 2010. Introduction to industrial polyethylene: properties, catalysts, and processes. John Wiley & Sons. 1–9.

Mazariego, M. 2023. Evolutionary analysis of plastic-degrading enzyme PETase found in the endophytic microbiome of marine viridiplantae for phytoremediation. Frontiers in Microbiology. 11: 2020.

McAdam, B., Brennan, F.M., McDonald, P. and Mojicevic, M. 2020. Production of polyhydroxybutyrate (PHB) and factors impacting its chemical and mechanical characteristics. Polymers. 12(12): 2908.

Miandad, R., Rehan, M., Barakat, M.A., Aburiazaiza, A.S., Khan, H., Ismail, I.M., et al. 2019. Catalytic pyrolysis of plastic waste: moving toward pyrolysis based biorefineries. Frontiers in Energy Research. 27.

Mihreteab, M., Stubblefield, B.A. and Gilbert, E.S. 2019. Microbial bioconversion of thermally depolymerized polypropylene by Yarrowia lipolytica for fatty acid production. Applied Microbiology and Biotechnology. 103: 7729–7740.

Mihreteab, M., Stubblefield, B.A. and Gilbert, E.S. 2021. Enhancing polypropylene bioconversion and lipogenesis by Yarrowia lipolytica using a chemical/biological hybrid process. Journal of Biotechnology. 332: 94–102.

Miliute-Plepiene, J., Fråne, A. and Almasi, A.M. 2021. Overview of polyvinyl chloride (PVC) waste management practices in the Nordic countries. Cleaner Environmental Technologies. 4: 100246.

Mohanan, N., Montazer, Z., Sharma, P.K. and Levin, D.B. 2020. Microbial and enzymatic degradation of synthetic plastics. Frontiers in Microbiology. 11: 580709.

Moog, D., Schmitt, J., Senger, J., Jan, K.H., Rexer, Z., Linne, T., et al. 2019. Using a marine microalga as a chassis for polyethylene terephthalate (PET) degradation. Microbial Cell Factories. 18: 1–15.

Munir, D. and Usman, M.R. 2022. Mesoporous HZSM-5 catalysts for the conversion of waste plastics to liquid fuels. Journal of Porous Materials. 29(3): 783–794.

Naima, A., Safia, A., Ali, M., Saadia, I., Bashir, A. and Robson, G. 2010. Isolation and identification of polystyrene biodegrading bacteria from soil. African Journal of Microbiology Research. 4(14): 1537–1541.

Narancic, T., Verstichel, S., Reddy, C., Morales-Gamez, L., Kenny, B., De Wilde, B., et al. 2018. Biodegradable plastic blends create new possibilities for end-of-life management of plastics but they are not a panacea for plastic pollution. Environmental Science and Technology. 52(18): 10441–10452.

Narancic, T., Cerrone, F., Beagan, N. and O'Connor, K.E. 2020. Recent advances in bioplastics: application and biodegradation. Polymers. 12(4): 920.

Nielsen, C., Rahman, A. and Rehman, A.U. 2017. Food waste conversion to microbial polyhydroxyalkanoates. Microbial Biotechnology. 10: 1338–1352.

Obruča, S., Dvořák, P., Sedláček, P., Koller, M., Sedlář, I., Pernicová, K., et al. 2022. Polyhydroxyalkanoates synthesis by halophiles and thermophiles: towards sustainable production of microbial bioplastics. Biotechnology Advances. 107906.

Ordan, J.L., Casem, D.T., Bradley, J.M., Dwivedi, A.K., Brown, E.N. and Jordan, C.W. 2016. Mechanical properties of low-density polyethylene. Journal of Dynamic Behavior of Materials. 2: 411–420.

Palmeiro-Sánchez, T., O'Flaherty, V. and Lens, P.N. 2022. Polyhydroxyalkanoate bio-production and its rise as biomaterial of the future. Journal of Biotechnology. 348: 10–25.

Pan, D., Su, F., Liu, C. and Guo, Z. 2020. Research progress for plastic waste management and manufacture of value-added products. Advances in Composite and Hybrid Materials. 3: 443–461.

Pandey, P., Dhiman, M., Kansal, A. and Subudhi, S.P. 2023. Plastic waste management for sustainable environment: techniques and approaches. Waste Disposal and Sustainable Energy. 1–18.

Perera, I.C. and Hemamali, E.H. 2022. Genetically modified organisms for bioremediation: Current research and advancements. pp. 163–186. *In*: Suyal, D.C. and Soni, R. (eds). Bioremediation of Environmental Pollutants: Emerging Trends and Strategies. Springer.

Perez-Puyana, V., Cuartero, P., Jiménez-Rosado, M., Martínez, I. and Romero, A. 2022. Physical crosslinking of pea protein-based bioplastics: effect of heat and UV treatments. Food Packaging and Shelf Life. 32: 100836.

Plastic Industry. 2023. A look at latest statistics. http://blog.gitnux.com.

Polzer, V.R. and Persson, K.M. 2015. Environmental and economical assessment of MSW management in Europe: an analysis between the landfill and WTE impacts. International Journal of Academic Research in Business and Social Sciences. 5(6): 2222–6990.

Puspitasari, N., Tsai, S.L. and Lee, C.K. 2021. Class I hydrophobins pretreatment stimulates PETase for monomers recycling of waste PETs. International Journal of Biological Macromolecules. 176: 157–164.

Qureshi, M.S., Oasmaa, A., Pihkola, H., Deviatkin, I., Tenhunen, J., Mannila, J., et al. 2020. Pyrolysis of plastic waste: Opportunities and challenges. Journal of Analytical and Applied Pyrolysis. 152: 104804.

Rafey, A. and Siddiqui, F.Z. 2021. A review of plastic waste management in India–challenges and opportunities. International Journal of Environmental Analytical Chemistry. 1–17.

Raghul, S.S., Bhat, S.G., Chandrasekaran, M., Francis, V. and Thachil, E.T. 2014. Biodegradation of polyvinyl alcohol-low linear density polyethylene-blended plastic film by consortium of marine benthic vibrios. International Journal of Environmental Science and Technology. 11: 1827–1834.

Raheem, A.B., Noor, Z.Z., Hassan, A., Abd Hamid, M.K., Samsudin, S.A. and Sabeen, A.H. 2019. Current developments in chemical recycling of post-consumer polyethylene terephthalate wastes for new materials production: a review. Journal of Cleaner Production. 225: 1052–1064.

Rahman, M.H. and Bhoi, P.R. 2021. An overview of non-biodegradable bioplastics. Journal of Cleaner Production. 294: 126218.

Rathour, R.K., Sharma, D., Sharma, N., Bhatt, A.K. and Singh, S.P. 2022. Engineered microorganisms for bioremediation. pp. 335–361. *In*: Joshi, S., Panday, A. and Park, S.H. (eds) Current Developments in Biotechnology and Bioengineering. Elsevier.

Rehan, M., Miandad, R., Barakat, M.A., Ismail, I.M., Almeelbi, T., Gardy, J., et al. 2017. Effect of zeolite catalysts on pyrolysis liquid oil. International Biodeterioration and Biodegradation. 119: 162–175.

Rendón-Villalobos, R., Lorenzo-Santiago, M.A., Olvera-Guerra, R. and Trujillo-Hernández, C.A. 2022. Bioplastic composed of starch and micro-cellulose from waste mango: mechanical properties and biodegradation. Polímeros. 32.

Reshmy, R.D., Thomas, E., Philip, S.A., Paul, A., Sindhu, R. and Binod, P. 2021. Bioplastic production from renewable lignocellulosic feedstocks: a review. Reviews in Environmental Science and BioTechnology. 20: 167–187.

Rosenboom, J.G., Langer, R. and Traverso, G. 2022. Bioplastics for a circular economy. Nature Reviews Materials. 7(2): 117–137.

Ru, J., Huo, Y. and Yang, Y. 2020. Microbial degradation and valorization of plastic wastes. Frontiers in Microbiology. 11: 442.

Saraswathi, G., Abirami, P., Kumar, P. and Meyyappan, N. 2022. Plastic roads as an alternative solution to solid waste management: a review. ECS Transactions. 107(1): 18799.

Schwarz, A.E., Ligthart, T.N., Bizarro, D.G., De Wild, P., Vreugdenhil, B. and Van Harmelen, T. 2021. Plastic recycling in a circular economy; determining environmental performance through an LCA matrix model approach. Waste Management. 121: 331–342.

Sekiguchi, T., Sato, T., Enoki, M., Kanehiro, H., Uematsu, K. and Kato, C. 2011. Isolation and characterization of biodegradable plastic degrading bacteria from deep-sea environments. JAMSTEC Report of Research and Development. 11: 33–41.

Shantini, K., Yahya, A.R.M. and Amirul, A.A. 2015. Influence of feeding and controlled dissolved oxygen level on the production of poly (3-Hydroxybutyrate-co-3-Hydroxyvalerate) Copolymer by Cupriavidus sp. USMAA2-4 and its characterization. Applied Biochemistry and Biotechnology. 176: 1315–1334.

Shen, L. and Worrell, E. 2024. Plastic recycling. pp. 497–510. *In*: Meskers, C., Worrell, E. and Reuter, M.A. (eds). Handbook of Recycling. Elsevier..

Shin, S.K., Um, N., Kim, Y.J., Cho, N.H. and Jeon, T.W. 2020. New policy framework with plastic waste control plan for effective plastic waste management. Sustainability. 12: 6049.

Singh, P. and Parma, N. 2011. Isolation and characterization of two novel polyhydroxybutyrate (PHB)-producing bacteria. African Journal of Biotechnology. 10: 4907–4919.

Statista. 2023. Annual production of plastics worldwide from 1950 to 2021. https://www.statista.com/statistics/282732/global-production-of-plastics-since-1950/ .

Tan, G.Y.A., Chen, C.L. and Li, L. 2014. Start research on biopolymer polyhydroxyalkanoate (PHA): a review. Polymers. 6: 706–754.

Thiagarajan, K. 2018. The man who paves India's roads with old plastic. A report in The Guardian. Retrieved 3 July 2020.

Trimbakwala, A. 2017. Plastic roads use of waste plastic in road construction. International Journal of Scientific Research Publications. 7(4): 137–139.

Tripathi, N., Misra, M. and Mohanty, A.K. 2021. Durable polylactic acid (PLA)-based sustainable engineered blends and biocomposites: recent developments, challenges, and opportunities. ACS Engineering Au/. 1(1): 7–38.

Tsui, J. 2022. Plastic roads: an innovative solution to reduce waste. A Report in Earth Organization. https://kids.earth.org/life-on-land/plastic-roads/.

Urbanek, A.K., Rymowicz, W. and Mirończuk, A.M. 2018. Degradation of plastics and plastic-degrading bacteria in cold marine habitats. Applied Microbiology and Biotechnology. 102: 7669–7678.

Urebanek, A.K., Kosiorowska, N.E. and Mironczuk, A.M. 2021. Current knowledge on polyethylene terephthalate degradation by genetically modified microorganisms. Frontiers in Bioengineering and Biotechnology. 9: 2021.

Vollmer, I., Jenks, M.J., Roelands, M.C., White, R.J., van Harmelen, T., de Wild, P., et al. 2020. Beyond mechanical recycling: giving new life to plastic waste. Angewandte Chemie International Edition. 59(36): 15402–15423.

Wang, Y., Dai, L., Fan, L., Cao, Y., Zhou, Y., Zhao, Y., et al. 2017. Catalytic co-pyrolysis of waste vegetable oil and high-density polyethylene for hydrocarbon fuel production. Waste Management. 61: 276–282.

Wittbrodt, B. and Pearce, J.M. 2015. The effects of PLA color on material properties of 3-D printed components. Additive Manufacturing. 8: 110–116.

Wojnowska-Baryła, I., Bernat, K. and Zaborowska, M. 2022. Plastic waste degradation in landfill conditions: the problem with microplastics, and their direct and indirect environmental effects. International Journal of Environmental Research and Public Health. 19(20): 13223.

Yee, M.S.L., Hii, L.W., Looi, C.K., Lim, W.M., Wong, S.F., Kok, Y.Y., et al. 2021. Impact of microplastics and nanoplastics on human health. Nanomaterials. 11(2): 496.

Yoshida, S., Hiraga, K., Takehana, T., Taniguchi, I., Yamaji, H., Maeda, Y., et al. 2016. A bacterium that degrades and assimilates poly (ethylene terephthalate). Science. 351(6278): 1196–1199.

Zheng, K., Wu, Y., Hu, Z., Wang, S., Jiao, X., Zhu, J., et al. 2023. Progress and perspective for conversion of plastic wastes into valuable chemicals. Chemical Society Reviews. 52: 8–29.

Zhou, X., Hao, Y., Zhang, X., He, X. and Zhang, C. 2021. Cellulose-based polymers. Physical Sciences Reviews. 000010151520200067.

Zhu, B., Ye, Q., Seo, Y. and Wei, N. 2022. Enzymatic degradation of polyethylene terephthalate plastics by bacterial curli display PETase. Environmental Science and Technology Letters. 9(7): 650–657.

Chapter **9**

Sustainable Management of E-Waste

Shivam Kumar[1], Saumya Shukla[2], Anil K. Singh[3], Jyoti Singh Jadaun[2] and Gyanendra Kumar*[4]

[1]CSIR-National Botanical Research Institute, Pratap Marg, Lucknow 226001, Uttar Pradesh, India.

[2]Department of Botany, Dayanand Girls PG College, Civil Lines, Kanpur Nagar 208001 UP, India.

[3]Department of Botany, PPN P.G. College, Kanpur 208001, U.P. India.

[4]Department of Physics, Government PG College, Hamirpur 210301 UP, India.

INTRODUCTION

The swift progression of technological development has transformed our lives, but it has also led to a concerning issue: the generation of electronic waste, commonly known as e-waste, encompasses discarded electronic gadgets like smartphones, laptops, televisions, and refrigerators (Table 9.1). These devices contain both valuable materials and harmful substances, presenting environmental and health dangers. Improper management of e-waste can contaminate soil and water with substances like lead, mercury, cadmium, and flame retardants, endangering ecosystems and human well-being. Furthermore, inefficient disposal and inadequate recycling contribute to resource depletion and perpetuate a linear production and consumption model (Ahirwar and Tripathi 2021). The magnitude of the e-waste

* For Correspondence: gyaniitk@gmail.com

Table 9.1 A list of different types of electronic wastes

S. No.	Categorization of Electronic Waste	Significant Electrical and Electronic Devices	Applications of E-Waste	References
1.	Major household appliances	• Refrigeration units • Laundry machines • Culinary equipment • Electric fans and air conditioning devices • Microwaves • Electric heaters • Cookers, etc.	Reduce environmental damage, protect precious resources, and ensure the safe management of harmful materials.	Yanagitani et al. 2000, Terazono et al. 2006, Yu et al. 2023
2.	Minor household appliances	• Hoover cleaners • Ironing equipment • Blenders, espresso machines, and bread makers • Hair clippers, electric razors, and toothbrushes • Watches, scales, and other timepieces.	Diminishes ecological footprint while also bolstering local economies, fostering creativity, and promoting educational endeavors.	Hoang et al. 2023, Islam and Huda 2019, Lou 2022
3.	Information technology and communication devices	• Mainframe computers • Calculators, desktop computers, mobile laptops, tablets, and data handling • Telephones, TVs, radio devices, fax machines, printer machine, copiers, and musical instruments.	Reuse, recycling, and secure data disposal	Khan et al. 2022, Garg and Adhana 2019
4.	Consumer devices	• Radios, TVs, video cameras • Cassette players amplifiers, and similar technology.	Artistic electronic items, responsible digital behavior, environmentally conscious technology products.	Ishchenko et al. 2019
5.	Illumination devices	• Fluorescent bulb-compatible lighting fittings • Fluorescent lighting, either linear or compact • Powerful discharge lamps • Sodium lamps	Minimize environmental impact, prevent the release of hazardous substances, and promote sustainable lighting solutions	Ravindra and Mor 2019, Rodríguez-Padrón et al. 2020

Table 9.1 (*Contd.*)

S. No.	Categorization of Electronic Waste	Significant Electrical and Electronic Devices	Applications of E-Waste	References
6.	Electric and electronic implements	• Cutting tools, grass cutters • Drills • Embroidery machines.	Repairing, reusing, recycling, educational efforts, and creative endeavors aimed at waste reduction and environmental preservation.	Barwood et al. 2015, Bailey-Van Kuren 2002
7.	Playthings, recreational, and athletic gear	• Electric train or race car kits • Gaming systems or platforms • Computers for activities like cycling, diving, jogging, rowing, etc. • Coin-operated machines	Initiatives for collecting e-waste that encourage and advance environmental well-being.	Miller et al. 2016
8.	Medical instruments	• Radiation therapy machines • Heart-related equipment • Kidney dialysis units • Respiratory ventilators • Sonogram equipment • Medical refrigeration units • Digital thermometers	Renovation, charitable giving, recycling, confidential data handling, and educational programs aimed at waste reduction and enhancing healthcare accessibility.	Bandyopadhyay et al. 2022, Banerjee et al. 2023.
9.	Equipment for surveillance and regulation	• Fire alarms • Sensing devices • Data collection systems • Temperature controllers • Temperature gauges	Reviving, recycling, ensuring secure data management, and educational efforts aimed at waste reduction, heightened security, and the promotion of responsible surveillance techniques.	Li et al. 2022, Kamel et al. 2022
10.	Automated vending machines	• Machines that dispense hot or cold beverages • Solid items, or cash.	Innovative efforts aimed at waste reduction, convenience improvement, and the advancement of sustainable vending practices.	Singh et al. 2021

problem is considerable with approximately 53.6 million metric tons generated globally in 2020, of which only 17.4% was properly collected and recycled. The majority of e-waste often ends up in landfills, incinerators, or informal recycling operations, exacerbating environmental consequences and health risks. Governments, international organizations, and stakeholders are actively seeking solutions to manage e-waste effectively, aiming to mitigate environmental and health effects while recovering valuable resources and promoting a circular economy principle applied to the electronics industry. This book chapter gives a summary of electronic waste management, addressing challenges, opportunities, and sustainable strategies. It explores the causes and impacts of e-waste, highlights responsible disposal and recycling practices, and emphasizes the importance of collaboration between governments, industry players, and consumers to establish effective e-waste management frameworks.

CAUSES OF E-WASTE

Multiple factors play a role in the worldwide generation of electronic waste, referred to as e-waste. These are as follows:

Technological progress and product obsolescence: The swift momentum of technological progress and innovation results in frequent enhancements and substitutions of electronic devices. With the introduction of newer models, older ones quickly become obsolete and are frequently disposed of, thereby contributing to the issue of electronic waste (Gurauskienė 2008).

Technological progress: The constant development and enhancement of technology is referred to as technological progress. It results in the development of new and innovative goods with better features, higher performance, and increased efficiency. Companies spend in R&D to remain competitive and meet consumer demand for cutting-edge technologies.

Product obsolescence: When an electronic item or technology becomes obsolete, inefficient, or incompatible with newer models or software, it is referred to as product obsolescence. This can occur for a variety of causes, such as the introduction of better technology, changes in industry standards, or the withdrawal of software or hardware support for older devices.

Increased consumerism: The increase in consumerism and the prevalence of a disposable culture have led to a heightened desire for electronic products. Individuals often acquire new devices even when their current ones are still operational, resulting in a greater accumulation of electronic waste.

Short product lifecycles: Planned obsolescence refers to a deliberate strategy employed in the design of certain electronic devices, where their lifespan is intentionally limited. This practice ensures that devices become obsolete or malfunction after a specific period, thereby encouraging consumers to purchase newer versions. Consequently, this approach generates more electronic waste as older devices are discarded and replaced.

Insufficient recycling infrastructure: In numerous regions, the lack of adequate infrastructure and awareness regarding proper e-waste recycling is a prevailing issue. Due to the absence of accessible recycling facilities, people frequently dispose of their electronic devices in the regular trash, resulting in these items ending up in landfills or being incinerated. This improper disposal contributes to environmental degradation and poses potential health hazards.

Illicit disposal and unauthorized export: E-waste occasionally undergo illegal dumping or is exported to developing countries with less stringent disposal regulations. This practice contributes to environmental pollution and poses health risks in those regions, as improper dismantling and recycling methods are commonly employed. Additionally, many electronic devices are designed in a manner that makes repairs challenging or costly. This discourages users from seeking repairs for faulty devices, resulting in an elevated production of electronic waste (Tsang 2002).

Knowledge gap and awareness deficiency: A lack of awareness regarding the environmental consequences of improper e-waste disposal, as well as the advantages of recycling, is widespread among many individuals. This inadequate education and awareness contribute to the improper handling and disposal of electronic devices (Figure 9.1). Effectively addressing these issues necessitates a comprehensive approach that includes sustainable product design, extended producer responsibility, enhanced recycling infrastructure, awareness campaigns, and stricter regulations pertaining to e-waste management (Ganguly 2016).

ENVIRONMENTAL IMPACTS OF E-WASTE

In today's technologically advanced world, the environmental consequences of electronic waste (e-waste) have become a growing concern. Inadequate disposal and mismanagement of electronic devices present substantial environmental risks. The existence of harmful elements like lead, mercury, cadmium, and brominated flame retardants in electronic waste is especially alarming. When not handled correctly, such toxic substances can pollute the soil, water sources, and the air, leading to severe health issues for both humans and wildlife. Additionally, the recovery of valuable metals from electronic waste often involves harmful practices like burning or acid leaching, which release toxic pollutants into the atmosphere and contribute to environmental pollution. Furthermore, the sheer volume of e-waste generated globally creates challenges for waste management, as a significant portion of it ends up in landfills or is exported to developing countries with inadequate infrastructure for proper handling. To mitigate these environmental impacts, it is crucial to promote responsible e-waste recycling, implement stricter regulations on disposal methods, encourage sustainable design practices in the electronics industry, and increase public awareness regarding the significance of appropriate e-waste management. Here are some of the key effects:

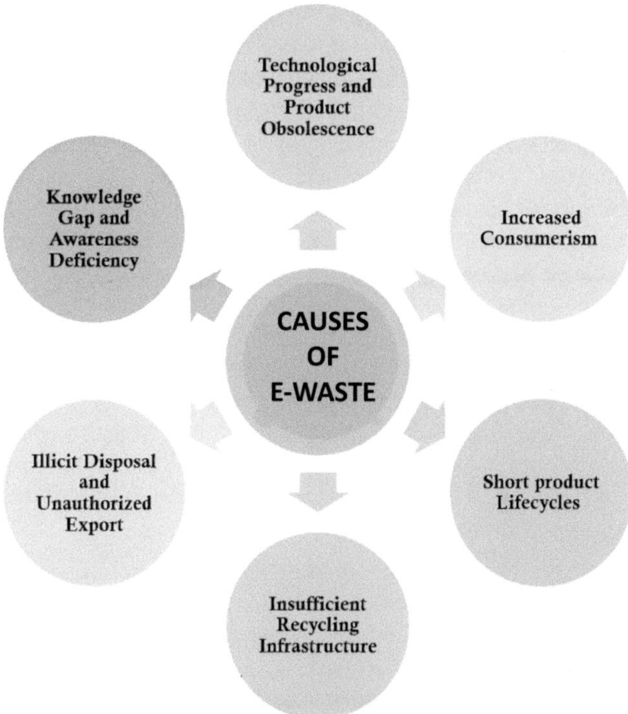

Figure 9.1 A schematic overview of different potential causes for generation of e-waste.

Soil and water pollution: Soil pollution caused by electronic waste, commonly known as e-waste, is a significant environmental concern in today's world. E-waste refers to discarded electronic devices such as computers, mobile phones, televisions, and other electronic appliances. These devices often contain hazardous materials like lead, mercury, cadmium, and brominated flame retardants, which can leach into the soil when improperly disposed of. When electronic waste is disposed of in landfills or incinerated, these toxic substances can contaminate the surrounding soil, resulting in enduring harm to ecosystems and posing a risk to human health. The soil becomes less fertile and loses its ability to support plant growth, disrupting the delicate balance of nutrients and microorganisms essential for a healthy ecosystem. Additionally, the toxic substances can infiltrate groundwater, further exacerbating the problem (Needhidasan et al. 2014). This form **of pollution** can have enduring impacts on the environment and pose significant risks to human health.

Air pollution: The practice of burning e-waste, commonly observed in informal recycling operations which results in to the release of detrimental pollutants into the atmosphere. The results release detrimental pollutants into the atmosphere. These pollutants include dioxins, furans, and brominated flame retardants, which can have severe health consequences when inhaled (Tsydenova and Bengtsson 2011). Additionally, this burning process contributes to air pollution and can contribute to climate change.

Human health risks: Inadequate management and disposal of e-waste, encompassing discarded electronic devices, present notable risks to human health. One primary concern is the release of hazardous substances, including lead, mercury, cadmium, and brominated flame retardants, commonly found in electronic devices. During the incineration of e-waste or dumped in landfills, these toxic materials can contaminate soil, water sources, and the air. Exposure to these substances can have severe health consequences. For example, lead exposure can result in neurological damage and developmental issues in children, mercury can cause nervous system harm and cognitive impairments, cadmium can lead to kidney damage and lung disorders, and brominated flame retardants have the potential to disrupt hormone regulation. Furthermore, the informal recycling and dismantling of e-waste, often carried out without proper safety measures in developing countries, pose additional risks. Workers involved in these processes face direct contact with harmful chemicals, which may lead to immediate and prolonged health issues, such as respiratory problems, skin ailments, organ damage, and a heightened cancer risk, often due to insufficient protective measures (Perkins et al. 2014).

Ecosystem damage: E-waste contamination poses a threat to diverse ecosystems, including freshwater bodies, soil organisms, and plant life (Thanomsangad et al. 2019). Improper disposal and management of electronic waste, commonly referred to as e-waste, can have a significant negative impact on ecosystems. E-waste encompasses discarded electronic devices like computers, mobile phones, televisions, and various other electronic equipment. One major concern associated with e-waste is the presence of toxic substances within these devices. Many electronic devices contain hazardous materials, including lead, mercury, cadmium, and brominated flame retardants. When e-waste is not disposed of properly, these toxic substances can seep into the soil and water, leading to the contamination of ecosystems. The contamination of ecosystems with these hazardous materials can have severe consequences for various forms of life, including plants, animals, and microorganisms. It can disrupt ecological balances, harm biodiversity, and even result in the decline or extinction of certain species. Additionally, e-waste often ends up in landfills or is incinerated, both of which contribute to air pollution. The incineration of e-waste releases harmful gases and particulate matter into the atmosphere, negatively impacting air quality and human health. The pollutants emitted during the incineration process can also settle on land and water bodies, exacerbating ecosystem damage. Therefore, it is crucial to ensure proper disposal and e-waste management to alleviate the adverse effects on ecosystems, safeguard the environment, and protect human health.

Balanced resource utilization and energy optimization: Manufacturing electronic devices requires extracting raw materials, notably valuable precious metals like gold, silver, and copper. Improper disposal of e-waste means that these valuable resources are not efficiently recovered and recycled (Kang and Schoenung 2005). Furthermore, the energy usage linked with the manufacturing and disposal processes of electronics contributes to carbon emissions and resource depletion.

Global ramifications: E-waste is frequently exported to developing countries, where unsafe recycling practices are prevalent due to lower costs and lenient regulations (Manhart 2011). This contributes to environmental degradation and poses risks to the health of workers engaged in dismantling and recycling operations.

TYPES OF E-WASTE

E-waste can be categorized based on the origin and purpose of electronic devices. According to Baldé et al. (2017), socioeconomic conditions, consumer trends, population size, and how heavily households and businesses rely on electronic and electrical equipment (EEE) can all have an impact on how e-waste is distributed among these categories. There are 10 different forms of recognised e-waste worldwide.

The predominant sources of e-waste are large household appliances, which make up 42.1% of the total, followed by IT and telecommunication equipment at 33.9%, consumer electronics at 13.7%, and minor household devices at 4.7% (Calliafas et al. 2012). While other categories like medical devices, lighting, electrical parts, and electronic tools contribute less than 2%, even fewer e-waste, under 1%, comes from automatic dispensers, toys, sports gear, monitoring systems, and control instruments. In developing nations, the primary e-waste items are TVs, computers, and mobile phones (Kalia et al. 2022).

STRATEGIES TO MANAGE E-WASTES

Various approaches for managing e-waste problems are evolving rapidly, and extensive research is currently underway to tackle e-waste management challenges on both a national and global scale. Various methodologies, such as Life Cycle Assessment (LCA) and Material Flow Analysis (MFA), have been developed and applied in the field of e-waste management. In developed nations, significant progress has been made in e-waste management, particularly with the implementation of waste electric and electronic equipment (WEEE) directives, aimed at reducing waste disposal and improving environmental conditions. Ongoing research efforts focus on sorting recyclable components and extracting rare and valuable metals.

Life Cycle Assessment

Designing products with environmental friendliness in mind is a preferable option and can potentially attract consumers. Life Cycle Assessment (LCA) serves as a robust tool for recognizing potential environmental effects, aiding in the creation of eco-friendly products like printers, desktop personal computers, heating and air conditioning devices, washing machines, and toys (Dauvergne and Lister 2013). Additionally, LCA offers a systematic approach to defining numerous environmental impact categories, including carcinogens, climate change, ozone layer depletion, ecotoxicity, acidification, eutrophication, and land use, aiming to enhance the environmental sustainability of products. LCA finds extensive application in e-waste management. In Europe, significant research has utilized LCA to assess the environmental impacts associated with end-of-life treatment of e-waste.

Material Flow Analysis

Material Flow Analysis (MFA) is a potent analytical tool employed to meticulously study the pathway of various materials, particularly focusing on the intricate journey of e-waste, as they traverse through recycling facilities, disposal sites, and material stocks, spanning both spatial and temporal dimensions. This methodological approach adeptly establishes vital connections between the inception of these materials, the intricate routes they take, and their eventual destinations, whether they be intermediate or final points in the flow (Brunner and Rechberger 2004). Serving as a robust decision support tool, MFA proves its mettle in the realms of environmental and waste handling, aiding in the strategic formulation of effective e-waste management approaches. It enables a thorough and holistic investigation into the flow of e-waste, facilitating a nuanced assessment that encompasses environmental impact, economic implications, and social considerations. By employing MFA, stakeholders are empowered to make informed decisions and implement sustainable practices, effectively navigating the complex landscape of e-waste management (Shinkuma and Huong 2009).

Multi Criteria Analysis (MCA)

Multi-Criteria Analysis (MCA) is a decision-making tool designed to address strategic decisions and solve intricate multi-faceted problems that encompass both qualitative and quantitative aspects of the given issue. MCA models have found application in environmental challenges, such as electronic waste management, enabling in identifying the optimal electronic waste management strategies (Wibowo and Deng 2015). In the context of End-of-Life (EoL) processing for coffee makers, MCA decision-making methodologies play a crucial role in weighing the compromises between economic gains and environmental benefits. A comprehensive six-step methodology is typically employed: defining EoL scenarios, specifying product models, creating an EoL evaluation model, formulating a multi-objective problem, generating solutions for the Pareto set, and constructing EoL strategy graphs that highlight optimal strategies minimizing environmental impacts and economic costs.

Extended Producer Responsibility

Extended Producer Responsibility (EPR) represents a paradigm where the responsibility for handling e-waste is allocated to the producers of electronic devices. In this framework, manufacturers bear the onus of managing the entire life cycle of their products, encompassing collection, recycling, and appropriate disposal, as noted by Monier et al. (2014). The central goal of EPR programs is to encourage sustainable product design by motivating manufacturers to create products that are environmentally conscious and easily recyclable. By imposing the responsibility for e-waste management on manufacturers, EPR instills greater accountability and drives the transition towards a circular and sustainable economy (McKerlie et al. 2006). EPR instills greater accountability and drives the transition towards a circular and sustainable economy. However, it's important to acknowledge that EPR programs can vary in terms of their specific requirements and implementations, differing across countries and regions. Some jurisdictions have adopted EPR initiatives tailored to specific product categories such as electronics, batteries, packaging materials, or automobiles. In these cases, the onus of managing these products is shifted from the consumer to the manufacturer.

METHODS FOR MANAGING E-WASTE

There are various methods for managing electronic waste, each with its own benefits and considerations (Figure 9.2). Here are some common types of e-waste management methods:

Recycling: Recycling is a vital method for management of electronic waste. It entails the collection and processing of electronic devices and components to extract valuable materials like metals, plastics, and glass. These materials can further be reused in the manufacturing of new products (Kang and Schoenung 2005). Recycling plays an important role in conserving resources, reducing the demand for raw material extraction, and preventing the discharge of hazardous chemical substances into the surroundings.

Refurbishment and resale: Refurbishment and resale are essential components of effective electronic waste (e-waste) management (Thapa et al. 2023). With the rapid advancement of technology, consumers often discard their old electronic devices in favor of newer models, leading to a significant amount of e-waste. However, not all discarded electronics are completely non-functional or obsolete. Many devices can be repaired, refurbished, and reintroduced into the market. The refurbishment process involves repairing and restoring electronic devices to their original working condition or enhancing their functionality. This may include addressing hardware issues, replacing faulty components, or upgrading software. By refurbishing e-waste, we can prolong the lifecycle of these devices, reducing the need for new manufacturing and minimizing the environmental impact associated with the extraction of raw materials. Once refurbished, these electronics can be resold in the second-hand market. Resale allows for the reuse of electronic devices, giving them a new lease on life and preventing premature waste generation. It provides an affordable option for individuals who may not have the means to purchase brand-new electronics. Moreover, it decreases the demand for new products, thereby reducing the resources and energy required for their production.

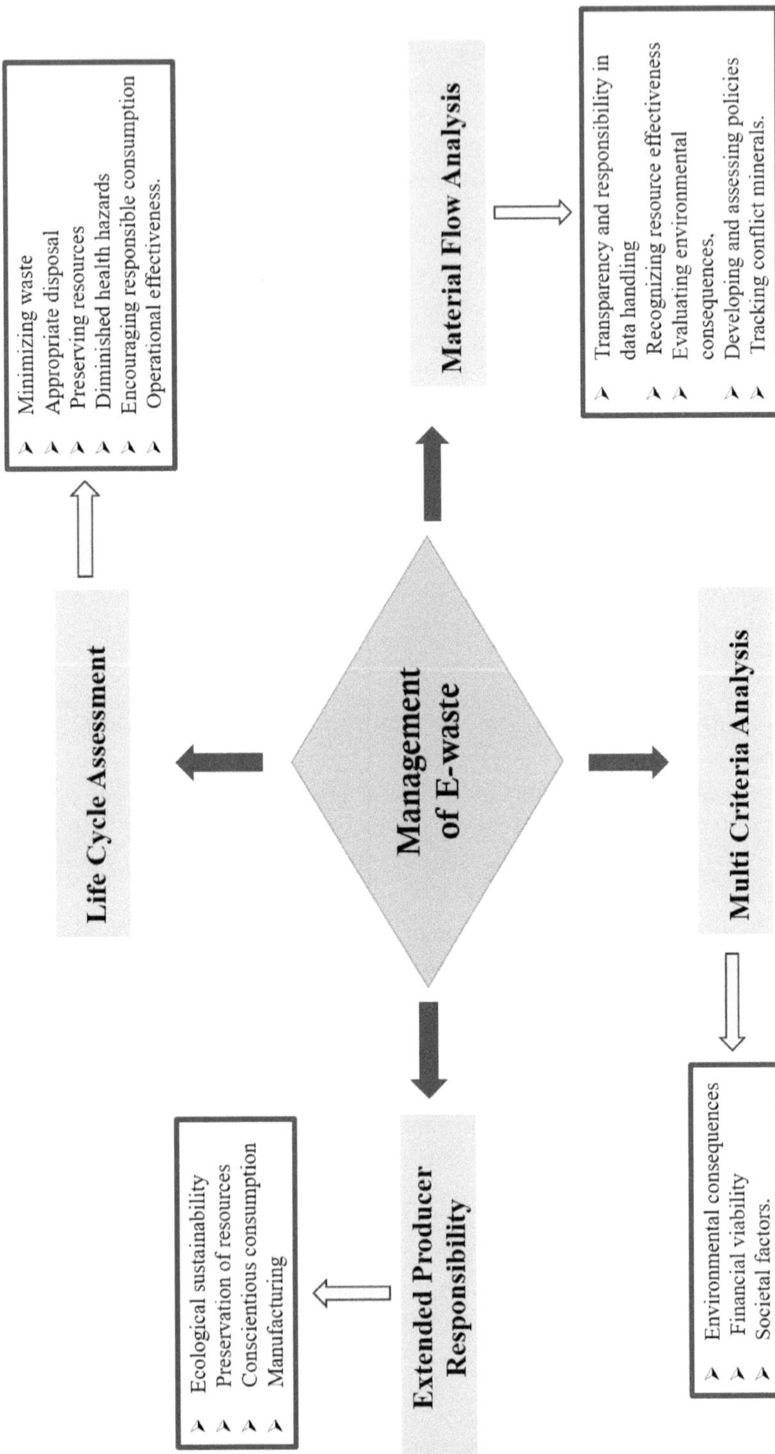

Figure 9.2 Different strategies for management of e-waste.

Donating and reuse: Donating electronic devices that are in good working condition to individuals or organizations in need is a favorable approach to e-waste management. The act of donating and reusing electronic devices plays a critical role in the effective management of e-waste. With technology advancing rapidly, the issue of electronic waste has become increasingly urgent. However, by donating functional electronics and promoting their reuse, we can significantly alleviate the environmental impact associated with e-waste. Donating electronic devices provides them with a second life, allowing others who may not have the means to purchase new equipment to benefit from their use. Many individuals, organizations, and educational institutions, particularly in developing regions with limited technology access, can benefit from donated electronics. By extending the lifespan of these devices, we reduce the necessity for manufacturing new ones, which in turn mitigates the extraction of raw materials and lowers energy consumption.

Confidential data erasure and destruction: Proper data destruction is a crucial component of e-waste management, especially for electronic devices such as smartphones, computers and storage devices (Perkins et al. 2014). Secure data destruction methods are employed to ensure that personal or sensitive information is permanently eradicated from these devices before disposal or recycling. This practice safeguards privacy and helps prevent potential data breaches.

Governance and regulatory frameworks: Governments worldwide are enacting rules and policies to address electronic waste management comprehensively. These regulations often entail compelling manufacturers to assume responsibility for their product at the conclusion of their lifespan. This includes establishing collection programs, setting recycling targets, and implementing penalties for improper disposal practices. These measures aim to encourage accountable handling of e-waste management and encourage sustainable practices throughout the electronics industry.

Extended producer responsibility (EPR): Extended Producer Responsibility (EPR) is an approach that assigns the responsibility for managing e-waste to producers of electrical devices. Under EPR, manufacturers are required to take responsibility for the entire life cycle of their products, which includes collections, re-cycling, and proper discarding or disposal (Monier et al. 2014). EPR programs aim to promote sustainable product design by incentivizing manufacturers to develop more environmentally friendly and recyclable products. By shifting the burden of e-waste management onto manufacturers, EPR encourages greater accountability and fosters the transition towards a more circular and sustainable economy. EPR programs differ across countries and regions in terms of their specific requirements and execution. Some jurisdictions have adopted EPR measures for particular product categories like electronics, batteries, packaging materials, or automobiles. Under these circumstances, the responsibility for handling these products is transferred from the consumer to the manufacturer. The intent of EPR is to support sustainable consumption and production, minimize waste generation, and promote efficient resource usage. By holding producers responsible for the ecological consequences of their products, EPR seeks to motivate the creation of environmentally conscious goods and establish a circular economy in which products are recycled and reused instead of being discarded as waste (Quinn and Sinclair 2006, Hickle 2013).

Disposal practices: When electronic devices reach the end of their functional or repairable lifespan, it is crucial to dispose of them properly to prevent environmental contamination. This involves utilizing authorized e-waste collection points or recycling facilities that are equipped to handle and process the waste in a safe and environmentally responsible manner. By utilizing such facilities, the harmful components and materials present in the electronic waste can be managed effectively, minimizing the environmental impact and promoting sustainable waste management practices. Indeed, e-waste management should prioritize environmentally friendly practices to achieve sustainable outcomes. This includes avoiding harmful practices such as incineration or the export of e-waste to developing countries with inadequate recycling facilities. Responsible e-waste management necessitates collaboration between individuals, businesses, governments, and recycling organizations. By working together, we can ensure that e-waste is handled and processed in a manner that minimizes environmental impact, promotes resource conservation, and protects the health of both people and the planet.

LEGAL AND REGULATORY FRAMEWORK FOR E-WASTE MANAGEMENT

The legal and regulatory framework for e-waste management varies across countries and regions. Many countries have recognized the need for proper management of electronic waste due to its environmental and health impacts. Here are some common elements found in the legal and regulatory frameworks for e-waste management.

Legislation and Regulations

Governments have implemented laws and regulations that specifically focus on the management of e-waste. These laws often encompass provisions related to the collection, transportation, recycling, treatment, and disposal of electronic waste (Joseph 2007). Additionally, they may address issues such as labelling requirements, extended producer responsibility, and restrictions on hazardous substances in electronic products.

Efficient Waste Management Systems

The legal framework often sets forth guidelines for the creation and upkeep of waste management infrastructure dedicated to e-waste. This encompasses the establishment of e-waste collection centres, recycling facilities, and appropriate disposal sites. Governments may also mandate the licensing or permitting of e-waste management facilities to ensure adherence to environmental and health regulations (Wath et al. 2010).

Global Cooperation and Harmonized Standards

Countries have the option to engage in international agreements or adopt international standards that pertain to e-waste management. An example is the Basel Convention on the Control of Transboundary Movements of Hazardous Wastes and their Disposal, which establishes a framework for regulating the transboundary movement of e-waste (Herat and Agamuthu 2012). Moreover, standards such as ISO 14001 for environmental management systems can assist organizations in implementing effective e-waste management practices.

Awareness and Education

Certain legal frameworks incorporate provisions for public awareness campaigns and education programs intended to educate consumers, businesses, and other stakeholders about the significance of proper e-waste management. These initiatives strive to promote responsible behavior and encourage the adoption of recycling and proper disposal practices.

PUBLIC EDUCATION AND AWARENESS REGARDING MANAGING E-WASTE

Public education and awareness regarding e-waste management have become increasingly important due to growing concerns about electronic waste's environmental and health impacts. Here are some key points to consider in promoting public awareness and education related to e-waste management:

Importance of management of e-waste: Start by educating the public about the significance of proper e-waste management. Explain the environmental hazards posed by electronic waste, such as the release of toxic substances when improperly disposed of or recycled.

Health risks: Raise awareness about the possible health risks which are associated with the e-waste, including exposure to hazardous chemicals like lead, mercury, and cadmium. Emphasize the importance of protecting human health by handling e-waste correctly (Awasthi et al. 2019).

Revitalizing and ethical discarding: Educate the public regarding the importance of responsible disposal and recycling of electronic waste. Explain how recycling can help recover valuable resources and reduce the environmental impacts of e-waste. Provide information about local e-waste recycling facilities or programs.

Consumer choices: Encourage individuals to make informed choices when purchasing electronic products. Promote the consideration of product durability, repairability, and recyclability when buying electronics. Inform them about eco-friendly certifications and labels to look for, such as energy star.

Extended producer responsibility (EPR): Educate the public about the concept of extending producer responsibility, that holds manufacturers responsible for the entire lifecycle of their products, including proper disposal and recycling. Highlight the importance of supporting companies that take responsibility for their e-waste management practices (McKerlie et al. 2006).

Donations and reuse initiatives: Promote the idea of donating or reusing electronics whenever possible. Educate the public on how to refurbish or repair electronic devices and the benefits of giving them a second life instead of disposing of them prematurely (Turban et al. 2004).

Community engagement: Organize awareness campaigns, workshops, or community events focused on e-waste management. Collaborate with local schools, universities, businesses, and organizations to spread the message effectively (Heacock et al. 2018). Encourage community members to take part in e-waste collection drives or recycling initiatives.

Online resources: Develop and provide easily accessible online resources such as informative websites, videos, and guides that explain e-waste management practices. Include step-by-step instructions on properly disposing of electronic waste and list local recycling options (Kapoor et al. 2021).

Government initiatives: Advocate for regulations and policies promoting proper e-waste management. Encourage the public to support and participate in these initiatives, such as government-led e-waste recycling programs or legislation related to responsible e-waste disposal (Wideman 2019).

Lifelong learning: Foster a culture of continuous learning by regularly updating the public about the most recent developments in e-waste management, re-cycling technologies and best practices. Encourage individuals to stay informed and share their knowledge with others. By promoting public awareness and education regarding e-waste management, we can encourage responsible consumer behavior, reduce the environmentally friendly impact of electronic waste, and protect human well-being.

TECHNOLOGICAL EVOLUTION: UNVEILING NEXT-GEN APPROACHES TO E-WASTE HANDLING

The field of e-waste management has undergone a transformation due to technological advancements, which have played a critical role in addressing the expanding challenges associated with electronic waste. Advanced recycling methods have enabled efficient separation and recovery of valuable elements from e-waste, leading to increased resource extraction and reduced waste output. Secure data wiping techniques ensure the protection of personal information and prevent data breaches. Technological advancements in design have also contributed to improved recyclability and sustainability of electronic gadgets. Tracking and management solutions for e-waste enhance transparency and accountability throughout the disposal process. Innovative resource recovery methods have been developed to capture valuable metals and minerals, reducing the demand for virgin resources. These trends and technologies reflect ongoing efforts to tackle the challenges of e-waste management, promote resource conservation, and mitigate environmental and social impacts associated with electronic waste. Some technologies are as follows:

Innovative techniques for sorting and separation: Innovations in sorting and separation technologies enable more efficient and precise reclaiming valuable materials from e-waste (Johnson et al. 2022). These technologies may include automated systems, robotics, machine learning algorithms, and optical sorting methods. In various industries and domains, innovative sorting and separation techniques play a vital role. These techniques are utilized in areas such as waste management, recycling, manufacturing, and logistics. They involve the application of advanced technologies and processes to achieve efficient and accurate sorting and separation of materials, objects, or components based on specific criteria. One significant benefit of these innovative techniques is their contribution to resource recovery and recycling. By effectively segregating and separating various types of materials, like plastic, metal, and paper, these techniques facilitate more efficient recycling processes. This not only reduces the pressure on natural resources but also fosters a closed-loop economy by maximizing the reuse of important materials and minimizing waste generation (Hsu et al. 2019).

Civic mining: The practise of removing valuable materials and resources from electronic waste (e-waste) produced in metropolitan settings is referred to as civic mining or urban mining. Discarded electronics like computers, smartphones, televisions, and other gadgets are referred to as "e-waste." These gadgets contain priceless metals that may be extracted and recycled, including rare earth elements, gold, silver, and copper (Arya and Kumar 2020). Due to the increase in global e-waste generation and the potential benefits of recycling electronic equipment for both the environment and the economy, urban mining has attracted more attention.

Responsible design and prolonged product lifespan: Innovations in product design are focused on creating electronics that are easier to repair, upgrade, and recycle. This includes modular designs, standardized components, and easily accessible repair manuals (Shahbazi and Jönbrink 2020). Additionally, initiatives promoting product life extensions, such as repair cafes and community-driven repair programs, are gaining traction. Energy efficiency is a crucial aspect of responsible design. By creating electronic devices that consume less power while in operation, the overall energy demand can be reduced, resulting in a decrease in carbon emissions. Energy-efficient designs not only have positive environmental impacts but also offer cost savings to users, which encourages the widespread adoption of sustainable technologies (Pacheco et al. 2012).

Sustainable resource circulation: Circular economy approaches are gaining attention as a means to minimize waste and maximize resource efficiency. This involves designing products with the intention of easy repair, reuse, and recycling, as well as implementing closed-loop supply chains and promoting product take-back programs (Mishra et al. 2023). The adoption and implementation of sustainable environmental practices for managing e-waste brings about several advantages. Firstly, it aids in the reduction of virgin material extraction from the Earth, which often involves destructive mining practices and contributes to habitat destruction and pollution. Through the recycle and re-use of e-waste, the demand for raw resources is minimized, leading to the conservation of natural resources and a decrease in associated ecological impacts. Secondly, sustainable resource circulation of e-waste helps prevent the release of hazardous substances into the environment. Many electronic devices contain hazardous components like cadmium, lead and mercury which can pollute soil and water if not handled correctly. By safely recycling e-waste, we ensure that these hazardous materials are appropriately managed, mitigating the risk they pose to human health and the environment.

Eco-friendly and renewable materials: Researchers and manufacturers are exploring the use of biodegradable and sustainable materials in electronic devices (Khalil et al. 2016). This includes bioplastics, bio-based circuit boards, and environmentally friendly alternatives to hazardous substances used in electronic components. The global trend towards sustainability has facilitated the embrace of eco-friendly and renewable materials in electronic manufacturing. With increasing awareness of the environmental impact, industries are actively seeking alternatives that reduce carbon emissions and waste generation. In the field of electronics, numerous innovative materials have emerged as transformative options. Among these materials is bamboo, a rapidly growing and renewable resource. Bamboo possesses impressive strength and flexibility, making it an excellent substitute for conventional plastics used in electronic casings. In addition to being biodegradable, bamboo requires considerably less energy for production, thereby further minimizing the carbon footprint associated with electronic devices.

Connected devices for e-waste management: IoT technologies can be utilized to enhance e-waste management processes. Smart tracking systems, sensors, and connected devices can help monitor and optimize e-waste collection, logistics, and recycling operations (Singh et al. 2021).

Distributed e-waste management: Instead of relying solely on large-scale centralized recycling facilities, decentralized processing methods are being explored (Rauland et al. 2015). This involves the establishment of smaller-scale recycling units in communities to reduce transportation and carbon footprint while promoting local job creation.

Eco-design and eco-labelling: Eco-design principles are being integrated into electronic product development to minimize the environmental impact at the design stage itself (Sanyé et al. 2014). Eco-labelling schemes are also being developed to provide consumers with information about the environmental performance of electronic devices, encouraging them to make sustainable choices.

Distributed ledger technology for transparency and traceability: Blockchain technology offers potential solutions for ensuring transparency and traceability in the management process of e-waste (Gheorghe and Ishii 2007). It can be used to track the entire lifecycle of electronic devices, from production to disposal, promoting accountability and reducing the risk of illegal e-waste dumping.

Joint ventures and public-private collaborations: Various collaborative initiatives and public-private partnerships are being formed to address the difficulties associated with e-waste management. These partnerships involve cooperation between governments, businesses, NGOs, and research institutions to share knowledge, resources, and expertise for developing sustainable solutions (Awasthi et al. 2019). The need for collaboration between joint ventures and the public and corporate entities in the field of e-waste management has become increasingly important in contemporary world. E-waste, which includes end-of-life electronic equipment and devices, poses a serious risk to the environment and health due to its dangerous components and improper disposal. Joint ventures and public-private collaborations play a crucial role in tackling the complexities associated with e-waste management. By pooling together resources, knowledge, and technological capacities, these partnerships enable the creation of eco-friendly solutions, mitigate environmental and health hazards, and ensure the appropriate discarding of disposal and recycling of electronic wastes. Through collaborative endeavours, we can strive towards a more sustainable and accountable approach to managing e-waste, safeguarding the surroundings and the well-being of future generations.

CONCLUSION

The management of electronic waste poses a formidable challenge that necessitates a comprehensive strategy encompassing technological progress, policy adjustments, and shifts in behavior. By embracing a comprehensive outlook and tackling e-waste management from its fundamental origins to its viable remedies, we can make significant strides toward a future that is both sustainable and environmentally mindful. Through collaborative endeavors involving stakeholders from different sectors, we can efficiently handle e-waste, minimize its ecological repercussions, and capitalize on its economic prospects, thereby safeguarding the well-being of future generations and promoting a healthier planet.

FUTURE PROSPECTS OF MANAGING E-WASTE

Robotic systems, automated processes, machine learning (MI) and artificial intelligence (AI) are examples of technological developments that can enhance the procedures for sorting, recycling, and identifying e-waste. Real-time monitoring of e-waste flows is made possible by advanced sensor technology for better tracking and collection. Approaches to the circular economy place a strong emphasis on resource recovery using cutting-edge technologies, expanded producer accountability, and sustainable product design. Advanced recycling techniques, ecologically friendly disposal methods, and the phase-out of hazardous compounds are all components of sustainable disposal and treatment. International cooperation, stricter enforcement, and financial incentives are only a few of the policy and regulatory approaches. Campaigns for public awareness and education encourage sustainable consumption and appropriate disposal. E-waste management also promotes entrepreneurship and innovation by producing green jobs and economic prospects.

REFERENCES

Ahirwar, R. and Tripathi, A.K. 2021. E-waste management: a review of recycling process, environmental and occupational health hazards, and potential solutions. Environmental Nanotechnology, Monitoring and Management. 15: 100409.

Arya, S. and Kumar, S. 2020. Bioleaching: urban mining option to curb the menace of E-waste challenge. Bioengineered. 11(1): 640–60.

Awasthi, A.K., Li, J. and Koh, L. 2019. Ogunseitan O.A. Circular economy and electronic waste. Nature Electronics. 2(3): 86–9.

Bailey-Van Kuren, M. 2002 (May). Automated demanufacturing studies in detecting and destroying threaded connections for processing electronic waste. *In*: Conference Record 2002 IEEE International Symposium on Electronics and the Environment (Cat. No. 02CH37273): 295–298. IEEE.

Baldé, C.P., Forti, V., Gray, V., Kuehr, R. and Stegmann, P. 2017. The global e-waste monitor 2017: quantities, flows and resources. United Nations University, International Telecommunication Union, and International Solid Waste Association.

Bandyopadhyay, A., Chatterjee, R. and Das, N. 2022. E-waste management in digital healthcare system and sustainability paradigm. pp. 157–166. *In*: Paul, P.K., Choudhury, A., Biswas, A. and Singh, B.K. (eds). Environmental Informatics: Challenges and Solutions. Singapore: Springer Nature Singapore.

Banerjee, D., Kushwaha, N., Ahmad, E., Quereshi, S. and Nigam, K.D.P. 2023. Eco-design strategies for recycling of e-waste. pages 26. *In*: Chauhan, G. and Sevda, S. (eds). Solid Waste Management. Boca Raton: CRC Press.

Barwood, M., Li, J., Pringle, T. and Rahimifard, S. 2015. Utilisation of reconfigurable recycling systems for improved material recovery from e-waste. Procedia CIRP. 29: 746–751.

Brunner, P.H. and Rechberger, H. 2004. Practical handbook of material flow analysis. The International Journal of Life Cycle Assessment. 9: 337–338.

Calliafas, P., Bates, M.P., Griffiths, G., Harding, A., Hawkes, A., Holloway, L., et al. 2012. Waste Electrical and Electronic Equipment (WEEE) regulations: individual producer responsibility (IPR) in a UK context.

Dauvergne, P. and Lister, J. 2013. Eco-Business: A Big-Brand Takeover of Sustainability. MIT Press.

Forti, V., Baldé, C.P., Kuehr, R. and Bel, G. 2020. The global e-waste monitor. United Nations University (UNU), International Telecommunication Union (ITU) and International Solid Waste Association (ISWA), Bonn/Geneva/Rotterdam. 120.

Ganguly, R. 2016. E-waste management in India–an overview. International Journal of Earth Sciences and Engineering. 9: 574–88.

Garg, N. and Adhana, D. 2019. E-waste management in India: a study of the current scenario. International Journal of Management, Technology and Engineering. 9. Available at SSRN: https://ssrn.com/abstract=3356877

Gheorghe, R. and Ishii, K. 2007. Eco-design value alignment: keys to success. *In*: ASME International Mechanical Engineering Congress and Exposition. 200: 267–277.

Gurauskienė, I. 2008. Behaviour of consumers as one of the most important factors in e-waste problem. Environmental Research, Engineering and Management. 46(4): 56–65.

Heacock, M., Trottier, B., Adhikary, S., Asante, K.A., Basu, N., Brune, M.N., et al. 2018. Prevention-intervention strategies to reduce exposure to e-waste. Reviews on Environmental Health. 33(2): 219–228.

Herat, S. and Agamuthu, P. 2012. E-waste: a problem or an opportunity? Review of issues, challenges and solutions in Asian countries. Waste Management and Research. 30(11): 1113–1129.

Hickle, G.T. 2013. Comparative analysis of extended producer responsibility policy in the United States and Canada. Journal of Industrial Ecology. 17(2): 249–261.

Hoang, A.Q., Tue, N.M., Tu, M.B., Suzuki, G., Matsukami, H., Tuyen, L.H., et al. 2023. A review of management practices, environmental impacts, and human exposure risks related to electrical and electronic waste in vietnam: findings from case studies in informal e-waste recycling areas. Environmental Geochemistry and Health. 45(6): 2705–2728.

Hsu, E., Barmak, K., West, A.C. and Park, A.H. 2019. Advancements in the treatment and processing of electronic waste with sustainability: a review of metal extraction and recovery technologies. Green Chemistry. 21(5): 919–936.

Ishchenko, V., Pohrebennyk, V., Kochanek, A. and Hlavatska, L. 2019. Waste electrical and electronic equipment management in Ukraine. *In*: Proceedings of International Conference on Geosciences. Saima Consult LTD. 197–204.

Islam, M.T. and Huda, N. 2019. Material flow analysis (MFA) as a strategic tool in e-waste management: applications, trends and future directions. Journal of Environmental Management. 244: 344–361.

Johnson, M., Khatoon, A. and Fitzpatrick, C. 2022. Application of AI and machine vision to improve battery detection and recovery in e-waste management. *In*: International Conference on Electrical, Computer, Communications and Mechatronics Engineering (ICECCME) IEEE. 1–6.

Joseph, K. 2007. Electronic waste management in India–issues and strategies. *In*: Eleventh International Waste Management and Landfill Symposium, Sardinia.

Kalia, P., Zia, A. and Mladenović, D. 2022. Examining country development indicators and e-waste under the moderating effect of country development levels and e-waste policy. International Journal of Quality and Reliability Management. 39(7): 1601–1616.

Kamel, R.M., Shahat, A., Atta, A.H. and Farag-Allah, M.M. 2022. Development of a novel and potential chemical sensor for colorimetric detection of Pd (II) or Cu (II) in e-wastes. Microchemical Journal. 172: 106951.

Kang, H.Y. and Schoenung, J.M. 2005. Electronic waste recycling: a review of US infrastructure and technology options. Resources, Conservation and Recycling. 45(4): 368–400.

Kapoor, N., Sulke, P. and Badiye, A. 2021. E-waste forensics: an overview. Forensic Science International: Animals and Environments. 1: 100034.

Khalil, H.A., Davoudpour, Y., Saurabh, C.K., Hossain, M.S., Adnan, A.S., Dungani, R., et al. 2016. A review on nanocellulosic fibres as new material for sustainable packaging: process and applications. Renewable and Sustainable Energy Reviews. 64: 823–836.

Khan, M.A.R., Motalib, M.A. and Saadat, A.H.M. 2022. Mobile phone and spent laptop waste: generation, dissemination and flow pattern analysis. Environmental Engineering and Management Journal. 21(8): 1339–1349.

Khetriwal, D.S., Kraeuchi, P. and Widmer, R. 2009. Producer responsibility for e-waste management: key issues for consideration–learning from the Swiss experience. Journal of Environmental Management. 90(1): 153–165.

Li, X., Vázquez-López, A., Sánchez del Río Sáez, J. and Wang, D.Y. 2022. Recent advances on early-stage fire-warning systems: mechanism, performance, and perspective. Nano-micro letters. 14(1): 197.

Lou, J.E. 2022. Developing an extended theory of planned behavior model for small e-waste recycling: an analysis of consumer behavior determinants. Journal of Environmental Science and Engineering. 11: 71–86.

Manhart, A. 2011. International cooperation for metal recycling from waste electrical and electronic equipment: an assessment of the "best-of-two-worlds" approach. Journal of Industrial Ecology. 15(1): 13–30.

Masud, M.H., Akram, W., Ahmed, A., Ananno, A.A., Mourshed, M., Hasan, M., et al. 2019. Towards effective e-waste management in Bangladesh: a review. Environmental Science and Pollution Research. 26(2): 1250–1276.

McKerlie, K., Knight, N. and Thorpe, B. 2006. Advancing extended producer responsibility in Canada. Journal of Cleaner Production. 14(6–7): 616–628.

Miller, G.Z., Tighe, M.E., Peaslee, G.F., Peña, K. and Gearhart, J. 2016. Toys, décor, and more: evidence of hazardous electronic waste recycled into new consumer products. Journal of Environmental Protection. 7(03): 341.

Mishra, A., Dutta, P., Jayasankar, S., Jain, P. and Mathiyazhagan, K. 2023. A review of reverse logistics and closed-loop supply chains in the perspective of circular economy. Benchmarking: An International Journal. 30(3): 975–1020.

Monier, V., Porsch, L., Hestin, M., Cavé, J., Laureysens, I., Watkins, E., et al. 2014. Development of Guidance on Extended Producer Responsibility (EPR). Final Report, European Commission – DG Environment; 40-76.

Needhidasan, S., Samuel, M. and Chidambaram, R. 2014. Electronic waste—an emerging threat to the environment of urban India. Journal of Environmental Health Science and Engineering. 12: 1–9.

Pacheco, R., Ordóñez, J. and Martínez, G. 2012. Energy efficient design of building: a review. Renewable and Sustainable Energy Reviews. 16(6): 3559–3573.

Perkins, D.N., Drisse, M.N., Nxele, T. and Sly, P.D. 2014. E-waste: a global hazard. Annals of Global Health. 80(4): 286–295.

Quinn, L. and Sinclair, A.J. 2006. Policy challenges to implementing extended producer responsibility for packaging. Canadian Public Administration. 49(1): 60–79.

Rauland, V. and Newman, P. 2015. Decarbonising Cities: Mmainstreaming Low Carbon Urban Development. Springer.

Ravindra, K. and Mor, S. 2019. E-waste generation and management practices in Chandigarh, India and economic evaluation for sustainable recycling. Journal of Cleaner Production. 221: 286–294.

Rodríguez-Padrón, D., ALOthman, Z.A., Osman, S.M. and Luque, R. 2020. Recycling electronic waste: prospects in green catalysts design. Current Opinion in Green and Sustainable Chemistry. 25: 100357.

Sanyé-Mengual, E., Lozano, R.G., Farreny, R., Oliver-Solà, J., Gasol, C.M. and Rieradevall, J. 2014. Introduction to the eco-design methodology and the role of product carbon footprint. Assessment of Carbon Footprint in Different Industrial Sectors. 1–24.

Shahbazi, S. and Jönbrink, A.K. 2020. Design guidelines to develop circular products: Action research on Nordic industry. Sustainability. 12(9): 3679.

Shinkuma, T. and Huong, N.T. 2009. The flow of e-waste material in the Asian region and a reconsideration of international trade policies on e-waste. Environmental Impact Assessment Review. 29(1): 25–31.

Singh, K., Arora, G., Singh, P. and Gupta, A. 2021. IoT-based collection vendor machine (CVM) for e-waste management. Journal of Reliable Intelligent Environments. 7: 35–47.

Terazono, A., Murakami, S., Abe, N., Inanc, B., Moriguchi, Y., Sakai, S.I., et al. 2006. Current status and research on e-waste issues in Asia. Journal of Material Cycles and Waste Management. 8: 1–12.

Thanomsangad, P., Tengjaroenkul, B., Sriuttha, M. and Neeratanaphan, L. 2019. Heavy metal accumulation in frogs surrounding an e-waste dump site and human health risk assessment. Human and Ecological Risk Assessment: An International Journal. 26(5): 1313–1328. https://doi.org/10.1080/10807039.2019.1575181

Thapa, K., Vermeulen, W.J., Deutz, P. and Olayide, O. 2023. Ultimate producer responsibility for e-waste management—a proposal for just transition in the circular economy based on the case of used European electronic equipment exported to Nigeria. Business Strategy and Development. 6(1): 33–52.

Tsang, A.H. 2002. Strategic dimensions of maintenance management. Journal of Quality in Maintenance Engineering. 8(1): 7–39.

Tsydenova, O. and Bengtsson, M. 2011. Chemical hazards associated with the treatment of waste electrical and electronic equipment. Waste Management. 31(1): 45–58.

Turban, E., King, D., Lee, J. and Viehland, D. 2004. Electronic Commerce: A Managerial Perspective. Pearson Education.

Wath, S.B., Vaidya, A.N., Dutt, P.S. and Chakrabarti, T. 2010. A roadmap for the development of sustainable e-waste management system in India. Science of the Total Environment. 409(1): 19–32.

Wibowo, S. and Deng, H. 2015. Multi-criteria group decision making for evaluating the performance of e-waste recycling programs under uncertainty. Waste Management. 40: 127–135.

Wideman, B.N. 2019. Grappling with the african e-waste pandemic: Contributing factors and future deterrence. Dissertations and Theses. Paper 4828. https://doi.org/10.15760/etd.6704

Yanagitani, K. and Kawahara, K. 2000. LCA study of air conditioners with an alternative refrigerant. The International Journal of Life Cycle Assessment. 5: 287–290.

Yu, Z., Gao, C., Yang, C. and Zhang, L. 2023. Insight into quantities, flows, and recycling technology of e-waste in China for a resource sustainable society. Journal of Cleaner Production. 393: 136222.

Chapter **10**

Implications of Microbes in Biorefinery and Curtailing Petroleum Industrial Wastes

Roxiette Heromina Siqueira, Savia Bernadette Botelho and Siddhi Kashinath Jalmi*

Dr. Siddhi Kashinath Jalmi, School of Biological Sciences and Biotechnology, Goa University, Taleigao Plateau, Goa, 403206, India.

INTRODUCTION

The high global demand for petroleum oil, a naturally occurring non-renewable resource, has resulted in its widespread use in various industrial processes. Extensive drilling, storage, distribution, and international transportation of petroleum and its derivatives have become common practices. However, these processes release contaminants that contribute to air, water, and soil pollution. Crude oil, diesel, and gasoline, collectively known as Total Petroleum Hydrocarbons (TPH), exhibit recalcitrant characteristics. Petroleum refineries and petrochemical plants are encountering challenges in effectively disposing of the waste and wastewater they generate. The wastewater discharged by petroleum industries contains a variety of organic and inorganic pollutants, including benzene, toluene, ethylbenzene, and xylene (BTEX), sulfides, hydrocarbons, phenol, polyaromatic hydrocarbons (PAHs), and heavy metals (He and Jiang 2008, Raza et al. 2019, Usman et al. 2012, Varjani 2017). These pollutants pose significant risks to both the environment

* For Correspondence: siddhi@unigoa.ac.in

and human health (Perera et al. 2012, Raza et al. 2019). The contamination of soil with petroleum hydrocarbons has far-reaching consequences for the local ecosystem, as the accumulation of pollutants within plant and animal tissue can lead to mutations and mortality. Petroleum refinery emissions pose significant risks of cancer, cardiovascular disorders, neurotoxic, respiratory, and reproduction disorders (Otitolaiye and Al-Harethiya 2022) (Figure 10.1).

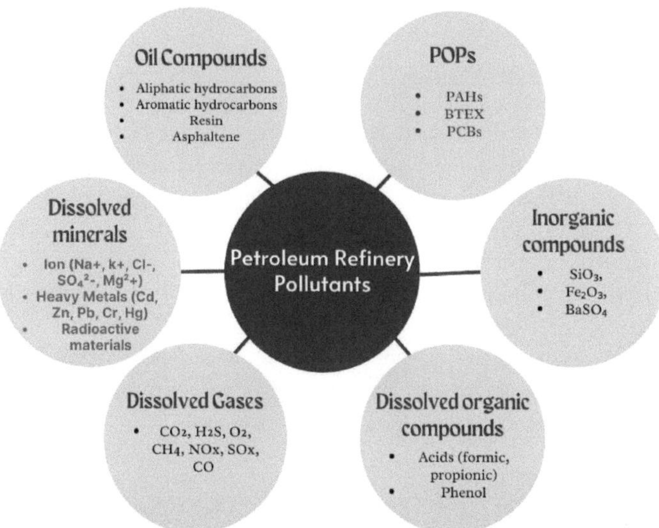

Figure 10.1 Major petroleum refinery pollutants found in waste/wastewater and refinery emissions, degraded by microbes converting them to non-toxic simpler compounds.

Several physical and chemical methods are employed to degrade these petroleum contaminants. However, these methods often come with high economic costs and may result in incomplete conversion of pollutants, potentially forming more harmful intermediates. In some cases, these methods merely transfer pollutants from one phase, such as soil, to another, like air, rather than effectively remediating the contamination. As a result, biodegradation-based remediation offers several advantages over conventional physicochemical technologies (viz. eco-friendly, economical, and beneficial concerning mineralization of contaminants) (Kumar and Khanna 2010, Suja et al. 2014). Microorganisms play a crucial role in the bioconversion and total breakdown of pollutants and other environmental xenobiotics. Even though the impacts of bacteria (microbes) on hydrocarbons have been understood for many years, bioremediation has emerged as a promising and highly effective method for treating contaminated sites. Its cost-effectiveness and ability to achieve complete mineralization make it a viable solution for thorough remediation (Speight and El-Gendy 2018). Bioremediation primarily relies on biodegradation, which can involve either the complete conversion of organic contaminants into carbon dioxide, water, cell protein, and inorganic compounds (mineralization), or the transformation of complex organic pollutants into simpler organic compounds that pose no harm to the environment (Obi et al. 2016).

The world population depends on the dwindling reserves of fossil fuels for energy and chemicals. The increasing use of fossil fuels negatively impacts the environment, leading to global warming, climate change, and adverse effects on human health (Bušić et al. 2018, Tursi 2019). Hence, it is imperative to identify alternate resources to produce these energies and chemicals to meet the population's growing demands (Takkellapati et al. 2018). Biomass can thus be used as an alternate source of energy. An increasing world population and rapid industrialization have resulted in the generation of enormous wastes like food waste, solid municipal waste, organic waste, agricultural residues, and industrial waste, such as pulp and paper mills. Biomass waste valorization is of very high importance as it offers a more sustainable solution for the conversion of harvested and waste biomass. These wastes can be converted into valuable renewable energy sources and chemicals through different technologies including biological approaches (Esposito and Antonietti 2015, Harirchi et al. 2022, Takkellapati et al. 2018). In this chapter, we will be discussing about the use of microbes as a biological approach in converting biomass to various forms of renewable energy. Also, the use of microbes in curtailing the petroleum industry waste will be discussed. First, let us get acquainted with the microbial processes involved in petroleum waste management and the role of microbes in producing some of the important renewable sources of energy from biomass.

BIOREMEDIATION OF PETROLEUM WASTE

Bioremediation is the utilization of biological organisms, including microscopic organisms like fungi, algae, and bacteria, to metabolically remove or neutralize environmental pollutants. The term "biological" refers to these organisms, while "remediation" pertains to the treatment of the environmental issue. Although microbes are the primary agents in this technology, the incorporation of plants can expedite the bioremediation process due to the beneficial conditions provided by the rhizosphere (Stout and Nüsslein 2010, Wenzel 2009). This approach, known as phytoremediation, harnesses the power of plants to assist in the remediation of polluted areas. Bioremediation offers flexibility in its application, allowing for both *ex-situ* and *in-situ* approaches based on various factors like cost, site characteristics, and the type and concentration of pollutants. It stands out as the most effective, cost-efficient, and environmentally friendly method for managing polluted environments. Several bioremediation treatments exist, encompassing the use of specific fungi species (mycoremediation), plants (phytoremediation, rhizofiltration), and microorganisms in general, along with their products or nutrients (bioaugmentation, biostimulation, biopiles, natural attenuation, and biosurfactants) (Leonardo et al. 2016, Njoku et al. 2017).

Bioremediation involves three fundamental principles: natural attenuation (NA) (relying solely on the inherent capacity of the native microbial community to degrade pollutants); biostimulation (BS) (involves providing organic or inorganic nutrients to enhance the function of the native community in a nutritionally imbalanced environment); and bioaugmentation (BA) (where specialized microbes are introduced into the targeted environment to aid in pollutant removal) (Li et al. 2009, Tyagi et al. 2011, Viñas et al. 2005, Yu et al. 2005).

Bacterial Bioremediation of Petroleum Waste

In the bioremediation of petroleum-contaminated soils, bacteria are the preferred microorganisms due to their abundance, fast growth, and ability to degrade a wide range of petroleum products (Wolicka et al. 2009). Several bacteria capable of breaking down petroleum hydrocarbons have been identified (Tremblay et al. 2017). Additionally, specific microorganisms play a crucial role in the degradation of petroleum hydrocarbons, as their abundance increases significantly after contamination (Yakimov et al. 2007). To enhance the degradation of petroleum, a combination of efficient indigenous bacteria with distinct degrading enzymes was used for remediating the contaminated soils. The synergistic effect of the native bacterial consortium and introduced exogenous bacteria significantly improves the efficiency of petroleum degradation (Tao et al. 2017). The major phyla of hydrocarbon degrading microorganisms found in refinery waste were *Proteobacteria*, *Actinobacteria*, *Acidobacteria*, *Bacteroidetes*, *Chloroflexi*, *Firmicutes*, *Spirochaetes*, *Thermotogae*, and *Caldiserica* (Table 10.1) (Das and Kazy 2014, Vasconcellos et al. 2010), which work in a coordinated manner. Through culture-based and enrichment studies, a diverse range of microbes capable of degrading complex hydrocarbons has been identified. These studies have also confirmed the selective ability of these microbes to utilize certain classes of hydrocarbons (Prince et al. 2018). Metabolic and genetic investigations have provided insights into the pathways used by these microbes to activate chemically inert hydrocarbons under aerobic and anaerobic conditions (Abbasian et al. 2015, Rojo 2009) (Figure 10.2).

Micro-Fungal Bioremediation of Petroleum Waste (Mycoremediation)

The organic matter present in wastes is utilized by fungi as a source of nutrients through their enzymatic activity. Fungi exhibit a notable ability to produce and release a wide range of extracellular enzymes into their surrounding environment, facilitating the degradation of diverse substrates into smaller molecules that can be absorbed and metabolized within their cells (Sardrood et al. 2013). Consequently, the process of biodegradation primarily occurs outside the fungal cell or its associated structures. The advantages of mycoremediation include the fungi's rapid growth rate, their capacity to effectively penetrate pollutants, and their comparatively higher survival rates even under stressful conditions such as low pH and nutrient scarcity (Davies and Westlake 1979).

Additionally, the presence of nutrient stress condition serves as a stimulus for the enzymatic machinery of fungi, which presents a signifiant advantage for bioremediation efforts (Aust et al. 2003, Mansur et al. 2003). Furthermore, fungi possess the ability to rapidly multiply their population due to their short life

Table 10.1 Bacterial species involved in curtailing different petroleum products

Microbial types	Petroleum pollutant	Reference
Bacteria		
Rhodococcus corynebacterioides	Crude oil	(Gentili et al. 2006)
Bacillus cereus	n-alkanes	(Liu et al. 2020)
Pseudomonas sp.	Phenol, PAH	(Xia et al. 2014)
Marinobacter sp.	PAH and alkane compounds.	(Dwinovantyo et al. 2016)
Bacillus megaterium	Oil sludge, aromatic compounds	(Cerqueira et al. 2011)
Sphingomonas sp.	Crude oil	(Adebusoye et al. 2007)
Microfungi		
Aspergillus niger	Alkanes, aromatics, nitrogen, sulfur, and oxygen (NSO)–containing compounds (PAHs)	(Okoro 2008)
Penicillium sp.	Crude oil	(Burghal et al. 2016)
Fusarium solani	Hydrocarbons	(Nasrawi 2012)
Candida tropicalis	Crude oil	(Farag and Soliman 2011)
Rhizopus stolonifer	Total Petroleum Hydrocarbon Content (TPHC)	(Okoro et al. 2013)
Microalgae		
Prototheca Zopfii	Resins, asphaltenes, aromatic hydrocarbons	(Walker et al. 1975a,b)
Chlorella vulgaris	Pyrene, fluoranthene, and their mixture, total petroleum hydrocarbons (TPH)	(Madadi et al. 2016)
Porphyridium sp.	Petroleum waste	(Vidyashankar and Ravishankar 2016)
Nostoc sp.	Petroleum wastes	(Ogidi and Njoku 2017)
Scenedesmus platydiscus	Fluoranthene and pyrene	(Lei et al. 2007)
Arthrospira platensis	Nickel, Copper	(Budi et al. 2020, Cepoi et al. 2022)
Rhizospheric microbes		
Rhizobium meliloti	PAH, crude oil, paraffin, toluene, naphthalene, phenanthrene	(Pourbabaee et al. 2021)
Pantoea sp.	Mixture of hydrocarbons	(Poddar et al. 2019, Yousaf et al. 2010)
Pseudomonas putida	PCBs	(Narasimhan et al. 2003)
Azospirillum lipoferum	Crude oil	(Muratova et al. 2005)
Enterobacter cloacae	Total petroleum hydrocarbons (TPHs)	(Huang et al. 2005)
Pseudomonas fluorescens	Trichloroethylene (TCE)	(Yee et al. 1998)

cycles and high rates of sporulation. Several fungal genera, including *Aspergillus, Saccharomyces, Candida hansenula* (Bartha and Atlas 1977), and *Cladosporium* (Walker et al. 1973) (Table 10.1) are utilized in the bioremediation of oil-contaminated areas due to their ability to metabolize hydrocarbons. Microfungi

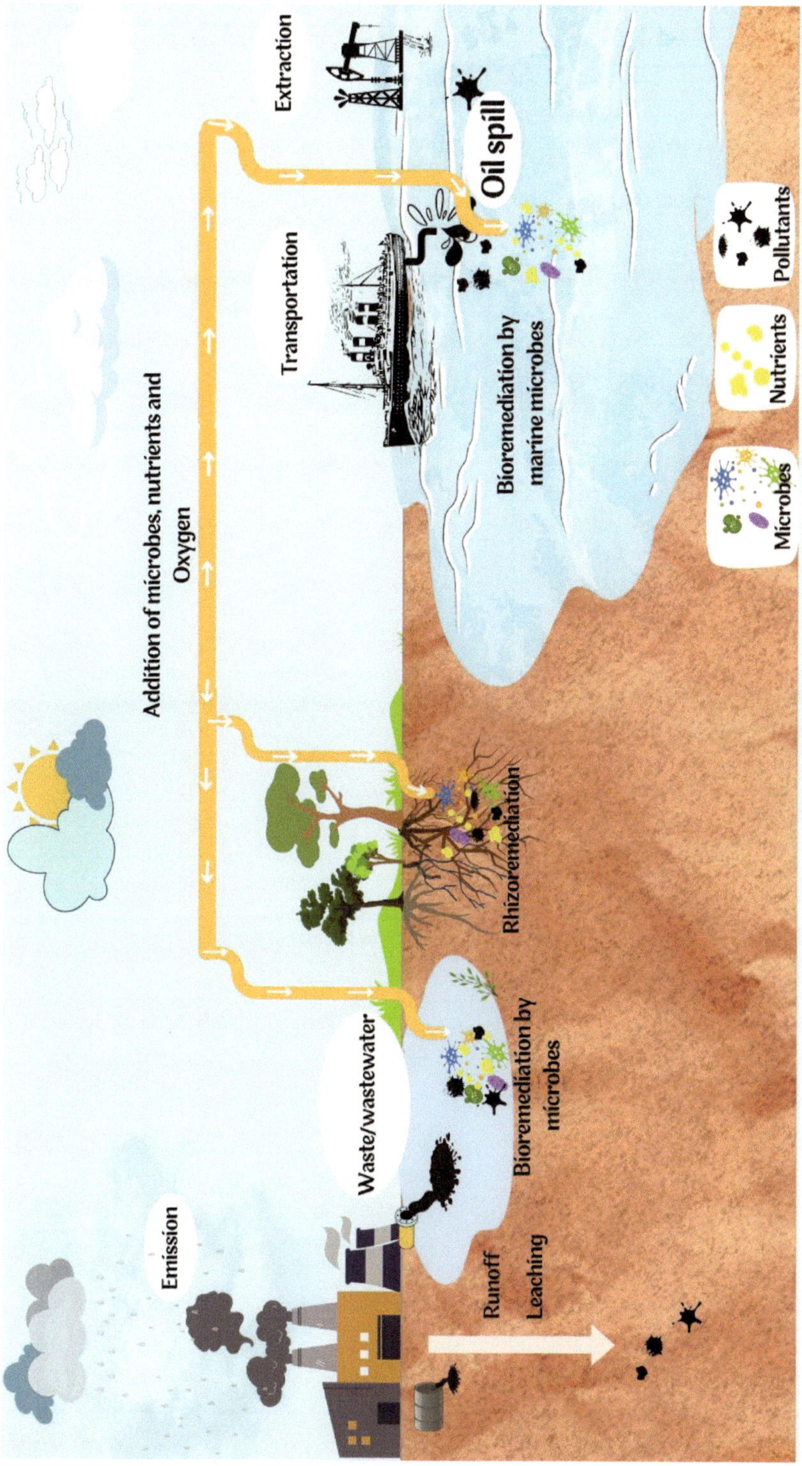

Figure 10.2 Bioremediation of petroleum pollutants scattered during the various processes of petroleum product recovery. Addition of specific microbes, nutrients and oxygen enhances the remediation process.

Table 10.2 Microbial Enzymes in the Bioremediation of Petroleum Pollutants

Enzyme	Applications	Microbes	References
Laccase	Detoxifies halogenated phenols, polyphenols, aromatic amines, nitro substituted compounds such as trinitrotoluene, hydroxy-indoles, and PAHs.	*Streptomyces* sp., *B. subtilis, Pseudomonas putida, Escherichia coli, Campylobacter jejuni, Thermus thermophiles. case (CueO), Aspergillus* sps., *Trichoderma reesei*	(Couto and Toca-Herrera 2007, Durán and Esposito 2000, Guan et al. 2018, Serrano-Posada et al. 2015, Silva et al. 2012, Strong and Claus 2011).
Mono-, Dioxygenase	Dehalogenation, desulfurization, denitrification, and hydroxylation of aromatic compounds, bioremediation of pentachlorophenol, phenol, p-nitrophenol, 2, 4, 6-Trichlorophenol, C10-C30 alkanes	*Pseudomonas putida, Nocardiodes* sp., *Mycobacterium* sp., *Acinetobacter* sp.	(Abbasian et al. 2015, Hamamura et al. 2001, Moody et al. 2001)
Cytochrome P450	Degrade xenobiotics like PAHs, polyhalogenated aromatics, linear alkanes	*Bacillus megaterium, Acinetobacter calcoaceticus, Rhodococcus* sp.	(Carmichael and Wong 2001) (Bhandari et al. 2021)
Dehalogenase	Mineralization of halogenated xenobiotics	*Bacillus* sp. *GZT, dehalococcoides mccartyi, Pseudomonas* sp., *Dehalobacter* sp., *Desulfitobacterium frappieri*	(Bhandari et al. 2021, Liu et al. 1994, Zu et al. 2012)
Dehydrogenase	Degrades TBP (tribromophenol) haloalkanoate. Polyethylene glycol, ethylbenzene	*Rhizobium* sp., *Sphingomonas* sp., *Azoarcus* sp.,	(Cairns et al. 1996, Kalimuthu et al. 2015, Sugimoto et al. 2001)
Hydrolase	Degradation of Phthalates	*P. aeruginosa, Acinetobacter* sp., *Micrococcus* sp., *Mycobacterium* sp., *Rhodococcus* sp.	(Casas-Godoy et al. 2012, Jiao et al. 2013, Verma et al. 2012)
Lipase	Degrades TPH	*Shewanella chilikensis, Bacillus firmus, Halomonas hamiltonii,*	(Suganthi et al. 2018)

such as Arbuscular mycorrhiza (AM) and yeast have been used to remediate petroleum-contaminated soils. The strain *Aspergillus fumigatus* has demonstrated the removal of Pb(II) ions from aquatic solutions. AM fungi, known for their extensive mycelia network, play a crucial role in the sequestration of heavy metals and creating a metal-free environment in the root zone through the release of glomalin protein (Muthukrishnan et al. 2018). The glomalin protein, present in the cell wall of AM fungi, is capable of binding various types of heavy metals (Muthukrishnan et al. 2018). These arbuscular mycorrhizal fungi provide significant benefits to scientists as they not only sequester heavy metals but also enhance soil nutrient content (Barea et al. 2005) (Figure 10.2).

Micro-Algal Bioremediation of Petroleum Waste

Cyanobacteria, referred to as blue-green algae, play a significant role in bioremediation. They accumulate and transform heavy metals as part of their metabolic activities, thereby reducing the presence of heavy metal contaminants in soils (Huertas et al. 2014). The high photosynthetic activity of many cyanobacterial species allows them to efficiently remove organic pollutants from aquatic environments, leading to the formation of substantial biomass (Chekroun et al. 2014, Semple et al. 1999). Cyanobacteria such as *Chlorella vulgaris, Scenedesmus quadricauda, S. capricornutum,* and *S. platydiscus* can efficiently oxidize PAHs present in crude oil pollutants (Cerniglia 1984) (Table 10.1). Due to microalgae's high lipid production potential, they can serve as alternative sources for biodiesel production and energy generation. For example, *Chlorella* sp. can degrade emulsified petroleum oil and can further be used for biodiesel production (Chisti 2007, Kuttiyathil et al. 2021). *Prototheca zopfii* exhibited significant potential for degradation and utilized crude oil or a mixture hydrocarbon substrates (Walker et al. 1975a,b) (Table 10.1). *Spirulina* sp. was used for biosorption of Chromium (Cr^{3+}), Cadmium (Cd^{2+}) and Copper (Cu^{2+}) ions (Chojnacka et al. 2005) (Table 10.1). In the presence of specific bacterial groups, certain microalgae have the ability to enhance the degradation of pyrene and fluoranthene. Mixotrophic strains of cyanobacteria and microalgae exhibit significant potential for bioremediation (Chekroun et al. 2014). This mode of nutrition offers numerous competitive advantages over other microbial groups such as fungi and bacteria, for the degradation of organic pollutants (Subashchandrabose et al. 2013) (Figure 10.2).

Bioremediation by Rhizosphere Microorganisms (Rhizoremediation)

The interaction between plants and their rhizospheric microorganisms can contribute to the degrdation of petroleum hydrocarbon. Plant root exudates provide carbon and nitrogen sources that support growth of microorganisms. This leads to significant increase in microbial activity within the rhizosphere, surpassing the population found in the surrounding soil by orders of magnitude. Enzymes

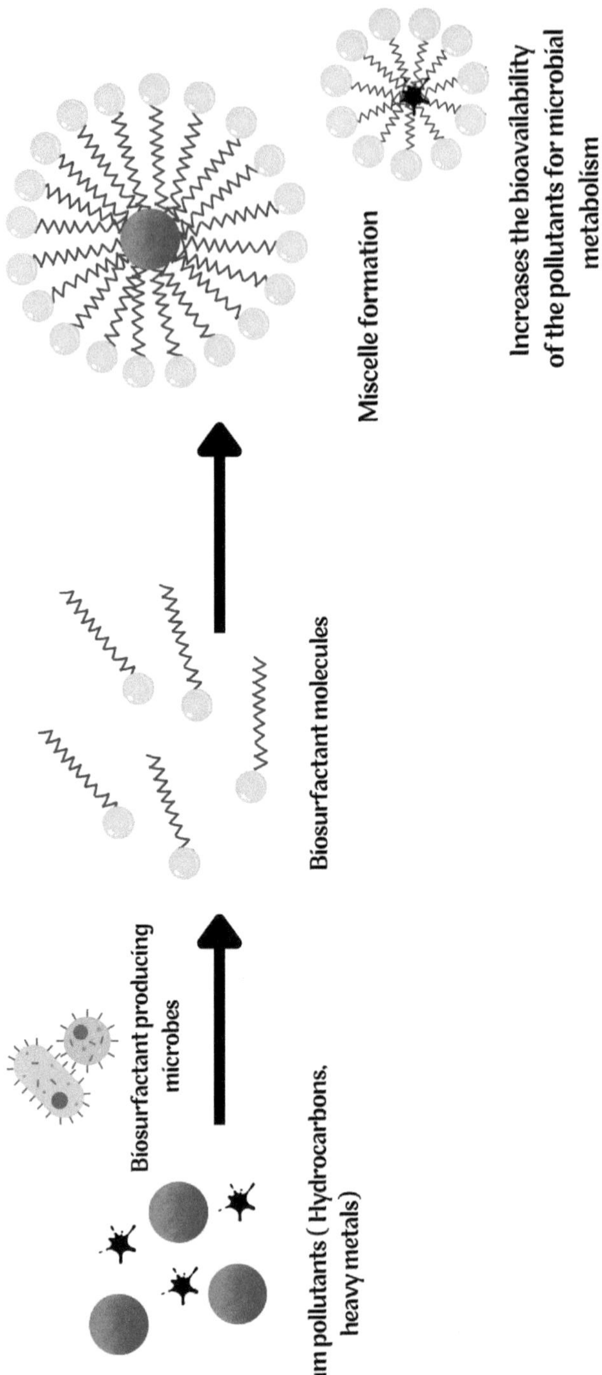

Figure 10.3 Biosurfactant bioremediation: Microbial biosurfactants attach to the pollutants forming miscelles, thereby reducing the surface tension of the contaminant making it readily available for microbial attack.

capable of degrading organic contaminants in large volumes of petroleum waste can be produced as a result of this plant-microorganism interaction (Van Hamme et al. 2003). The Plant growth-promoting rhizobacteria (PGPR) and arbuscular mycorrhizal fungi are most extensively utilized in the bioremediation of contaminated soils (Ma et al. 2011). Adding *Rhizobium meliloti* strain to Alfalfa plant significantly decreases PAH concentrations in contaminated soil (Teng et al. 2011). The combination of *Zea mays* and *Pseudomonas* sp. was employed to achieve the successful bioremediation of phenanthrene and pyrene (Chouychai et al. 2009). The plant *Lotus corniculatus,* with the support of the rhizobacteria *Pantoea* sp. successfully degraded diesel oil (Yousaf et al. 2010). In their study, Yu et al. employed the bacterium *Acinetobacter* sp. along with the plant *Lolium multiflorum* to remove various PAHs (Yu et al. 2011) (Table 10.1). The rhizospere regions were found to contain *Flavobacterium, Pseudomonas, Streptomyces, Agromyces,* and *Serratia,* which demonstrated the ability to absorb Zinc and Cadmium (Ghasemi et al. 2018) (Figure 10.2).

Petroleum Waste Bioremediation with Biosurfactants

Many microbes possess the ability to degrade hydrocarbons through the secretion of biosurfactants. These biosurfactants comprise of both hydrophilic and hydrophobic components, that can reduce surface tension and increase solubility. As a result, they facilitate increased chance of direct contact between the bacteria and oil droplets (Das and Chandran 2011) (Figure 10.3). Chemical surfactants have been traditionally used; however, there is now a widespread adoption of amphiphilic biological molecules that can reduce surface and interfacial tension as a greener and economically viable alternative to chemical surfactants (Banat et al. 2000). Biosurfactants possess significant surface and emulsifying activity (Figure 10.3). Living cells are capable of synthesizing surface-active substances that are environmental friendly, highly selective, and diverse in nature (Rahman et al. 2002). Thus, biosurfactants reduce the hydrophobicity of target pollutants, and enhance their bioavailability (Ron and Rosenberg 2002).

Several researchers have reported the biodegradation of (PAHs) through the utilization of surfactants produced by bacteria and fungi (Batista et al. 2010, Jacques et al. 2005, Wong et al. 2004). Some of these studies have specially investigated the capabilities of different *Pseudomonas* strains in degrading BTEX (benzene, toluene, ethylene, and xylene), naphthalene and total petroleum hydeocarbons (TPH). They have also explored the effects of these strains on surface tension reduction of petroleum oil, biofilm formation in hydrocarbon-supplemented media, and the production of surfactants such as rhamnolipids and glycolipids (Ye et al. 2011). Pseudomonads are well-known bacteria renowned for their ability to utilize hydrocarbons as carbon and energy sources while producing biosurfactants. Among these, *Pseudomonas aeruginosa,* a hydrocarbon-degrading gram-negative bacteria, is best known for biosurfactant production that produces rhamnolipids (a glycolipid surfactant). Studies have reported that *P. putida* and *P. chlororaphis* produce glycolipid-type biosurfactants (Das and Chandran 2011). *Bacillus* sp. have been found to produce "'surfactins' (Whang et al. 2008),

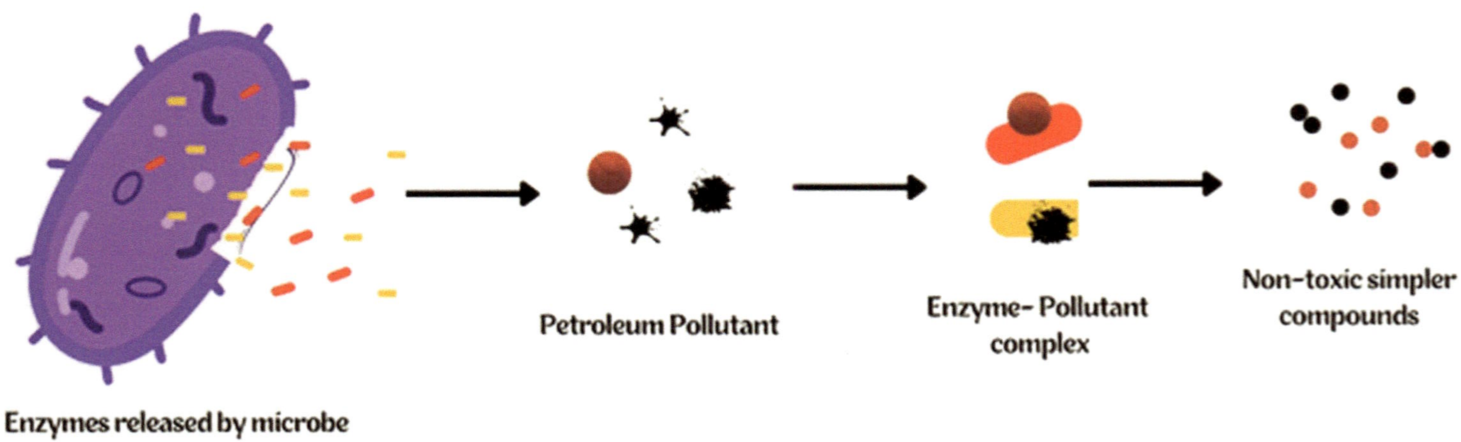

Figure 10.4 Enzymatic Bioremediation: Microbial enzymes utilize petroleum pollutants as substrates forming an enzyme-pollutant complex and converting toxic compounds to non-toxic simpler compounds that are environmental friendly.

with *B. subtilis* being particularly notable for surfactin production (Bais et al. 2004). *Acinetobacter venetianus* RAG has been identified as a producer of a lipopolysaccharide biosurfactant (Fondi et al. 2016). Additionally, *Mycobacterium* sp. and *Rhodococcus erythropolis* are known to synthesize trehalose lipids (White et al. 2013, Kügler et al. 2015). The yeast *Candida* sp. has also been recognised for producing another promising group of biosurfactants known as sophorolipids (Elshafie et al. 2015).

Petroleum Waste Bioremediation with Enzymes

Microbial enzymes play a crucial role in the degradation of various organic pollutants, as they possess high specificity towards a wide range of substrates (pollutants) and can be utilized in the bioremediation of environmental contaminants from industrial wastes (Wainwright 1999). These enzymes offer several advantages, including their ability to function under extreme conditions where microbes may struggle, their effectiveness at low pollutant concentrations, enhanced activity in the presence of microbial metabolism inhibitors, and their high mobility due to their small size. Specific enzymes such as dehydrogenase, oxygenase, and ligninolytic enzymes are involved in the degradation of PAHs. These biodegradation reactions are facilitated by enzymes that are responsible for activating specific energy-yielding metabolic pathways. Certain species are highly adaptable and can thrive in both oxygen-rich (aerobic) and oxygen-limited (anaerobic) conditions, using molecular oxygen or nitrate as the final electron acceptor in the electron transport chain. On the other hand, fermentative microbes do not rely on external electron acceptors, and generate energy through fermentative processes (Pandey and Fulekar 2012). Moreover, some microorganisms secrete enzymes that are functional only in the extracellular environment (Figure 10.4).

Laccases, known as benzenediol oxygen oxidoreductases, are extracellular enzymes that contain multiple copper ions. These enzymes are found in plants, bacteria, and fungi and exist in the form of monomeric, dimeric, and tetrameric glycoproteins (Shraddha et al. 2011). Various microbes produce laccases, such as *Streptomyces* sp. (Guan et al. 2018), *B. subtilis*, *Pseudomonas putida*, etc. Laccase, either on its own or in combination with redox mediators, has been documented to oxidize many dangerous compounds. These include phenols, halogenated phenols, aromatic amines, nitro-substituted compounds like trinitrotoluene, hydroxy-indoles, polycyclic aromatic hydrocarbons, chlorinated hydroxybiphenyls, bisphenol A and nonylphenol (Durán and Esposito 2000, Strong and Claus 2011). Cytochrome P450 is a superfamily of ubiquitous heme enzymes found in microbes. P450 enzymes possess inherent capabilities to degrade xenobiotics (Anzenbacher and Anzenbacherová 2001) through chemical transformations, including aliphatic hydroxylations and epoxidations, dealkylations, dehalogenation, and various mechanism-based inactivations. These enzymatic activities play a crucial role in bioremediation chemistry. *Bacillus megaterium*, CYP102A1 (P450BM3), demonstrated its ability to oxidize PAHs such as phenanthrene, fluoranthene, and pyrene, yielding a mixture of phenols and quinones. Some microbial P450s showed promise in the bioremediation of polyhalogenated aromatics (Carmichael and Wong 2001). Microbial dehalogenase (EC 3.8.1.5) is involved in the mineralization of halogenated xenobiotics. *Bacillus* sp. GZT simultaneously debrominates and mineralizes tribromophenol (Zu et al. 2012). *Dehalococcoides mccartyi* can dechlorinate widely distributed and highly toxic environmental pollutants like pentachlorophenol and other chlorinated phenols (Men et al. 2017) (Table 10.2).

MICROBES IN CONVERSION OF BIOMASS TO RENEWABLE ENERGY SOURCE

In earlier sections, we have seen the implication of different microbes in degrading the variety of toxic pollutants generated in petroleum refineries. However, considering the source of clean energy, biomass conversion to renewable energy sources is earning great importance. Biomass has been identified as a candidate for renewable resources. It can be used to produce a wide range of energy services (like "bioheat", "biofuel, "biopower", etc.) and biomaterials sustainably (Takkellapati et al. 2018). Hence, the use of microbes in biological biomass conversion will be an essential area to look upon. Here, we have made efforts to detail different types of microbial conversion involved in reducing biomass and producing renewable energy sources.

An article by Jacobsson and Johnson (2000) stated that "every organic substance deriving directly or indirectly from the photosynthesis process is considered biomass" (Jacobsson and Johnson 2000, Tursi 2019). The definition of biomass varies depending on its origin and usage. Hence, a more generalized definition is given by McKendry (2002), "Biomass is a combination of naturally derived materials, originating from plants such as shrubs, trees, algae, crops as well as all the materials composed of organic matrix except for plastics originating from petrochemical and fossil materials" (McKendry 2002, Tursi 2019). Many biomass sources exist, mainly crops and residues, animal residues, industrial waste, forestry crops and residues, and municipal waste (Figure 10.5).

There are several advantages of using biomass as a source of renewable energy. Primarily, it is geographically evenly distributed (Bušić et al. 2018). The availability of plant biomass is reliable to a greater extent because it can quickly grow in variable environmental conditions compared to other energy sources like wind, solar, and hydropower (Chauhan 2020). In contrast, fossil fuels make them more feasible for storing and transportation (Chauhan 2020). Biomass is chemically similar to fossil fuels; hence, it can be easily used as a substitute for products derived from petrochemicals like thickeners, stabilizers, textiles, polymers, etc. (Chaturvedi 2004).

Moreover, the greenhouse emissions are very low or zero for biofuel derived from biomass (Bušić et al. 2018, Chaturvedi 2004). Quality of human life is improved due to less exposure to harmful emissions in comparison to fossil fuels (Bušić et al. 2018, Chaturvedi 2004, Takkellapati et al. 2018). Several factories can be set up close to farms, or small-scale units can be developed to meet the local needs for fuel, electricity, etc. Thus, it also helps in rural development by generating employment opportunities (Chaturvedi 2004). Finally, using products developed from biomass can be more economic in comparison to imported fossil fuels (Chaturvedi 2004).

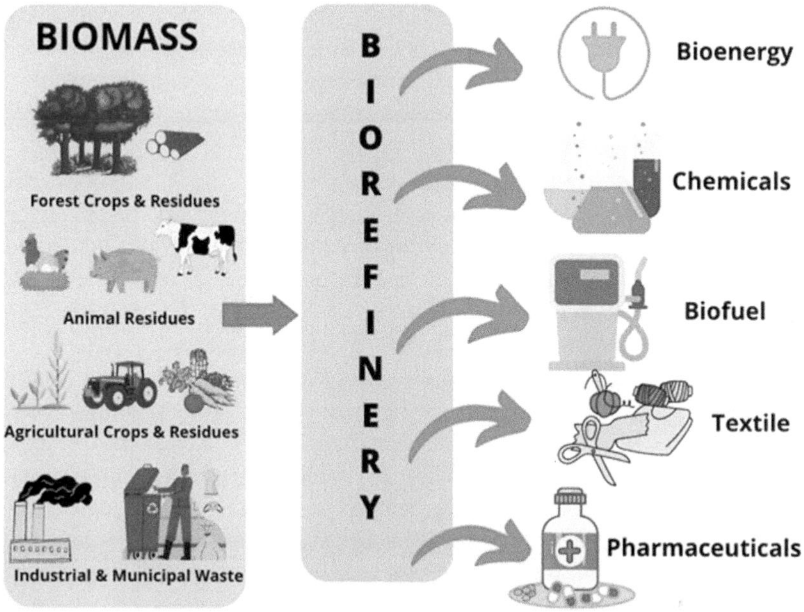

Figure 10.5 Biomass Valorization: This diagram explains how different biomass sources like forest crops and residues, animal residues, agricultural crops and residues, industrial waste, and municipal waste are upgraded by biorefineries to products like bioenergy, chemicals, biofuels, textiles, and pharmaceuticals.

Biorefineries, Types, and use in Biomass Degradation

Biomass can be converted to biofuels or biomaterials using biorefineries which use existing and novel technologies to develop these products. The National Renewable Energy Laboratory (NREL) defines biorefinery as a facility that integrates biomass conversion processes and equipment to produce power, fuel, and chemicals from biomass (Hingsamer and Jungmeier 2019). The most alluring part of the biorefinery is the production of high-added-value compounds and large volumes of biofuels with competitive market prices while lessening waste disposal and energy costs (Paes and Almeida 2014). Also, biorefineries always consider the process's sustainability and its indirect impacts on biodiversity and soil, use of water resources, and competition for food (Paes and Almeida 2014). Biorefinery can be a single machine or a polygeneration plant integrated with many industries and energy systems to upgrade/refine the biomass into one or more environment-friendly products like biofuels and biochemical (Hingsamer and Jungmeier 2019) (Figure 10.5).

Biorefineries can be classified into three different types based on the kind of feedstocks utilized, the nature of intermediates produced (syngas or sugar), the conversion processes used (e.g. thermochemical, biochemical, two platforms), and the status of technology execution (conventional, advanced, etc.) (Takkellapati et al. 2018). The types of biorefineries are discussed below.

(A) **Phase 1 Biorefinery:** It uses only a single feedstock material to produce a single product with the help of fixed/limited processing capabilities. Some examples are ethanol production from corn grain and biodiesel from vegetable oil like rapeseed oil and pulp and paper mill (Clark and Deswarte 2008, Takkellapati et al. 2018).

(B) **Phase 2 Biorefinery:** It also uses a single feedstock material; however, various products like chemicals, energy, and materials can be developed. Examples include the production of various chemicals like biodegradable polyesters from corn starch and carbohydrate derivatives like polyol, sweeteners, and bioethanol from cereal grains (Clark and Deswarte 2008, Takkellapati et al. 2018).

(C) **Phase 3 Biorefinery:** It uses several feedstock types and processing strategies to produce various products. There are four classes of Phase III Biorefineries. These are identified as: (1). Whole-crop biorefinery, (2). Lignocellulosic biorefinery, (3). Green biorefinery, (4). Two-platform concept biorefinery (Takkellapati et al. 2018). As the name suggests, the entire crop is used as raw material in a whole-crop biorefinery. For example, when cereal grains are used as raw material, the cereal grain is initially split up into corn and straw, where the former will be converted into starch. Hydrolysis of this starch results in the production of glucose which is used as a substrate for various chemicals, whereas the straw is processed in a lignocellulosic biorefinery (Naik et al. 2010, Takkellapati et al. 2018). The primary substrate for lignocellulosic biorefinery is lignocellulosic feedstock, available in different forms like forest residues, herbaceous and woody energy crops, and agricultural residues (rice straw and husk, banana waste, sugarcane bagasse, crop residue, etc.) (Prasad et al. 2019). These lignocellulosic materials are composed of cellulose (32–54%) and hemicellulose (11–37%) held together by lignin (17–32%) (Galbe and Zacchi 2002, Kumar et al. 2019, Naik et al. 2010). Upgradation of materials like these leads to the manufacturing of high-value products such as ethanol, acetic acid, acetone, butanol, syngas, biofuel, etc. (Kumar et al. 2019, Naik et al. 2010). Green biorefinery makes use of natural wet feedstocks derived from untreated products, like grass from the cultivation of permanent grassland, nature preserve, closure fields, and green crops such as clover, lucerne, and immature cereals from extensive land cultivation as input to produce chemicals like levulinic acid and also for conversion to syngas and hydrocarbons (Kamm Kamm 2004, Naik et al. 2010, Takkellapati et al. 2018). The two-platform concept biorefinery introduced by NREL includes two types of platforms: the sugar platform and the syngas platform (Clark and Deswarte 2008).

In the sugar platform, biochemical conversion methodologies are employed, while thermochemical methods are utilized in the syngas platform (Clark and Deswarte 2008, Takkellapati et al. 2018). Both energy and chemicals can be obtained as the end products in two-platform concept biorefinery (Takkellapati et al. 2018) (Figure 10.6).

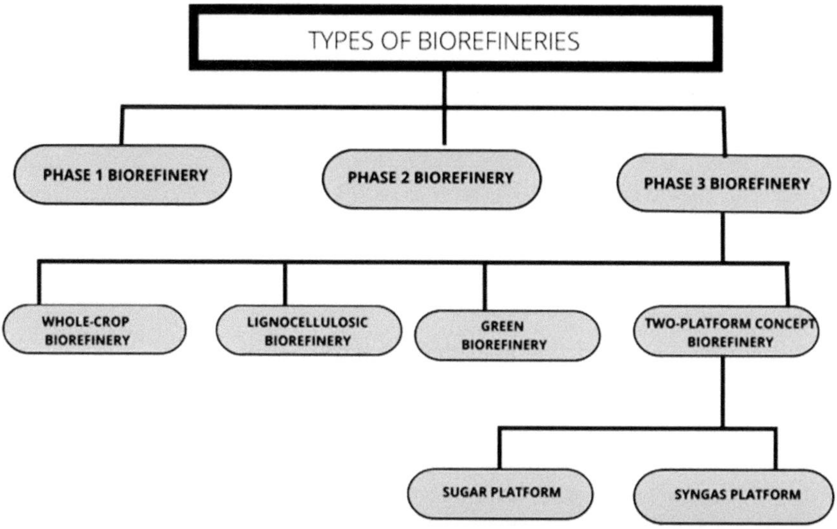

Figure 10.6 Types of Biorefineries: Biorefineries can be divided into three groups based on the source of biomass, the process involved, and products developed: Phase 1 biorefinery, Phase 2 biorefinery, and phase 3 biorefinery. Phase 3 biorefinery is divided into four types based primarily on the source: whole crop biorefinery, lignocellulosic biorefinery, green biorefinery, and two platform biorefinery. The latter is classified into sugar platform and syngas platform.

The processes involved in biomass conversion can be classified into three main types:

1. Biochemical conversions (Anaerobic digestion, Fermentation, and Composting)
2. Physio-chemical conversions (Esterification or Transesterification)
3. Thermo-chemical (Combustion, Pyrolysis, Gasification, and Liquefaction) (Fatih Demirbas 2009, Tursi 2019).

Biochemical conversion processes either utilize microorganisms or enzymes, or chemicals derived from them. Hence, microorganisms are considered the protagonist in industrial bioprocesses, as they are directly accountable for producing the desired chemical or indirectly providing intermediate compounds needed for the processes (Paes and Almeida 2014). Currently, microbes are used all across the globe in several industrial processes, like the production of bioethanol from sugarcane in Brazil (Paes and Almeida 2014). Since this book chapter focuses on microbes playing a role in biorefineries, we will be dealing with biochemical conversion processes.

In the biochemical conversion of biomass, enzymes obtained from bacteria or other microorganisms are utilized to break down biomass through processes like anaerobic digestion (anaerobic respiration, fermentation, and composting to develop valuable products in an eco-friendly and sustainable manner) (Sam and Barik 2019, Williams et al. 2020).

Several products can be developed from biomass, such as ethanol, hydrogen, biogas, and organic acids (like lactate, citrate, pyruvate, mannitol, isobutanol, and xanthan gum). Different microorganisms have to be selected for biochemical conversion depending on the product to be developed (Ding et al. 2016).

Compared to the other two technologies, biochemical conversion technologies are more clean, efficient, moderate, and eco-friendly (Ding et al. 2016, Sam and Barik 2019). By using enzymes or microbes individually or in a consortium, a wide variety of products or intermediate substances of industrial value can be developed (Ding et al. 2016). Hence, biochemical conversion technologies of biomass have much scope in recent years to help build a clean and sustainable future.

Biomass Degradation by Anaerobic Digestion

Anaerobic digestion, also known as anaerobic respiration or biomethanation, is a process in which organic materials (like livestock manure, food wastes, sewage sludge, crop residues, agricultural byproducts, and the organic fraction of municipal solid wastes) are digested in the absence of molecular oxygen (Tursi 2019, Lim et al. 2020, Harirchi et al. 2022). This technique is employed for the production of methane, bio-hydrogen, and volatile fatty acids. Also, anaerobic digestion helps in waste management and pathogen deactivation, among others (Harirchi et al. 2022).

The anaerobic digestion process involves the stepwise degradation of complex biomass by a diverse range of microorganisms interacting with each other to produce methane (Lv et al. 2010, Amin et al. 2021). The four steps include hydrolysis, acidogenesis, acetogenesis, and methanogenesis (Tursi 2019, Lim et al. 2020).

Hydrolysis: It is the initial phase of anaerobic digestion which makes use of extracellular hydrolases (e.g. cellulase, xylanase, pectinase, amylase, lipase, and protease) secreted by hydrolytic bacteria to breakdown complex polymers of carbohydrates, lipids, and proteins into monomers or oligomers, such as glucose, cellobiose, glycerol, fatty acids, and amino acids, respectively (Lim et al. 2020). The examples of microbial genera used in this process include *Aminobacterium, Bacillus, Caldanaerobacter, Clostridium, Eubacterium, Lactobacillus, Streptococcus, Trichococcus, Aspergillus, Humicola, Penicillium* and *Trichoderma* (Harirchi et al. 2022). Further decomposition of the molecules is done in the next phase, i.e. acidogenesis.

Acidogenesis: During the second phase, the acidogenic bacteria, also called fermentative bacteria, degrade the products of hydrolysis, thereby producing volatile fatty acids (like acetic acid, propionic acid, formic acid, and butyric acid-precursors for methane production), NH_3, CO_2, H_2, H_2S, alcohols, etc. (Harirchi et al. 2022, Lim et al. 2020, Tursi 2019). The acidogenic bacteria used in this phase can either be aerobes, facultative anaerobes, or strict anaerobes. *Klebsiella,*

Bacteroidia, Enterobacter, Lactobacilli, Clostridia, Escherichia, Advenella, Alkalitalea, Christensenella, Corynebacterium, Dechloromon, Ferroplasma, Geofilum, Lutispora, Mariniphaga and *Sedimentibacter* members aid in this process (Amin et al. 2021, Lim et al. 2020).

Syntrophic acetogenesis: Some of the products obtained from the previous phase, like acetate, formate, H_2, and CO_2, can be directly utilized by methanogens to produce biogas, etc. (Lim et al. 2020, Tursi 2019). However, other products like propionate, butyrate, isobutyrate, valerate, and isovalerate must first be broken down into acetate, H_2, and CO_2 by acetogens via syntrophic acetogenesis. A syntrophic relationship occurs between acetate-forming bacteria (acetogens) and methanogens, as the former bacteria are responsible for providing the primary substrates (acetate and hydrogen) for methane production by methanogens (Lim et al. 2020). Acetogens can be divided into two groups:

(1) Those that produce hydrogen and acetate from organic acids and carbohydrates (e.g. *Anaerovorax odorimutans, Macellibacteroides fermentans, Saccharofermentans acetigenes, Proteiniphilum acetatigenes, Levilinea saccharolytica, Hydrogenispora ethanolica*, and *Hydrogenophaga carboriunda*) (Amin et al. 2021);

(2) Those that consume hydrogen and CO_2 to produce acetate, such as *Acetobacterium wieringae, Acetobacterium woodii, Acetogenium kivui, Clostridium aceticum, Clostridium thermoautotrophicum*, and *Sulfurovum riftia*e (Amin et al. 2021).

Methanogenesis: Methanogens produces methane through the process of methanogenesis. The two main enzymes involved in this process are coenzyme M (CoM) and methyl–coenzyme M reductase (MCR), which are involved in three different pathways for methane production and these pathways include (1) methylotrophic methanogenesis; (2) acetoclastic methanogenesis; and (3) hydrogenotrophic methanogenesis through CO_2 reduction (Lim et al. 2020). The first two pathways are mainly found in *Methanosarcinales,* while the latter is seen in most methanogens (Harirchi et al. 2022) (Figure 10.7).

Biomass Degradation by Fermentation

The fermentation process involves a series of biochemical reactions that convert simple sugars (hexoses and pentoses) into ethanol and CO_2 under anaerobic conditions by microorganisms (Tursi 2019). The main product of fermentation is bioethanol. Apart from bioethanol, glycerol and carboxylic acids are also produced. Bioethanol can be produced from different sources, i.e. feedstock, lignocellulosic biomass, and algal biomass (Prasad et al. 2019). Based on the sources, there are four different ways of generating biofuels.

Among the different sources available, using lignocellulosic material for bioethanol production is advantageous as (1) it is renewable, (2) does not interfere with food supply, (3) abundant in nature, (4) geographically widely distributed, (5) cheaper/inexpensive, (6) bioethanol developed from lignocellulosic material is associated with considerably lower greenhouse emission (Bušić et al. 2018, Mahalaxmi and Williford 2016, Prasad et al. 2019)

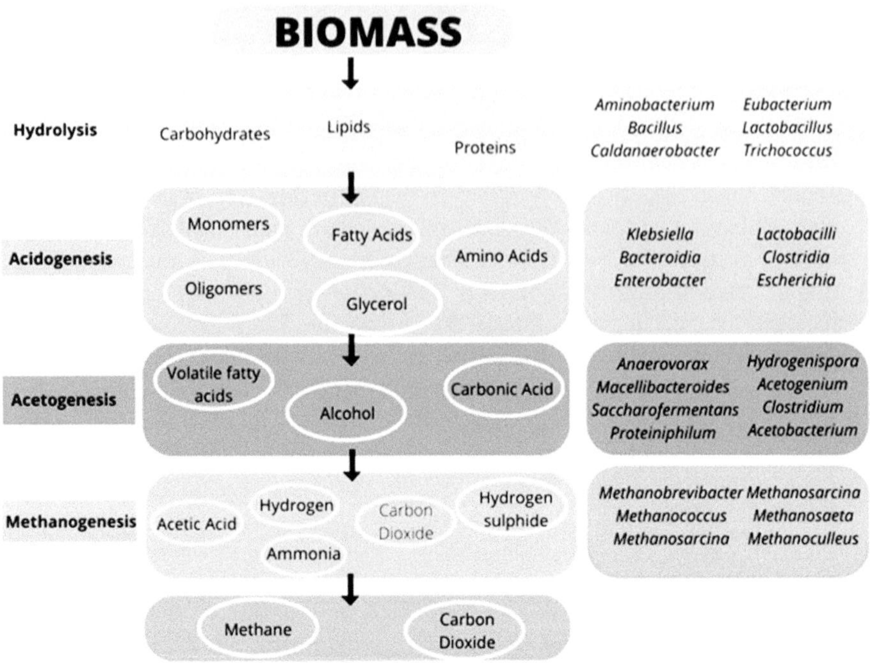

Figure 10.7 Stages of Anaerobic Digestion: Stage 1 is hydrolysis which involves the breakdown of complex compounds. Acidogenesis is the 2nd stage, where acidogenic bacteria degrade hydrolysis products to produce volatile fatty acids, NH_3, CO_2, H_2, H_2S, alcohols, etc. Stage 3 is syntrophic acetogenesis, producing acetate and hydrogen for the final stage. Stage 4 is methanogenesis, which results in the production of methane and carbon dioxide.

The lignocellulose containing raw materials for bioethanol production can be placed into six main groups: (1) crop residues (like rice hulls, corn stover, cane and sweet sorghum bagasse, different straw types, olive stones, and pulp); (2) hardwood (which includes aspen, poplar); (3) softwood (for instance pine, spruce); (4) cellulose wastes (e.g. recycled paper sludge and waste paper); (5) herbaceous biomass (switchgrass, alfalfa hay, and other types of grasses) and finally (6) municipal solid wastes (Bušić et al. 2018).

Production of bioethanol from lignocellulosic material involves three main steps, which include:

(i) Cellulose and hemicellulose pretreatment to make it more accessible in the subsequent steps,
(ii) Polysaccharide hydrolysis (either acid or enzymatic) to produce simple sugars (like hexoses and pentoses),
(iii) Finally, fermentation of the simple sugars produced in the previous step using microbes to yield ethanol (Bušić et al. 2018, Kumar et al. 2019, Prasad et al. 2019).

Pretreatment

The pretreatment step is essential to expose the cellulose or modify the pores in the raw materials. This allows the enzymes to penetrate the fibers and thus hydrolyze the cellulose to monomeric sugars. This pretreatment of biomass can be done using several methods, like physical, physio-chemical, chemical and biological methods.

The physical methods include milling, fragmentation, shearing, or grinding of the biomass (Kumar et al. 2019, Prasad et al. 2019).

Physio-chemical treatments consist of steam explosion (autohydrolysis), microwave-chemical pretreatment, CO_2 explosion, ammonia fiber explosion (AFEX), steam explosion with the addition of SO_2, liquid hot water pretreatment, etc. (Bušić et al. 2018, Prasad et al. 2019, Takkellapati et al. 2018).

Chemical treatment methods include wet oxidation alkaline treatment, sulphite pretreatment, acidic treatment, ionic liquids (green solvents), organic solvent pretreatment, oxidizing agents, etc. (Bušić et al. 2018, Kumar et al. 2019, Takkellapati et al. 2018).

The biological treatment is done by enzymatic degradation (hydrolysis) of lignocellulosic material (Bušić et al. 2018). These specific enzymes are obtained from either bacteria or fungi (Kumar et al. 2019, Takkellapati et al. 2018). Kumar et al. (2019) have reported several white, brown, and soft rot fungi for the degradation of lignin and hemicelluloses and very little cellulose. Furthermore, Alvira et al. (2010) have also reported several white-rot fungi like *Pycnoporus cinnarbarinus, Pleurotus ostreaus, Phanerochaete chrysosporium, Cyathus stercolerus, Ceriporia lacerata*, and *Ceriporiopsis subvermispora*, for their lignin degrading efficiency. Biological treatment is cheaper, more efficient, and eco-friendly (Prasad et al. 2019).

Hydrolysis

In the succeeding pretreatment, the cellulose is hydrolyzed with the help of cellulases. Cellulase enzyme production is typical among fungi and can be found in many species, including *Trichoderma, Penicillium,* and *Aspergillus* (Bušić et al. 2018). According to Galbe and Zacchi, *Trichoderma reesei* is the most efficient cellulose-hydrolyzing organism (Galbe and Zacchi 2002).

Fermentation

It can occur in three ways: separate enzymatic hydrolysis and fermentation (SHF) process, simultaneous saccharification and fermentation (SSF), and consolidated bioprocessing (CBP).

In SHF, hydrolysis of pretreated lignocellulosic feedstock to glucose and its fermentation to ethanol occurs in separate bioreactors. The pretreatment of lignocellulosic raw materials is followed by the separation of the solid phase from the liquid phase, mainly comprising of sugars (pentose and some hexoses). This is succeeded by treatment with dilute acid, which results in the residual solid phase with primarily lignin and cellulose. The cellulolytic enzymes aid in the hydrolysis of the cellulose (Bušić et al. 2018).

Simultaneous saccharification and fermentation (SSF): Cellulose hydrolysis and hexose fermentation occur in a single reactor in SSF. The sugars released from the former process are acted upon by microbes immediately. Hence, cellulase inhibition is prevented by the low amount of sugar in the broth (Bušić et al. 2018, Prasad et al. 2019). Most often, *T. reesei* and *S. cerevisiae* are used in this process (Bušić et al. 2018).

Consolidated bioprocessing (CBP), as the name suggests, is a consolidation of four biological reactions that occur in one step: (1) the production of enzymes like cellulases and hemicellulases, (2) the hydrolysis of carbohydrates into sugars, (3) fermentation of hexoses like galactose, glucose, and mannose, and (4) fermentation of pentoses, for example, xylose and arabinose (Bušić et al. 2018). CBP microbial consortium of *Clostridium thermocellum* and *C. thermosaccharolyticum* can be used, where *Clostridium thermocellum* produces cellulase for cellulose hydrolysis and ferment glucose into ethanol, while *C. thermosaccharolyticum* is used to ferment pentose sugars into ethanol (Cardona and Sánchez 2007). Another example of a microbial consortium that can be used is *Trichoderma reesei*, *Saccharomyces cerevisiae*, and *Scheffersomyces stipites*. Here *T. reesei* is utilized for enzyme production to hydrolyze lignocellulosic biomass, *S. cerevisiae* is used to ferment hexose sugars, and *S. stipites* is responsible for the fermentation of pentose sugar (Kumar et al. 2019).

Biomass Degradation by Composting

Composting is an aerobic, biochemical, and microbial process involving hydrolysis of organic matter into humus (Wei et al. 2017). This process is aided by microbes that decompose the organic matter by utilizing the nitrogen and carbon as energy sources along with oxygen and water, thereby producing soil-enriching compost and also heat and carbon dioxide as byproducts (Białobrzewski et al. 2015, Rastogi et al. 2020).

The process of composting can be divided into three major phases:

(i) During the initial mesophilic phase, simple compounds like sugars, amino acids, etc., are degraded by mesophilic bacteria and fungi, increasing temperature and carbon dioxide production (Rastogi et al. 2020). Mesophilic microbes include *Bacillus, Streptomyces rectus,* etc.

(ii) During the second thermophilic phase, mesophilic microbes are replaced by thermophiles like *Therinomnonospora curvata, Actinobifida chromagena* that degrade organic matter like fats, cellulose, hemicellulose, and lignin (Pan et al. 2011). At this stage, the rapid increase in temperature facilitates killing the pathogens, thereby rendering the compost free from pathogens and safe for use (Pan et al. 2011, Rastogi et al. 2020).

(iii) Finally, the cooling phase is represented by a decreased temperature and microbial activity. During this time, mesophilic bacteria recolonize the compost mass and degrade the cellulose and hemicellulose, residual sugars, materializing humic-like substances.

Apart from indigenous microbes, efficient microbe cultures of dominant and non-dominant microbes can be added to the compost. Cao et al. (2013) reported that inoculating the compost with *Aspergillus fumigatus* F12 strain helped to degrade cellulose before other biopolymers. Tran et al. (2019) demonstrated the role of *Pediococcus* and *Weissella* in producing lactic acid that aided in accelerating the composting process and activating the indigenous microorganisms within the compost. Furthermore, studies show that inoculation of bacterial consortium (anti-acidification), which consisted of *Bacillus*, *Pseudomonas, Lactobacillus*, and others, helped to obtain superior quality compost with a higher humic acid content (Ding et al. 2016).

CONCLUSION

Microbes have been long studied for curtailing different waste materials and, more prominently, toxic wastes that are hazardous to humans. This chapter discusses various bioremediation methods in curtailing petroleum refinery waste and microbes' role in biorefineries for biomass conversion as a renewable energy source. Bioremediation is an emerging technology that shows great promise. It directly uses microbes or their byproducts to detoxify and breakdown petroleum-based organic pollutants. It plays a crucial role in the biogeochemical process, aiding in the restoration and recycling of minerals within polluted ecosystems. Microbes play an essential role in biorefineries as they are directly responsible for producing the desired biochemical. Biorefineries use microbial enzymes and other chemicals derived from microbes to generate biofuels that have advantages over fossil fuels. Bioenergy production offers a viable alternative to petroleum as a feedstock, thus contributing to a cleaner and low-carbon environment.

REFERENCES

Abbasian, F., Lockington, R., Mallavarapu, M. and Naidu, R. 2015. A comprehensive review of aliphatic hydrocarbon biodegradation by bacteria. Applied Biochemistry and Biotechnology. 176(3): 670–699. https://doi.org/10.1007/s12010-015-1603–5

Adebusoye, S.A., Ilori, M.O., Amund, O.O., Teniola, O.D. and Olatope, S.O. 2007. Microbial degradation of petroleum hydrocarbons in a polluted tropical stream. World Journal of Microbiology and Biotechnology. 23(8): 1149–1159. https://doi.org/10.1007/s11274-007-9345-3.

Alvira, P., Tomás-Pejó, E., Ballesteros, M. and Negro, M.J. 2010. Pretreatment technologies for an efficient bioethanol production process based on enzymatic hydrolysis: a review. Bioresource Technology. 101(13): 4851–4861. https://doi.org/10.1016/J.BIORTECH.2009.11.093

Amin, F.R., Khalid, H., El-Mashad, H.M., Chen, C., Liu, G. and Zhang, R. 2021. Functions of bacteria and archaea participating in the bioconversion of organic waste for methane production. The Science of The Total Environment. 763: 143007. https://doi.org/10.1016/J.SCITOTENV.2020.143007

Anzenbacher, P. and Anzenbacherová, E. 2001. Cytochromes P450 and metabolism of xenobiotics. Cellular and Molecular Life Sciences. 58(5): 737–747. https://doi.org/10.1007/PL00000897.

Aust, S.D., Swaner, P.R. and Stahl, J.D. 2003. Detoxification and metabolism of chemicals by White-Rot Fungi. 3–14. https://doi.org/10.1021/bk-2004-0863.ch001.

Bais, H.P., Fall, R. and Vivanco, J.M. 2004. Biocontrol of *Bacillus subtilis* against infection of Arabidopsis roots by *Pseudomonas syringae* is Facilitated by Biofilm Formation and surfactin Production. Plant Physiology. 134(1): 307–319. https://doi.org/10.1104/pp.103.028712.

Banat, I.M., Makkar, R.S. and Cameotra, S.S. 2000. Potential Commercial Applications of Microbial Surfactants. Applied Microbiology and Biotechnology. 53(5): 495–508. https://doi.org/10.1007/s002530051648.

Barea, J.-M., Pozo, M.J., Azcón, R. and Azcón-Aguilar, C. 2005. Microbial co-operation in the rhizosphere. Journal of Experimental Botany. 56(417): 1761–1778. https://doi.org/10.1093/jxb/eri197.

Bartha, R. and Atlas, R.M. 1977. The microbiology of aquatic oil spills. Advances in Applied Microbiology. 22: 225–266. https://doi.org/10.1016/S0065-2164(08)70164-3.

Batista, R.M., Rufino, R.D., Luna, J.M., de Souza, J.E.G. and Sarubbo, L.A. 2010. Effect of medium components on the production of a biosurfactant from *Candida tropicalis* applied to the removal of hydrophobic contaminants in soil. Water Environment Research. 82(5): 418–425. https://doi.org/10.2175/106143009X12487095237279.

Bhandari, S., Poudel, D.K., Marahatha, R., Dawadi, S., Khadayat, K., Phuyal, S., et al. 2021. Microbial enzymes used in bioremediation. Journal of Chemistry. 2021: 1–17. https://doi.org/10.1155/2021/8849512.

Białobrzewski, I., Mikš-Krajnik, M., Dach, J., Markowski, M., Czekała, W. and Głuchowska, K. 2015. Model of the sewage sludge-straw composting process integrating different heat generation capacities of mesophilic and thermophilic microorganisms. Waste Management. 43: 72–83. https://doi.org/10.1016/J.WASMAN.2015.05.036.

Budi, R.M.S., Rahardja, B.S. and Masithah, E.D. 2020. Potential concentration of heavy metal copper (cu) and microalgae growth *Spirulina plantesis* in culture media. IOP Conference Series: Earth and Environmental Science. 441(1): 012147. https://doi.org/10.1088/1755-1315/441/1/012147.

Burghal, A.A., Abu-Mejdad, N.M.J.A. and Al-Tamimi, W.H. 2016. Mycodegradation of crude oil by fungal species isolated from petroleum contaminated soil. International Journal of Innovative Research in Science. 5(2). https://doi.org/10.15680/IJIRSET.2016.0502052.

Bušić, A., Marđetko, N., Kundas, S., Morzak, G., Belskaya, H., Šantek, M.I., et al. 2018. Bioethanol production from renewable raw materials and its separation and purification: a review. Food Technology and Biotechnology. 56(3): 289–311. https://doi.org/10.17113/ftb.56.03.18.5546.

Cairns, S.S., Cornish, A. and Cooper, R.A. 1996. Cloning, sequencing and expression in *Escherichia coli* of two *Rhizobium* sp. genes encoding haloalkanoate dehalogenases of opposite stereospecificity. European Journal of Biochemistry. 235(3): 744–749. https://doi.org/10.1111/j.1432-1033.1996.t01-1-00744.x.

Cao, Y., Chang, Z., Wang, J., Ma, Y. and Fu, G. 2013. The fate of antagonistic microorganisms and antimicrobial substances during anaerobic digestion of pig and dairy manure. Bioresource Technology. 136: 664–671. https://doi.org/10.1016/J.BIORTECH.2013.01.052

Cardona, C.A. and Sánchez, Ó.J. 2007. Fuel ethanol production: process design trends and integration opportunities. Bioresource Technology. 98(12): 2415–2457. https://doi.org/10.1016/J.BIORTECH.2007.01.002

Carmichael, A.B. and Wong, L.-L. 2001. Protein engineering of *Bacillus megaterium* CYP102. European Journal of Biochemistry. 268(10): 3117–3125. https://doi.org/10.1046/j.1432-1327.2001.02212.x.

Casas-Godoy, L., Duquesne, S., Bordes, F., Sandoval, G. and Marty, A. 2012. Lipases: an overview. Methods in Molecular Biology. 861: 3–30. https://doi.org/10.1007/978-1-61779-600-5_1.

Cepoi, L., Zinicovscaia, I., Rudi, L., Chiriac, T., Djur, S., Yushin, N., et al. 2022. Assessment of metal accumulation by *Arthrospira platensis* and its adaptation to iterative action of nickel mono- and polymetallic synthetic effluents. Microorganisms. 10(5): 1041. https://doi.org/10.3390/microorganisms10051041.

Cerniglia, C.E. 1984. Microbial metabolism of polycyclic aromatic hydrocarbons. Advances in Applied Microbiology. 30: 31–71. https://doi.org/10.1016/S0065-2164(08)70052-2.

Cerqueira, V.S., Hollenbach, E.B., Maboni, F., Vainstein, M.H., Camargo, F.A.O., Peralba, M. do C.R., et al. 2011. Biodegradation potential of oily sludge by pure and mixed bacterial cultures. Bioresource Technology. 102(23): 11003–11010. https://doi.org/10.1016/j.biortech.2011.09.074.

Chaturvedi, P. 2004. Biomass-the fuel of the rural poor in developing countries. Bioenergy options for a cleaner environment: *In*: Developed and Developing Countries, 161–181. https://doi.org/10.1016/B978-008044351-5/50008-8.

Chauhan, A.K. 2020. Biofuel: types and process overview. *In*: Advances in Bioenergy. 1–28. https://doi.org/10.1007/978-981-13-9333-4_1.

Chekroun, K.B., Sánchez, E. and Baghour, M. 2014. The role of algae in bioremediation of organic pollutants. International Journal of Public and Environmental Health. 1(2): 19–32.

Chisti, Y. 2007. Biodiesel from microalgae. Biotechnology Advances. 25(3): 294–306. https://doi.org/10.1016/j.biotechadv.2007.02.001.

Chojnacka, K., Chojnacki, A. and Górecka, H. 2005. Biosorption of Cr^{3+}, Cd^{2+} and Cu^{2+} ions by blue–green algae *Spirulina* sp.: kinetics, equilibrium and the mechanism of the process. Chemosphere. 59(1): 75–84. https://doi.org/10.1016/j.chemosphere.2004.10.005.

Chouychai, W., Thongkukiatkul, A., Upatham, S., Lee, H., Pokethitiyook, P. and Kruatrachue, M. 2009. Plant-enhanced phenanthrene and pyrene biodegradation in acidic soil. Journal of Environmental Biology. 30(1): 139–144.

Clark, J.H. and Deswarte, F.E.I. 2008. The biorefinery concept–an integrated approach. *In*: Introduction to Chemicals from Biomass. 1–20. https://doi.org/10.1002/9780470697474.CH1

Couto, S.R. and Toca-Herrera, J.L. 2007. Laccase production at reactor scale by filamentous fungi. Biotechnology Advances. 25(6): 558–569. https://doi.org/10.1016/j.biotechadv.2007.07.002.

Das, N. and Chandran, P. 2011. Microbial degradation of petroleum hydrocarbon contaminants: an overview. Biotechnology Research International. 2011: 1–13. https://doi.org/10.4061/2011/941810.

Das, R. and Kazy, S.K. 2014. Microbial diversity, community composition and metabolic potential in hydrocarbon contaminated oily sludge: prospects for in situ bioremediation. Environmental Science and Pollution Research. 21(12): 7369–7389. https://doi.org/10.1007/s11356-014-2640-2.

Davies, J.S. and Westlake, D.W.S. 1979. Crude oil utilization by fungi. Canadian Journal of Microbiology. 25(2): 146–156. https://doi.org/10.1139/m79-023.

Ding, L., Cheng, J., Xia, A., Jacob, A., Voelklein, M. and Murphy, J.D. 2016. Co-generation of biohydrogen and biomethane through two-stage batch co-fermentation of macro- and micro-algal biomass. Bioresource Technology. 218: 224–231. https://doi.org/10.1016/J.BIORTECH.2016.06.092.

Durán, N. and Esposito, E. 2000. Potential applications of oxidative enzymes and phenoloxidase-like compounds in wastewater and soil treatment: a review. Applied Catalysis B: Environmental. 28(2): 83–99. https://doi.org/10.1016/S0926-3373(00)00168-5.

Dwinovantyo, A., Prartono, T., Rahmaniar, R., Prabowo, N.W. and Susanti, S. 2016. Oil spill biodegradation by bacteria isolated from Jakarta Bay Marine Sediments. ILMU KELAUTAN: Indonesian Journal of Marine Sciences. 21(1). https://doi.org/10.14710/ik.ijms.21.1.29-36.

Elshafie, A.E., Joshi, S.J., Al-Wahaibi, Y.M., Al-Bemani, A.S., Al-Bahry, S.N., Al-Maqbali, D., et al. 2015. Sophorolipids Production by *Candida bombicola* ATCC 22214 and its potential application in microbial enhanced oil recovery. Frontiers in Microbiology. 6. https://doi.org/10.3389/fmicb.2015.01324.

Esposito, D. and Antonietti, M. 2015. Redefining biorefinery: the search for unconventional building blocks for materials. Chemical Society Reviews. 44: 5821. https://doi.org/10.1039/c4cs00368c.

Farag, S. and Soliman, N.A. 2011. Biodegradation of crude petroleum oil and environmental pollutants by *Candida tropicalis* strain. Brazilian Archives of Biology and Technology. 54(4): 821–830. https://doi.org/10.1590/S1516-89132011000400023

Fatih Demirbas, M. 2009. Biorefineries for biofuel upgrading: a critical review. Applied Energy. 86(SUPPL. 1): S151–S161*. https://doi.org/10.1016/J APENERGY.2009.04.043.

Fondi, M., Maida, I., Perrin, E., Orlandini, V., La Torre, L., Bosi, E., et al. 2016. Genomic and phenotypic characterization of the species *Acinetobacter venetianus*. Scientific Reports. 6(1): 21985. https://doi.org/10.1038/srep21985.

Galbe, M. and Zacchi, G. 2002. A review of the production of ethanol from softwood. Applied Microbiology and Biotechnology. 59(6): 618–628. https://doi.org/10.1007/S00253-002-1058-9.

Gentili, A.R., Cubitto, M.A., Ferrero, M. and Rodriguéz, M.S. 2006. Bioremediation of crude oil polluted seawater by a hydrocarbon-degrading bacterial strain immobilized on chitin and chitosan flakes. International Biodeterioration and Biodegradation. 57(4): https://doi.org/10.1016/j.ibiod.2006.02.009.

Ghasemi, Z., Ghaderian, S.M., Rodríguez-Garrido, B., Prieto-Fernández, Á. and Kidd, P.S. 2018. Plant species-specificity and effects of bioinoculants and fertilization on plant performance for nickel phytomining. Plant and Soil. 425(1–2): 265–285. https://doi.org/10.1007/s11104-017-3553-x.

Guan, Z.B., Luo, Q., Wang, H.R., Chen, Y. and Liao, X.R. 2018. Bacterial laccases: promising biological green tools for industrial applications. Cellular and Molecular Life Sciences. 75(19): 3569–3592. https://doi.org/10.1007/s00018-018-2883-z.

Hamamura, N., Yeager, C.M. and Arp, D.J. 2001. Two Distinct Monooxygenases for Alkane Oxidation in *Nocardioides* sp. Strain CF8. Applied and Environmental Microbiology. 67(11): 4992–4998. https://doi.org/10.1128/AEM.67.11.4992-4998.2001.

Harirchi, S., Wainaina, S., Sar, T., Nojoumi, S.A., Parchami, M., Parchami, M., et al. 2022. Microbiological insights into anaerobic digestion for biogas, hydrogen or volatile fatty acids (VFAs): a review. Bioresource Technology. 13(3): 6521–6557. https://doi.org/10.1080/21655979.2022.2035986.

He, Y. and Jiang, Z.-W. 2008. Technology review: treating oilfield wastewater. Filtration and Separation. 45(5): 14–16. https://doi.org/10.1016/S0015-1882(08)70174-5.

Hingsamer, M. and Jungmeier, G. 2019. Biorefineries. *In*: The Role of Bioenergy in the Emerging Bioeconomy: Resources, Technologies, Sustainability and Policy. 179–222. https://doi.org/10.1016/B978-0-12-813056-8.00005-4.

Huang, X.D., El-Alawi, Y., Gurska, J., Glick, B.R. and Greenberg, B.M. 2005. A multi-process phytoremediation system for decontamination of persistent total petroleum hydrocarbons (TPHs) from soils. Microchemical Journal. 81(1). https://doi.org/10.1016/j.microc.2005.01.009.

Huertas, M., López-Maury, L., Giner-Lamia, J., Sánchez-Riego, A. and Florencio, F. 2014. Metals in cyanobacteria: analysis of the copper, nickel, cobalt and arsenic homeostasis mechanisms. Life. 4(4): 865–886. https://doi.org/10.3390/life4040865.

Jacobsson, S. and Johnson, A. 2000. The diffusion of renewable energy technology: an analytical framework and key issues for research. Energy Policy. 28(9): 625–640. https://doi.org/10.1016/S0301-4215(00)00041-0.

Jacques, R.J.S., Santos, E.C., Bento, F.M., Peralba, M.C.R., Selbach, P.A., Sá, E.L.S., et al. 2005. Anthracene biodegradation by *Pseudomonas* sp. isolated from a petrochemical sludge landfarming site. International Biodeterioration and Biodegradation. 56(3): 143–150. https://doi.org/10.1016/j.ibiod.2005.06.005.

Jiao, Y., Chen, X., Wang, X., Liao, X., Xiao, L., Miao, A., et al. 2013. Identification and characterization of a cold-active phthalate esters hydrolase by screening a metagenomic library derived from biofilms of a wastewater treatment plant. PLoS ONE. 8(10): e75977. https://doi.org/10.1371/journal.pone.0075977.

Kalimuthu, P., Heider, J., Knack, D. and Bernhardt, P.V. 2015. Electrocatalytic hydrocarbon hydroxylation by ethylbenzene dehydrogenase from *Aromatoleum aromaticum*. The Journal of Physical Chemistry B. 119(8): 3456–3463. https://doi.org/10.1021/jp512562k.

Kamm Kamm, B.M. 2004. MINI-REVIEW. Applied Microbiology and Biotechnology. 64: 137–145. https://doi.org/10.1007/s00253-003-1537-7.

Kumar, M. and Khanna, S. 2010. Diversity of 16S rRNA and dioxygenase genes detected in coal-tar-contaminated site undergoing active bioremediation. Journal of Applied Microbiology. 108(4): 1252–1262. https://doi.org/10.1111/j.1365-2672.2009.04523.x

Kumar, A., Singh, J. and Baskar, C. 2019. Lignocellulosic Biomass for bioethanol production through microbes: strategies to improve process efficiency. *In*: Prospects of Renewable Bioprocessing in Future Energy Systems, 357–386. https://doi.org/10.1007/978-3-030-14463-0_14.

Kuttiyathil, M.S., Mohamed, M.M. and Al-Zuhair, S. 2021. Using microalgae for remediation of crude petroleum oil–water emulsions. Biotechnology Progress. 37(2). https://doi.org/10.1002/btpr.3098.

Lei, A.-P., Hu, Z.-L., Wong, Y.-S. and Tam, N.F.-Y. 2007. Removal of fluoranthene and pyrene by different microalgal species. Bioresource Technology. 98(2): 273–280. https://doi.org/10.1016/j.biortech.2006.01.012.

Leonardo, T., Farhi, E., Pouget, S., Motellier, S., Boisson, A.-M., Banerjee, D., et al. 2016. Silver accumulation in the green microalga *Coccomyxa actinabiotis*: toxicity, *in situ* speciation, and localization investigated using synchrotron XAS, XRD, and TEM. Environmental Science and Technology. 50(1): 359–367. https://doi.org/10.1021/acs.est.5b03306.

Li, X., Lin, X., Li, P., Liu, W., Wang, L., Ma, F., et al. 2009. Biodegradation of the low concentration of polycyclic aromatic hydrocarbons in soil by microbial consortium during incubation. Journal of Hazardous Materials. 172(2–3). https://doi.org/10.1016/j.jhazmat.2009.07.044.

Lim, J.W., Park, T., Tong, Y.W. and Yu, Z. 2020. The microbiome driving anaerobic digestion and microbial analysis. *In*: Advances in Bioenergy. 5: 1–61. https://doi.org/10.1016/BS.AIBE.2020.04.001.

Liu, J.Q., Kurihara, T., Hasan, A.K., Nardi-Dei, V., Koshikawa, H., Esaki, N., et al. 1994. Purification and characterization of thermostable and nonthermostable 2-haloacid dehalogenases with different stereospecificities from *Pseudomonas* sp. strain YL. Applied and Environmental Microbiology. 60(7): 2389–2393. https://doi.org/10.1128/aem.60.7.2389-2393.1994.

Liu, Y., Wan, Y.Y., Wang, C., Ma, Z., Liu, X. and Li, S. 2020. Biodegradation of n-alkanes in crude oil by three identified bacterial strains. Fuel. 275: 117897. https://doi.org/10.1016/j.fuel.2020.117897.

Lv, W., Schanbacher, F.L. and Yu, Z. 2010. Putting microbes to work in sequence: recent advances in temperature-phased anaerobic digestion processes. Bioresource Technology. 101(24): 9409–9414. https://doi.org/10.1016/J.BIORTECH.2010.07.100.

Ma, Y., Prasad, M.N.V., Rajkumar, M. and Freitas, H. 2011. Plant growth promoting rhizobacteria and endophytes accelerate phytoremediation of metalliferous soils. Biotechnology Advances. 29(2): 248–258. https://doi.org/10.1016/j.biotechadv.2010.12.001.

Madadi, R., Pourbabaee, A.A., Tabatabaei, M., Zahed, M.A. and Naghavi, M.R. 2016. Treatment of petrochemical wastewater by the green algae *Chlorella vulgaris*. International Journal of Environmental Research. 10(4).

Mahalaxmi, S. and Williford, C. 2016. Biochemical conversion of biomass to fuels. *In*: Handbook of Climate Change Mitigation and Adaptation, 2 Ed. (Vol. 3). https://doi.org/10.1007/978-3-319-14409-2_26.

Mansur, M., Arias, M.E., Copa-Patiño, J.L., Flärdh, M. and González, A.E. 2003. The white-rot fungus *Pleurotus ostreatus* secretes laccase isozymes with different substrate specificities. Mycologia. 95(6): 1013–1020. https://doi.org/10.1080/15572536.2004.11833017.

McKendry, P. 2002. Energy production from biomass (part 1): overview of biomass. Bioresource Technology. 83(1): 37–46. https://doi.org/10.1016/S0960-8524(01)00118-3.

Men, Y., Yu, K., Bælum, J., Gao, Y., Tremblay, J., Prestat, E., et al. 2017. Metagenomic and metatranscriptomic analyses reveal the structure and dynamics of a dechlorinating community containing *Dehalococcoides mccartyi* and corrinoid-providing microorganisms under cobalamin-limited conditions. Applied and Environmental Microbiology. 83(8). https://doi.org/10.1128/AEM.03508-16.

Moody, J.D., Freeman, J.P., Doerge, D.R. and Cerniglia, C.E. 2001. Degradation of phenanthrene and anthracene by cell suspensions of *Mycobacterium* sp. strain PYR-1. Applied and Environmental Microbiology. 67(4): 1476–1483. https://doi.org/10.1128/AEM.67.4.1476-1483.2001.

Muratova, A.Y., Turkovskaya, O.V., Antonyuk, L.P., Makarov, O.E., Pozdnyakova, L.I. and Ignatov, V.V. 2005. Oil-oxidizing potential of associative rhizobacteria of the genus *Azospirillum*. Microbiology. 74(2). https://doi.org/10.1007/s11021-005-0053-4.

Muthukrishnan, G., Gopalasubramaniam, S.K. and Perumal, P. 2018. Prospects of arbuscular mycorrhizal fungi for heavy metal-polluted soil management. *In*: Microorganisms for Green Revolution. 91–113. https://doi.org/10.1007/978-981-10-7146-1_5.

Naik, S.N., Goud, V.V., Rout, P.K. and Dalai, A.K. 2010. Production of first and second generation biofuels: a comprehensive review. Renewable and Sustainable Energy Reviews. 14(2): 578–597. https://doi.org/10.1016/J.RSER.2009.10.003.

Narasimhan, K., Basheer, C., Bajic, V.B. and Swarup, S. 2003. Enhancement of plant-microbe interactions using a rhizosphere metabolomics-driven approach and its application in the removal of polychlorinated biphenyls. Plant Physiology. 132(1). https://doi.org/10.1104/pp.102.016295.

Nasrawi, H. Al. 2012. Biodegradation of crude oil by fungi isolated from Gulf of Mexico. Journal of Bioremediation and Biodegradation. 03(04). https://doi.org/10.4172/2155-6199.1000147.

Njoku, K., Yussuf, A., Akinola, M., Adesuyi, A., Jolaoso, A. and Adedokun, A. 2017. Mycoremediation of petroleum hydrocarbon polluted soil by *Pleurotus pulmonarius*. Ethiopian Journal of Environmental Studies and Management. 9(1): 865. https://doi.org/10.4314/ejesm.v9i1.6s.

Obi, L.U., Atagana, H.I. and Adeleke, R.A. 2016. Isolation and characterisation of crude oil sludge degrading bacteria. SpringerPlus. 5(1): 1946. https://doi.org/10.1186/s40064-016-3617-z.

Ogidi, O.I. and Njoku, O.C. 2017. A Review on the possibilities of the application of bioremediation methods in the oil spill clean-up of ogoni land. *In*: International Journal of Biological Sciences and Technology. 9(6).

Okoro, C.C. 2008. Biodegradation of hydrocarbons in untreated produce water using pure fungal cultures. African Journal of Microbiology Research. 2: 217–223.

Okoro, B.C., Nwadike, O.A. and Agunwamba, J.C. 2013. Prediction of remediation rates of microbes in polluted crude oil soil samples. Environment and Natural Resources Research. 3(3). https://doi.org/10.5539/enrr.v3n3p89.

Otitolaiye, V.O. and Al-Harethiya, G.M. 2022. Impacts of petroleum refinery emissions on the health and safety of local residents. Journal of Air Pollution and Health. https://doi.org/10.18502/japh.v7i1.8921.

Paes, B.G. and Almeida, J.R. 2014. Genetic improvement of microorganisms for applications in biorefineries. Chemical and Biological Technologies in Agriculture. 1(1): 1–10. https://doi.org/10.1186/S40538-014-0021-1/TABLES/1.

Pan, I., Dam, B. and Sen, S.K. 2011. Composting of common organic wastes using microbial inoculants. 3 Biotech. 2(2): 127–134. https://doi.org/10.1007/S13205-011-0033-5.

Pandey, B. and Fulekar, M.H. 2012. Bioremediation technology: a new horizon for environmental clean-up. Biology and Medicine.

Perera, F.P., Tang, D., Wang, S., Vishnevetsky, J., Zhang, B., Diaz, D., et al. 2012. Prenatal polycyclic aromatic hydrocarbon (PAH) exposure and child behavior at age 6–7 years. Environmental Health Perspectives. 120(6): 921–926. https://doi.org/10.1289/ehp.1104315.

Poddar, K., Sarkar, D. and Sarkar, A. 2019. Construction of potential bacterial consortia for efficient hydrocarbon degradation. International Biodeterioration and Biodegradation. 144. https://doi.org/10.1016/j.ibiod.2019.104770.

Pourbabaee, A.A., Khazaei, M., Alikhani, H.A. and Emami, S. 2021. Root nodulation of alfalfa by *Ensifer meliloti* in petroleum contaminated soil. Rhizosphere. 17. https://doi.org/10.1016/j.rhisph.2021.100305.

Prasad, R.K., Chatterjee, S., Mazumder, P.B., Gupta, S.K., Sharma, S., Vairale, M.G., et al. 2019. Bioethanol production from waste lignocelluloses: a review on microbial degradation potential. Chemosphere. 231: 588–606. https://doi.org/10.1016/J.CHEMOSPHERE.2019.05.142.

Prince, R.C., Amande, T.J. and McGenity, T.J. 2018. Prokaryotic hydrocarbon degraders. *In*: Taxonomy, Genomics and Ecophysiology of Hydrocarbon-Degrading Microbes. 1–41. Springer International Publishing. https://doi.org/10.1007/978-3-319-60053-6_15-1.

Rahman, K.S.M., Rahman, T.J., McClean, S., Marchant, R. and Banat, I.M. 2002. Rhamnolipid biosurfactant production by strains of *Pseudomonas aeruginosa* using low-cost raw materials. Biotechnology Progress. 18(6): 1277–1281. https://doi.org/10.1021/bp020071x.

Rastogi, M., Nandal, M. and Khosla, B. 2020. Microbes as vital additives for solid waste composting. Heliyon. 6(2): e03343. https://doi.org/10.1016/J.HELIYON.2020.E03343.

Raza, W., Lee, J., Raza, N., Luo, Y., Kim, K.-H. and Yang, J. 2019. Removal of phenolic compounds from industrial wastewater based on membrane-based technologies. Journal of Industrial and Engineering Chemistry. 71: 1–18. https://doi.org/10.1016/j.jiec.2018.11.024.

Rojo, F. 2009. Degradation of alkanes by bacteria. Environmental Microbiology. 11(10): 2477–2490. https://doi.org/10.1111/j.1462-2920.2009.01948.x.

Ron, E.Z. and Rosenberg, E. 2002. Biosurfactants and oil bioremediation. Current Opinion in Biotechnology. 13(3): 249–252. https://doi.org/10.1016/S0958-1669(02)00316-6.

Sam, A. and Barik, D. 2019. Toxic waste from municipality. Energy from Toxic Organic Waste for Heat and Power Generation. 7–16. https://doi.org/10.1016/B978-0-08-102528-4.00002-X.

Sardrood, B.P., Goltapeh, E.M. and Varma, A. 2013. An Introduction to Bioremediation. *In*: Fungi as Bioremediators. 3–27. https://doi.org/10.1007/978-3-642-33811-3_1.

Semple, K.T., Cain, R.B. and Schmidt, S. 1999. Biodegradation of aromatic compounds by microalgae. FEMS Microbiology Letters. 170(2): 291–300. https://doi.org/10.1111/j.1574-6968.1999.tb13386.x.

Serrano-Posada, H., Centeno-Leija, S., Rojas-Trejo, S.P., Rodríguez-Almazán, C., Stojanoff, V. and Rudiño-Piñera, E. 2015. X-ray-induced catalytic active-site reduction of a multicopper oxidase: structural insights into the proton-relay mechanism and O_2-reduction states. Acta Crystallographica Section D Biological Crystallography. 71(12): 2396–2411. https://doi.org/10.1107/S1399004715018714.

Shraddha, Shekher, R., Sehgal, S., Kamthania, M. and Kumar, A. 2011. Laccase: microbial sources, production, purification, and potential biotechnological applications. Enzyme Research. 2011: 1–11. https://doi.org/10.4061/2011/217861.

Silva, C.S., Durão, P., Fillat, A., Lindley, P.F., Martins, L.O. and Bento, I. 2012. Crystal structure of the multicopper oxidase from the pathogenic bacterium *Campylobacter jejuni*CGUG11284: characterization of a metallo-oxidase. Metallomics. 4(1): 37–47. https://doi.org/10.1039/C1MT00156F.

Speight, J.G. and El-Gendy, N.S. 2018. Introduction to petroleum biotechnology. *In*: Introduction to Petroleum Biotechnology. https://doi.org/10.1016/b978-0-12-805151-1.00003-5.

Stout, L. and Nüsslein, K. 2010. Biotechnological potential of aquatic plant–microbe interactions. Current Opinion in Biotechnology. 21(3): 339–345. https://doi.org/10.1016/j.copbio.2010.04.004.

Strong, P.J. and Claus, H. 2011. Laccase: a review of its past and its future in bioremediation. Critical Reviews in Environmental Science and Technology. 41(4): 373–434. https://doi.org/10.1080/10643380902945706.

Subashchandrabose, S.R., Ramakrishnan, B., Megharaj, M., Venkateswarlu, K. and Naidu, R. 2013. Mixotrophic cyanobacteria and microalgae as distinctive biological agents for organic pollutant degradation. Environment International. 51: 59–72. https://doi.org/10.1016/j.envint.2012.10.007.

Suganthi, S.H., Murshid, S., Sriram, S. and Ramani, K. 2018. Enhanced biodegradation of hydrocarbons in petroleum tank bottom oil sludge and characterization of biocatalysts and biosurfactants. Journal of Environmental Management. 220: 87–95. https://doi.org/10.1016/j.jenvman.2018.04.120.

Sugimoto, M., Tanabe, M., Hataya, M., Enokibara, S., Duine, J.A. and Kawai, F. 2001. The first step in polyethylene glycol degradation by sphingomonads proceeds via a flavoprotein alcohol dehydrogenase containing flavin adenine dinucleotide. Journal of Bacteriology. 183(22): 6694–6698. https://doi.org/10.1128/JB.183.22.6694-6698.2001.

Suja, F., Rahim, F., Taha, M.R., Hambali, N., Rizal Razali, M., Khalid, A., et al. 2014. Effects of local microbial bioaugmentation and biostimulation on the bioremediation of total petroleum hydrocarbons (TPH) in crude oil contaminated soil based on laboratory and field observations. International Biodeterioration and Biodegradation. 90: 115–122. https://doi.org/10.1016/j.ibiod.2014.03.006.

Takkellapati, S., Li, T. and Gonzalez, M.A. 2018. An overview of biorefinery-derived platform chemicals from a cellulose and hemicellulose biorefinery. Clean Technologies and Environmental Policy. 20(7): 1615–1630. https://doi.org/10.1007/S10098-018-1568-5/FIGURES/1.

Tao, K., Liu, X., Chen, X., Hu, X., Cao, L. and Yuan, X. 2017. Biodegradation of crude oil by a defined co-culture of indigenous bacterial consortium and exogenous *Bacillus subtilis*. Bioresource Technology. 224: 327–332. https://doi.org/10.1016/j.biortech.2016.10.073.

Teng, Y., Shen, Y., Luo, Y., Sun, X., Sun, M., Fu, D., et al. 2011. Influence of *Rhizobium meliloti* on phytoremediation of polycyclic aromatic hydrocarbons by alfalfa in an aged contaminated soil. Journal of Hazardous Materials. 186(2–3): 1271–1276. https://doi.org/10.1016/j.jhazmat.2010.11.126.

Tran, Q.N.M., Mimoto, H., Koyama, M. and Nakasaki, K. 2019. Lactic acid bacteria modulate organic acid production during early stages of food waste composting. Science of The Total Environment. 687: 341–347. https://doi.org/10.1016/J.SCITOTENV.2019.06.113.

Tremblay, J., Yergeau, E., Fortin, N., Cobanli, S., Elias, M., King, T.L., et al. 2017. Chemical dispersants enhance the activity of oil- and gas condensate-degrading marine bacteria. The ISME Journal. 11(12): 2793–2808. https://doi.org/10.1038/ismej.2017.129.

Tursi, A. 2019. A review on biomass: importance, chemistry, classification, and conversion. Biofuel Research Journal. 6(2): 962–979. https://doi.org/10.18331/BRJ2019.6.2.3.

Tyagi, M., da Fonseca, M.M.R. and de Carvalho, C.C.C.R. 2011. Bioaugmentation and biostimulation strategies to improve the effectiveness of bioremediation processes. Biodegradation. 22(2): 231–241. https://doi.org/10.1007/s10532-010-9394-4.

Usman, M., Faure, P., Hanna, K., Abdelmoula, M. and Ruby, C. 2012. Application of magnetite catalyzed chemical oxidation (Fenton-like and persulfate) for the remediation of oil hydrocarbon contamination. Fuel. 96: 270–276. https://doi.org/10.1016/j.fuel.2012.01.017.

Van Hamme, J.D., Singh, A. and Ward, O.P. 2003. Recent advances in petroleum microbiology. Microbiology and Molecular Biology Reviews. 67(4): 503–549. https://doi.org/10.1128/MMBR.67.4.503-549.2003.

Varjani, S.J. 2017. Microbial degradation of petroleum hydrocarbons. Bioresource Technology. 223: 277–286. https://doi.org/10.1016/j.biortech.2016.10.037.

Vasconcellos, S.P. de, Angolini, C.F.F., García, I.N.S., Martins Dellagnezze, B., Silva, C.C. da, Marsaioli, A.J., et al. 2010. Screening for hydrocarbon biodegraders in a metagenomic clone library derived from Brazilian petroleum reservoirs. Organic Geochemistry. 41(7): 675–681. https://doi.org/10.1016/j.orggeochem.2010.03.014.

Verma, S., Saxena, J., Prasanna, R., Sharma, V. and Nain, L. 2012. Medium optimization for a novel crude-oil degrading lipase from *Pseudomonas aeruginosa* SL-72 using statistical approaches for bioremediation of crude-oil. Biocatalysis and Agricultural Biotechnology. 1(4): 321–329. https://doi.org/10.1016/j.bcab.2012.07.002.

Vidyashankar, S. and Ravishankar, G.A. 2016. Algae-based bioremediation. *In*: Bioremediation and Bioeconomy. 457–493. Elsevier. https://doi.org/10.1016/B978-0-12-802830-8.00018-6.

Viñas, M., Sabaté, J., Espuny, M.J. and Solanas, A.M. 2005. Bacterial community dynamics and polycyclic aromatic hydrocarbon degradation during bioremediation of heavily creosote-contaminated soil. Applied and Environmental Microbiology. 71(11): 7008–7018. https://doi.org/10.1128/AEM.71.11.7008-7018.2005.

Wainwright, M. 1999. Bioremediation. *In*: An Introduction to Environmental Biotechnology. 37–61. Springer US. https://doi.org/10.1007/978-1-4615-5251-2_5.

Walker, J.D., Jr., L.C. and Cooney, J.J. 1973. Microbial petroleum degradation: the role of cladosporium resinae. International Oil Spill Conference Proceedings. 1973(1): 821–825. https://doi.org/10.7901/2169-3358-1973-1-821.

Walker, J.D., Colwell, R.R. and Petrakis, L. 1975a. Degradation of petroleum by an alga, *Prototheca zopfii*. Applied Microbiology. 30(1): 79–81. https://doi.org/10.1128/am.30.1.79-81.1975.

Walker, J.D., Colwell, R.R., Vaituzis, Z. and Meyer, S.A. 1975b. Petroleum-degrading achlorophyllous alga *Prototheca zopfii*. Nature. 254(5499): 423–424. https://doi.org/10.1038/254423a0.

Wei, Y., Li, J., Shi, D., Liu, G., Zhao, Y. and Shimaoka, T. 2017. Environmental challenges impeding the composting of biodegradable municipal solid waste: a critical review. Resources, Conservation and Recycling. 122: 51–65. https://doi.org/10.1016/J.RESCONREC.2017.01.024.

Wenzel, W.W. 2009. Rhizosphere processes and management in plant-assisted bioremediation (phytoremediation) of soils. Plant and Soil. 321(1–2): 385–408. https://doi.org/10.1007/s11104-008-9686-1.

Whang, L.-M., Liu, P.-W.G., Ma, C.-C. and Cheng, S.-S. 2008. Application of biosurfactants, rhamnolipid, and surfactin, for enhanced biodegradation of diesel-contaminated water and soil. Journal of Hazardous Materials. 151(1): 155–163. https://doi.org/10.1016/j.jhazmat.2007.05.063.

White, D.A., Hird, L.C. and Ali, S.T. 2013. Production and characterization of a trehalolipid biosurfactant produced by the novel marine bacterium *Rhodococcus* sp., strain PML026. Journal of Applied Microbiology. 115(3): 744–755. https://doi.org/10.1111/jam.12287.

Williams, C.L., Dahiya, A. and Porter, P. 2020. Introduction to bioenergy and waste to energy. Bioenergy: Biomass to Biofuels and Waste to Energy. 5–44. https://doi.org/10.1016/B978-0-12-815497-7.00001-4.

Wolicka, D., Suszek, A., Borkowski, A. and Bielecka, A. 2009. Application of aerobic microorganisms in bioremediation in situ of soil contaminated by petroleum products. Bioresource Technology. 100(13): 3221–3227. https://doi.org/10.1016/j.biortech.2009.02.020.

Wong, J.W.C., Fang, M., Zhao, Z. and Xing, B. 2004. Effect of surfactants on solubilization and degradation of phenanthrene under thermophilic conditions. Journal of Environmental Quality. 33(6): 2015–2025. https://doi.org/10.2134/jeq2004.2015.

Xia, W., Du, Z., Cui, Q., Dong, H., Wang, F., He, P., et al. 2014. Biosurfactant produced by novel *Pseudomonas* sp. WJ6 with biodegradation of n-alkanes and polycyclic aromatic hydrocarbons. Journal of Hazardous Materials. 276: 489–498. https://doi.org/10.1016/j.jhazmat.2014.05.062.

Yakimov, M.M., Timmis, K.N. and Golyshin, P.N. 2007. Obligate oil-degrading marine bacteria. Current Opinion in Biotechnology. 18(3): 257–266. https://doi.org/10.1016/j.copbio.2007.04.006.

Ye, J.-S., Yin, H., Qiang, J., Peng, H., Qin, H.-M., Zhang, N., et al. 2011. Biodegradation of anthracene by *Aspergillus fumigatus*. Journal of Hazardous Materials. 185(1): 174–181. https://doi.org/10.1016/j.jhazmat.2010.09.015.

Yee, D.C., Maynard, J.A. and Wood, T.K. 1998. Rhizoremediation of trichloroethylene by a recombinant, root-colonizing *Pseudomonas fluorescens* strain expressing toluene ortho-monooxygenase constitutively. Applied and Environmental Microbiology. 64(1). https://doi.org/10.1128/aem.64.1.112-118.1998.

Yousaf, S., Andria, V., Reichenauer, T.G., Smalla, K. and Sessitsch, A. 2010. Phylogenetic and functional diversity of alkane degrading bacteria associated with Italian ryegrass (*Lolium multiflorum*) and Birdsfoot trefoil (*Lotus corniculatus*) in a petroleum oil-contaminated environment. Journal of Hazardous Materials. 184(1–3): 523–532. https://doi.org/10.1016/j.jhazmat.2010.08.067.

Yu, K.S.H., Wong, A.H.Y., Yau, K.W.Y., Wong, Y.S. and Tam, N.F.Y. 2005. Natural attenuation, biostimulation and bioaugmentation on biodegradation of polycyclic aromatic hydrocarbons (PAHs) in mangrove sediments. Marine Pollution Bulletin. 51(8–12): 1071–1077. https://doi.org/10.1016/J.MARPOLBUL.2005.06.006.

Yu, X.Z., Wu, S.C., Wu, F.Y. and Wong, M.H. 2011. Enhanced dissipation of PAHs from soil using mycorrhizal ryegrass and PAH-degrading bacteria. Journal of Hazardous Materials. 186(2–3): 1206–1217. https://doi.org/10.1016/j.jhazmat.2010.11.116.

Zu, L., Li, G., An, T. and Wong, P.-K. 2012. Biodegradation kinetics and mechanism of 2,4,6-tribromophenol by Bacillus sp. GZT: a phenomenon of xenobiotic methylation during debromination. Bioresource Technology. 110: 153–159. https://doi.org/10.1016/j.biortech.2012.01.131.

Chapter **11**

Management of Hazardous Waste

Narinderpal Kaur*[1], Balraj Saini[2] and Shiwali Sharma[1]

[1]Chitkara University School of Pharmacy, Chitkara University, Himachal Pradesh 174103, India.

[2]Chitkara College of Pharmacy, Rajpura, Punjab, India

INTRODUCTION

Disasters can happen due to the natural as well as man-made activities for different purpose (Khare 2001). A hazard is an erratic or extreme incident that has an undesirable impact on the activity, property and overall life of humans. The term "hazardous" waste is differently used in various countries. It can be broadly defined as non-household waste that also includes hazardous chemicals. Any material when thrown away without taking into account the wages it can give for its intrinsic value is said to be waste (Misra et al. 2005) and if the management of these wastes is not done in a proper way, then it may present a potential hazard to the health of the human or to the environment by contaminating air, soil or water. Hazardous wastes are usually non-biodegradable wastes that are tenacious in the environment and are venomous to the natural resources or to the health of human. Industries are the major part of present society and production of this hazardous waste is an inexorable outcome of all these progressive activities and industrialization. The fast progression of technology and the intensification of the need of consumers all over the world causes greater depletion of natural resources (Srivastav et al. 2023). Even the majority of human history has seen

* For Correspondence: narinder.sonia@gmail.com

minimal waste production because of low population densities and overuse of natural resources. In premodern times, the majority of trash created was ashes and biodegradable human waste, which was locally reburied with little effect on the ecosystem. Wooden and metal tools were typically passed down through the generations or repurposed (Barbalace et al. 2013)

Nonetheless, compared to other civilizations, some have been more wasteful than others. Specifically, the Maya of Central America observed a set monthly ceremony in which the village's residents would assemble and burn their trash in expansive landfills. Hazardous wastes are defined as materials that are considered to be non-biodegradable, harmful, especially to human health and natural resources, and persistent to the environment. These materials include litter, trash, rubble, and a combination of these that could pose a serious risk to human health or safety (RCRA 2016).

One of the most significant parts of municipal garbage is domestic waste. Food scraps, paper, glass, metals, plastics, textiles, and so forth are examples of household garbage. Vegetable, fruit peel, bone, and meat waste, as well as the waste from chickens and fish, make up a sizable portion of household garbage. These wastes are categorized as wet wastes. Waste newspapers, books, notebooks, wrapping paper, cardboard, and paper products also account for a significant amount of the trash. Toys, plastics, glass and metal cans, and throwaway plates are further types of household garbage. The components of municipal solid trash have been described using a variety of definitions and categories in books and publications (Dehghani et al. 2021).

There are some common possible forms of domestic waste management (Vallero 2019).

(a) The whole public has to be aware of and engaged in the issues surrounding HHW.
(b) Making local collection and disposal plans and guidelines public and offering a sufficient justification for the plans that have been chosen.
(c) The management of household waste should be publicly announced by both national and local efforts, such as flyers, websites, public announcements, news articles, and so on, all of which should convey the same clear message.
(d) Frequent door-to-door collections that take into account homeowners who are unable to drive themselves.
(e) There should be phone support lines and an easily accessible internet available.
(f) Access to recycling facilities for home garbage that are prominently marked.
(g) Retail stores may have taken-back programs for WEEE (Waste from Electrical and Electronic Equipment), appliances that contain CFCs (Chlorofluorocarbons) in the home, and other items.
(h) Legislation or voluntary programs to compel manufacturers to reveal the contents of any potentially dangerous components.

Along with all these, the process of management of hazardous waste comprises of the collection, transference, treatment, reprocessing, clearance, and monitoring of the various sites of disposal of the wastes. However, simply gathering and disposing of these wastes without the necessary processing or treatment leads to a large amount of this type of waste remaining unattended at various collection sites, riverbanks, and roadside areas, in addition to the waste from several large and small-scale industries that typically dispose of their waste in open spaces or areas close to water sources, which pollutes the environment and poses health risks to the general public (Kumar et al. 2017) and people living proximate to old and abandoned sites of disposal of waste may be in the most endangered state (Nathanson 2023). Presently, in many of the emerging nations, these wastes are usually disposed of comprehensively that creates a risk to health as well as to the environment. In addition to this, the governments and various international agencies are endeavouring to resist the emerging problems of the substances that are hazardous to the environment and that seems to be difficult to manage because of a number of sources from which they can be released (Amadi et al. 2017). These substances can cause minor to serious effects on health of human such as discomfort for a shorter duration of time, headache, nausea or may lead to serious health issues like birth defects, various types of cancers or major accidents that may cause serious injury or even death. So, the proper disposal or management of hazardous waste is an important issue all over the world as the probability of diseases related to exposure to this kind of waste, specifically in the countries known as middle- or low-income countries, are cumulating without acknowledging adequately (Landrigan et al. 2015), and management of domestic and menacing waste is a matter of concern (Nweke et al. 2009). It is crucial that hazardous waste be disposed of or managed properly anywhere in the globe. Hence, this chapter gives emphasis on hazardous wastes, types, various sources of wastes, disposal and management of this kind of waste.

TYPES OF HAZARDOUS WASTES

Hazardous wastes can be classified based on physical, biological or chemical properties. Depending on their physical properties, hazardous waste can be classified as (i) solid wastes, (ii) liquid wastes (iii) gaseous wastes (iv) sludge wastes. Further, it can be classified based on the source of the waste such as radioactive wastes, biomedical waste, chemical waste, flammable waste, explosive waste, household hazardous waste, etc. There may be different collection and treatment approaches for each category of waste. Identification and segregation of specific type of waste can help in developing strategies for their proper management.

SOURCES OF HAZARDOUS WASTES

There are a number of sources from where hazardous wastes can be produced, the most common source being the industries. Other sources may include wastes from agriculture, household or domestic, minerals and mines, health care facilities, processing sites, commercial sites, agro industry, sites of disposal of solid waste, institutional and industrial wastes, building materials, contaminated sites, and radioactive elements involving activities. A few of these hazardous substances are categorised as apparently carcinogenic to humans and are allied with augmented rates of damage of organ, formation of tumour and sometimes even death in humans (Roger et al. 1989). Some of the sources of hazardous waste are discussed below:

Hazardous Waste from Small Scale Industry

Various sources like processing of textile, different techniques of metal treatment such as etching, electroplating, galvanizing and anodizing, printing, photofinishing and leather tanning can generate a number of hazardous waste materials.

Hazardous Waste from Large Scale Industry

Various sources like manufacturing of pharmaceuticals, chemicals, petrochemical, refining of oil and production of etc. produces oily wastes, spent catalysts, red muds, solvent mercury and tarry residues as hazardous waste.

Hazardous Waste from Agricultural Industry

Agricultural industry like farming etc. utilises fertilizers, pesticides and a number of veterinary product wastes that comes under the hazardous waste.

Hazardous Waste from Healthcare Industry

Different sources such as clinics, medical laboratories and hospitals generate infectious needles, contaminated blood and pathological wastes as a source of hazardous waste.

Hazardous Waste from Building and Roofing Industries

Material used in manufacturing of roofs and buildings, specifically material made up of asbestos and copper, are common sources of generation of hazardous waste.

Miscellaneous Waste

Some other type of hazardous wastes like halogenated solvents, oils, polychlorinated biphenyls and hydraulic fluids etc. are produced from processing and mining of uranium, repair of automobiles, dry cleaning, gasoline stations, and workshops, airports, electrical transformers, and municipal parks (Figure 11.1). Some of the hazardous substances are categorised as potential carcinogens to humans and are linked to the increased rate of organ damage, tumour formation and even death.

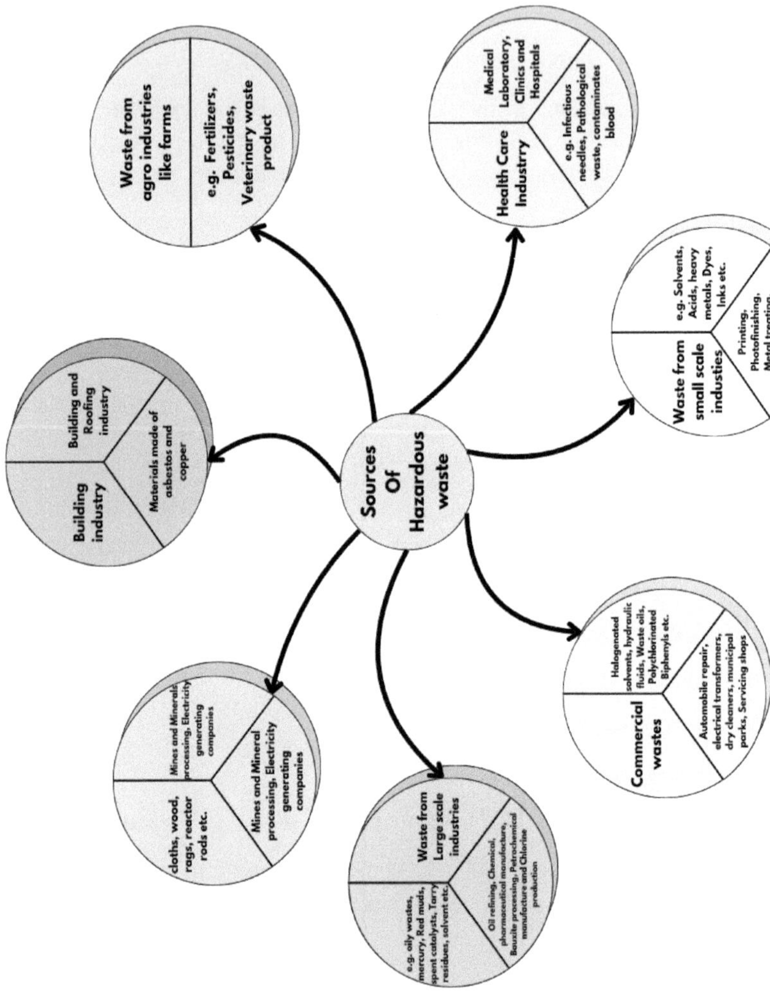

Figure 11.1 Various sources of hazardous waste.

Although it is a common assumption that maximum hazardous waste comes from industries only, our daily household waste also comprises of many types of dangerous and toxic waste substances such as floor cleaners, glass, insecticides, anti-clogs etc. Evidently, the amount of this household waste is comparable to the waste from industries (Exposto et al. 2021).

These hazardous wastes show different type of effects on environment which are deleterious in general. Now a days, increased industrial development is the main cause of generation of hazardous waste and due to lack of proper disposal, this becomes the main area of focus to manage these wastes because these wastes are harmful in terms of mental, physical and social health of humans (Etim 2003). Other than this, dumping of hazardous wastes without proper treatment causes deterioration of soil quality, which may lead to reduction in vegetation that is again a matter of concern (Dawsin 2000) Along with this, increased production of hazardous wastes and disposal of this waste in open spaces can cause deterioration of water and air quality. The harmful chemicals that are disposed of in the form of waste in open spaces usually enter the groundwater in one or the other way and causes contamination of groundwater. Our general water supply contains these chemicals and sometimes it is susceptible to contain particles of asbestos and heavy metals such as cadmium, lead, nitrates, arsenic and a number of chemicals which are considered as carcinogenic in nature. Certain chemical substances are required for normal body functioning, although in minute quantities, primarily for synthesis of various enzymes, whereas in enormous amounts these chemicals can be toxic (Bond 2016). Another way of getting impacted by the adverse effects of hazardous wastes is by inhalation, ingestion or directly coming in contact with skin. Exposure to these can occur even when people breathe in the air contaminated with hazardous chemical vapours. If something comes directly in contact with skin, it causes exposure to skin and ingestion can occur as a secondary route of exposure such as when skin contact has occurred, or if someone put his or her contaminated hands in his or her own mouth, which leads to transfer of the hazardous chemicals from their hands to mouth. In addition to this, exposure to hazardous chemicals can occur through the food which is consumed in the form of contaminated fish and other secondary sources (Duan et al. 2011).

CHARACTERISTICS OF HAZARDOUS WASTE

There are a number of common features of hazardous waste substances like (Krishnaswamy Kanagamani et al. 2020) corrosivity, toxicity, reactivity and gnitability. These characteristics of hazardous waste are briefly described below:

Corrosivity

Usually, liquid waste materials having pH values too high or too low are considered to be corrosive. Generally, if the pH of the material is equal to or less than 2 or if it is equal to or greater than 12.5, the waste is called as corrosive hazardous waste. Materials with high pH e.g. sodium hydroxide and low pH e.g. hydrochloric acid is habitually used in many industries for cleaning purpose or to degrease the different parts made up of metals. Prior to painting, disposal of solvents devoid of any management leads to production of corrosive hazardous waste.

Ignitability

Any waste material is called as an ignitable hazardous waste if the flash point of this waste is not more than 60°C, that eagerly catches fire and burns robustly as a hazard, or is a compressed combustible gas or an oxidizer. E.g. adhesives, epoxy resins, naphtha, oil based paints, lacquer thinner etc.

Toxicity

Toxicity is an important factor to be taken into consideration for evaluation of the waste that is hazardous in nature. It can be resolved by selecting random samples of the item in question and subjecting them to a variety of tests in a licensed laboratory to identify any harmful properties.

Reactivity

Any substance is said to be reactive hazardous waste if it is not stable, viciously reacts with water, and leads to generation of toxic gases in particular when it is exposed to any material that is corrosive, water or explodes when exposed to flame or heat; e.g. sodium metal, gunpowder waste or wastes containing sulphides or cyanides. Due to all these characteristic features, it becomes necessary to manage the hazardous wastes by disposing of the hazardous waste after proper treatment, recycling etc.

MANAGEMENT OF HAZARDOUS WASTE

The management, disposal, and dumping of this type of garbage is a global issue. Indigenous people are sadly impacted by outdated, subpar, and illegal urban hazardous waste disposal methods in practically every nation. This generally takes into account the illicit trans boundary trading, typically from industrialized countries (Marsili et al. 2009). As a result, the risk of illnesses linked to garbage exposure is rising, especially in middle- and low-income nations, which is not even fully recognized (Landrigan et al. 2015). It is one of the biggest challenges of the 21st century to maintain quality of environment of the entire world (Chakraborty and Maity 2020) due to rapid, unplanned urbanization and excessive industrialization, which result in a massive demand for natural resources (Kanwar et al. 2023). Improper handling and disposal of hazardous material can have serious repercussions. Hazardous gaseous materials can be released into the air if they are disposed of incorrectly, increasing illness and mortality (Marfe 2020). In general, management of hazardous waste focuses on various techniques to decrease the amount of production of hazardous waste, decrease the toxicity of hazardous waste by involving various treatments, and to reduce exposure to this type of waste through the use of various technical controls (Vallero et al. 2022).

Conventionally, the disposal of hazardous wastes usually involves the dumping of the waste and burning of this waste in open public space. But this burning and dumping of the waste in open space causes soil, water and even contamination of air in the form of evolution of gases, deleterious odours etc. Mismanaged disposal of residues of solid waste can be the cause of remarkable water pollution. The process of incineration by municipal corporations with the help of sophisticated energy recovery systems were widely used in American and European cities, but vanished because of the high cost of operation. In current years, this technique of incineration for management of hazardous solid waste has reduced its popularity due to risk related to increased air pollution control requirements. Due to increasing industrial culture, the management of hazardous waste becomes the most important challenge in urban areas all over the world. Because of lack of proper management of waste program, remains produced as a result of countless human activities, including both household and industrial garbage, can give rise to a number of health related issues and can have a deleterious influence on the environment as well. Because of this, the concern towards the management of hazardous waste is continuously increasing (Iyyanki and Vali 2017). In the process of management, handling of hazardous waste is another issue. It is suggested to take care of the person who is involved in handling of hazardous wastes in terms of wearing protective measures related to skin, eyes, respiration and to wear protective clothing so as to diminish the consequences of the contact, because these kind of wastes can be the cause of asthma, tightening of the chest and dermatitis on exposure for a longer duration of time. We can manage hazardous wastes by diminishing the source of these kind of waste, by recycling and recovery or by the treatment of hazardous waste (Figure 11.2).

But, the final phase of management system is the disposal of hazardous waste. Appropriate pre-treatment is required in almost all the possible methods of disposal (Visvanathan 1996). Traditionally, waste can be managed by various techniques like deep well, bedrock disposal, secure landfill etc. Although adoption of various techniques of minimization of waste material while designing of any process is the most competent way to overcome the problem of generation of waste (Yadav et al. 2021), the qualities of the trash being treated determine which technique is most effective (Fox and Tuchman 1996).

Deep-well Disposal

This disposal process is particularly used for disposal of liquid waste predominantly from industry in the form of injection directly deep into the wells. This process of disposal comprise use of injection wells for placing managed or unmanaged waste in liquid form into geologic formations which shows almost zero probability to permit the relocation of pollutants into the possible and potable water aquifers. These wells for disposal of waste are very deep, generally thousands or hundreds of meters deep, that is far away from the outer layer. and beneath the provincial table of water. Indigenous well water has to be monitored for detection of any inadvertent leaks. The flow of ground water that is deep inside is generally considered to be slow; in addition to this, it is assumed that the chemicals that are lethal in nature have drifted far enough sideways to spread into an operational aquifer or any other water body and the concentration of these ingredients has become adequately less so as it will no longer remain endangered. This technique is moderately inexpensive and it requires a small amount or no pre-treatment of the waste, but it eliminated the risk of this type of garbage leaking, which eventually contaminates subterranean water supplies (Linday 2012).

Figure 11.2 Management of Hazardous waste processes.

Bedrock Disposal

This type of disposal is used for management of solid hazardous waste. It basically includes the concept of multi barrier around the waste in solid form with numerous diverse forms of materials to construct manifold of obstacles to the leakage of waste or incursion by ground water. The most extensively recognised strategies are for those which are concealed in the deep inside with stability like rock structures. A lot of geological formations, for instance salt, volcanic tuff, granite, thick basalts (like the shale or Columbia River plateau basalt) will be suitable (Salma 2012).

Secure Landfills

Traditionally, some of the wastes that are hazardous were discarded in systematic landfills. But this leads to seepage of hazardous materials into the ground and in this way, these chemicals can enter into the natural hydrologic systems. This is the reason that currently many landfills require countermeasures in contrast to the contamination of groundwater. A landfill is the facility of disposal of hazardous wastes where these waste material are retained and stowed in the soil. The purpose of landfills is to reframe or to bury the waste material so that these materials no longer remain hazard either to environment or to public health (Encyclopedia Britannica 2016).

Besides all these techniques, biotechnology plays an important role in management of hazardous waste now a days. In recent years, biotechnological application in management of hazardous wastes, along with the concerns of biosafety towards both health of human and environment, has shown a remarkable increase. Biotechnological applications are the application of various technologies that includes traditional biotechnology along with the modern biotechnology which is based on bioinformatics, recombinant DNA technology etc. These various techniques are used for clean-up of the environment, detoxification of the hazardous wastes that are toxic, generation of the materials that are biodegradable and to develop eco-friendly and safe processes of manufacturing as well as for disposal of the hazardous waste. Generally, biotechnology counts on the usage of biological processes that take into account the microbial reactions for degradation and detoxification of the hazardous wastes. A number of microorganisms are there which can be used for biological process of treatment of hazardous wastes such as alga, fungi, bacteria and protozoa (Singh 2010).

Recombinant DNA Technique

Genetic engineering is another name for this technique. Genetic engineering gives emphasis particularly on alteration of the genetic material of an organism so as to alter the expression of that genetic material at a desired level. This technique involves the isolation and identification of the gene of interest and then inclusion of this desired gene into the other organism. Then this gene of interest gives its expression in the organism in which it is integrated. With the help of this technique of recombinant DNA technology, microorganisms can be modified genetically to remediate the hazardous wastes. For example, genetic modification of *Deinococcus radiodurans,* which is tremendously radiant-resistant bacterium, is being done with the help of biodegradation gene so as to render it suitable for the treatment of mixed waste (Thakur 2006). Another good example is of cotton, which is commercially grown in many countries and the growth of cotton gets affected by the pests. To overcome this problem and to minimize the use of pesticides, genetic modification of cotton has been done by inclusion of insecticidal proteins from *Bacillus thuringiensis*. This genetically modified cotton known as *Bacillus thuringiensis* cotton has found a good acceptance rate all over the world.

Bioremediation Technique

This technique is globally recognised as an effective biotechnological method used for the environmental clean-up by the use of microbes for detoxification of hazardous waste. This technique basically gives stress on degradation or transformation of toxic material or to render this toxic material as harmless. This technique generally involves indigenous microbial population with or without supplementation of nutrients, or it can involve inoculation of exogenous organisms into the site. The process is called bio-augmentation when exogenous organisms are involved. In both the cases, the objective is to defuse the toxic chemicals, that is, without the formation of new toxins (Bennett et al. 2002). A number of bioremediation approaches have been established for detoxification of the hazardous wastes. Bioremediation can broadly be divided into two categories: one is phytoremediation and another is microbial bioremediation. Phytoremediation involves the use of plants to detoxify the water and soil, whereas microbial bioremediation involves usage of living microorganisms for treatment of complex toxic compounds and convert it into the harmless by products of cellular metabolism such as CO_2 and H_2O (Thakur 2006).

CONCLUSION

It is important to make people aware about proper storage, transport, treatment and disposal of hazardous waste so as to protect the environment and health of humans from harmful effects of these wastes generated by various sources. Industries have to focus on various technologies that are environment friendly for producing different products or government has to stringent and make some strict laws for management or disposal of hazardous wastes.

REFERENCES

Amadi, C.C., Okeke, O.C. and Amadi, D.C. 2017. Hazardous waste management: a review of principles and methods. International Journal of Advanced Academic Research Sciences, Technology and Engineering. 3(8). ISSN: 2488–9849.

Barbalace, R.C. 2003. The History of Waste. EnvironmentalChemistry.com. Retrieved from https://environmentalchemistry.com/yogi/environmental/wastehistory.html.

Bennett, J.W., Wunch, K.G. and Faison, B.D. 2002. Use of fungi in biodegradation. *In*: C.J. Hurst (ed.). Manual of Environmental Microbiology. ASM Press. 960–971.

Bond, O. 2016. Signs and symptoms of selenium poisoning. Retrieved from www.livestrong.com/article/367955-signs-symptoms-of-selenium-poisoning.

Chakraborty, I. and Maity, P. 2020. COVID-19 outbreak: migration, effects on society, global environment and prevention. Science of the Total Environment. 728: 138882. https://doi.org/10.1016/j.scitotenv.2020.138882.

Dawsin, D.A. 2000. Alternative measures to waste disposal. Journal of Hazardous Substance. 12(1–2): 79–91.

Dehghani, M.H., Omrani, G.A. and Karri, R.R. 2023. SSolid waste—sources, toxicity, and their consequences to human health. pp. 205-213. *In*: Karri, R.R., Ravindran, G. and Dehghani, M.H. (eds). Soft Computing Techniques in Solid Waste and Wastewater Management. Elsevier.

Duan, V., Chen, Q., Ye et al. 2011. Hazardous incidents. Journal of Hazardous Materials. 186(2–3): 1489–1494. https://doi.org/10.1016/B978-0-12-824463-0.00013-6.

Encyclopedia Britannica. 2016. Hazardous waste disposal. Retrieved from https://www.britannica.com/technology/hazardous-waste-management.

Etim, I.U. 2003. Basic Principles of Environmental Science. Etiliew International Publishers. Uyo, Nigeria.

Exposto, L.A.S. and Sujaya, I.N. 2021. The impacts of hazardous and toxic waste management: a systematic review. Interdisciplinary Social Studies. 1(2): 103–123.

Fox, R. and Tuchman, M. 1996. The assessment and remediation of contaminated sediments (ARCS) program. Journal of Great Lakes Research. 22(3): 493–494.

Iyyanki, V.M. and Vali, M. 2017. Science and Engineering for Industry: Environmental Management 1st Edition. Elsevier Inc.

Kanwar, V.S., Sharma, A., Rinku et al. 2023. An overview for biomedical waste management during pandemic like COVID-19. International Journal of Environmental Science and Technology. 20: 8025–8040. https://doi.org/10.1007/s13762-022-04287-5

Khare, D. 2001. Warning precursors in disaster management. *In*: Proceedings of Conference on Disaster Management: Lessons learnt from Orissa Cyclone and Gujarat-Rajasthan Drought, BITS, Pilani. 5–7: 12–13.

Krishnaswamy, K., Geethamani, P. and Narmatha, M. 2020. Hazardous waste management: Environmental Issues and Sustainable Development. DOI: http://dx.doi.org/10.5772/intechopen.94080.

Kumar, S., Smith, S.R., Fowler, G., Velis, C., Kumar, S.J., Arya, S., et al. 2017. Challenges and opportunities associated with waste management in India. Royal Society Open Science. 4: 160764. https://doi.org/10.1098/rsos.160764.

Landrigan, P.J., Wright, R.O., Cordero, J.F., Eaton, D.L., Goldstein, B.D., Hennig, B., et al. 2015. The NIEH superfund research program: 25 years of translational research for public health. Environmental Health Perspectives. 123: 909–918. doi:10.1289/ehp.1409247

Linday, D.P. 2012. Deep Well injection waste disposal technology. Retrieved from http://www.theguardian.com.

Marfe, G. 2020. Hazardous Waste Management and Health Risks (1st Ed.). ProQuest: Bentham Science Publishers. Retrieved December 11, 2022.

Marsili, D., Fazzo, L. and Comba, P. 2009. Health risks from hazardous waste disposal: the need for international scientific cooperation. European Journal of Oncology. 14: 151–159.

Misra, V. and Pandey, S.D. 2005. Hazardous waste, impact on health and environment for development of better waste management strategies in the future in India. Environment International. 31(3): 417–435. https://doi.org/10.1016/j.envint.2004.08.005.

Nathanson, J.A. 2023. Hazardous-waste management. Encyclopedia Britannica. https://www.britannica.com/technology/hazardous-waste-management. Accessed 1 July 2023.

Nweke, O.C. and Sanders, W.H. III. 2009. Modern environmental health hazards: a public health issue of increasing significance in Africa. Environmental Health Perspectives. 117: 863–870. doi:10.1289/ehp.0800126.

Roger, B., James, E.S. and David, D. 1989. The safe disposal of hazardous wastes: the special needs and problems of developing countries. World Bank Technical Paper 1 (93)154. Washington D.C: World Bank.

Salma. 2012. Is zero waste possible? Retrieved from http://www.global.britannica.com/topic/secure-landfill.

Singh, M.P. 2010. Biotechnology in hazardous waste management. Recent Trends in Biotechnology. 1. Nova Science Publishers, New York, USA.

Srivastav, A.L., Markandeya, Patel, N., Pandey, M., Pandey, A.K., Dubey, A.K., et al. 2023. Concepts of circular economy for sustainable management of electronic wastes: challenges and management options. Environmental Science and Pollution Research. 30: 48654–48675. https://doi.org/10. 1007/s11356-023-26052-y.

Thakur, I.S. 2006. Environmental Biotechnology: Basic Concepts and Applications. I.K. International Pvt. Ltd., New Delhi.

The Resource Conservation and Recovery Act (RCRA). 2016. Hazardous wastes regulations, Chapters 10 through 32 of Division 4.5 of Title 22 of the California Code of Regulations. Retrieved from http://www.dtsc.ca.gov/HazardousWaste/upload/HWMP_DefiningHW111.pdf.

Vallero, D.A. 2019. Hazardous wastes. pp. 585–630. *In*: Letcher, T.M. and Vallero, D.A. (eds). Waste: A Handbook for Management (2nd ed.). Academic Press. https://doi.org/10.1016/B978-0-12-815060-3.00031-1.

Vallero, D.A. 2022. Hazardous wastes and the environment. pp. 3–32. *In*: Yadav, D., Kumar, P., Singh, P. and Vallero, D.A. (eds). Hazardous Waste Management: An Overview of Advanced and Cost-effective Solutions. Elsevier Inc. https://doi.org/10.1016/B978-0-12-824344-2.00016-1.

Visvanathan, C. 1996. Hazardous waste disposal. Resources, Conservation and Recycling. 16: 201–212. https://doi.org/10.1016/0921-3449(95)00057-7.

Yadav, D., Kumar, P., Singh, P. and Vallero, D.A. (eds). 2021. Hazardous Waste Management: An Overview of Advanced and Cost-effective Solutions. Elsevier Inc. Paperback ISBN: 9780128243442, eBook ISBN: 978032385.

Chapter **12**

Biosensors in Environmental Monitoring

Jyoti Singh Jadaun[1#], Pradeep Kumar Yadav[2#], Saumya Shukla[1], Sudhir Pratap Singh[3] and Gyanendra Kumar*[4]

[1]Department of Botany, Dayanand Girls' PG College, Civil Lines, Kanpur Nagar-208001, UP, India

[2]Department of Botany, University of Allahabad, Prayagraj, UP, India

[3]Gujarat Biotechnology University, GIFT City, Gandhinagar- 382355, Gujarat, India

[4]Department of Physics, Government PG College, Hamirpur-210301, UP, India

INTRODUCTION

Human life is very much connected with the environment and nature gives blessing to the humans and other living being in various forms. Science and technology are achieving new milestones on a daily basis in all the domains starting from nuclear dimensions to astronomical dimensions. To fulfill these promises, the manufacturing, industrialization, infrastructure, fossil fuels and technical advancements are the key players. These processes release lots of toxicity in the environment in the forms of industrial waste, residual wastewater, greenhouse gases, etc. Some other pollutants are in the form of heavy metals, pesticides, polychlorinated biphenyls (PCBs) etc. (Wang et al. 2022, Cui et al. 2022). The density of these pollutants in the environment is expediting on a daily basis, and hence the environmental monitoring has become imperative. To protect the environment, detection and

# Equal contribution; *For Correspondence: gyaniitk@gmail.com

sensing have pivotal role. For example, in our daily life, we frequently notice the sensors in the microwaves, air conditioners, smoke detectors, refrigerators, computers, mobile phones, etc. Such sensors are quite precise, robust and protect these electronics devices. Similarly, several biosensors have been developed, for instance, today one can easily measure blood sugar in couple of minutes at their home, blood pressure can be measured by a compact electronic device, several blood parameters by using ELISA kits. Recently, when whole world was struggling with the corona catastrophe, thousands of doctors and scientists sacrificed their lives to develop the corona detection kits. From this, we realized that the sensors should be accurate, compact, fast, economic and do not need lot of expertise. For the environment monitoring, though there are existing conventional methods, they are expensive, time consuming and also need skilled manpower. To overcome these limitations, biosensor can be good candidates. In this direction, lots of work is going on to develop the biosensors for continuous environmental monitoring (Justino et al. 2017). Since the world community is going through lots of technical advancement, it is need of hour to do the continuous environmental monitoring. To synchronize with the fast-paced world, practically the monitoring needs to be rapid, sensitive, user friendly, robust, affordable and require less instrumentation. In this chapter, we will elaborate the working principle of a biosensor and also discuss various kind of sensors related to the environmental monitoring.

BASIC STRUCTURE OF BIOSENSORS

The biosensors are basically designed to detect the minute alteration in the properties of a molecule or cell. These changes may happen in the shape, size, structure, chemical composition, DNA etc. of the cells or in its other properties. In principle, a biosensor is a receptor-transducer-based device, which is utilised for the interpretation of the biological and chemical properties of the medium (Figure 12.1). In these sensors, due to presence of biological materials, they are able to detect specific biological substances such as proteins, enzymes and other molecules etc. in the medium (Wang et al. 2006). The key components of the biosensors are as follows.

Biochemical components: Particular biochemical component should has an interaction with the targeted analyte present in sample. Therefore, the selection of a bio-component for a specific analyte has a pivotal role for its success. Whole sensing mechanism depends on the type of specific biochemical present in the specimen. This biocomponent can be any biological material or by product which is produced during the sensing process.

Transducers: In the sensing process, the energy needs to convert from one form to another. The component of the sensor which performs this process is called transducer. Depending on the transducer, the biological signal interacts with the biological material and the analyte and it is converted in the form of an electrochemical, optical or biochemical signal, etc. Accordingly, the transducers are broadly classified as thermal, optical, electrical or photochemical transducers.

The choice of transducer mainly depends on interaction type and produced biological signal (e.g. if the pH changes after the interaction, the difference in mass, the activation and/or inactivation of enzymes, etc.). Combinations of interaction types and transducers describe the principles of bioanalysis of specific biosensors.

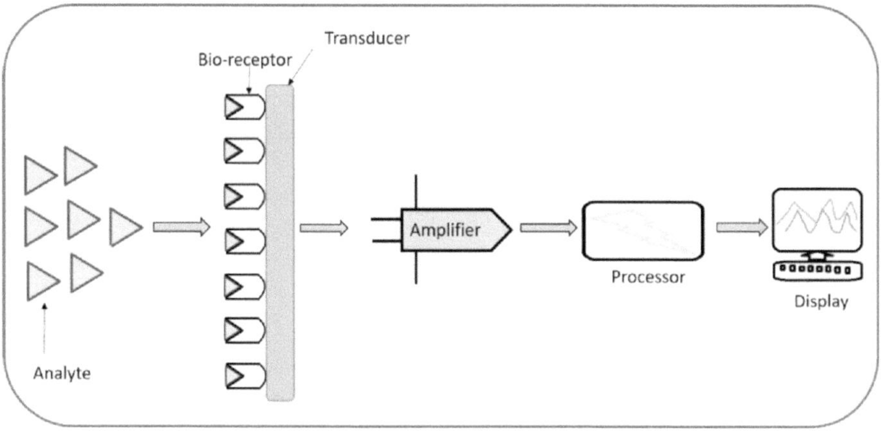

Figure 12.1 Schematic diagram of biosensor.

Amplifiers: In the sensors, amplifiers have significant role. Sometimes, significantly less amount of biochemical is present in the sample, which is going to be used for sensing. In such cases, direct measurement of the biochemical becomes complicated. In these cases, amplifiers either localise the concentration of the biochemical or with some chemical processes produce quite sufficient amount of by product, which is required for sensing purpose.

Detectors: Detector is a read out device of results. Most of the detectors have a display in which the signal can be analysed. Depending upon the type of the sensors, different detectors are utilised. For instance, for electrical signals, oscilloscope, light emitting diodes etc. are used. For optical signals, optical spectrum analyser, fibre-optic spectrofluorometer etc. are used. For thermal signals, thermocouples, pyroelectric detectors, bolometers etc. are used.

The main advantages of biosensors over the existing technologies (Rogers and Gerlac 1999) are measurement speed, high specificity, high sensitivity, small quantity of reagent required for calibration, portability etc.

TYPES OF BIOSENSORS

One can classify the biosensors depending upon the transduction principle such as **electrochemical** (impedance biosensors, amperometric biosensors), **optical** (surface plasmon resonance-based biosensors, optical fibre based biosensors), and **piezoelectric** (quartz crystal based biosensors) (Gavrilas et al. 2022). Biosensors can also be classified on the basis of their recognition elements such as **microbial fuel cell-based** biosensors, **aptamer**-based biosensors, **enzyme**-based biosensors, **whole cell**-based biosensors, **antibody**-based biosensors.

Types of Biosensors Depending Upon the Transduction Principle

Electrochemical biosensor: The principle of an electrochemical biosensor involves using the electrical properties of biomolecular interactions to detect and quantify specific analytes. A recognition element (e.g. antibody, enzyme, DNA) immobilized on an electrode surface interacts with the target analyte, causing a measurable change in the electrical signal (current or potential) (Shanbhag et al. 2023). This change is directly proportional to the concentration of the analyte, enabling rapid and sensitive detection for various applications, including medical diagnostics, environmental monitoring, and food safety. Amperometric, potentiometric, impedimetric and voltametric biosensors are examples of electrochemical biosensors (Wu et al. 2023).

Applications

Medical Diagnostics: For detecting biomarkers associated with diseases such as diabetes, cardiovascular disorders, and infectious diseases.

1. **Environmental Monitoring:** To detect pollutants, heavy metals, and toxins in air, water, and soil, contributing to environmental protection and public health.
2. **Food Safety:** Identifies contaminants, allergens, and pathogens in food products, ensuring their safety and quality.
3. **Biotechnology and Drug Development:** For screening drug candidates, studying enzyme kinetics, and monitoring cellular responses in drug development.
4. **Point-of-Care Testing:** They enable rapid on-site testing, making them valuable tools in resource-limited or remote areas.

Advantages

1. **High Sensitivity:** Enables the detection of low concentrations of analytes.
2. **Quantitative Analysis:** Many electrochemical biosensors provide quantitative results, allowing accurate determination of analyte concentrations.
3. **Real-Time Monitoring:** Electrochemical methods often allow real-time monitoring of reactions and interactions, providing dynamic insights into biological processes.
4. **Minimal Sample Volume:** Electrochemical biosensors often require small sample volumes, conserving precious samples.
5. **Specificity:** Properly designed recognition elements offer high specificity for target analytes, reducing false positives.
6. **Ease of Miniaturization:** Electrochemical biosensors can be miniaturized, enabling the development of portable and point-of-care devices.

Disadvantages

1. **Instrumentation Complexity:** Some electrochemical techniques require specialized equipment for signal measurement and analysis.
2. **Interference:** Background signals and matrix effects can interfere with accurate measurements.
3. **Electrode Fouling:** Biological samples can lead to electrode fouling, affecting sensor performance and requiring maintenance.
4. **Expertise Required:** Designing and optimizing electrochemical biosensors might necessitate specialized knowledge.
5. **Limited Analyte Range:** Some electrochemical methods are specific to certain types of analytes and may not be suitable for all applications.
6. **Electrode Stability:** Electrode materials can deteriorate over time, affecting the sensor's long-term performance.

Immunosensors: Immunosensors operate based on the principle of immunorecognition, where the highly specific binding affinity between antibodies and antigens is utilized. In this type of biosensor, an immunochemical reaction is operated through fixation of an antigen or antibody as a bio identification component on the surface of transducer (Luppa et al. 2001). These types of biosensors are utilized for detection of pathogens such as virus and bacteria, proteins, cancerous particles and LDLs, due to their high specificity. They can detect biomolecule in the concentration ranges from nanomolar to femtomolar levels.

There are two types of immunosensors generally: non-labelled and labelled immunosensors.

Direct measurement of physical changes during the formation of immunocomplex is determined in the non-labelled immunosensors while in labelled immunosensor, a detectable label is incorporated.

Applications

1. **Clinical Diagnostics:** Detection of biomarkers for diseases (e.g. cardiac markers, cancer markers), monitoring of infectious agents (e.g. viruses, bacteria), point-of-care testing for rapid diagnosis.
2. **Environmental Monitoring:** Detection of pollutants and toxins (e.g. pesticides, heavy metals), monitoring of water quality, air quality, and soil contamination.
3. **Food and Beverage Industry:** Detection of allergens, pathogens, and contaminants in food products, authentication and quality control of food products.
4. **Biomedical Research:** Studying protein-protein interactions and binding kinetics, quantification of specific molecules in biological samples.
5. **Pharmaceutical Industry:** Monitoring the production process and quality control of biopharmaceuticals, detection of residual impurities in pharmaceutical products.

6. **Veterinary Diagnostics:** Detection of animal diseases and pathogens, monitoring of animal health markers.
7. **Security and Defense:** Detection of biothreat agents in security settings, screening for illicit drugs and explosives.

Advantages

1. **High Sensitivity and Selectivity:** Immunosensors can detect target molecules at very low concentrations, making them highly sensitive. Additionally, their selectivity arises from the specificity of antibody-antigen interactions, allowing them to distinguish between closely related molecules.
2. **Rapid Detection:** Immunoreactions are typically fast, leading to rapid detection times. This is especially useful for applications requiring real-time or near-real-time analysis, such as medical diagnostics and environmental monitoring.
3. **Wide Range of Applications:** Immunosensors have applications in various fields, including medical diagnostics (detection of biomarkers, viruses, and bacteria), food safety (detection of allergens, pathogens, and contaminants), environmental monitoring (detection of pollutants and toxins), and more.
4. **Portability:** Many immunosensors are designed to be portable and easy to use in the field, allowing for on-site testing and reducing the need for complex laboratory equipment.
5. **Low Sample Volume:** Immunosensors often require only small sample volumes, which can be advantageous when dealing with limited or precious samples.

Disadvantages

1. **Complexity:** The design and fabrication of immunosensors can be technically complex, requiring expertise in biotechnology, chemistry, and sensor fabrication.
2. **Cost:** Developing and manufacturing immunosensors can be expensive due to the need for specialized materials, antibodies, and equipment.
3. **Stability and Shelf Life:** Immobilized antibodies on sensor surfaces may degrade over time, affecting the sensor's performance and shelf life. Proper storage conditions are crucial to maintaining sensor stability.
4. **Matrix Effects:** Complex sample matrices (like blood, serum, food, etc.) can interfere with sensor performance due to nonspecific binding or other interactions, leading to false positive or false negative results.
5. **Cross-reactivity:** Some antibodies might exhibit cross-reactivity with similar molecules, potentially leading to inaccurate results if the target is present alongside a structurally similar interfering substance.

6. **Regeneration Challenges:** Some immunosensors can be difficult to regenerate (clean and reuse) after use, which can limit their practicality in some applications.
7. **Quantification Challenges**: Converting sensor signals into quantitative measurements may require calibration curves and standardization procedures.
8. **Limited Shelf Stability of Reagents:** Antibodies and other reagents used in immunosensors might have limited shelf lives, potentially leading to variability in results over time.
9. Despite these challenges, ongoing research aims to address these disadvantages and improve the performance, robustness, and practicality of immunosensor technologies. As a result, immunosensors continue to find broader applications and contribute significantly to fields such as food safety, healthcare, and environmental monitoring.

Optical sensor: An optical sensor transforms light rays into electronic signals. It captures the intensity of light and processes it into a format interpretable by an instrument. It comprises of biological equipment, detector element and transducer associated with both parts. Typically, the biological component is a protein target which is fixed to the surface of the transducer (Singh et al. 2023).

Applications

1. **Medical Diagnostics:** Used to detect specific protein, DNA/ RNA detection making them valuable in diagnosing diseases like cancer, HIV, and cardiac markers and also in the identification of genetic mutations, sequence variations, and nucleic acid targets, aiding in genetic disease screening and pathogen detection.
2. **Drug Development and Pharmacology:** Optical biosensors enable rapid screening of compounds for drug discovery by assessing binding interactions with drug targets.
3. **Enzyme Kinetics:** They are employed to study enzyme-substrate interactions, allowing researchers to understand enzymatic reactions and optimize drug compounds.
4. **Environmental Monitoring:** Optical biosensors can detect pollutants, heavy metals, and toxins in environmental samples, helping assess water and soil quality.
5. **Pathogen Detection:** They are used to identify microbial contaminants in water supplies and other environmental samples.
6. **Allergen Detection:** Optical biosensors detect allergens in food products, aiding in quality control and ensuring consumer safety.
7. **Pathogen Identification:** They are utilized to identify pathogens like bacteria and viruses in food, reducing the risk of foodborne illnesses.

8. **Molecular Interactions:** Optical biosensors provide insights into biomolecular interactions, such as protein-protein interactions and ligand-receptor binding.
9. **Cellular Analysis:** They facilitate studies of cell adhesion, migration, and signaling processes
10. **Point-of-Care Testing:** Rapid optical biosensors are employed for on-site testing, enabling quick results and timely interventions.
11. **Remote Areas:** Portable optical biosensors are valuable in resource-limited settings where access to traditional laboratories is restricted.
12. **Toxin Sensing:** They can detect chemical and biological toxins, aiding in homeland security efforts.

Advantages

1. They enable the direct, real-time and label-free identification of many biological and chemical compounds.
2. High specificity, small size, sensitivity, and cost effectiveness
3. Safe, no electrical device interacts with the body.
4. Good biocompatability

Disadvantages

1. High cost of instrumentation
2. May be invasive
3. Fluorescent signal may not be good enough

Physical Biosensors

In physical biosensors, a biological event happens which follow the changes in physical parameters such as fluorescence, mass, resonance frequency, refractive index etc. They are basically two types: one is thermometric biosensor and another is piezoelectric biosensor. Thermometric biosensors gauge the temperature shift in the solution containing the analyte caused by these enzymatic reactions. Physical biosensors measure changes in physical properties, which result due to interaction between a biological molecule (analyte) and a recognition element (such as enzymes, DNA probes or antibodies). These changes are converted into detectable signals, which are then quantified to determine the presence and concentration of the analyte (Naresh and Lee 2021).

Applications

1. **Biomedical Diagnostics:** Physical biosensors are used for detecting biomarkers associated with diseases, monitoring glucose levels in diabetes, detecting cardiac markers and cancer, etc. (Kaur et al. 2022).
2. **Drug Discovery and Development:** They play a role in screening potential drug candidates, studying interactions between molecules, and monitoring enzymatic reactions.

3. **Environmental Monitoring:** Physical biosensors can detect pollutants, toxins, and contaminants in air, water, and soil.
4. **Food and Beverage Industry:** These biosensors are employed to monitor food quality, detect pathogens, and assess the presence of allergens.
5. **Biotechnology and Research:** They are used to study molecular interactions, enzyme kinetics, and cellular responses in research settings.

Advantages

1. **High Sensitivity:** Physical biosensors often offer high sensitivity, allowing for the detection of low concentrations of analytes.
2. **Quantitative Analysis:** Many physical biosensors provide quantitative measurements, enabling precise determination of analyte concentrations.
3. **Label-Free Detection:** Some physical biosensors work without the need for labeling molecules, which simplifies the assay and reduces potential interference.
4. **Real-Time Monitoring:** Some physical biosensors enable real-time monitoring of interactions, reactions, or changes in biological systems.
5. **Wide Range of Analytes:** These biosensors can be designed to detect a wide range of analytes, including proteins, nucleic acids, small molecules, and more.

Disadvantages

1. **Complex Instrumentation:** Physical biosensors often require specialized equipment for signal detection and analysis, which can limit their use in certain settings.
2. **Cost:** The instrumentation and setup for some physical biosensors can be expensive.
3. **Expertise Required:** Designing, developing, and interpreting results from physical biosensors might require specialized knowledge and skills.
4. **Specificity Concerns:** Ensuring high specificity can be challenging, as some physical changes might be influenced by non-specific interactions.
5. **Interference:** Environmental factors, sample matrices, and background signals can interfere with the accuracy of measurements.

Paper-based Biosensor

Paper-based biosensors work on immobilizing a biological recognition element (such as enzymes, antibodies, or DNA probes) onto a paper substrate. When the target molecule (analyte) is present in a sample, it interacts with the recognition element, leading to a detectable signal. This signal can be visualized using various methods, such as colour changes, fluorescence, or electrochemical reactions (Kuswandi et al. 2022).

Applications

1. **Healthcare Diagnostics:** Paper-based biosensors are used for rapid detection of various biomarkers, including proteins, nucleic acids, and pathogens. They find applications in point-of-care testing for diseases like HIV, malaria, and COVID-19.
2. **Environmental Monitoring:** On-site detection of pollutants, toxins, and contaminants in water, soil, and air.
3. **Food Safety:** To detect pathogens, allergens, and spoilage indicators in food products.
4. **Biological Research:** They are employed in research settings to study molecular interactions, enzyme activity, and other biochemical processes.
5. **Forensics:** For detecting traces of substances like drugs or explosives.

Advantages

1. **Cost-Effective:** Paper is an inexpensive and widely available material, making paper-based biosensors cost-effective to produce.
2. **Portability:** These biosensors are lightweight, portable, and require minimal equipment, enabling field use and point-of-care testing, especially in resource-limited settings.
3. **Simplicity:** Paper-based biosensors are user-friendly and do not require specialized training to operate, making them suitable for non-experts.
4. **Rapid Results:** Many paper-based biosensors provide rapid results within minutes, allowing for timely decisions and interventions.
5. **Low Sample Volume:** They often require small sample volumes, which is especially useful when dealing with limited sample availability.
6. **Stability:** Some paper substrates can provide stability to biomolecules, preserving their activity even without refrigeration.

Disadvantages

1. **Limited Sensitivity:** Paper-based biosensors may have lower sensitivity compared to some other advanced detection methods.
2. **Quantitative Limitations:** Quantitative measurements can be challenging due to variations in signal intensity and difficulty in precise quantification.
3. **Interference:** Environmental factors like humidity, temperature, and contaminants can interfere with the performance of paper-based biosensors.
4. **Lack of Multiplexing:** Multiplexing (detecting multiple analytes in a single test) can be limited on paper-based platforms due to the potential for signal crossover.
5. **Storage Concerns:** Long-term stability of reagents and recognition elements on paper can be a concern.

Biosensors on the Basis of their Recognition Elements

Microbial fuel cell-based biosensors: A technology that was initially developed for electricity generation from the organic waste with the help of microorganism is known as Microbial fuel cell (MFC) (Lovley 2006, Zhao et al. 2009, Yong et al. 2013). Although with time MFC came to be used in numerous potential applications, it got quite significant attention in the biosensing technology for systematic environmental monitoring (Figure 12.2). MFC based biosensors are of two types: single-chamber and dual-chambers. Single-chamber MFC has anodic chamber and cathode is made of air while dual-chamber MFC has distinct anode and cathode, separated by proton exchange membrane (Hernández-Fernández et al. 2015).

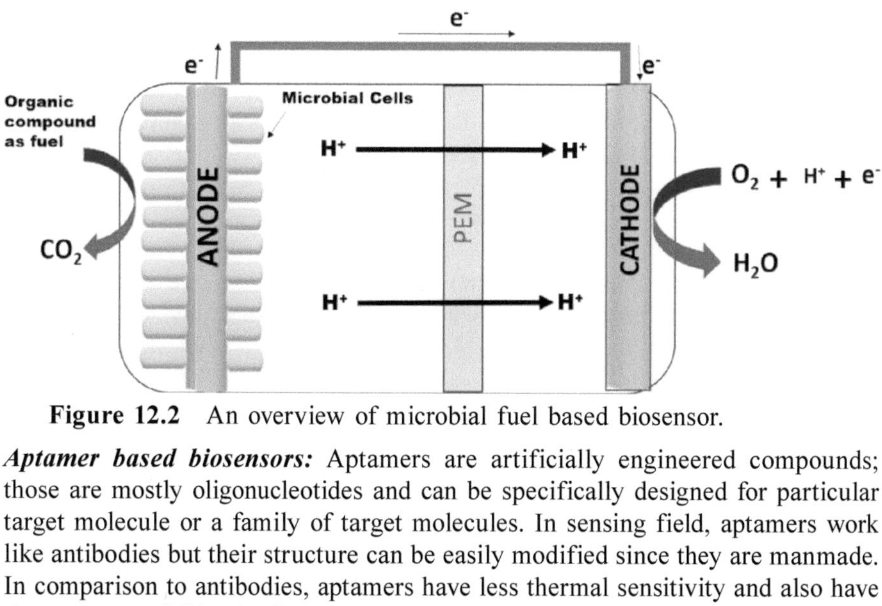

Figure 12.2 An overview of microbial fuel based biosensor.

Aptamer based biosensors: Aptamers are artificially engineered compounds; those are mostly oligonucleotides and can be specifically designed for particular target molecule or a family of target molecules. In sensing field, aptamers work like antibodies but their structure can be easily modified since they are manmade. In comparison to antibodies, aptamers have less thermal sensitivity and also have thermal reversibility (Kelly et al. 2021).

Enzymes based biosensors: Enzymes are the kind of proteins that work as a catalyst in biological reactions. In enzyme-based biosensors, the sensing element is made of specific enzyme and is attached with a transducer which converts the signal in an appropriate form that can be measured with the corresponding detector (Rahimi and Joseph 2019). Heat emission, variation in proton concentration, light absorption/emission, oxygen intake or release etc. are the few outcomes of the enzymatic reactions. The attached (immobilized) transducer converts it in the form of physically measurable quantities such as voltage, current, heat, pressure etc. (Asal et al. 2018). In enzyme-based sensors, the enzyme immobilization with the transducer surface is most important and can be achieved by various techniques. Adsorption is one simple and economic technique to achieve immobilization, in which enzymatic activity is less destructive and also no additional reagent is required. Entrapment is another immobilization technique, which provides high stability to the enzymes (Nguyen et al. 2017). Another immobilization technique is microencapsulation, which provides a better protection to the enzyme from the environmental conditions and also safeguards the structural and acting integrities of the enzymes. To achieve higher efficiency and stable binding of the enzyme, cross-linking immobilization technique can be a good choice (Nguyen et al. 2019). To achieve even more stability to the enzyme and enzyme-support complexes, the covalent binding technique can be an appropriate one. With a period of time, several authors have reported the use of enzyme-based sensors for environmental monitoring.

Whole cell-based biosensors: In the whole cell-based biosensors, a real cell or a genetically modified microorganism (virus, fungi, protozoa, bacteria or algae) is used as an active biocomponent. It interferes with the analyte and generates a signal, which is picked up by transducer. Transducer transforms that signal into a physically measurable quantity (Chang et al. 2017). Whole cell-based biosensors are also known as microbial sensors. Microbial sensors are the improvised version of the enzyme-based sensors since their working principle is similar. Both of them work on immobilization techniques to attach the transducer with the active biocomponent (Bilal et al. 2019). Microbial sensors have edge over the other molecular biosensors in the process of immobilization and bio-fabrication, since these sensors involve the whole cells. On the other hand, in terms of sensitivity to the environmental signals, microbial sensors are less sensitive in comparison to molecular biosensors but their sensitivity can be improved by simple genetic engineering approach (Xu and Ying 2011). From last one decade, several articles reported the utilization of microbial sensors for the detection of heavy metals, pesticides, phenols and other pollutants in the environment (Xu and Ying 2011, Gupta et al. 2019, Chang et al. 2021, Do et al. 2021).

Antibody based biosensors: A wider class of glycoproteins produced by particular cells are called antibodies, and also known as immunoglobulins. Antibodies are a part of immune system and they detect the antigens (chemicals and microorganisms). Antigens are detected and restrained by antibodies to form antigen-antibody complexes (Sharma et al. 2016, Felix and Angnes 2018). In the antibody-based biosensors, the specific interaction between antigens and immunoglobulins are used for detection. Unlike other sensors, in antibody-based biosensors various kind of transducing mechanisms such as electrochemical, optical, colorimetric and microgravimetric are used. These sensors can be categorized as labelled and non-labelled sensors (Fraka et al. 2017, Fang et al. 2020). In labelled sensors, a detectable marker is used to detect the analyte or bioreceptor. These are made of enzyme, nanoparticles, electroactive materials and fluorescent dyes, etc. (Jia et al. 2017). On the other hand, in non-labelled sensors antibody-antigen complex is quantified with the physical changes (Naresh and Lee 2021). With time, lot of work has been done in development and improvement of the antibody-based sensors. Selected antibody-based biosensors with their potential applications are discussed here.

APPLICATIONS OF BIOSENSORS FOR ENVIRONMENTAL MONITORING

Biosensors have tremendous applications in several areas but our main focus is environmental monitoring, hence we will limit our discussion to the sensors useful for environment, such as the detection of water contaminations, heavy metal detection, toxin detection, pathogen detection, air quality monitoring etc.

Biosensor for organic carbon measurement of waste water: Water is essential for the survival of living beings, so it becomes imperative to have a track of its availability and purity. Due to excessive industrialization, water and air are the two natural resources that face maximum contaminations. In contaminated water, the pollutants are mainly organic matters that comprises amino acids, proteins, lipids, nitrogenous compounds, etc. and they cause environmental problems like eutrophication (Kim et al. 2006). To control such unwanted threats to the environment, it is the need of the hour to have continuous monitoring of the contaminated water. The monitoring of individual water pollutant is quite sophisticated. Genrally, we measure the bulk parameters such as chemical oxygen demand (COD) and biochemical oxygen demand (BOD) of the water (Bourgeois et al. 2001). Though there are conventional methods to measure these bulk parameters, in recent years MFC based biosensors have attracted people's interest (Zhou et al. 2017, ElMekawy et al. 2018). In 2009, Lorenzo et al. had measured BOD with the linear range up to 350 mg/L by using single chamber MFC biosensor in which anode is made of carbon cloth and have air cathode. Further, mediator-less MFC biosensor, which incorporated a mixture of six bacteria strains utilized for determination of BOD concentration, was reported by Hsieh and Chung (2014). This has successfully measured the BOD concentration below 240 mg/L of actual wastewater. Later on, several scientists have improved the performance of the MFC biosensors for BOD concentration determination. Yamashita et al. (2016) had designed a novel electrochemical open type biosensor for *in situ* BOD monitoring during intermittent aeration. Similar results were obtained with this biosensor under aerating and non-aerating environments and it measured the BOD concentration up to 250 mg/L. Further, for continuous BOD monitoring of the real wastewater, Kharkwal et al. (2017) reported a MFC based biosensor. This sensor is cost effective and gives similar results for artificial and real wastewaters.

Biosensor in toxin detection of waste water: Presence of toxic substances in water affects the life of living beings. It is not only necessary to check the quality of drinking water supply but it is also required to analyze the quality of waste water disposed by various industries. If we analyse the water quality before disposing, we can minimize the toxicity level at source point and MFC based biosensors proved suitable for this purpose. Presence of toxic substances affect the metabolism of microorganisms; it hampers the speed of electron transfer towards the electrodes which resulted in reduced electricity production. Kim and his colleagues (2007) developed a MFC based biosensor applicable for detection of toxic substance. In this biosensor, MFC output was decreased up to 61%,

46%, 28% or 38% after incorporation of organophosphorus, lead, mercury, or polychlorinated biphenyls in comparison to control, respectively. Presence of formaldehyde can be detected by using culture of *Shewanella oneidenesis* MR-1 in a single chambered bioelectrochemical system (BES) and concentration range varies from 0.01 to 0.10% (Wang et al. 2013). A dual-chamber MFC can be utilized to detect diazinon and polychlorinated biphenyls (PCBs) in the range of 10 mg L^{-1} and 1 to 5 mg L^{-1}, respectively (Kim et al. 2007). Enzyme based biosensor was developed to detect the organophosphorus insecticides using carbon nanotube based thick film strip electrode. The lowest detection limit of the sensor was reported as 0.5 nM with stability and reproducibility. The monitoring of organophosphorus insecticides in real water sample was demonstrated which insured its feasibility for on-site detection (Joshi et al. 2005).

Biosensors for heavy metal detection: Mercury, lead, arsenic, silver, cadmium, etc. are the heavy metals those are dissolved in the water and contaminate it. Such contaminated water is hazardous for human life as well as aquatic life (Yadav et al. 2019). There are several biosensors that have been developed for detection of heavy metals (Table 12.1). Electrochemical methods are mostly used for heavy metals detection (Rapini and Marrazza 2017, Mishra et al. 2018, Malik et al. 2019). Few authors have also explored the mechanical and optical methods (Khoshbin et al. 2018, Zhang et al. 2018, Kudłak and Wieczerzak 2020). Aptamer based biosensor for mercury detection is well known and also widely documented (Zhou et al. 2017). For mercury detection, the concentration of Hg^{2+} in water is monitored. Lu et al. (2018) developed an aptamer modified SiO_2@Au nanoparticles for the detection of Hg^{2+} concentration of lake water. In this modified apatmer, HS–$(CH_2)_6$– is conjugated with the surface of nanoparticles. By using Surface Enhanced Raman Spectroscopy (SERS), they detected the Hg^{2+} concentration ranging from 10 nM to 1 mM. Tian et al. (2021) demonstrated an aptasensor based on surface enhanced Raman spectroscopy using dual recycling amplification for mercury ion detection in water environment. They had reported a low detection limit of 0.11 fM with a linear range of 0.2–125 fM.

Silver has a variety of applications from daily use jewelry to top end medical and electrical technologies due to its good electrical and thermal conductivities, ductility and high resistance from oxidation. Due to its extensive uses across all the fields, there are significant probabilities of silver contamination to air and water. Even though such contaminations are small in amount, it can cause serious health issues. Like other metals, silver is also detected by emission spectroscopy or mass spectroscopy and atomic absorption. Zhang and Wang (2019) designed a biosensor to detect the Ag^+ in water by using the non-canonical interaction property of Ag^+ with DNA. Similar to mercury aptamer, G-Ag^+-G interactions are utilized for binding Ag^+ ions. In this, the rotational spectra of silver bound aptamer is distinct from the non-bound aptamer rotational spectra. Above mentioned sensor has a probe which can detect the Ag^+ concentration ranging from 2–100 nM with the resolution of 0.5 nM. This sensor is also capable of sensing silver in the presence of 14 interfering metals and showed the recovery of Ag^+ from 71.8 to 83.6% in river water and 97.6 to 98.8% in spiked drinking water.

Lead is a widely used metal in batteries, electric cables, gasolines etc., and due to excessive use, it is releases in the environment and pollutes the water bodies, ground water and soil. Lead is a very toxic metal for human beings; hence, it is necessary to detect it with high sensitivity and accuracy. To keep in mind the toxicity of lead, Environmental Protection Agency (EPA), United States, has set a maximum concentration level of lead contamination as 72 nM (Yang et al. 2017). Even though there are traditional methods such as mass spectrometry and atomic absorption spectroscopy, they need costly and sophisticated instrumentations. To easily detect lead, aptamer-based biosensors can be a good option. Yang et al. (2017) published a review article stating the recent advancement of aptamer-based sensors for lead detection. Liu et al. (2021) had exhibited a fluorescent aptasensor for lead ion (Pb^{2+}) detection using integrated gold nanoflowers associated with complementary DNA of sulfhydryl group in the tea and tab water. With the optimized parameters, this aptasensor delivered the lowest detection limit of 0.285 nM with a quite good selectivity of Pb^{2+} in the concentration range of 0.5 nM to 1 μM (Liu et al. 2021).

Arsenic is a pollutant which is found in air, water, food and soil. There are natural sources of human exposure to the arsenic such as mineralization of ground water, volcanic ash etc. All the plants absorb the arsenic but the absorption is more in rice, leafy vegetables, apple, seafoods etc. Like other pollutant metals, the detection and quantification of arsenic is also imperative. Detection of arsenic with the aptamer-based biosensors (by electrochemical and optical methods) is well explored and first time reported by Kim et al. 2009. The detection range of various reports ranges in the order of magnitude 10^{-3} to 10^{-6} nM. Webster et al. (2014) reported an arsenic specific microbial biosensor by using bioelectrochemical system. In this, they used genetically engineered *Shewanella oneidensis* in such a manner that on increasing the amount of arsenic in the sample, the current increases. The reported linear range of detection is up to 100 μM arsenite with the detection limit of ~40 μM.

For detection of heavy metal toxicity, a self-powered MFC biosensor was developed by Yu et al. (2017); they noted an inhibition ratio up to 12.56%, 13.99%, 8.81%, 9.29%, 5.59% and 1.95% in the presence of heavy metals like Cu^{2+}, Hg^{2+}, Zn^{2+}, Cd^{2+}, Pb^{2+} and Cr^{3+}, respectively. Similarly, other MFC based biosensors have been developed to trace the presence of other metals like Fe (Tran et al. 2015), Ni^{++} (Stein et al. 2012), and Cu^{++} (Shen et al. 2013). A chromium sensing device was developed by using the *Ochrobactrum anthropi* YC152 as anodic microorganism catalyst (Wang et al. 2016a). Rathnayake et al. demonstrated a whole cell bacterial biosensor by using green fluorescent protein. The used bacterial strain is sensitive to multiple heavy metals (Cd, Cu and Zn) and appropriate to the soil sample. Its name is *Bacillus megaterium* VR1 and was optimized for pH range 5–8.5. The linear detection ranges for Cd, Cu and Zn are 0–10 mg/L, 0–20 mg/L and 0–100 mg/L, respectively. The lowest detection limits for all the three are 1.42×10^{-4} mg/L, 3.16×10^{-4} mg/L and 2.42×10^{-4} mg/L, respectively (Rathnayake et al. 2021).

Table 12.1 Biosensors applicable for heavy metal detection.

S. No.	Heavy metal	Biosensors (Biocatalyst)	Diagnostic methods	Limit of detection	Stability	References
1.	I. Mercury II. Arsenic III. Cadmium IV. Zinc V. Cupper	Urease enzyme	Electrochemical	I. 7.2 mg L^{-1} II. 0.08 μg L^{-1} III. 0.3 mg L^{-1} IV. 0.2 mg L^{-1} V. 8.5 mg L^{-1}	21–23°C for 2 year	Ilangovan et al. 2006 Dugo et al. 2005 Rodriguez et al. 2004
2.	I. Mercury II. Arsenic III. Cadmium IV. Zinc V. Cupper VI. Lead	Glucose oxidase	Electrochemical, Optical	I. 0.5 nM II. Not available III. 1 μg L^{-1} –19 μg L^{-1} IV. 6 μg L^{-1} V. 3 μg L^{-1} VI. 0.91 nM	37–60°C, 80% of the enzyme's activity was lost after 0.5 hr incubation.	Ghica et al. 2008 Tang et al. 2019
3.	I. Mercury II. Cadmium III. Zinc IV. Cupper	Glutathione S-transferase	Electrochemical, Optical	I. 10 mM^{-1} II. 10 $μm^{-1}$ III. 10 $μm^{-1}$ IV. 100 nm^{-1}	Stable for several months at –20 °C	Ahmed et al. 2017 Saatçi 2007
4.	I. Mercury II. Arsenic III. Cadmium IV. Zinc V. Cupper VI. Lead VII. Chromium	Alkaline phosphatase	Electrochemical, Optical	I. 0.017 mM II. 0.033 mM III. 0.021 mM IV. 1.30 mM V. 1.36 mM VI. 2.75 mg L^{-1} VII. 2.03 mg L^{-1}	Denatures when subjected to temperatures of approximately 72°C for at least 15 seconds.	Neale et al. 1965 Islam et al. 2017 Yorganci, and Akyilmaz 2011 Fahiminia et al. 2017

5.	I. Mercury II. Zinc	Lactate dehydrogenase	Electrochemical, Optical	I. 1 µM II. 0.2–0.8 µmol L^{-1}	–20°C for up to 45 days	Baustad and Tollersrud 1969 Rathee et al. 2016 Tollersrud S 1970
6.	I. Lead II. Chromium	Acid phosphatase	Electrochemical, Optical	I. 2.75 mg L^{-1} II. 2.03 mg L^{-1}	Enzyme was stable up to 55°C. Further heating of the enzyme resulted in about 45% loss in enzyme activity at 60 °C with a total loss in activity at 80°C	Mobley et al. 1984 Fahiminia et al. 2017

For Detection of Pathogens: An electrochemical genosensor coupled with AuNP mercaptosilane- modified screen-printed electrodes was developed for detection of *Salmonella typhi* (Das et al. 2014). A virtual biosensor is developed for the detection of both Gram (+) and Gram (–) bacteria by using the characteristic of manan binding lectin (MBL) for recognizing the pathogen associated molecular patterns (Naseri et al. 2022). MBL protein was immobilized on carboxyl-functionalized polystyrene microparticles (PSM). This biosensor was applied to Gram negative bacteria like *Escherichia coli, Klebsiella pneumoniae, Acinetobacter baumannii,* and *Pseudomonas aeruginosa* as well as Gram positive bacteria like *Enterococcus faecalis* and *Staphylococcus aureus.* This biosensor is equally effective for detection of bacterial contamination in real life water samples.

E. coli is considered as the most potent pathogen found in contaminated food and water samples. Various biosensors were developed for rapid detection of *E. coli.* These biosensors include surface-enhanced Raman spectroscopy, surface plasmon resonance, colorimetric, localized surface plasmon resonance, fluorescence and chemiluminescence (Pebdeni et al. 2022). Recently, paper based biosensor has emerged as an alternate with low cost and high efficiency; these sensors are successfully used for detection of bacterial cells (Rengaraj et al. 2018). A biosensor was developed for detection of food borne pathogen *L. monocytogenes* in seafood by using laminated lab-on-paper chip. Base for this assay is the loop mediated DNA amplification (LAMP) of hly (listeriolysin O precursor) gene. After that, a fluorescent signal was produced which is detected by using SYBR Gold^{TM} (Pisamayarom et al. 2017)

Biosensor for detection of herbicide and pesticides: Belkhamssa et al. (2016) has developed a non-labelled antibody-based biosensor to detect the atrazine. When the sensor is exposed to atrazine, it absorbs the specific antibodies. With the appropriate conditions, the detection ranges from 0.001 to 10 ng m/L with the resolution of 0.001 ng m/L in real water. This sensor can also be useful for the detection of atrazine in riverine water and seawater. Antunes et al. (2018) also developed a biosensor with a graphene-based field effect transistor to detect the okadaic acid (it is a tumour promotor marine toxin and also inhibitor of various kinds of cancers) in seawater. The reproducibility was reported to be 0.54–2.19% with the detection limit of 0.05 ngmL^{-1}. For the herbicides (specially diuron and atrazine) detection, a microbial biosensor was developed by photocurrent inhibition of cyanobacteria *Anabaena variabilis* using amperometric technique. They have immobilized bacterial cells on a carbon felt electrode through entrapping method using alginate polymer. For atrazine. the detection range is up to 0.56 μM with the sensitivity of –24.6 μA μM^{-1} cm^{-2} and for diuron only turn on and turn off occurs since it is very strong photosynthesis inhibitor (Tucci et al. 2019). From last two decades, so many biosensors have been reported in the literature which have shown great potential for use in pollution control (Table 12.2 and Table 12.3).

Table 12.2 List of biosensors used for detection of pesticides

Sl. No	Pesticide	Enzyme	Method	Limit of Detection	Reference
1.	I. Paraoxon Fluorescence	Organophosphate hydrolase	Amperometry	20 μM	Besombes et al. 1995 Touloupakis et al. 2012 LeDoux 2011
2.	I. Parathion Carbaryl II. Atrazine	Tyrosinase	Amperometry	I. 0.005 ppb II. 0.3 ppm	Zamaleeva et al. 2011
3.	I. Methomyl II. Pirimicarb III. Formetanate	Laccase	Square wave voltammetry	I. 2.35 × 10 × 7 M II. 1.8 × 10 × 7 M III. 95 nM	Moro et al. 2018 Rasmussen et al. 2014
4.	I. Glyphosate II. Dichlofenthion	Heme-containing enzymes	I. Square wave voltammetry II. Amperometry	I. 30 μg L^{-1} II. 24 μM	Oliveira et al. 2012
5.	I. Atrazine II. Glyphosate	Urease	I. Enzyme Field Effect Capacitive System II. Potentiometric	I. 0.12 μM II. 0.5 ppm	EPEC 2000
6.	Aldicarb	AChE	Fluorescence	3.3 μg L^{-1}	Wang et al. 2016b
7.	I. Metham-sodium II. Tetradifon III. Fenitrothion	Alkaline phosphatase	Fluorimetry	I. 36.5 μM II. 4.1 μM III. 45.5 μM	Besombes et al. 1995
8.	Acetamiprid	AChE and ChO	UV-visible absorbance spectroscopy	2.66 ng mL^{-1}	Wu et al. 2021

Table 12.3 Biosensors used for detection of insecticides.

S. No.	Insecticides	Method of detection	Biosensor system	LOD (μg L^{-1})	References
1.	Carbendazim	DNA Aptamer based biosensor immobilized on gold surface	Glutathione S-transferases	2	Borah et al. 2018
2.	Deltamethrin	A biosensor featuring polyaniline immobilized catalase	Catalase	177	Paluzar and Özcan 2017
3.	Molinate	GST affixed to a glassy carbon electrode using a covalent attachment method involving aminosilane and glutaraldehyde	Glutathione S-transferases	64	Almeida and De Souza 2023
4.	Captan	GST is anchored onto a self-assembled monolayer-modified gold electrode	Glutathione S-transferases immobilized	1000	Samal et al. 2023
5.	Permethrin	Biosensor based on AChE integrated with SPCEs	Acetylcholinesterase (AChE)	3169	Quiros 2023
6.	Fenobucarb	GST tethered to a platinum electrode.	Glutathione S-transferases	2	Alwin 2023
7.	Chlorpyrifos	Acetylcholinesterase (AChE) immobilized on Fc-F hydrogel support	Glutathione S-transferases biosensor using graphene oxide.	60	Akpe. 2023
8.	Ethion	Butyrylcholinesterase biosensor based on multiwalled carbon nanotube–polyvinyl chloride (MWNT–PVC) composite	Glutathione S-transferases biosensor using graphene oxide	100	Li 2023

CONCLUSION AND FUTURE PERSPECTIVES

Even though the biosensing field is evolving in all the dimensions, lots of promises need to be fulfilled such as the simultaneous detection of various pollutants and the advancement in such a way that they can be used like a cell phone. Environmental monitoring through satellite technology is another area which has a wide scope of sensing of various environmental parameters such as availability of water, concentration of different kind of gases, temperature, pressure, weather forecasting etc. Few advance countries are using unmanned aerial vehicles fitted with sensors for surveillance. In this field, a lot of improvisation and technical advancement is possible to make it robust and available across the world. Recently, the whole world faced the horrible corona virus catastrophe. We can develop the sensors which will be able to detect the presence of corona virus. With the time, a lot of progress has been made for development of biosensors but still we need to develop portable biosensors which can produce results on site.

REFERENCES

Ahmed, S., Sohail, A., Khatoon, S., Khan, S. and Saifullah, M.K. 2017. Partial purification and characterization of glutathione S-transferase from the somatic tissue of *Gastrothylax crumenifer* (Trematoda: Digenea). Veterinary World. 10(12): 1493.

Akpe, M.A., Louis, H., Gber, T.E., Chima, C.M., Brown, O.I. and Adeyinka, A.S. 2023. Modeling of Cu, Ag, and Au-decorated Al12Se12 nanostructured as sensor materials for trapping chlorpyrifos insecticide. Computational and Theoretical Chemistry. 1226: 114218.

Almeida, E.M.F. and De Souza, D. 2023. Current electroanalytical approaches in the carbamates and dithiocarbamates determination. Food Chemistry. 135900.

Alwin, S., Baby, D., Rajakani, S.R., Karpagavinayagam, P., and Chinnapiyan, V. 2023. Development of sustainable electrochemical sensors. pp. 341–366. *In:* Recent Developments in Green Electrochemical Sensors: Design, Performance, and Applications. American Chemical Society.

Antunes, J., Justino, C., da Costa, J.P., Cardoso, S., Duarte, A.C. and Rocha-Santos, T. 2018. Graphene immunosensors for okadaic acid detection in seawater. Microchemical Journal. 138: 465–471.

Asal, M., Ozen, O., Sahinler, M. and Polatoglu, I. 2018. Recent developments in enzyme, DNA and immuno-based biosensors. Sensors. 18: 1924.

Baustad, B. and Tollersrud, S. 1969. Isoenzymes of lactate dehydrogenase in swine. Stability during storage at different temperatures and by heat treatment. Acta Veterinaria Scandinavica. 10: 372–381.

Belkhamssa, N., Justino, C.I.L., Santos, P.S.M., Cardoso, S., Lopes, I., Duarte, A.C., et al. 2016. Label-free disposable immunosensor for detection of atrazine. Talanta. 146: 430–434.

Besombes, J.-L., Cosnier, S., Labbé, P. and Reverdy, G. 1995. A biosensor as a warning device for the detection of cyanide, chlorophenols, atrazine, and carbamate pesticides. Analytica Chimica Acta. 311: 255–263.

Bilal, M. and Iqbal, H.M.N. 2019. Microbial-derived biosensors for monitoring environmental contaminants: Recent advances and future outlook. Process Safety and Environmental Protection. 124: 8–17.

Borah, H., Gogoi, S., Kalita, S. and Puzari, P. 2018. A broad-spectrum amperometric pesticide biosensor based on glutathione S-transferase immobilized on graphene oxide-gelatin matrix. Journal of Electroanalytical Chemistry. 828: 116–23.

Bourgeois, W., Burgess, J.E. and Stuetz, R.M. 2001. On-line monitoring of wastewater quality: a review. Journal of Chemical Technology and Biotechnology. 76: 337–348.

Chang, H.J., Voyvodic, P.L., Zuniga, A. and Bonnet, J. 2017. Microbially derived biosensors for diagnosis, monitoring, and epidemiology. Microbial Biotechnology. 10: 1031–1035.

Chung, T.H., Meshref, M.N.A. and Dhar, B.R. 2021. A review and roadmap for developing microbial electrochemical cell-based biosensors for recalcitrant environmental contaminants, emphasis on aromatic compounds. Chemical Engineering Journal. 424: 130245.

Cui, M., Xu, S., Song, W., Ye, H., Huang, J., Liu, B., et al. 2022. Trace metals, polycyclic aromatic hydrocarbons and polychlorinated biphenyls in the surface sediments from Sanya River, China: distribution, sources and ecological risk. Environmental Pollution. 294: 118614.

Das, R., Sharma, M.K., Rao, V.K., Bhattacharya, B.K., Garg, I., Venkatesh, V., et al. 2014. An electrochemical genosensor for Salmonella typhi on gold nanoparticles-mercaptosilane modified screen printed electrode. Journal of Biotechnology. 188: 9–16.

Deng, H., Jiang, Y., Zhou, Y., Shen, K. and Zhong, W. 2015. Using electrical signals of microbial fuel cells to detect copper stress on soil microorganisms. European Journal of Soil Science. 66: 369–377.

Do, M.H., Ngo, H.H., Guo, W., Chang, S.W., Nguyen, D.D., Pandey, A., et al. 2021. A dual chamber microbial fuel cell-based biosensor for monitoring copper and arsenic in municipal wastewater. Science of the Total Environment. 811: 152261.

Dugo, G., La Pera, L., Turco, V.L. and Di Bella, G. 2005. Speciation of inorganic arsenic in alimentary and environmental aqueous samples by using derivative anodic stripping chronopotentiometry (dASCP). Chemosphere. 61(8): 1093–1101.

ElMekawy, A., Hegab, H.M., Pant, D. and Saint, C.P. 2018. Bio-analytical applications of microbial fuel cell–based biosensors for onsite water quality monitoring. Journal of Applied Microbiology. 124(1): 302–313.

European Union. 2000. Directive 2000/60/EC of the European Parliament and of the Council of 23 October 2000 establishing a framework for the community action in the field of water policy (water framework directive). Official Journal of the European Communities Series. 327: 1–7.

Fahiminia, M., Rahmanifar, B., Koohpaei, A., Dehaghi, S.M. and Behnamipour, S. 2017. Heavy metal monitoring by a new tool: acid phosphatase/chitosan nanoparticles thin film biosensor. Fresenius Environmental Bulletin. 26(2 A): 1579–1584.

Fang, L., Liao, X., Jia, B., Shi, L., Kang, L., Zhou, L., et al. 2020. Recent progress in immunosensors for pesticides. Biosensors and Bioelectronics. 164: 112255.

Farka, Z., Juriik, T., Kovaar, D., Trnkova, L. and Sklaadal, P. 2017. Nanoparticle-based immunochemical biosensors and assays: recent advances and challenges. Chemical Reviews. 117: 9973–10042.

Felix, F.S. and Angnes, L. 2018. Electrochemical immunosensors—a powerful tool for analytical applications. Biosensors and Bioelectronics. 102: 470–478.

Gavrilaş, S., Ursachi, C.Ş., Perţa-Crişan, S. and Munteanu, F.D. 2022. Recent trends in biosensors for environmental quality monitoring. sensors (Basel, Switzerland). 22(4): 1513.

Ghica, M.E. and Brett, C.M.A. 2008. Glucose oxidase inhibition in poly (neutral red) mediated enzyme biosensors for heavy metal determination. Microchimica Acta. 163: 185–193.

Gupta, N., Renugopalakrishnan, V., Liepmann, D., Paulmurugan, R. and Malhotra, B.D. 2019. Cell-based biosensors: Recent trends, challenges and future perspectives. Biosensors and Bioelectronics. 141: 111435.

Hernández-Fernández, F.J., Pérez de los Ríos, A., Salar-García, M.J., Ortiz-Martínez, V.M., Lozano-Blanco, L.J., Godínez, C., et al. 2015. Recent progress and perspectives in microbial fuel cells for bioenergy generation and wastewater treatment. Fuel Processing Technology. 138: 284–297.

Hsieh, M.C. and Chung, Y.C. 2014. Measurement of biochemical oxygen demand from different wastewater samples using a mediator-less microbial fuel cell biosensor. Environmental Technology. 35(17): 2204–2211.

Ilangovan, R., Daniel, D., Krastanov, A., Zachariah, C. and Elizabeth R. 2006. Enzyme-based biosensor for heavy metal ions determination. Biotechnology and Biotechnological Equipment. 20(1): 184–9.

Islam, M.S., Sazawa, K., Hata, N., Sugawara, K. and Kuramitz, H. 2017. Determination of heavy metal toxicity by using a micro-droplet hydrodynamic voltammetry for microalgal bioassay based on alkaline phosphatase. Chemosphere. 188: 337–344.

Jia, M.X., Liao, X.F., Fang, L., Jia, B.Y., Liu, M., Li, D.H., et al. 2021. Recent advances on immunosensors for mycotoxins in foods and other commodities. TrAC—Trends in Analytical Chemistry. 136: 116193.

Joshi, K.A., Tang, J., Haddon, R., Wang, J., Chen, W. and Mulchandani, A. 2005. A disposable biosensor for organophosphorus nerve agents based on carbon nanotubes modified thick film strip electrode. Electroanalysis. 17: 54–58.

Kaur, B., Kumar, S. and Kaushik, B.K. 2022. Recent advancements in optical biosensors for cancer detection. Biosensors and Bioelectronics. 197: 113805.

Kelly, L., Maier, K.E. and Yan, A. 2021. A comparative analysis of cell surface targeting aptamers. Nature Communications. 12: 6275.

Kharkwal, S., Tan, Y.C., Lu, M. and Ng, H.Y. 2017. Development and long-term stability of a novel microbial fuel cell BOD sensor with MnO_2 Catalyst. International Journal of Molecular Sciences. 18(2): 276.

Kim, B.H., Chang, I.S. and Moon, H. 2006. Microbial fuel cell-type biochemical oxygen demand sensor. Encyclopedia of Sensors. X: 1–12.

Kim M., Sik Hyun M., Gadd G.M. and Joo Kim, H. 2007. A novel biomonitoring system using microbial fuel cells. Journal of Environmental Monitoring. 9: 1323–1328.

Kim, M., Um, H.J., Bang, S., Lee, S.H., Oh, S.J., Han, J., et al. 2009. Arsenic removal from Vietnamese groundwater using the arsenic-binding DNA aptamer. Environmental Science & Technology. 43(24): 9335–9340.

Kuswandi, B., Hidayat, M.A. and Noviana, E. 2022. Paper-based sensors for rapid important biomarkers detection. Biosensors and Bioelectronics: X. 12: 100246.

LeDoux, M. 2011. Analytical methods applied to the determination of pesticide residues in foods of animal origin: a review of the past two decades. Journal of Chromatography A. 1218: 1021–1036.

Li, G., Huang, X., Peng, C. and Sun, F. 2023. Highly sensitive fluorescence detection of three organophosphorus pesticides based on highly bright DNA-templated silver nanoclusters. Biosensors. 13(5): 520.

Liu, R., He, B., Jin, H. and Suo, Z. 2021. A fluorescent aptasensor for Pb^{2+} detection based on gold nanoflowers and RecJf exonuclease induced signal amplification. Analytica Chimica Acta. 1192: 339329.

Lovley, D.R. 2006. Bug juice: harvesting electricity with microorganisms. Nature Reviews Microbiology. 4: 497–508.

Luppa, P.B., Sokoll, L.J. and Chan, D.W. 2001. Immunosensors—principles and applications to clinical chemistry. Clinica Chimica Acta. 314(1–2): 1–26.

Mobley, D.M., Chengappa, M.M., Kadel, W.L. and Stuart, J.G. 1984. Effect of pH, temperature and media on acid and alkaline phosphatase activity in "clinical" and "nonclinical" isolates of *Bordetella bronchiseptica*. Canadian Journal of Comparative Medicine. 48(2): 175.

Moro, L., Pezzotti, G., Turemis, M., Sanchís, J., Farré, M., Denaro, R., et al. 2018. Fast pesticide pre-screening in marine environment using a green microalgae-based optical bioassay. Marine Pollution Bulletin. 129: 212–221.

Naresh, V. and Lee, N. 2021. A review on biosensors and recent development of nano-structured materials-enabled biosensors. Sensors. 21: 1109.

Naseri, M., Maliha, M., Dehghani, M., Simon, G.P. and Batchelor, W. 2022. Rapid detection of gram-positive and-negative bacteria in water samples using mannan-binding lectin-based visual biosensor. ACS Sensors. 7(4): 951–959.

Neale, F.C., Clubb, J.S., Hotchkis, D. and Posen, S. 1965. Heat stability of human placental alkaline phosphatase. Journal of Clinical Pathology. 18(3): 359–363.

Nguyen, H.H. and Kim, M. 2017. An overview of techniques in enzyme immobilization. Applied Sciences Convergence Technology. 26: 157–163.

Nguyen, H.H., Lee, S.H., Lee, U.J., Fermin, C.D. and Kim, M. 2019. Immobilized enzymes in biosensor applications. Materials. 12: 121.

Oliveira, G.C., Moccelini, S.K., Castilho, M., Terezo, A.J., Possavatz, J., Magalhães, M.R.L., et al. 2012. Biosensor based on atemoya peroxidase immobilised on modified nanoclay for glyphosate biomonitoring. Talanta. 98: 130–136.

Paluzar, H. and Özcan, H.M. 2017. Determination of deltametry with a catalase based PANI biosensor. Journal of Suleyman Demirel University Institute of Science and Technology. 21(2): 644–651.

Pebdeni, A.B., Roshani, A., Mirsadoughi, E., Behzadifar, S. and Hosseini, M. 2022. Recent advances in optical biosensors for specific detection of *E. coli* bacteria in food and water. Food Control. 135: 108822.

Pisamayarom, K., Suriyasomboon, A. and Chaumpluk, P. 2017. Simple screening of listeria monocytogenes based on a fluorescence assay via a laminated lab-on-paper chip. Biosensors. 7(4): 56.

Quiros, M.B. 2023. Metallic Nanoparticles Modification of Epinephrine Biosensor Improves its Amperometric Response. Chemical Sciences Journal. 14: 2–8.

Rahimi, P. and Joseph, Y. 2019. Enzyme-based biosensors for choline analysis: a review. TrAC Trends in Analytical Chemistry. 110: 367–374.

Rasmussen, M., Wingersky, A. and Minteer, S.D. 2014. Comparative study of thylakoids from higher plants for solar energy conversion and herbicide detection. Electrochimica Acta. 140: 304–308.

Rathee, K., Dhull, V., Dhull, R. and Singh, S. 2016. Biosensors based on electrochemical lactate detection: a comprehensive review. Biochemistry and Biophysics Reports. 5: 35–54.

Rathnayake, I.V.N., Megharaj, M. and Naidu, R. 2021. Green fluorescent protein-based whole cell bacterial biosensor for the detection of bioavailable heavy metals in soil environment. Environmental Technology Innovation. 23: 101785.

Rengaraj, S., Cruz-Izquierdo, Á., Scott, J.L. and Di Lorenzo, M. 2018. Impedimetric paper-based biosensor for the detection of bacterial contamination in water. Sensors and Actuators B: Chemical. 265: 50–58.

Rodriguez, B.B., Bolbot, J.A. and Tothill, I.E. 2004. Development of urease and glutamic dehydrogenase amperometric assay for heavy metals screening in polluted samples. Biosensors and Bioelectronics. 19(10): 1157–1167.

Saatçi, E., Nistor, M., Gaspar, S., Csöregi, E. and İşcan, M. 2007. Comparison of two glutathione S-transferases used in capacitive biosensors for detection of heavy metals. International Journal of Environmental and Analytical Chemistry. 87(10-11): 745–754.

Samal, S., Mohanty, R.P., Mohanty, P.S., Giri, M.K., Pati, S. and Das, B. 2023. Implications of biosensors and nanobiosensors for the eco-friendly detection of public health and agro-based insecticides: a comprehensive review. Heliyon. 9(5): e15848.

Shanbhag, M.M., Manasa, G., Mascarenhas, R.J., Mondal, K. and Shetti, N.P. 2023. Fundamentals of bio-electrochemical sensing. Chemical Engineering Journal Advances. 16.

Sharma, S., Byrne, H. and O'Kennedy, R.J. 2016. Antibodies and antibody-derived analytical biosensors. *In*: Estrela, P. (ed.). Biosensor Technologies for Detection of Biomolecules. Essays in Biochemistry. Portland Press: South Portland, ME, USA. 60: 9–18.

Singh, A.K., Mittal, S., Das, M., Saharia, A. and Tiwari, M. 2023. Optical biosensors: a decade in review. Alexandria Engineering Journal. 67: 673–691.

Tang, D., Xia, B., Tang, Y., Zhang, J. and Zhou, Q. 2019. Metal-ion-induced DNAzyme on magnetic beads for detection of lead (II) by using rolling circle amplification, glucose oxidase, and readout of pH changes. Microchimica Acta. 186. 1–9.

Tian, C., Zhao, L., Zhu, J. and Zhang, S.S. 2021. Ultrasensitive detection of trace Hg^{2+} by SERS aptasensor based on dual recycling amplification in water environment. Journal of Hazardous Materials. 416: 126251.

Tollersrud S. 1970. Heat stability of serum lactate dehydrogenase and its isoenzymes in young and adult cattle and sheep: evaluation of a relative heat stability test and serum determination of α-hydroxy-butyrate dehydrogenase in diagnostic work. Acta Veterinaria Scandinavica. 11: 510–524.

Touloupakis, E., Boutopoulos, C., Buonasera, K., Zergioti, I. and Giardi, M.T. 2012. A photosynthetic biosensor with enhanced electron transfer generation realized by laser printing technology. Analytical and Bioanalytical Chemistry. 402: 3237–3244.

Tran, P.H., Luong, T.T., Nguyen, T.T., Nguyen, H.Q., Duong, H.V., Kim, B.H., et al. 2015. Possibility of using a lithotrophic iron-oxidizing microbial fuel cell as a biosensor for detecting iron and manganese in water samples. Environmental Science: Processes and Impacts. 17(10): 1806. doi: 10.1039/C5EM00099H.

Tucci, M., Grattieri, M., Schievano, A., Cristiani, P. and Minteer, S.D. 2019. Microbial amperometric biosensor for online herbicide detection: photocurrent inhibition of *Anabaena variabilis*. Electrochimica Acta. 302: 102–108.

Wang, J.X., Sun, X.W., Wei, A., Lei, Y., Cai, X.P., Li, C.M., et al. 2006. Zinc oxide nanocomb biosensor for glucose detection. Applied Physics Letters. 88(23): 233106.

Wang, X., Gao, N. and Zhou, Q. 2013. Concentration responses of toxicity sensor with *Shewanella oneidensis* MR-1 growing in bioelectrochemical systems. Biosensors and Bioelectronics. 43: 264–267.

Wang, G.H., Cheng, C.Y., Liu, M.H., Chen, T.Y., Hsieh, M.C. and Chung, Y.C. 2016a. Utility of *Ochrobactrum anthropi* YC152 in a microbial fuel cell as an early warning device for hexavalent chromium determination. Sensors. 16: 1272. doi: 10.3390/s16081272.

Wang, X., Hou, T., Dong, S., Liu, X. and Li, F. 2016b. Fluorescence biosensing strategy based on mercury ion-mediated DNA conformational switch and nicking enzyme-assisted cycling amplification for highly sensitive detection of carbamate pesticide. Biosensors & Bioelectronics. 77: 644–649.

Wang, Z., Luo, P., Zha, X., Xu, C., Kang, S., Zhou, M., et al. 2022. Overview assessment of risk evaluation and treatment technologies for heavy metal pollution of water and soil. Journal of Cleaner Production. 379: 134043. doi: 10.1016/j.jclepro.2022.134043.

Webster, D.P., TerAvest, M.A., Doud, D.F.R., Chakravorty, A., Holmes, E.C., Radens, C.M., et al. 2014. An arsenic-specific biosensor with genetically engineered *Shewanella oneidensis* in a bioelectrochemical system. Biosensors and Bioelectronics. 62: 320–324.

Wu, L., Zhou, M., Liu, C., Chen, X. and Chen, Y. 2021. Double-enzymes mediated Fe^{2+}/Fe^{3+} conversion as magnetic relaxation switch for pesticide residues sensing. Journal of Hazardous Materials. 403: 123619.

Wu, J., Liu, H., Chen, W., Ma, B. and Ju, H. 2023. Device integration of electrochemical biosensors. Nature Reviews Bioengineering. 1(5). https://doi.org/10.1038/s44222-023-00032-w.

Xu, X. and Ying, Y. 2011. Microbial biosensors for environmental monitoring and food analysis. Food Reviews International. 27(3): 300–329.

Yadav, R., Kushwah, V., Gaur, M.S., Bhadauria, S., Berlina, A.N. and Zherdev, A.V. 2019. Electrochemical aptamer biosensor for As^{3+} based on apta deep trapped Ag-Au alloy nanoparticles-impregnated glassy carbon electrode. International Journal of Environmental Analytical Chemistry. 100: 623–634.

Yamashita, T., Ookawa, N., Ishida, M., Kanamori, H., Sasaki, H., Katayose, Y., et al. 2016. A novel open-type biosensor for the *in-situ* monitoring of biochemical oxygen demand in an aerobic environment. Scientific Reports. 6: 38552.

Yang, D., Liu, X., Zhou, Y., Luo, L., Zhang, J., Huang, A., et al. 2017. Aptamer-based biosensors for detection of lead (ii) ion: a review. Analytical Methods. 9(13): 1976–1990.

Yong, Y.C., Yu, Y.Y., Li, C.M., Zhong, J.J. and Song, H. 2013. Bioelectricity enhancement via overexpression of quorum sensing system in *Pseudomonas aeruginosa*-inoculated microbial fuel cells. Biosensors and Bioelectronics. 30: 87–92.

Yorganci, E. and Akyilmaz, E. 2011. Alkaline phosphatase based amperometric biosensor immobilized by cysteamine-glutaraldehyde modified self-assembled monolayer. Artificial Cells, Blood Substitutes, and Biotechnology. 39(5): 317–323.

Zamaleeva, A.I., Sharipova, I.R., Shamagsumova, R.V., Ivanov, A.N., Evtugyn, G.A., Ishmuchametova, D.G., et al. 2011. A whole-cell amperometric herbicide biosensor based on magnetically functionalized microalgae and screen-printed electrodes. Analytical Methods. 3: 509–513.

Zhao, F., Slade, R.C.T. and Varcoe, J.R. 2009. Techniques for the study and development of microbial fuel cells: an electrochemical perspective. Chemical Society Reviews. 38: 1926–1939.

Zhou, T., Han, H., Liu, P., Xiong, J., Tian, F. and Li, X. 2017. Microbial fuel cell-based biosensor for toxicity detection: a review. Sensors (Basel, Switzerland). 17(10): 2230

Zhou, S., Huang, S., Li, Y., Zhao, N., Li, H., Angelidaki, I., et al. 2018. Microbial fuel cell-based biosensor for toxic carbon monoxide monitoring. Talanta. 186: 368–371.

Chapter 13

Nano-technology as a Tool for Environment Management

Shital Desai and Siddhi Kashinath Jalmi*

Dr. Siddhi Kashinath Jalmi, School of Biological Sciences and Biotechnology, Goa University, Taleigao Plateau, Goa, 403206, India.

INTRODUCTION

Nano-technology is one of the rapidly growing fields of applied science. It involves multidisciplinary approach which deals with manipulation of matter at nanoscale to get the beneficial products. The prefix "nano" depicts something very small, as small as one thousand millionth of a meter. The concept of nano-technology was introduced by American physicist Richard Feynman in 1959 and nano-technology was coined by Norio Taniguchi (1974) to describe the manufacturing of materials at the nanometre scale. Nano-technology is actively helping in improving and revolutionizing almost every field of science including, material science, medical science, computer science, environmental science and also in different industries (Bayda et al. 2019). Among many others, the contribution of nano-technology in the field of environmental science and its management is noteworthy. The unique ability to handle and manipulate the nanosized particles has opened up a new possibility in a different aspect of environmental management (Rani and Sridevi 2017).

The direct and indirect effects of rapid industrialization, burning activities, cutting down the trees, release of agricultural and household wastes in water bodies, application of chemicals for harvesting aquatic fauna, have led to undesirable

* For Correspondence: siddhi@unigoa.ac.in

changes in air, land, and water qualities. The problem of environmental pollution persists from decades, it remains the world's greatest problem faced by human society (Ukaogo et al. 2020). Most of the contaminants are found suspended in the atmosphere so we need a technology that helps to monitor and also clean the contaminants present in the environment. In this context, nano-technology plays a crucial role in creating new tool, develop novel methods and appropriate techniques to deal with various environmental problems. It can be employed for environmental remediation, pollution sensing and detection, prevention of pollution and also to improve the quality of existing environment.

Environmental management can be defined as the set of actions undertaken by society, or part of it, in order to protect the environment. It aims at modifying a current undesired situation to another desired one. It includes all the activities, means and techniques which aims at conserving the different elements of ecosystem and the ecological relationships between them, especially when there are alterations due to human action (Alexis 2021). It is a continuous process of prevention, restoration, conservation and sustainable development to maintain the ecological balance (Bayda et al. 2019).

This chapter summarizes the methods, tools and techniques of nano-technology employed to deal with different environmental issues like water pollution, waste water treatment, water scarcity, air pollution, solid waste management, and nuclear waste management. It includes information about different types of nanomaterials and their applications in different aspects of environmental management.

NANOMATERIALS FOR ENVIRONMENT MANAGEMENT

The materials employed for bio-remediation should not cause additional harm to the environment and turn out to be other pollutants. Biodegradable materials provide safer and greener options for environmental remediation. They do not generate any of the waste material that needs to be disposed of. These facts about biodegradable materials make them a better choice for use in dealing with different environmental issues. Many studies are being undertaken to develop nanomaterials that could be used to deal with challenges faced during environmental remediation (Bhawana and Fulekar 2012). Unique properties like greater surface area, stability, and different chemical functionalization make them ideal choices over traditional methods for addressing environmental challenges. The creation of nanomaterials like nanotubes, nano filters, nanocomposites, and nanoparticles are examples of nano-technology products used to build systems to deal with environmental issues (Baig et al. 2021).

The complex nature of different compounds makes the detection and management of environmental pollutants challenging; therefore, many approaches are being explored to fulfill this purpose. Current research is focused on the production of different nanomaterials for the development of new techniques for environmental remediation. Table 13.1 provides the list of different nanomaterials and their uses in the field of environment. Advances in the field of nano-technology are making differences in various areas including resource management, environmental protection, and remediation through the development of smart materials and connected devices (Kuhn et al. 2022).

Table 13.1 Different types of materials used for environmental remediation

	Nanomaterial	Type of remediation	References
		Graphene based nanomaterials	
1	Pristine graphene	Water remediation (Fluoride)	Li et al. 2011
2	Graphene Oxide	Water remediation (pesticides, heavy metals) and gas remediation (Ammonia, Sulphur dioxide), radionuclides in water	Song et al. 2016, Seredych and Bandosz 2007
3	Zn oxide graphene or Cadmium sulphide	Water and heavy metal remediation	Zhang et al. 2013
4	Titanium dioxide graphene	Gas remediation	Zhang et al. 2010
		Silica based nanomaterials	
1	Amine modified silica aerogels	CO_2 and H_2S remediation	Huang et al. 2003
2	Amine modified porous silica and aluminosilicates	Aldehyde, Ketone and CO_2 remediation	Nomura and Jones 2013
3	Amino functionalised mesoporous silica	Heavy metals (waste water)	Wang et al. 2015
4	Carboxylic acid functionalised mesoporous silica	Heavy metals and cationic dyes (waste water)	Bruzzoniti et al. 2007
5	Thiol functionalised mesoporous silica	Heavy metals (waste water)	Arencibia et al. 2010
		Metal based nanomaterials	
1	Titanium based nanomaterial (Titanium oxide, Titanate)	Microbial disinfection, Water and soil-aromatic hydrocarbon, Gas-Nitric oxide, radionuclides	Cho et al. 2005, Li et al. 2014
2	Iron based nanomaterials	Water-heavy metals, organic solvents, chlorinated solvents, radionuclides	Ebrahim et al. 2016
3	Bimetallic nanomaterials	Chlorinated solvents, brominated solvents in soil and water, radionuclides	Zhu et al. 2010, Lien and Zhang 2005

Remediation of various environmental media is done with the help of various technologies employing absorption, filtration, photocatalysis, chemical reactions, etc. High surface area to volume ratio, which provides high reactivity, makes nanoparticles excellent materials for remediation of contaminants from the environment. Three main categories of nanomaterials can be used for environmental remediation: carbon-based, polymeric-based, and inorganic (Guerra et al. 2018).

In contrast to metal-based nanoparticles, carbon-based particles show specific electronic, chemical, and physical features (Paul 2019). Single-walled nanotubes, multi-walled nanotubes, graphene, fullerene C_{60}, and fullerene C_{540} are some of the different configurations developed as a result of changeable hybridization states. These nanomaterials have been the interest of many studies. The adsorption properties of these materials make them particularly useful for the removal of organic and inorganic pollutants from the air as well as from large volumes of aqueous solution. Carbon-based nanomaterials are also employed to remediate contaminants through photocatalytic approaches. Pristine, graphene, or its modified form has been widely used in environmental remediation (Li et al. 2011, Song et al. 2016).

Environment-friendly Synthesis of Nanoparticles

Nanoparticles are considered as basic building blocks of nano-technology (Ijaz et al. 2020). Different strategies are employed for the synthesis of nanoparticles which includes physical, biological chemical, or a combination of techniques. Since nanoparticles act as starting material for many nanostructures materials and devices, their synthesis forms a crucial step in development of green products (Rani and Sridevi 2017).

"Green Nano Technology" involves production of nanomaterials and products without harming the environment and human health. It is a branch of green technology that utilizes the concepts and principles of green chemistry and green engineering. It cut downs the use of energy and fuel by using minimum and renewable material wherever possible (Verma et al. 2019). Green synthesis utilizes plant extract and microorganisms such as bacteria, actinomycetes, algae, fungi or their by-products like proteins and lipids for the production of nanoparticles (Ijaz et al. 2020, Gour and Jain 2019). Green nano-technology provides a cost effective, eco-friendly route for the synthesis of nanoparticles. It uses bottom-up approach similar to chemical reduction but in this, expensive chemical agent are replaced by natural extract from the microorganisms and plants (Joglekar et al. 2011, Lee et al. 2011, Gopinath et al. 2014, Hussain et al. 2016, Pal et al. 2019). Hence, nanoparticles synthesized by using biological method are free of toxic chemicals.

Top-down and bottom-up are the two basic approaches used for synthesis of nanoparticles. In top-down approach, suitable bulk material is subjected to size reduction to form small fine particles. It uses various physical and chemical methods and various techniques like mechanical/chemical grinding, milling, sputtering, thermal/laser ablation, etc., whereas in bottom-up approach NPs are synthesized using chemical and biological methods which involves self-assembly of atoms to new nuclei, and these nuclei are further converted to nanoscale dimensions (Singh et al. 2010, Hussain et al. 2016).

Conventional chemical and physical methods used for synthesis of nanoparticles exert stress on environment. The large amount of heat generated during physical mode of synthesis leads to increase in environmental temperature around the source material. Also, it requires large amount of space and is very time consuming. On the other hand, the chemical method uses toxic solvents and chemicals, which cause more harm to the environment (Gour and Jain 2019, Hussain et al. 2016).

Plants possess variety of metabolites like enzymes, vitamins, antioxidant, and nucleotides which can be easily utilized for green synthesis of nanoparticles. Different nanoparticles have been synthesized using plant and microbial extract like Gold, Copper, Copper oxide (CuO) Titanium dioxide (TiO_2), Zinc oxide (ZnO) etc. Table 13.1 shows list of nanoparticles synthesized by using plant and microbial source.

Nano-technology in Improving Air Quality

Air pollution resulting from human activities and natural processes has become a major problem at global level. The increased concentration of primary and secondary air pollutants negatively affects the air quality. Combustion of fossil fuel and burning of biomass are the main sources of airborne pollutants including oxides of sulphur, nitrogen, carbon monoxide, unburned hydrocarbons and particulate matter (Gomez-Garcia et al. 2005, Skalska et al. 2010). Of these, NO_2 is photochemically reactive and listed as the priority gas pollutant declared by the WHO. This is due to the acute injuries that it can cause to pulmonary tissues. Burning of biomass and motor vehicle exhaust are the most important sources of volatile compounds (VOCs) emission (Li et al. 2017). VOCs include more than 300 types of carbon-based chemicals that share similar characteristics of low boiling point, high vapour pressure, and strong reactivity (Zhang et al. 2017). It is very well known that many of these VOCs cause a variety of threats to the environment including ozone depletion, production of photochemical smog and some of the VOCs are carcinogens and mutagens (Manisalidis et al. 2020).

Current developments in the area of nano-technology offers new opportunities for removing fine particles and gaseous pollutants from air. Different methods like nanoscale based absorption of gases, particle filtration, bacteria disinfection and catalytic processes have been components of various types of purification devices used to reduce particulate matter, gaseous precursors and other toxic pollutants (Figure 13.1). Some of these compounds have been discussed in Table 13.1.

Nanomaterials in Monitoring of Air Pollution

The first important step to control air pollution is continuous monitoring. Nanosensors are an excellent choice for the detection of chemical and biological reactions (Jimenez-Cadena et al. 2007, Aguilar-Pérez et al. 2020). They are able to detect and respond to physical stimuli at the nanometer scale. Nanosensors are made up of zero-dimensional nanomaterials (nanoparticles and quantum data), one-dimensional nanomaterials (nanorods, nanotubes, and nanowires), two-dimensional nanomaterials (nanosheets), and three-dimensional nanomaterials (hard-core nanostructures) (Beni and Jabbari 2022).

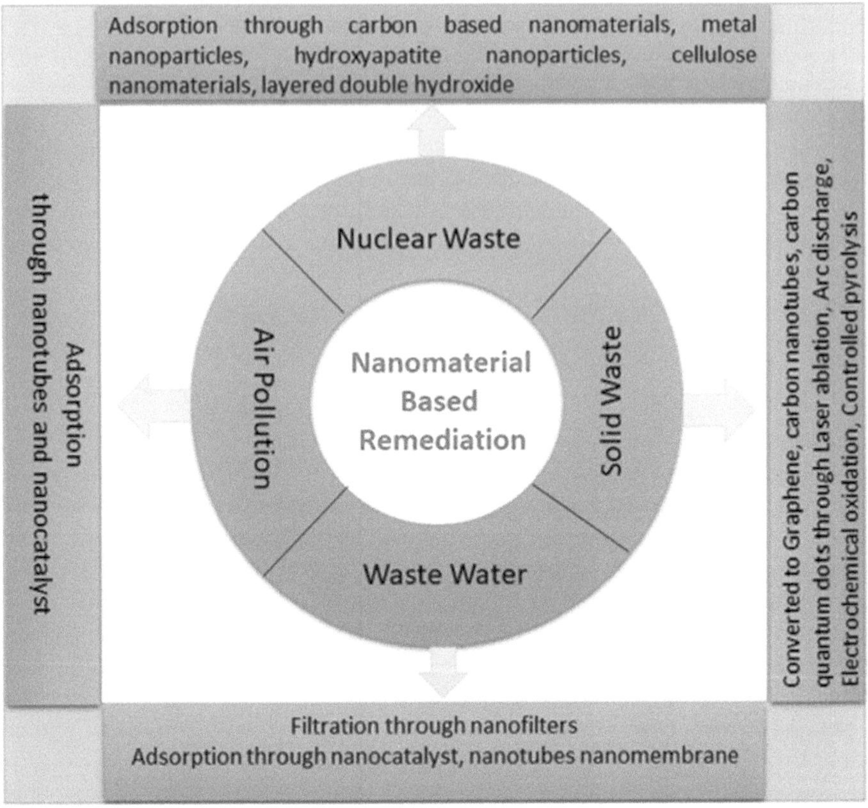

Figure 13.1 Nanomaterials based remediation of various environmental pollutants: Different ways by which nanomaterials are adopted for the use in remediation of pollutants in air, waste water solid waste and nuclear waste are shown in the figure.

The carbon nanotube sensors are made of single-layer nanotubes about 1 nm thick and can absorb toxic gas molecules (Beni and Jabbari 2022). Some nanosensors are also able to detect different types of gaseous compounds, dioxin, sulphur dioxide, carbon dioxide, nitrogen oxides, and ammonia.

They possess more accuracy and speed than conventional sensors. Quantum dots synthesized by using nanomaterials are advanced and highly accurate and are used for the detection of heavy metals (Beni and Jabbari 2022).

Nanomaterial as Adsorbents for Toxic Gases

The process of adsorption can be used to remove gaseous pollutants from air under ambient conditions (Cao et al. 2021). Scientists have developed carbon nanotubes (CNTs) which are useful for the removal of toxic gases from the environment. CNTs are made up of hexagonally placed carbon atoms in graphene sheets that surround the tube axis. CNTs can be both single-walled nanotubes or multi-walled nanotubes. Both of them display unique macromolecules that have a one-dimensional structure, thermal stability, and exceptional chemical properties. CNTs are also modified with gold or platinum nanoparticles (Zhang et al. 2017). Saleem et al. noted that carbon nanotubes and zinc oxides-based nanoadsorbents show good adsorption efficiency for the elimination of toxic gases (CO_2, CH_3OH, and H_2S). Also, the modified forms of the CNTS have also proved to have better adsorption ability as compared to the unmodified forms. Strong oxidation resistance of CNTs has also been beneficial for the regeneration of the adsorbent at high temperatures (Yunus et al. 2012). CNTs are also used for controlling the discharge of CO_2 and methane from the industrial chimney exhausts as well as vehicle exhausts. These materials also have the ability to trap greenhouse gases from coal mines and power generation (Saleem et al. 2022).

Dioxins are toxic by-product of many industrial processes and are categorized as persistent environmental pollutants; they can remain in the environment for many years (Das et al. 2015). Traditionally, activated carbon adsorption is widely used to remove dioxins from incinerators (Wei and Li 2021). CNT showed much higher efficiency in removal of dioxin than other adsorbents such as activated carbon, clay, Al_2O_3 and zeolites (Aguilar-Pérez et al. 2020). CNTs could be used as an adsorbent for the removal of NOx (Long and Yang 2001).

Volatile organic compounds (VOCs) can also be removed from air. A manganese oxide-based catalyst coated with gold nanoparticles is developed by the researchers which offers a highly porous space and can decompose acetaldehyde, toluene and hexane and remove it from room air. SWNTs/NaClO had the best performance to adsorb isopropyl alcohol (IPA) followed by SWNTs/HNO_3 from air (Yunus et al. 2012).

Nano-catalysts in Air Purification

Nano-technology can be effective in reducing the entry of air pollutants by manufacturing exhaust catalysts to purify the exhaust of gases from industrial processes, automobiles and cars. Nano-catalysts with increased surface area for gaseous reactions transform harmful vapours from cars and industrial plants into harmless gases. Catalysts made up of NP of manganese oxide can remove volatile organic compounds from industrial smokestacks. Gold NP offers a significant catalytic activity in converting highly toxic air pollutants into harmless CO_2 and H_2O (Chen and Goodman 2006). Extensive use of nanocomposites can reduce the consumption of 1.5 billion litres of gasoline in the one-year life of vehicles and reduce carbon dioxide emissions by more than 5 billion kg per year (Beni and Jabbari 2022).

Nano-technology in Water Purification

Water is a vital requirement for existence of life on this Earth. Potable water supply is a basic human need. One of the biggest challenges faced by human society is the supply of safe and pure water. Various human activities like, domestic and livestock operations, food processing waste, insecticides and herbicides from agriculture runoff, industrial effluents, sewage water, VOCs, heavy metals from electronic wastes, chemical waste, and medical wastes, oil spills etc. introduces pollutants into water. Leaching of naturally occurring ores, which are rich in heavy metals like arsenic, lead also cause pollution of water bodies (Briffa et al. 2020, Akhtar et al. 2021, Saha et al. 2021). The contaminants may be of inorganic, organic or biological origin. Inorganic pollutants include metals anions, acids and alkalis. Pesticides phenolic compounds and various dyes are the major organic pollutants, whereas viruses, bacteria and protozoans constitute the biological pollutants. Heavy metals like arsenic, nickel, zinc, cadmium, mercury chromium, lead, nickel and others show toxicities (Ismail et al. 2019). Many of these pollutants accumulate in the food chain and affect the food chain at different levels (Chatelet et al. 2015).

Remediation of environmental contaminants may be done *in-situ* or *ex-situ*. *In-situ* method involves treatment of contaminants in same place whereas in *ex-situ* treatment, contaminated material is transferred to a more convenient location Nano-technology helps in developing tools that can perform *in-situ* remediation at inaccessible area such as crevices and aquifers. This eliminates cost effective operations like pumping and heat necessary for *ex situ* treatment (Hussain et al. 2021). Different processes like adsorption, use of nanomembranes, photocatalysis, and antimicrobial activities are used for treatment of waste water. These processes make use of inorganic, organic and polymer-based nanoparticles depending on nature of pollutants (Matei et al. 2022). NPs possess very high flexibility for both *in-situ* and *ex-situ* purification of waste water and treatment of industrial effluents (Hussain et al. 2021).

Nanoadsorption for Treatment of Water

Adsorption is a surface phenomenon, wherein the pollutants are adsorbed on a solid surface. Use of adsorbent materials is one of the most efficient technologies in water and wastewater treatment. Working efficiency of traditional adsorbents may be restricted by their surface area. Due to very large surface area, multiple sorption sites, low-temperature modification, short intraparticle diffusion distance, tunable pore size and surface chemistry, nanoparticles act as good adsorbents (Ali et al. 2020). Different types of nanoparticles like carbon nanotubes, metal oxides, fullerenes, organic nanomaterials and zeolites have been developed to use as nanoadsorbents. The characteristics for ideal nano adsorbent materials for water and wastewater remediation purposes are as follows:

1. The material used as nanoadsorbents should be non-toxic.
2. The materials should have high adsorption capacity and accuracy for removing the low concentration of contaminants.
3. The surface of the nanoadsorbent materials should be easily reactivated (Sadegh et al. 2017).

Carbon-based Nanoadsorbents such as carbon nanotubes (CNTs) are used as substitutes for activated carbon. Due to their hydrophobic nature, CNTs form loose aggregates in liquid medium and these aggregates provide a high energy site for the adsorption of organic contaminants present in the water (Zhang et al. 2009, Ren et al. 2011). Studies have shown that CNTs are very good adsorbing nanomaterials for heavy metal ions, such as copper, lead, cadmium and zinc (Aguilar-Pérez et al. 2020). CNTs provide high solute uptake speed and this feature is widely used for removal of lead ions from waste water (Gul and Nasreen 2018). CNTs can also be used for removal of oil spills. A sponge made up of CNTs and boron showed a very good adsorbing capacity for oil from water (Gui et al. 2011). CNTs can be regenerated by altering the pH towards acidic condition. The regeneration will not affect the adsorbing capacity for metal ions (Toński et al. 2021) (Figure 13.1).

Nanofilters as Tool for Water Purification

Water filtration involves removal of contaminants such as suspended particles, microorganisms and other detrimental biological and chemical pollutants from water to make it safe and clean for drinking and other applications. Nanofilters can be used in water and wastewater treatment. Nanofilters are able to remove a wide variety of bacteria, viruses, calcium and magnesium salts, pesticides, organic pollutants, and heavy metals from the water (Shon et al. 2013, Yaqoob et al. 2020, Singh et al. 2020). As no chemicals are used for water softening, its adverse environmental effects are less compared to traditional chemical methods of water treatment (OboteyEzugbe and Rathilal 2020). Nanofilters possesses properties between non-porous reverse osmosis membranes and porous ultrafiltration (UF) membranes (Shon et al. 2013). Nanofilters are made up of using high surface energy nanoparticles like zinc and silver (Harish et al. 2022). The important advantage of nanofilters compared to conventional filters is their selectivity power in removing the ions. They can be used in such a way that the minerals necessary for human health can be retained in the water, and harmful contaminants can be removed (OboteyEzugbe and Rathilal 2020) (Figure 13.1).

Nano Catalysts for Water Treatment

Different categories of nano-catalyst used for remediating both organic and inorganic contaminants are chemical oxidants, electrocatalyst, photocatalyst and Fenton based catalyst (Figure 13.1). The catalytic reactions of nanomaterials are based on light and are hence called photocatalyst. The nano-catalyst is made up of semiconductor material showing ability to degrade wide variety of halogenated and non-halogenated organic pollutants as well as heavy metals in water (Adeleye et al. 2016). The mechanism used by semiconductor nano catalyst is photoexcitation of electron in the catalyst and generation of hydroxyl radicals, which act as strong oxidizing agents for oxidation of organic pollutants into water and gaseous degradation products (Anjum et al. 2016). Wide number of research is taken up to elevate the photocatalytic activity of nano-catalyst. The methods used for doping or modifying catalyst include production of hybrid nanoparticles, doping metal impurities, and dye sensitization. The nanomaterials like ZnO and TiO_2 having wide band gap energy have been studied extensively for doping of their activity. Both these catalysts can absorb a small portion of UV region decreasing their efficiency. Further modification by loading of metals on their surface increases the efficiency by decreasing band energy (Chen and Zhou 2004). The doping elements used for such reactions are Chromium, Silicone, Cobalt, Magnesium, Manganese, Iron, Aluminium, Galium, etc. Apart from metallic dopants, anionic dopants like Nitrogen are also considered to be more feasible and cost effective (Qu et al. 2013). These elements increases surface area of metal oxide nano-catalyst and also protects them from change in morphology, shape and size reduction (Jamal et al. 2012). Apart from ZnO and TiO_2, many other nano-catalyst like Nickel, ZnS:Mn, ZnS:Cu, ZnSe:Mn, ZnS:Pb, CdS:Eu, and CdS:Mn have been doped (Chandrakar et al. 2015).

Nano-technology Soil Health

Decreasing soil nutrition and soil quality is a major concern in agriculture. The prevailing conditions of climate change, depletion of water resources and use of chemical fertilizers are the major causes of soil degradation. Nanomaterials play an important and promising role in overcoming agricultural problems and maintain soil health. Nanomaterials are used as carriers for release of fertilizer or controlled release of vectors as a part of smart delivery system (Nair et al. 2010). The greater surface area and smaller size allows to adsorb more fertilizer and can easily penetrate the soil and plant. Nano-fertilizers can act as both source of essential nutrients and carriers of traditional fertilizers or help in effective delivery of DNA and chemicals into target plant (Ghafariyan et al. 2013). Nano-fertilizers do not pose any threat to the environment as different vegetative or reproductive parts of the plant are used for their extraction. Fe oxide, silica, carbon coated iron, nano-clays and zeolites are the most efficient naturally occurring compounds used as nanocarriers. Kaolin and other polymeric biocompatible nanomaterials like biodegradable polymeric chitosan have also been used for the slow delivery of different NPK (Corradini et al. 2010). Nano-fertilizers are developed by two ways: either by encapsulating nutrients inside them or by adsorbing the nutrients on the surface of nanomaterials (Rai et al. 2012). Nanofertilizers are classified as macronutrients NFs and micronutrients NFs, depending on type of nutrients used. Various macronutrients like Ca, P, N, Mg, S are used for preparation of nanosized fertilizers which are required in larger amounts, while micronutrient nano fertilizers are required in less than 100 ppm which includes Iron, Boron, Manganese, Zinc, Copper, Molybdenum, Chloride, Nickel, etc. (Kottegoda et al. 2011, Delfan et al. 2014).

Different nanomaterials have been used in different aspects of plant growth and development. Carbon nanotubes have been used in seed germination of seedlings, nanonutrients have function in plant nutrition, nanopesticides are used in plant protection, nanoscale carriers are used for more effective application of pesticides and fertilizers, and nanosensors are used for detection of nutrients and contaminants (Adhikari 2021).

Nano-technology in Solid Waste Management

The rapid increase in human population and increase in global productivity and economy has resulted in overproduction of solid waste which results in adverse effects on human health and the entire ecosystem (Johnson et al. 1999). The better solution in management of this solid waste is the use of nanomaterials. The best practice followed is the conversion of solid waste into valuable carbon-based nanomaterials like graphene, carbon nanotubes, carbon quantum dots. The presence of high carbon content makes them very suitable for the production of nanotubes. These solid waste derived carbon nanomaterials can further be used for various industrial applications and in field of energy like supercapacitors, solar cells, fuel cells and even in drug delivery and bioimaging (Sun et al. 2013).

Solid waste mainly consists of plastic, paper, polymer, municipal, industrial and agricultural waste, etc, which provide structural framework that is carbonic skeleton for the synthesis of nanomaterials. The carbonic skeleton consisting of millions of layers of atom acts as precursor for conversion into graphene. Graphene is characterized by single layer of graphite and is the thinnest form of carbon which provides higher conductivity and increased surface area (El Essawy et al. 2017). Pyrolysis unit can be used to convert solid waste carbon nanotubes (CNT); with the help of pyrolysis unit, CNTs can be made into single-walled or multi-walled carbon nanotubes as defined by its structure. Single-walled CNT has a structure of one cylindrical rolled graphene sheet. For this conversion of plastic waste to CNT, it requires use of catalyst like Nickel, Molybdenum, Magnesium oxide, etc. Like graphene and carbon nanotubes, carbon quantum dots (CQD) are also formed from solid waste. CQD exhibits properties like optical stability, photoluminescence, biocompatibility, low toxicity, etc. (Peng et al. 2012). CQDs can be made by using methodologies like laser ablation, arc discharge, electrochemical oxidation, microwave heating, deposition of chemical vapour, physical vapour and pyrolytic heating, and hydrothermal approach (Figure 13.1). Controlled pyrolysis is used for conversion of plastic waste like PET, polypropelene, carbonic waste into carbon quantum dots, while hydrothermal approach is used for agricultural and food waste (Prasannan and Imae 2013, Kamali et al. 2019). The structural and morphological properties that has to be considered in conversion reaction of solid waste to nanomaterials are surface framework, extent of carbonization and so on. Doping of these solid waste-derived nanomaterials with various metallic and non-metallic elements can signify surface porosity and morphology.

Nano-technology in Nuclear Waste

Nuclear plants are considered as alternative of the energy production to meet the high demand and decreasing non-renewable sources of energy. Nuclear energy is preferred form of energy due to less green-house gas emission, clean form and important source of renewable energy. However, in the process of energy production, lot of nuclear waste is produced as a by-product of nuclear power generation. This waste is considered to be the most hazardous material to all life forms and environment. The radioisotopes in aquatic habitats are significantly more hazardous due to their long-term radiological and chemical toxicities. This waste is being managed by only segregation and storage; however, a sustainable method is required for treating this waste (Blowers 1999). Also, the traditional waste water treatment techniques are not useful for removing nuclear waste since the radioisotopes are present in low concentration as compared to other components in waste water. Hence, efficient methods for selective removal are necessary. Several conventional methods like chemical precipitation, adsorption, evaporation, ion exchange have been employed to remove the radioactive compounds of which adsorption is widely employed. The advancement in nanotechnology and development of wide range of nanomaterials have provided positive opportunities in the remediation of radioactive waste from waste water (Liu et al. 2019, Teow and Mohammad 2019). Extensive efforts have been put up for employing these nanomaterials in decontaminating radioactive water waste in countries like U.S., Russia, Japan, Germany and France (Zhang and Liu 2020). There are various categories of nanomaterials used for this purpose like carbon-based nanomaterials, metal nanoparticles, hydroxyapatite nanoparticles, cellulose nanomaterials, layered double hydroxides, etc. (Figure 1). Carbon based nanomaterials, graphene, and carbon nanotubes are very much used for cleaning radioactive waste (Li et al. 2012). In contrast to organic exchange resins, carbon nanotubes have higher thermal and radiation resistance making them more stable in highly acidic media like nuclear waste water (Omidi et al. 2017). The first reported of use of carbon nanotubes as absorbent was for element Americium Am-(III) from aqueous solution. Also, multi-walled carbon nanotubes are effective for removing Uranium (Wang et al. 2005). Graphene oxide possess properties of higher surface area and functional groups with oxygen, which act as better anchoring sites for radionuclides. Modified graphene oxide was used for removal of toxic radioactive materials like Americium, Plutonium, Uranium, Thorium, Strontium, etc. from polluted water (Wang et al. 2005).

In the category of metal nanomaterials, nanoscale zero-valent irons (NZVI), silver nanoparticles, gold nanoparticles, nano-sized metal oxides, silver oxides, iron oxides are popular for removing radionuclides from waste water. NZVI are environment friendly and highly efficient nanomaterial with high reactivity and abundant active sites for reacting with radioactive materials. They have been widely used for removing heavy metals and showed very high ability to adsorb, reduce and precipitate radionuclides like Uranium (Li et al. 2013). Gold and silver nanoparticles have been used for adsorption and removal of iodine nanoparticles from aqueous solutions due to the high affinity between gold/silver and iodine (Choi et al. 2017). Silver nanoparticles immobilized with dextran gel and Titanium modified with cellulose, and chitin have been also used for removal of Strontium, Cobalt, Caesium from aqueous solutions (Pospěchová et al. 2016). Likewise, the oxides of metals like Titanium oxides, and Iron oxides which are stable nanostructures under thermal, chemical, radiation and mechanical conditions, are used for adsorption of Uranium, Molybdenum, Technetium, Cobalt, Cesium, Europium, etc. (Borai et al. 2015, Zhang et al. 2018).

CONCLUSION

Nano-technology is one of the rapidly growing fields of applied science. Nano-technology is actively helping in improving and revolutionizing almost every field of science including, material science, medical science, computer science, environmental science and also different industries. The direct and indirect effects of various development processes have led to undesirable changes in the environment. Nano-technology is seen as a promising technology to solve major environmental issues such as water pollution, wastewater treatment, water scarcity, air pollution, solid waste management, and nuclear waste management. Recent developments in the field of nano-technology offer a new solution for removing gaseous pollutants and particles from the air, wastewater, and even nuclear waste. Research has been carried out to convert solid waste to nanomaterials which can further be used for remediation. The remediation of waste by various nanomaterials is mainly achieved through adsorption, filtration, chemical precipitation, pyrolysis, electrochemical oxidation, etc. The importance of this area in maintaining quality of the environment and life on Earth demands more research. Also, more efforts are needed to extend the research on nanomaterials from laboratory to the field application.

REFERENCES

Adeleye, A.S., Conway, J.R., Garner, K., Huang, Y., Su, Y. and Keller, A.A. 2016. Engineered nanomaterials for water treatment and remediation: Costs, benefits, and applicability. Chemical Engineering Journal. 286: 640–662.

Adhikari, T. 2021. Nanotechnology in environmental soil science. Soil Science: Fundamentals to Recent Advances. 297–309.

Aguilar-Pérez, K.M., Heya, M.S., Parra-Saldívar, R. and Iqbal, H.M. 2020. Nano-biomaterials in-focus as sensing/detection cues for environmental pollutants. Case Studies in Chemical and Environmental Engineering. 2. 100055.

Ajith, M.P., Aswathi, M., Priyadarshini, E. and Rajamani, P. 2021. Recent innovations of nanotechnology in water treatment: a comprehensive review. Bioresource Technology. 342. 126000.

Akhtar, N., SyakirIshak, M.I., Bhawani, S.A. and Umar, K. 2021. Various natural and anthropogenic factors responsible for water quality degradation: a review. Water. 13(19): 2660.

Alexis, A.P.F. 2021. The environmental management tools and instruments of aplication. American Journal of Environmental Protection. 10(6): 120–126.

Ali, I. 2012. New generation adsorbents for water treatment. Chemical Reviews. 112(10): 5073–5091.

Ali, M.E., Hoque, M.E., Safdar Hossain, S.K. and Biswas, M.C. 2020. Nanoadsorbents for wastewater treatment: next generation biotechnological solution. International Journal of Environmental Science and Technology. 17: 4095–4132.

Anjum, M., Al-Makishah, N.H. and Barakat, M.A. 2016. Wastewater sludge stabilization using pre-treatment methods. Process Safety and Environmental Protection. 102: 615–632.

Arencibia, A., Aguado, J. and Arsuaga, J.M. 2010. Regeneration of thiol-functionalized mesostructured silica adsorbents of mercury. Applied Surface Science. 256(17): 5453–5457.

Baig, N., Kammakakam, I. and Falath, W. 2021. Nanomaterials: a review of synthesis methods, properties, recent progress, and challenges. Materials Advances. 2(6): 1821–1871.

Bayda, S., Adeel, M., Tuccinardi, T., Cordani, M. and Rizzolio, F. 2019. The history of nanoscience and nanotechnology: from chemical-physical applications to nanomedicine. Molecules (Basel, Switzerland). 25(1): 112.

Beni, A.A. and Jabbari, H. 2022. Nanomaterials for environmental applications. Results in Engineering. 15: 100467.

Bhawana, P. and Fulekar, M. 2012. Nanotechnology: remediation technologies to clean up the environmental pollutants. Research Journal of Chemical Sciences. 2(2): 90–96.

Blowers, A. 1999. Nuclear waste and landscapes of risk. Landscape Research. 24(3): 241–264.

Borai, E.H., Breky, M.M.E., Sayed, M.S. and Abo-Aly, M.M. 2015. Synthesis, characterization and application of titanium oxide nanocomposites for removal of radioactive cesium, cobalt and europium ions. Journal of Colloid and Interface Science. 450: 17–25.

Brenner, A., Persson, K.M., Russell, L., Rosborg, I. and Kozisek, F. 2015. Technical and mineral level effects of water treatment. pp. 103–117. *In*: Rosborg, I. (ed.). Drinking Water Minerals and Mineral Balance. Springer.

Briffa, J., Sinagra, E. and Blundell, R. 2020. Heavy metal pollution in the environment and their toxicological effects on humans. Heliyon. 6(9): e04691.

Bruzzoniti, M.C., Prelle, A., Sarzanini, C., Onida, B., Fiorilli, S. and Garrone, E. 2007. Retention of heavy metal ions on SBA-15 mesoporous silica functionalised with carboxylic groups. Journal of Separation Science. 30(15): 2414–2420.

Cao, J.J., Huang, Y. and Zhang, Q. 2021. Ambient air purification by nanotechnologies: from theory to application. Catalysts. 11(11): 1276.

Chandrakar, R.K., Baghel, R.N., Chandra, V.K. and Chandra, B.P. 2015. Synthesis, characterization and photoluminescence studies of Mn doped ZnS nanoparticles. Superlattices and Microstructures. 86: 256–269.

Chatelet, A., Fournier, A., Jurjanz, S., Lerch, S., Toussaint, H., Delannoy, M., et al. 2015. Transfer of organic and inorganic pollutants in the food chain after spreading of residual-origin fertilizing materials on grasslands. INRA Productions Animales. 28(5): 383–397.

Chen, C.Z. and Zhou, Z.W. 2004. The preparation of nano-ZnO and its middle infrared-ultraviolet-visible light absorption properties. Journal of Functional Materials. 35(1): 97–98.

Chen, M. and Goodman, D.W. 2006. Catalytically active gold: from nanoparticles to ultrathin films. Accounts of Chemical Research. 39(10): 739–746.

Cho, M., Chung, H., Choi, W. and Yoon, J. 2005. Different inactivation behaviors of MS-2 phage and *Escherichia coli* in TiO_2 photocatalytic disinfection. Applied and Environmental Microbiology. 71(1): 270–275.

Choi, M.H., Jeong, S.W., Shim, H.E., Yun, S.J., Mushtaq, S., Choi, D.S., et al. 2017. Efficient bioremediation of radioactive iodine using biogenic gold nanomaterial-containing radiation-resistant bacterium, *Deinococcus radiodurans* R1. Chemical Communications. 53(28): 3937–3940.

Cui, H., Zhang, X., Chen, D. and Tang, J. 2018. Geometric structure and SOF_2 adsorption behavior of Ptn (n = 1 – 4) clustered (8, 0) single-walled CNT using density functional theory. Journal of Fluorine Chemistry. 211: 148–153.

Das, S., Sen, B. and Debnath, N. 2015. Recent trends in nanomaterials applications in environmental monitoring and remediation. Environmental Science and Pollution Research. 22(23): 18333–18344.

Delfani, M., BaradarnFirouzabadi, M., Farrokhi, N. and Makarian, H. 2014. Some physiological responses of black-eyed pea to iron and magnesium nanofertilizers. Communications in Soil Science and Plant Analysis. 45(4): 530–540.

Ebrahim, S.E., Sulaymon, A.H. and SaadAlhares, H. 2016. Competitive removal of Cu^{2+}, Cd^{2+}, Zn^{2+}, and Ni^{2+} ions onto iron oxide nanoparticles from wastewater. Desalination and Water Treatment. 57(44): 20915–20929.

El-Essawy, N.A., Ali, S.M., Farag, H.A., Konsowa, A.H., Elnouby, M. and Hamad, H.A. 2017. Green synthesis of graphene from recycled PET bottle wastes for use in the adsorption of dyes in aqueous solution. Ecotoxicology and Environmental Safety. 145: 57–68.

El-Sayed, M.E. 2020. Nanoadsorbents for water and wastewater remediation. Science of the Total Environment. 739: 139903.

Ghafariyan, M.H., Malakouti, M.J., Dadpour, M.R., Stroeve, P. and Mahmoudi, M. 2013. Effects of magnetite nanoparticles on soybean chlorophyll. Environmental Science and Technology. 47(18): 10645–10652.

Gomez-Garcia, M.A., Pitchon, V. and Kiennemann, A. 2005. Pollution by nitrogen oxides: an approach to NOx abatement by using sorbing catalytic materials. Environment International. 31(3): 445–467.

Gopinath, K., Karthika, V., Gowri, S., Senthilkumar, V., Kumaresan, S. and Arumugam, A. 2014. Antibacterial activity of ruthenium nanoparticles synthesized using *Gloriosa superba* L. leaf extract. Journal of Nanostructure in Chemistry. 4(1): 1–6.

Gour, A. and Jain, N.K. 2019. Advances in green synthesis of nanoparticles. Artificial Cells, Nanomedicine, and Biotechnology. 47(1): 844–851.

Guerra, F.D., Attia, M.F., Whitehead, D.C. and Alexis, F. 2018. Nanotechnology for Environmental Remediation: Materials and Applications. Molecules (Basel, Switzerland). 23(7): 1760.

Gui, X., Li, H., Wang, K., Wei, J., Jia, Y., Li, Z., et al. 2011. Recyclable carbon nanotube sponges for oil absorption. ActaMaterialia. 59(12): 4798–4804.

Gul, H. and Nasreen, S. 2018. Heavy metal uptake from contaminated water using carbon nanotubes: A Review. Reviews of Environmental Contamination and Toxicology. 1: 04–08.

Gupta, V., Moradi, O., Tyagi, I., Agarwal, S., Sadegh, H., Shahryari-ghoshekandi, R., et al. 2016. Study on the removal of heavy metal ions from industry waste by carbon nanotubes: effect of the surface modification: a review. Critical Reviews in Environmental Science and Technology. 46(2): 93–118.

Harish, V., Ansari, M.M., Tewari, D., Gaur, M., Yadav, A.B., García-Betancourt, M.L., et al. 2022. Nanoparticle and Nanostructure synthesis and controlled growth methods. Nanomaterials. 12(18): 3226.

Hashim, D.P., Narayanan, N.T., Romo-Herrera, J.M., Cullen, D.A., Hahm, M.G., Lezzi, P., et al. 2012. Covalently bonded three-dimensional carbon nanotube solids via boron induced nanojunctions. Scientific Reports. 2: 363. https://doi.org/10.1038/srep00363

Huang, H.Y., Yang, R.T., Chinn, D. and Munson, C.L. 2003. Amine-grafted MCM-48 and silica xerogel as superior sorbents for acidic gas removal from natural gas. Industrial and Engineering Chemistry Research. 42(12): 2427–2433.

Huang, Z.H., Yin, Y.N. and Zhang, Y. 2016. Preparation of a novel positively charged nanofiltration composite membrane incorporated with silver nanoparticles for pharmaceuticals and personal care product rejection and antibacterial properties. Water Science and Technology. 73: 1910–1919. https://doi.org/10.2166/wst.2016.037.

Hussain, I., Singh, N.B., Singh, A., Singh, H. and Singh, S.C. 2016. Green synthesis of nanoparticles and its potential application. Biotechnology letters. 38(4): 545–560.

Hussain, M.M., Farooqi, Z.U.R., Mohy-Ud-Din, W., Younas, F., Shahzad, M.T.T., Ghani, M.U.M.U., et al. 2021. Application of bioremediation as sustainable approach to remediate heavy metal and pesticide polluted environments. Plant and Environment. 2(01): 62–92.

Hussain, A., Rehman, F., Rafeeq, H., Waqas, M., Asghar, A., Afsheen, N., et al. 2022. *In-situ, Ex-situ,* and nano-remediation strategies to treat polluted soil, water, and air–a review. Chemosphere. 289: 133252.

Ijaz, I., Gilani, E., Nazir, A. and Bukhari, A. 2020. Detail review on chemical, physical and green synthesis, classification, characterizations and applications of nanoparticles. Green Chemistry Letters and Reviews. 13(3): 223–245.

Ismail, M., Akhtar, K., Khan, M.I., Kamal, T., Khan, M.A., M Asiri, A., et al. 2019. Pollution, toxicity and carcinogenicity of organic dyes and their catalytic bio-remediation. Current Pharmaceutical Design. 25(34): 3645–3663.

Jain, K., Patel, A.S., Pardhi, V.P. and Flora, S. 2021. Nanotechnology in wastewater management: a new paradigm towards wastewater treatment. Molecules (Basel, Switzerland). 26(6): 1797. https://doi.org/10.3390/molecules26061797.

Jamal, A., Rahman, M.M., Khan, S.B., Faisal, M., Akhtar, K., Rub, M.A., et al. 2012. Cobalt doped antimony oxide nano-particles based chemical sensor and photo-catalyst for environmental pollutants. Applied Surface Science. 261: 52–58.

Jimenez-Cadena, G., Riu, J. and Rius, F.X. 2007. Gas sensors based on nanostructured materials. Analyst. 132(11): 1083–1099.

Joglekar, S., Kodam, K., Dhaygude, M. and Hudlikar, M. 2011. Novel route for rapid biosynthesis of lead nanoparticles using aqueous extract of *Jatropha curcas* L. latex. Materials Letters. 65(19–20): 3170–3172.

Johnson, B.L., Hicks, H.E., Cibulas, W., Faroon, O., Ashizawa, A.E., De Rosa, C.T., et al. 1999. Public health implications of exposure to polychlorinated biphenyls (PCBs). Washington, DC: Agency for Toxic Substances and Disease Registry (USDHHS) and US Environmental Protection Agency.

Kamali, A.R., Yang, J. and Sun, Q. 2019. Molten salt conversion of polyethylene terephthalate waste into graphene nanostructures with high surface area and ultra-high electrical conductivity. Applied Surface Science. 476: 539–551.

Khin, M.M., Nair, A.S., Babu, V.J., Murugan, R. and Ramakrishna, S. 2012. A review on nanomaterials for environmental remediation. Energy and Environmental Science. 5(8): 8075–8109.

Kobya, M., Demirbas, E., Senturk, E. and Ince, M. 2005. Adsorption of heavy metal ions from aqueous solutions by activated carbon prepared from apricot stone. Bioresource Technology. 96(13): 1518–1521.

Kottegoda, N., Munaweera, I., Madusanka, N. and Karunaratne, V. 2011. A green slow-release fertilizer composition based on urea-modified hydroxyapatite nanoparticles encapsulated wood. Current Science. 73–78.

Kuhn, R., Bryant, I.M., Jensch, R. and Böllmann, J. 2022. Applications of environmental nanotechnologies in remediation, wastewater treatment, drinking water treatment, and agriculture. Applied Nano. 3(1): 54–90.

Lee, H.J., Lee, G., Jang, N.R., Yun, J.H., Song, J.Y. and Kim, B.S. 2011. Biological synthesis of copper nanoparticles using plant extract. Nanotechnology. 1(1): 371–374.

Li, Y., Zhang, P., Du, Q., Peng, X., Liu, T., Wang, Z., et al. 2011. Adsorption of fluoride from aqueous solution by graphene. Journal of colloid and Interface Science. 363(1): 348–354.

Li, Z., Chen, F., Yuan, L., Liu, Y., Zhao, Y., Chai, Z. and Shi, W. 2012. Uranium (VI) adsorption on graphene oxide nanosheets from aqueous solutions. Chemical engineering Journal. 210: 539–546.

Li, X., Zhang, M., Liu, Y., Li, X., Liu, Y., Hua, R., et al. 2013. Removal of U (VI) in aqueous solution by nanoscale zero-valent iron (nZVI). Water Quality, Exposure and Health. 5: 31–40.

Li, D., Cui, F., Zhao, Z., Liu, D., Xu, Y., Li, H. and Yang, X. 2014. The impact of titanium dioxide nanoparticles on biological nitrogen removal from wastewater and bacterial community shifts in activated sludge. Biodegradation. 25: 167–177.

Li, M., Liu, H., Geng, G., Hong, C., Liu, F., Song, Y., et al. 2017. Anthropogenic emission inventories in China: a review. National Science Review. 4(6): 834–866.

Lien, H. L. and Zhang, W.X. 2005. Hydrodechlorination of chlorinated ethanes by nanoscale Pd/Fe bimetallic particles. Journal of Environmental Engineering. 131(1): 4–10.

Liu, Y., Zeng, X., Hu, X., Hu, J. and Zhang, X. 2019. Two-dimensional nanomaterials for photocatalytic water disinfection: recent progress and future challenges. Journal of Chemical Technology and Biotechnology. 94(1): 22–37.

Long, R.Q. and Yang, R.T. 2001. Carbon Nanotubes as Superior Sorbent for Dioxin Removal. Journal of the American Chemical Society. 123(9): 2058–2059.

Manisalidis, I., Stavropoulou, E., Stavropoulos, A. and Bezirtzoglou, E. 2020. Environmental and health impacts of air pollution: a review. Frontiers in Public Health. 14.

Matei, E., Predescu, A.M., Râpă, M., Ţurcanu, A.A., Mateş, I., Constantin, N., et al. 2022. Natural polymers and their nanocomposites used for environmental applications. Nanomaterials. 12(10): 1707.

Nair, R., Varghese, S.H., Nair, B.G., Maekawa, T., Yoshida, Y. and Kumar, D.S. 2010. Nanoparticulate material delivery to plants. Plant Science. 179(3): 154–163.

Nomura, A. and Jones, C.W. 2013. Amine-functionalized porous silicas as adsorbents for aldehyde abatement. ACS Applied Materials and Interfaces. 5(12): 5569–5577.

ObotеyEzugbe, E. and Rathilal, S. 2020. Membrane technologies in wastewater treatment: a review. Membranes. 10(5): 89.

Omidi, M.H., Azad, F.N., Ghaedi, M., Asfaram, A., Azqhandi, M.H.A. and Tayebi, L. 2017. Synthesis and characterization of Au-NPs supported on carbon nanotubes: application for the ultrasound assisted removal of radioactive UO_2^{2+} ions following complexation with Arsenazo III: spectrophotometric detection, optimization, isotherm and kinetic study. Journal of Colloid and Interface Science. 504: 68–77.

Oubagaranadin, J.U.K. and Murthy, Z.V.P. 2009. Adsorption of divalent lead on a montmorillonite—illite type of clay. Industrial and Engineering Chemistry Research. 48(23): 10627–10636.

Pal, G., Rai, P. and Pandey, A. 2019. Green synthesis of nanoparticles: A greener approach for a cleaner future. pp. 1–26. *In*: Shukla, A.K. and Iravani, S. (eds). Green Synthesis, Characterization and Applications of Nanoparticles. Elsevier.

Paul, R. 2019. Prospects of carbon nanomaterials for energy storage and conversion. pp. 423–430. *In*: Paul, R., Etacheri, V., Wang, Y. and Lin, C.-T. (eds). Carbon Based Nanomaterials for Advanced Thermal and Electrochemical Energy Storage and Conversion. Elsevier.

Peng, J., Gao, W., Gupta, B.K., Liu, Z., Romero-Aburto, R., Ge, L., et al. 2012. Graphene quantum dots derived from carbon fibers. Nano Letters. 12(2): 844–849.

Pokrajac, L., Abbas, A., Chrzanowski, W., Dias, G.M., Eggleton, B.J., Maguire, S., et al. 2021. Nanotechnology for a sustainable future: Addressing global challenges with the international network4sustainable nanotechnology. ACS Nano. 15(12): 18608–18623.

Pospěchová, J., Brynych, V., Štengl, V., Tolasz, J., Langecker, J.H., Bubeníková, M., et al. 2016. Polysaccharide biopolymers modified with titanium or nickel nanoparticles for removal of radionuclides from aqueous solutions. Journal of Radioanalytical and Nuclear Chemistry. 307: 1303–1314.

Prasannan, A. and Imae, T. 2013. One-pot synthesis of fluorescent carbon dots from orange waste peels. Industrial and Engineering Chemistry Research. 52(44): 15673–15678.

Qu, X., Alvarez, P.J. and Li, Q. 2013. Applications of nanotechnology in water and wastewater treatment. Water Research. 47(12): 3931–3946.

Rani, K. and Sridevi, V. 2017. An overview on role of nanotechnology in green and clean technology. Austin Environmental Sciences. 2(3): 1026.

Ren, X., Chen, C., Nagatsu, M. and Wang, X. 2011. Carbon nanotubes as adsorbents in environmental pollution management: a review. Chemical Engineering Journal. 170(2–3): 395–410.

Ronen, A., Semiat, R. and Dosoretz, C.G. 2013. Antibacterial efficiency of composite nano-ZnO in biofilm development in flow-through systems. Desalination. Water Treat. 51: 988–996. https://doi.org/10.1080/19443994.2012.693698.

Sadegh, H., Ali, G.A., Gupta, V.K., Makhlouf, A.S.H., Shahryari-Ghoshekandi, R., Nadagouda, M.N., et al. 2017. The role of nanomaterials as effective adsorbents and their applications in wastewater treatment. Journal of Nanostructure in Chemistry. 7: 1–14.

Saha, L., Kumar, V., Tiwari, J., Rawat, S., Singh, J. and Bauddh, K. 2021. Electronic waste and their leachates impact on human health and environment: Global ecological threat and management. Environmental Technology and Innovation. 24: 102049.

Saleem, H., Zaidi, S.J., Ismail, A.F. and Goh, P.S. 2022. Advances of nanomaterials for air pollution remediation and their impacts on the environment. Chemosphere. 287: 132083.

Seredych, M. and Bandosz, T.J. 2007. Removal of ammonia by graphite oxide via its intercalation and reactive adsorption. Carbon (New York, NY). 45(10): 2130–2132.

Shon, H.K., Phuntsho, S., Chaudhary, D.S., Vigneswaran, S. and Cho, J. 2013. Nanofiltration for water and wastewater treatment–a mini review. Drinking Water Engineering and Science. 6(1): 47–53.

Singh, S.C., Mishra, S.K., Srivastava, R.K. and Gopal, R. 2010. Optical properties of selenium quantum dots produced with laser irradiation of water suspended Se nanoparticles. The Journal of Physical Chemistry A. 114: 17374–17384.

Singh, R., Bhadouria, R., Singh, P., Kumar, A., Pandey, S. and Singh, V.K. 2020. Nanofiltration technology for removal of pathogens present in drinking water.

pp. 463–489. *In*: Prasad, M.N.V. and Grobelak, A. (eds). Waterborne Pathogens. Butterworth-Heinemann.

Skalska, K., Miller, J.S. and Ledakowicz, S. 2010. Trends in NOx abatement: a review. Science of the Total Environment. 408(19): 3976–3989.

Song, B., Zhang, C., Zeng, G., Gong, J., Chang, Y. and Jiang, Y. 2016. Antibacterial properties and mechanism of graphene oxide-silver nanocomposites as bactericidal agents for water disinfection. Archives of Biochemistry and Biophysics. 604: 167–176.

Sun, C., Qu, R., Ji, C., Wang, Q., Wang, C., Sun, Y., et al. 2006. A chelating resin containing S, N and O atoms: synthesis and adsorption properties for Hg (II). European Polymer Journal. 42(1): 188–194.

Sun, L., Tian, C., Li, M., Meng, X., Wang, L., Wang, R., et al. 2013. From coconut shell to porous graphene-like nanosheets for high-power supercapacitors. Journal of Materials Chemistry A. 1(21): 6462–6470.

Taran, M., Safaei, M., Karimi, N. and Almasi, A. 2021. Benefits and application of nanotechnology in environmental science: an overview. Biointerface Research in Applied Chemistry. 11(1): 7860–7870.

Teow, Y.H. and Mohammad, A.W. 2019. New generation nanomaterials for water desalination: a review. Desalination. 451: 2–17.

Tońśki, M., Paszkiewicz, M., Dołżonek, J., Flejszar, M., Bielicka-Giełdoń, A., Stepnowski, P., et al. 2021. Regeneration and reuse of the carbon nanotubes for the adsorption of selected anticancer drugs from water matrices. Colloids and Surfaces A: Physicochemical and Engineering Aspects. 618: 126355.

Ukaogo, P.O., Ewuzie, U. and Onwuka, C.V. 2020. Environmental pollution: causes, effects, and the remedies. pp. 419–429. *In*: Chowdhary, P., Raj, A., Verma, D., and Akhter, Y. (eds). Microorganisms for Sustainable Environment and Health. Elsevier.

Verma, A., Gautam, S.P., Bansal, K.K., Prabhakar, N. and Rosenholm, J.M. 2019. Green nanotechnology: advancement in phytoformulation research. Medicines. 6(1): 39.

Wang, X., Chen, C., Hu, W., Ding, A., Xu, D. and Zhou, X. 2005. Sorption of 243Am (III) to multiwall carbon nanotubes. Environmental Science and Technology. 39(8): 2856–2860.

Wang, X., Guo, Y., Yang, L., Han, M., Zhao, J. and Cheng, X. 2012. Nanomaterials as sorbents to remove heavy metal ions in wastewater treatment. Journal of Environmental Analytical Toxicology. 2(7): 154–158.

Wang, S., Wang, K., Dai, C., Shi, H. and Li, J. 2015. Adsorption of Pb^{2+} on amino-functionalized core–shell magnetic mesoporous SBA-15 silica composite. Chemical Engineering Journal. 262: 897–903.

Wei, X. and Li, T. 2021. Wooden activated carbon production for dioxin removal via a two-step process of carbonization coupled with steam activation from biomass wastes. ACS omega. 6(8): 5607–5618.

World Health Organization (WHO). 2010. WHO guidelines for indoor air quality: selected pollutants. World Health Organization. Regional Office for Europe.

Yaqoob, A.A., Parveen, T., Umar, K. and Mohamad Ibrahim, M.N. 2020. Role of nano-materials in the treatment of wastewater: a review. Water. 12(2): 495.

Yunus, I.S., Harwin, Kurniawan, A., Adityawarman, D. and Indarto, A. 2012. Nanotechnologies in water and air pollution treatment. Environmental Technology Reviews. 1(1): 136–148.

Zhang, S., Shao, T., Bekaroglu, S. S. K. and Karanfil, T. 2009. The impacts of aggregation and surface chemistry of carbon nanotubes on the adsorption of synthetic organic compounds. Environmental Science and Technology. 43(15): 5719–5725.

Zhang, Y., Tang, Z.R., Fu, X. and Xu, Y.J. 2010. TiO_2–graphene nanocomposites for gas-phase photocatalytic degradation of volatile aromatic pollutant: is TiO_2–graphene truly different from other TiO_2–carbon composite materials? ACS Nano. 4(12): 7303–7314.

Zhang, N., Yang, M.Q., Tang, Z.R. and Xu, Y.J. 2013. CdS–graphene nanocomposites as visible light photocatalyst for redox reactions in water: a green route for selective transformation and environmental remediation. Journal of Catalysis. 303: 60–69.

Zhang, X., Gao, B., Creamer, A.E., Cao, C. and Li, Y. 2017. Adsorption of VOCs onto engineered carbon materials: A review. Journal of Hazardous Materials. 338: 102–123.

Zhang, H., Dai, Z., Sui, Y., Xue, J. and Ding, D. 2018. Adsorption of U (VI) from aqueous solution by magnetic core–dual shell Fe_3O_4@ PDA@ TiO_2. Journal of Radioanalytical and Nuclear Chemistry. 317: 613–624.

Zhang, X. and Liu, Y. 2020. Nanomaterials for radioactive wastewater decontamination. Environmental Science: Nano. 7(4): 1008–1040.

Zhu, N., Luan, H., Yuan, S., Chen, J., Wu, X. and Wang, L. 2010. Effective dechlorination of HCB by nanoscale Cu/Fe particles. Journal of Hazardous Materials. 176(1–3): 1101–1105.

Chapter **14**

Agricultural Waste Management for Bacterial Cellulose Production

Susana Calderón-Toledo*[1], Amparo Iris Zavaleta[1] and Guillermo R. Castro[2]

[1]Laboratorio de Biología Molecular, Facultad de Farmacia y Bioquímica, Universidad Nacional Mayor de San Marcos, Lima, Perú.

[2]Nanomedicine Nucleus, Center for Natural and Human Sciences, Federal University of ABC, Sao Paulo, Brazil.

INTRODUCTION

The sustainable production of biopolymers is one of the areas that has gained more attention in recent years due to the high demand in daily and specialized use, so it is key to reduce production costs. Among these, cellulose is a homogeneous polymer, constituted by 9 000 to 10 000 β-D-glucopyranose units polymerized by hydroxyl groups. The most widely used cellulose today is from plant origin, which requires purification processes because it comprises natural lignocellulosic fibers with 30 to 100 individual molecules packed together depending on its biosynthesis (Ng et al. 2015). This can also have a bacterial origin, which stands out for its high purity and advantageous physicochemical characteristics compared to vegetable (Lahiri et al. 2021). Vegetable cellulose is the most abundant; however, its low sustainability has prompted the search for renewable sources.

* For Correspondence: susana.calderon2@unmsm.edu.pe

The physicochemical properties possessed by the bacterial source have given it great applicability in the biomedical, electronic engineering, electrical, textile, cosmetic and food industries (Ashrafi et al. 2019, Cazón and Vázquez 2021b, Guan and Guo 2022, Iqbal 2013, Luo et al. 2019a, Padhi et al. 2023, Phukon et al. 2024). Likewise, the availability of hydroxyl groups allows their functionalization for the addition of compounds by *in situ* or *ex situ* modifications (Cazón and Vázquez 2021b). Its production has limitations due to the high cost of its culture medium, which restricts its scaling up. For this purpose, strategies such as the cheapening of carbon and nitrogen sources, the use of transgenic cells and bioreactor designs have been applied.

On the other hand, it has been observed that in the last 50 years, agricultural production has increased more than three times, so it is necessary to provide approximately 24 million tons of food worldwide (Duque-Acevedo et al. 2020, FAO 2017). The processing of these products requires different stages in which solid wastes accumulate according to the type of product, production technique and utility. However, the use of these wastes contributes to their reintroduction into the production cycle, giving them an added value and generating a decrease in agricultural losses, since the current management of these wastes has minimal impact on the economic retribution to agricultural producers (Ogbu et al. 2023). In addition to contributing to the reduction of greenhouse gas emissions, there has been an increase in the number of research studies related to the production of bacterial cellulose (BC) using different types of agricultural residues (Adejumo et al. 2020).

PHYSICOCHEMICAL PROPERTIES

BC is composed of 7–8 nm nanofibrils that are gradually bundled into fibrils smaller than 100 nm (Gorgieva and Trček 2019). Its physicochemical properties are due to the uniformity in the arrangement of the nanofibrils, which will then be packed in a two-dimensional and parallel manner due to van de Waals forces (de Olyveira et al. 2011, Lahiri et al. 2021). These microfibers possess a fully crystalline core that is surrounded by less crystalline cellulose that will be arranged in 3D reticular structure interrupted by an amorphous arrangement (Rodriguez et al. 2022). These possess high purity (98–99%) as they do not possess lignin and hemicellulose, compared to vegetable cellulose (Zhang et al. 2011). Also, its high water retention capacity (more than 100 times its weight), high porosity and biocompatibility have allowed it to have high acceptance in tissue engineering. Water retention capacity is indirectly correlated with tensile strength and stiffness, which is due to the amorphous and crystalline cellulose composition. Crystalline cellulose is impermeable to hydration and its adsorption is only superficial, while amorphous can hydrate in all its domains (Bodea et al. 2021, Mazeau 2015).

BC possesses high flexibility, and high tensile strength (200–300 MPa), which are related to the crystalline structure of nano- and microfibrils (Ruka et al. 2015). Additionally, it possesses uniform porosity, high water vapor transmission rate, absence of toxicity and thermal stability (Naomi et al. 2020, Sriplai et al. 2018). It is highly resistant to oxygen permeability, which is altered by water swelling, breaking hydrogen bridges and causing expansion of amorphous regions and increased movement of microfibers (Cazón et al. 2020, Naomi et al. 2020). It exerts a protective effect against the UV-light region and its permeability to water vapor is directly correlated to the higher moisture content (Cazón et al. 2020). BC is easier to eliminate than the plant source due to the structure and composition of the latter, whereas BC can be discarded using alkaline solutions or in case it is found in an organism based on its hydrophilic capacity or with lysosomes (Naomi et al. 2020).

The BC has hydroxyl groups that allow its functionalization, whereas that of microbial origins, having a greater number than the vegetable one, increases its reactivity due to the greater availability of oxygen to form hydrogen bonds (Pogorelova et al. 2020). Its high reactivity for cyanoethylation and carboxymethylation reactions stands out due to its high crystallinity (Keshk and Nada 2003). This abundance of hydroxyl groups gives it hydrophilic properties that can be modeled by *in situ* or *ex situ* chemical modifications (Choi and Shin 2020). Based on the above, the physicochemical properties of cellulose can be altered according to the culture medium, fermentation conditions and the microorganism employed, in addition to post-culture chemical modifications, which gives it different applicability (Ashrafi et al. 2019, Tsouko et al. 2015).

Applications

Electrical and Electronic Applicability

BC has had great acceptability in this field due to its porosity, flexibility and strength, which are important for the design of nanocomposites, since a smaller size would give it greater strength and ductility, with tensile strength higher than carbon nanotubes. The high density in the hydrogen bridge arrangement contributes to energy release, which also improves strength (Guan and Guo 2022). Graphene sheets have been realized using BC, which allows synthesizing flexible and tough membranes, with low weight loss at values below 70°C and evidencing high electrical conductivity (218 MPa) (Luo et al. 2019a).

Cellulose has also contributed to give elasticity to conventional supercapacitors, which have a rigid structure. These have evidenced high specific capacitance (302 F/g at 0.5 A/g), which is due to the increased surface area and structural porosity of carbon-activated cellulose (Sezali et al. 2021). Lithium batteries have been developed based on the tensile strength, thermostability and modifiability of BC, which can be reinforced with polyethylene oxide, that provides flexibility and conductivity, and has led to the formation of batteries with high resistance to explosion, short circuit at high temperatures and a life of 1160 hours without polarization (Li et al. 2021). Also, genetic manipulation tools have been applied, as well as the modification of the fermentation process and its processing in order to obtain membranes that by binding with lithium obtain a retention of 80% of activity after 1000 cycles of use (Gwon et al. 2019).

Toxics Removal

In recent years, the applicability of BC for the removal of toxics has been related to its mesh structure, which allows the adsorption of these compounds, being functionalized with the entrapment of various chemicals that have improved its removal capacity. Such is the case of BC aerogels associated with aramid nanofibers, which possess hydrophilicity and negative surface charges, giving it adsorptive capacity of cationic dyes and therefore its application in wastewater treatment (Yi et al. 2021). Likewise, BC nanocrystals with keratin allow the removal of metals by adsorption such as cadmium and crystal violet, which is ideal for water treatment (Peiravi-Rivash et al. 2023). This dye trapping was also evidenced with biofilms with magnetite, under visible light in short periods of time (from 100 to 110 min) and the immobilization of silver and nickel oxide on its membrane allows high levels of adsorption of cationic and anionic dyes (~90%), with reusability for up to six consecutive times (Kamal et al. 2022, Maruthupandy et al. 2021).

It has also been applied in the design of nanocomposites based on cellulose films together with metal oxide to avoid the corrosive effect of steel against carbon (Kreydie and Al-Abdaly 2021). Based on its functionalization technique, it has been shown that *ex situ* modifications of cellulose with iron ions have better efficiency for the degradation of textile dyes, compared to *in situ* ones, since the cellulose produced by the *ex situ* method reflects a higher degree of porosity, which favors the storage of ions in its structure (Gomes Silva et al. 2023).

Biomedical Applications

The morphology of BC is important in the biomedical area, as its surface topography stimulates cell surface interactions, which would be desirable for its tissue integration (Aditya et al. 2022). The 3D mesh structure prevents microbial attack and magnetite entrapment forms a film with growth inhibitory activity of *Escherichia coli* and *Proteus mirabilis* similar to ceftazidime, which has earned its application for tissue engineering (Lahiri et al. 2021, Maruthupandy et al. 2021). Its nanostructures have been shown to be antibiotic loading matrices, which can be released in a controlled mode when associated with minerals (Meneguin et al. 2021).

Its modification capacity allows its functionalization, by which functional groups, nanoparticles and other polymers have been incorporated to obtain blood vessel grafts, bone grafts, artificial skin, dental implants, artificial corneas, biosensors, among others (Aditya et al. 2022). Due to its low endotoxin level (<20 EU per port), its biomedical applicability has been approved by the FDA, evidenced by the absence of epidermal irritation compared to other materials (Picheth et al. 2017). Hydrogels with silica nanoparticles encapsulating a pH indicator have been developed, which act as pH sensors in their application as skin grafts in order to monitor possible infections (Eskilson et al. 2023).

The adherence of cellulose to bone tissue promotes its regeneration and proliferation, as it simulates an extracellular matrix. This has promoted the addition of compounds such as hydroxyapatite, chitosan in oxidized BC, which induces bone formation and improves its mineralization capacity with good cytocompatibility (Cao et al. 2022). Functionalization with metal nanoparticles inhibits microbial colonization and biofilm formation in wounds due to alteration of porosity, conductivity, surface charge and bacterial cell membrane composition (Mallick et al. 2022). Immobilization of silver ions evidences antioxidant activity and accelerates treatment by enhancing fibroblast propagation and proliferation (Shaheen et al. 2022). Additionally, aggregation of chitosan to cellulose films with silver ions allows a sustained release of the ions, which has inhibitory effect on *E. coli* and *Staphylococcus aureus*, and aggregation of graphene oxide and silver nanoparticles complements the inhibitory effect on Gram negative and positive growth, as evidenced for *E. coli*, *S. enterica*, *B. cereus* and *S. aureus* (Barjasteh et al. 2023, Luo et al. 2019a, Mariadoss et al. 2023).

Food Applications

BC has interesting applications in the food industry, either as an additive or in food packaging. Its properties such as aqueous absorption, water vapor permeability, visible and UV light barrier of cellulose have been combined with the presence of plasticizing agents, which increases elasticity and resistance to rupture, and decreases tension, which is an advantageous alternative in food packaging (Cazón and Vázquez 2021). The addition of polyvinyl alcohol can be applied to protect food from microbial activity for more than 10 days, with the ability to degrade in soil after this time (Naik et al. 2022). It can also be reinforced with chitosan to obtain smooth fibers of low diameter that increase the surface area, to be then functionalized with prodigiosin with which a synergistic antimicrobial effect is exerted and an attractive coloration is given to the consumer (Amorim et al. 2022). The immobilization of silver ions, D-pentaacetate and agarose has been applied for the identification of biogenic amines, which would identify possible microbial contamination during the storage process (Nath et al. 2023).

BC is a product generally recognized as safe (GRAS), a source of fiber that promotes intestinal transit. Its rheological, physical and textural characteristics are applicable as a replacement for fats linked to food preparation, giving them volume, softness and palatability (Azeredo et al. 2019). Their consumption also decreases serum cholesterol levels by adsorbing them at gastrointestinal tract level, thus preventing liver damage or atherosclerosis (Zhang et al. 2022). Additionally, its highly porous matrix can serve to immobilize beneficial lactic acid bacteria, which allows encapsulation of probiotics and their protection against gastric juices and bile salts (Fijałkowski et al. 2016).

BACTERIAL SOURCES

BC can be produced in static or agitated form, being in the former its presence evident at the liquid-medium interface due to its low density, by which it allows colonization in the interface (Spiers et al. 2003). Among the different bacterial genera described are *Komagataeibacter, Agrobacterium, Aerobacter, Rhizobium, Sarcina, Pseudomonas, Alcaligenes, Rhizobium, Azotobacter*, among others (Augimeri et al. 2015, Römling and Galperin 2015).

Acetic acid bacteria are strict Gram-negative bacteria belonging to the α-Proteobacteria, among which *Komagataeibacter xylinum, K. medellinensis, K. nataicola, K. oboediens, K. rhaeticus, K. maltaceti, K. saccharivorans, K. hansenii* and *K. pomaceti* stand out for their high levels of cellulose production (Li et al. 2019, Škraban et al. 2018, Thongwai et al. 2022, Vigentini et al. 2019, Zhang et al. 2017). Among these, *K. xylinum* has been extensively studied due to its high production from mainly glucose, which has been complemented by the publication of genomic sequences linked to cellulose production (Lahiri et al. 2021, Römling and Galperin 2015). Comparison of genomes of acetic acid bacteria has determined that they differ in operon composition, which generates an effect on synthesis, extracellular transport and assembly of nanofibers (Gorgieva and Trček 2019).

Cellulose synthesis starts from hexoses, pyruvate, dihydroxyacetone and glycerol (Benziman and Burger-Rachamimov 1962, Bricotte et al. 2023, Lahiri et al. 2021). If the substrate is glucose, the mechanism of synthesis in *K. xylinus* is based on the isomerization of glucose 6-phosphate to glucose 1-phosphate due to phosphoglucomutase, after which UDPG-pyrophosphorylase promotes the synthesis of UDP-glucose, and subsequently cellulose synthase, composed of four subunits (BcsA, BcsB, BcsC and BcsD), performs the polymerization of cellulose from UDP-glucose, which will be released into the extracellular space (Lustri et al. 2015). The BcsC subunit is responsible for exporting cellulose, while others would have effect on physicochemical properties, such as the catalytic subunit BcsA, which determines the nanofibers' length and BcsD would be responsible for crystallinity (Sajadi et al. 2017, Jacek et al. 2019, Yang et al. 2019).

Bioprocessing

Cellulose production depends on the selection of the fermenting bacteria, the components of the medium and the culture conditions. Regarding bacteria, several highly BC-producing microbial sources have been isolated from acidic media, such as *Lactiplantibacillus plantarum*, isolated from rotting fruits, which would possess metabolic adaptability mechanisms to acidic organic substrates (Shoda and Sugano 2015, Saleh et al. 2022). Likewise, *Gluconacetobacter* sp. RV28, *Enterobacter* sp. RV11 and *Pseudomonas* sp. RV14 have been isolated from rotten sweet potato, rotten potato and rotten pomegranate, the first being the one with which the highest production has been obtained (4.70 g/L) (Rangaswamy et al. 2015). Also, recombinants have been developed, with the introduction of the *cmcax*, *ccp*Ax, *bcs*A, *bcs*B, *bcs*C and *bcs*D genes *of K. hansenii*, which facilitates cellulose production in less time compared to the native strain (Buldum et al. 2018).

The traditional culture medium is Hestrim-Schramm, which has a high cost due to the fact that its composition includes glucose, yeast extract, peptone, among others (Hestrin and Schramm 1954). Likewise, in order to improve their production, other carbon sources such as sucrose, fructose and mannitol have been employed to increase production (Ruka et al. 2012, Molina-Ramírez et al. 2017). Regarding the nitrogen source, peptone, casein, tryptone, meat extract and ammonium chloride have been used to improve production, depending on the metabolism of the strain used (Çoban and Biyik 2011, Chand Singh et al. 2016, Gopu and Govindan 2018, Bagewadi et al. 2020). The search for higher production levels has led to increasing quantities of these sources, thus making it expensive and uneconomical to produce them on a larger scale and limiting their applicability (Rani and Appaiah 2011).

Additionally, magnesium ions and vitamin C are described to have increased cellulose production (Zeng et al. 2011, Keshk 2014). The acidity level is an important factor, as acetic acid bacteria produce between pH 3.5 to 7.5, with best levels obtained between pH 4–5 (Zahan et al. 2015, Kuo et al. 2016). Acetic acid supplementation to a conventional medium increases BC production by 17%, with increased tensile strength and decreased elongation at break and crystallinity (Yassine et al. 2016, Agustin and Padmawijaya 2018). The presence of ethanol would also increase production in *Komagataeibacter* sp. PAP1, but high values (>15 g/L) would have opposite effect on *Komagataeibacter xylinus* subsp. *sucrofermentans* BPR3OOlA (Naritomi et al. 1998, Suwanposri et al. 2014). In addition, ethanol would have effect on the properties of BC as it increases its swelling index, moisture content and elasticity due to the random assembly of fibrils, which gives it higher porosity (Agustin and Padmawijaya 2018). This balance between production and applicability of BC has driven the search for alternative sources, among which agricultural wastes possess potential applicability because they comprise high amounts of carbohydrates, proteins, phenolic compounds, among others (Ravindran et al. 2018).

Among the most influential culture conditions are temperature, incubation time, inoculum size and culture method, either by agitation or static (Rani and Appaiah 2011, Zeng et al. 2011, Aswini et al. 2020). Bacteria of the genus *Komagataeibacter* produce higher levels of cellulose at longer incubation times; however, at times longer than 15 days, productivity slowly decreases because as cellulose is produced, it is disposed at the liquid-air interface, resulting in decreased cellular oxygen availability (Gomes et al. 2018, Photphisutthiphong and Vatanyoopaisarn 2020). In relation to temperature, this genus presents better production between 27 and 30°C, and minimum production exceeding 35°C because glucose would be converted to gluconic acid via direct oxidation instead of being converted to cellulose, which was contrary to what was reported for *Acetobacter senegalensis* MA1, which obtained a maximum production at 37°C, so it would have metabolic adaptability mechanisms for biosynthesis at higher temperatures (Rani and Appaiah 2011, Zahan et al. 2015, Aswini et al. 2020).

Inoculum size exerts effect on production according to the cellular metabolic system, since for *Komagataeibacter* sp. RV28 the maximum production is obtained with a cell density of 5% (v/v), contrary to what was obtained for *K. xylinus* LKN6, which produced up to 13.85 g/L of cellulose with an inoculum of 25% (v/v), after which the production level decreased (Rangaswamy et al. 2015, Yanti et al. 2018).

However, values higher than 20% would lead to competition for nutrients and thus interference with cellulose production (Photphisutthiphong and Vatanyoopaisarn 2020). Also, the stirred culture mode can be used, which can be characterized at the bioreactor level by the outgassing method. The bioreactor configuration has importance on BC production and carbon source yield on BC, which will be k_L a dependent, as mixing and oxygen transfer is increased. Likewise, cellulose production is mainly dependent on the rate of consumption of sugars and high-energy components derived from respiration (Kouda et al. 1997).

PRETREATMENT OF AGRICULTURAL RESIDUES

The utilization of agricultural residues generally involves a pretreatment, which can be a structural reduction or chemical treatment in order to expose the cellulosic material for subsequent hydrolysis. The composition of the residue will influence the type of treatment applied in order to improve the rate of sugars released in the subsequent hydrolysis, as well as to decrease the loss of sugars in the pretreatment and reduce costs (El-Gendi et al. 2022). The lignocellulosic material, of which most agricultural residues are composed, comprises cellulose (30–35%, w/w), hemicellulose (20–35%, w/w) and lignin (15–20%, w/w), and its variation will depend on genetic variables, environmental influence, soil type and growing climate (Nashiruddin et al. 2020). Among the chemical methods, acid, alkaline or ozonation pretreatment stand out, as well as particle size reduction, which improves the surface area of contact with the subsequent treatment (Palacios et al. 2017).

Physical Pretreatment

These methods are widely applied in accompaniment of chemical pretreatment, which comprise vaporization, milling, irradiation, temperature and pressure application, leading to the increase of surface area and pore size. Grinding is one of the most widely applied, where size can be reduced to about 0.2 mm (Baruah et al. 2018). Microwave-induced particle size reduction has also been applied, which is based on molecular collisions caused by dielectric polarization, and generates thermal energy that breaks the lignocellulosic structure (Aguilar-Reynosa et al. 2017). Also, ultrasonic pretreatment allows structural breakdown and extraction of cellulose, hemicellulose and lignin through cavitation generated by ultrasonic radiation (Ravindran and Jaiswal 2016). These processes that require the application of energy can facilitate the breakdown of lignocellulosic material; however, they are not cost-effective to scale up due to their high cost.

Acid Pretreatment

Acid pretreatment usually involves the use of sulfuric acid, nitric acid or hydrochloric acid, by which hemicellulose is removed and cellulose is exposed for subsequent hydrolysis (Nashiruddin et al. 2020). Sulfuric acid pretreatment of fruit peels allows the release of a maximum content of 34.8% of cellulose with pectin decrease from 17.10 to 6.73%, which is higher in direct relation to the exposure time (John et al. 2017). Pretreatment of banana peel with sulfuric acid decreases the hemicellulosic content as temperature increases (Palacios et al. 2017). Likewise, with 4% nitric acid, an increase from 54% to 74.3% of cellulosic material and a decrease in lignin was also obtained (Skiba et al. 2020).

These processes require concentrated acids between 30 to 70% at low temperatures (<100°C) or dilute acids (0.1–10%) at high temperatures (100 –250°C). The latter treatment can generate intermediates such as furfurans, phenolic acids and aldehydes, which would have negative effects on enzymatic hydrolysis and subsequent fermentation (Zhao et al. 2007, Palacios et al. 2017, Sarkar et al. 2022). In addition, this method requires subsequent detoxification processes, use of anticorrosive equipment and application of energy for acid recovery (Nashiruddin et al. 2020).

Alkaline Pretreatment

Alkaline pretreatment comprises the removal of lignin by its solubilization in the alkaline solution without carbohydrate degradation, the most commonly used alkaline solutions being NaOH, KOH, NH_3 and $Ca(OH)_2$. Alkaline peroxide has been used prior to enzymatic hydrolysis for the removal of more than 98% of lignin to improve conversion to reducing sugars; however, this pretreatment removes 49% of hemicellulose and about 25% of cellulose, but increases glucose by about 210 times compared to untreated fiber (Palamae et al. 2017).

Fruit peels, which show low content of soluble sugars, are pretreated with NaOH 1%, having better results with the help of microwaves so that a higher release of reducing sugars (0.16 g/g) is evidenced at higher intensity (600 W) (Tiwari et al. 2017). The removal of acetyl groups and uronic acid substitutions in generated hemicelluloses increase the area available for enzymatic hydrolysis; however, a disadvantage of this method consists in the recovery of the added alkalis (Baruah et al. 2018).

Other Techniques

Other recently applied techniques are the use of ionic liquids, cationic and anionic solvents, which solubilize cellulose and lignin, but with the advantage of being recoverable, reusable and non-toxic. Among the most used are those based on imidazolium ($[(C_3N_2)X_n]^+$), pyridinium ($[(C_5N)X_n]^+$), pyrrolidinium ($[C_4N)X_n]^+$, ammonium $[NX_4]^+$, among others, where the first group is the most used (Baruah et al. 2018). An example is the pretreatment of wheat straw with the ionic liquid 1-allyl-3- methylimidazolium chloride at 100°C for 2 h prior to saccharification, resulting in a lignocellulosic material yield on sugars of 32.5% (Chen et al. 2013).

Also, a process of delignification and solubilization of hemicellulose has been developed that leads to the increase of pore volume and surface area by pretreatment with organic solvents or aqueous solutions, such as ethanol, methanol, acetone, among others. This process has the addition of catalysts such as mineral acids that enhance the delignification process (Zhang et al. 2016, Baruah et al. 2018).

WASTE HYDROLYSIS

This step allows the release of sugars from the pretreated waste, which is done by acid or enzymatic hydrolysis.

Acid Hydrolysis

Acid hydrolysis of vegetable fiber allows the release of pentoses from hemicellulose and hexoses from cellulose, but with the presence of co-products that can have inhibitory effect on the growth of microorganisms (Ho et al. 2014). To reduce the formation of these toxic compounds, additional detoxification processes are required (El-Gendi et al. 2022). Among the most commonly used acids is sulfuric acid, which was applied for the release of glucose, xylose and arabinose from elephant grass for BC production (Yang et al. 2013).

Enzymatic Hydrolysis

Enzymatic hydrolysis can be carried out with endoglucanases, exoglucanases, xylanases, glucosidases, galactosidases, pectinases, among others, isolated from bacteria or fungi, which perform a hydrolytic and oxidative role (Ballesteros 2010, Phukon et al. 2020, Kumari et al. 2021). Cellulases have been applied at 50°C for 24 h for the treatment of wheat straw, improving 3.3 times the waste yield on sugars compared to the untreated fraction (Chen et al. 2013). Likewise, sugar cane bagasse treated with cellulases for 48 h allowed a cellulose to glucose conversion of 81% (Dhar et al. 2019). Due to the robust nature of the lignocellulosic material, high enzyme concentrations would be needed to enhance the conversion of glucans to glucose, but this increase would be due to cellulases binding to lignin, causing non-productive binding (Luo et al. 2019b). In addition, the specific rate of cellulose hydrolysis decreases with increasing enzyme concentration (Ballesteros 2010). In general, the treatment of these residues with enzymes has advantages due to their absence of toxicity and loss of sugars; however, some sugars released can inhibit the catalytic process, so their removal is required, which is not economically profitable (El-Gendi et al. 2022, Stickel et al. 2014).

PRODUCTION USING AGRICULTURAL RESIDUES

Agricultural residues are produced at each stage of the processing chain, so they can be classified according to their origin and type of processing. Table 14.1 summarizes the production of BC using different agricultural wastes as alternative substrates.

Table 14.1 Alternative substrates and conditions used for bacteria cellulose production

Alternative substrate	Bacteria	Culture conditions	BC (g/L)	Reference
Pineapple juice	*K. xylinus* TISTR 893	30°C, 14 day, static	576.66*	Kongruang 2008
Wheat straw	*K. xylinus* ATCC 700178	29°C, 7 day, 175 rpm	5.20	Dahman et al. 2010
Grape skins	*K. sacchari*	30°C, 4 day, static	0.63	Carreira et al. 2011
Pineapple skin/ sugar cane molasses	*K. swingsii* sp.	28°C, 13 day, static	2.80	Castro et al. 2011
Maple syrup	*K. xylinus* BPR 200	25°C, 3 day, 135 rpm	1.51	Zeng et al. 2011
Cotton based textiles	*K. xylinus* ATCC 23770	30°C, 14 day, static	10.80	Hong et al. 2012
Wheat straw	*K. xylinus* ATCC 23770	30°C, 7 day, static	8.30	Chen et al. 2013
Coffee cherry husk	*K. hansenii* UAC09	27°C, 14 day, static	8.20	Rani and Appaiah 2013
Elephant grass	*K. xylinus* CH001	28°C, 14 day, static	6.40	Yang et al. 2013
Rotten fruits	*K. xylinus* ATCC 53582	30°C, 4 day, static	60.20*	Jozala et al. 2014
Corncob acid	*K. xylinus* CH001	28°C, 14 day, static	4.10	Huang et al. 2014
Waste distillation	*K. xylinus* CH001	28°C, 7 day, static	1.34	Huang et al. 2015
Cashew juice	*K. hansenii* CCT 1431	30°C, 7 day, static	0.28	Duarte et al. 2015
Sisal juice	*K. hansenii* CCT 1431	30°C, 7 day, static	0.54	Duarte et al. 2015
Carob and haricot bean (1:2)	*K. xylinus* ATCC 700178	30°C, 9 day, static	7.46	Bilgi et al. 2016
Lipid fermentation	*K. xylinus* CH001	28°C, 5 day, static	0.66	Huang et al. 2016
Sisal juice	*K. hansenii* ATCC 23769	30°C, 5 day, static	2.60	Lima et al. 2017
Pawpaw juice	*Acetobacter pasteurianus* PW1	28°C, 15 day, static	7.70	Adebayo-Tayo et al. 2017
Oil Palm Shoot Juices and Coconut Medium (1:1)	*K. xylinus* TISTR 086	30°C, 10 day, static	4.97	Phruksaphithak et al. 2017
Cider waste and sugar cane (1.5:2.3)	*K. medellinensis* ID13488	28°C, 14 day, static	2.50	Urbina et al. 2017

(*Contd.*)

Alternative substrate	Bacteria	Culture conditions	BC (g/L)	Reference
Molasses	*A. pasteurianus* HBB6	30°C, 7 day, static	0.31	Biyik and Coban 2017
Corn steep liquor	*K. hansenii* UCP1619	30°C, 10 day, static	5.80	Costa et al. 2017
Molasses sugar cane/corn steep liquor	*K. xylinus* C18	31°C, 7.2 day, static	4.43	Singh et al. 2017
Apple juice	*K. rhaeticus* P 1463	30°C, 14 day, static	9.49	Semjonovs et al. 2017
Wheat thin stillage	*K. sucrofermentans* B-11267	28°C, 3 day, 250 rpm	6.19	Revin et al. 2018
Tomato juice	*Acetobacter pasteurianus* RSV-4	30°C, 7 day, static	7.80	Kumar et al. 2018
Cane molasses	*Acetobacter pasteurianus* RSV-4	30°C, 7 day, static	3.60	Kumar et al. 2018
Orange pulp	*Acetobacter pasteurianus* RSV-4	30°C, 7 day, static	2.80	Kumar et al. 2018
Rotten banana	*K. medellinensis* NBRC 3288	28°C, 10 day, static	4.81	Molina-Ramirez et al. 2018
Rotten mango	*K. medellinensis* NBRC 3288	28°C, 10 day, static	1.95	Molina-Ramirez et al. 2018
Cheese whey	*K. medellinensis* NBRC 3288	28°C, 10 day, static	2.37	Molina-Ramirez et al. 2018
Polysaccharide fermentation	*K. xylinus* BC-11	30°C, 7 day, static	1.18	Zhao et al. 2018
Grapefruit juice	*K. sucrofermentans* DSM 15973	30°C, 13 day, static	6,70	Andritsou et al. 2018
Rice washing drainage	*K. nataicola* Li1	25°C, 13 day, static	0.20	Sudying et al. 2019
Coconut water	*K. xylinus* 0416	27°C, 5 day, 150 rpm	1.13	Esa et al. 2019
Potato peel	*K. xylinus* ATCC 10245	35°C, 6 day, static	4.70	Abdelraof et al. 2019
Sugarcane molasses	*K. saccharivorans* MD1	28°C, 7 day, static	3.90	Abol-Fotouh et al. 2020
Banana peel	*K. nataicola* TISTR 2661	30°C, 9 day, static	0.89	Moukamnerd et al. 2020
Passion fruit peel	*K. nataicola* TISTR 2661	30°C, 9 day, static	0.31	Moukamnerd et al. 2020
Date syrup and cheese whey (1:1)	*K. xylinus* PTCC 1734	28°C, 10 day, static	18.80	Raiszadeh-Jahromi et al. 2020

Alternative substrate	Bacteria	Culture conditions	BC (g/L)	Reference
Mango pulp waste	*K. xylinus* DSMZ 2004	30°C, 16 day, static	6.32	García-Sanchez et al. 2020
Mango extract	*K. rhaeticus* SU12	30°C, 21 day, static	25.34	Calderón-Toledo et al. 2021
Grape pomace	*K. medellinensis* ID13488	28°C, 7 day, 130 rpm	0.91	Diaz-Ramirez et al. 2021
Sugarcane bagasse	*K. xylinus* ATCC 23769	28°C, 9 day, static	1.88	Lin et al. 2022
Lotus rhizome	*A. pasteurianus* MGC-N8819	30°C, 14 day, static	2.10	Nie et al. 2022
Sugarcane bagasse	*Komagataeibacter* sp. CCUG73629	30°C, 10 day, 100 rpm	1.20	Akintunde et al. 2022
Corncob	*Komagataeibacter* sp. CCUG73629	30°C, 10 day, 100 rpm	1.60	Akintunde et al. 2022
Mango peel	*Achromobacter* S3	28°C, 7 day, static	1.22	Hasanin et al. 2023

*wet cellulose

Waste from Vegetable and Animal Processing

Residues derived from the processing of various plant products have been mostly applied, such as the fermenting waste of acid corn hydrolysate (pH 6), which was used as the sole source of carbon and nitrogen (without the addition of trace elements) by *K. xylinus*, producing 0.66 g/L BC at 5 days, after which it decreased due to the deficiency of carbon source in the medium (Huang 2016). Corn cooking liquor was used as a nitrogen source, which produced 5.80 g/L BC of stability at higher temperatures compared to HS medium and similar crystallinity index (Costa et al. 2017). This has been used as a nitrogen source (20 mL/L), adding wheat straw, corn fibers or distillers dried grains as carbon sources (40 g sugars/L), wheat straw being the medium that produced 5.20 g/L BC (Dahman et al. 2010). Likewise, the hydrolyzate of beer yeast waste pretreated with ultrasonication allowed the production of 7.02 g/L of cellulose using *K. hansenii* CGMCC 3917 with an alteration in the arrangement of nanofibers, which is evidenced in the irregularity of the porosity and increases its water holding capacity (Lin et al. 2014).

K. sucrofermentans B-11267 produced up to 6.19 g/L of cellulose using thin wheat stillage (pH 3.95) as the sole carbon and nitrogen source, which was three times compared to that produced on HS medium (2.14 g/L) by stirred fermentation (Revin et al. 2018). Likewise, date syrup and cheese whey have been combined in similar amounts as carbon sources, but require yeast extract and peptone as a factor for bacterial growth and cellulose production (Raiszadeh-Jahromi et al. 2020). Cheese whey contains about 23 g/L of glucose, besides being a source of K^+, Mg^{2+} and Ca^{2+}, which was employed to produce 2.37 g/L of BC. This cellulose evidenced decreased crystallinity probably due to the presence of other polysaccharides, proteins and lipids, which would hinder microfiber packing, as well as maximum degradation temperature similar to HS medium due to the presence of hydrogen bridges in the matrix (Molina-Ramírez et al. 2018).

The refining of raw sugar at the industrial level generates as a by-product; a viscous liquid known as cane molasses, which due to its high composition of reducing sugars, organic acids and minerals, has been applied as an alternative substrate in cellulose production (Palmonari et al. 2020). This is able to produce cellulose (0.31 g/L), not very distant to that produced with a glucose and yeast extract medium (0.45 g/L) using *A. pasteurianus* HBB6 (Biyik and Coban 2017). The application of 20 g/L of sugar cane molasses, composed of 17.2% reducing sugars and 61.3% total sugars, as the sole carbon source produces 6.34 g/L of cellulose using *K. hansenii*, while adding 5 g/L of glucose increases this to 19.16 g/L, which would evidence the preference of this bacteria for reducing sugars (de Costa et al. 2018). The 10% sugar cane molasses would generate the maximum BC production (3.4 g/L) as the sole carbon source using *K. xylinus* C18, as amounts higher than 13% would lead to co-production of gluconic acid or lactic acid and consequently lower BC production (Singh et al. 2017). However, the use of this waste as a carbon source would lead to a rough surface morphology, but would have no effect on fiber diameter, where a matrix with low water release rate is generated (Abol-Fotouh et al. 2020).

Cane bagasse is produced by milling sugar cane and up to 2011 represented about 279 million metric tons of waste per year (Chandel et al. 2012). This waste has a complex composition based on cellulose, hemicellulose and lignin and therefore requires a process of aqueous extraction and acid hydrolysis, which releases glucose, xylose and fructose, obtaining 1.88 g/L BC without alteration in the morphology of the nanofibers, but different arrangement of the microfibrils due to the incorporation of impurities to the medium. Enzymatic hydrolysates of this residue (14.50 g/L of hexoses and 7.80 g/L of pentoses) and corn discards (20.60 g/L of hexoses and 12.60 g/L of pentoses) were applied as sole carbon source, by which 1.70 and 2.60 g/L were obtained, respectively, with better stability at elevated temperatures compared to HS medium, but lower elasticity and tensile strength (Akintunde et al. 2022).

The waste generated by oil palm logging has generated a search to reuse its sugars by hydrolysis with α-amylase. Its combination with coconut juice (1:1) produced up to 4.94 g/L of cellulose, but with decreased molecular weight and internal porosity (Phruksaphithak et al. 2017). On the other hand, corn oil residues contain high rates of reducing sugars (50 g/L), which diluted to 50% (v/v) as the sole carbon source, produced up to 4 g/L of high crystallinity cellulose (95.2%) (Huang et al. 2014). Likewise, corn steeping liquor residues have a buffering effect on the pH of the medium and has induced maximum production with 12.4% of this residue, while with higher amounts, decreased production is observed when applied in conjunction with sugarcane molasses due to a possible interaction between the substrates (Singh et al. 2017).

Soy whey, a by-product of a tofu company, as a source of carbon and nitrogen would have an accelerating effect on cell growth and BC production, since it stimulated the production of 4.14 g/L of BC, 3.6 times higher than that obtained with HS medium (1.15 g/L). This cellulose presents nanofibers in the form of thicker ribbons than those produced in the conventional medium, with a reticulate structure closely intertwined, and elongation at break 3.4 times lower than that produced using HS medium (Suwanposri et al. 2014). One of the most abundant wastes during rice processing is the rice wash water, which was used as a carbon source for *K. nataicola* Li1 as it contains 6.12 g/L of total sugars. Through this waste, 0.2 g/L of cellulose was obtained, similar to a sucrose medium, but with a higher degree of purity (Sudying et al. 2019).

Grape skins are an important source of sugars (46 g/L) with which 0.63 g/L of BC was obtained, considerably lower than that obtained with glucose and galactose 20 g/L, but with similar structural quality, so these agroindustrial residues could present inhibitors of cell growth and BC production (Carreira et al. 2011). In this aspect, the presence of polyphenolic compounds could have an inhibitory effect on BC production (Diaz-Ramirez et al. 2021). These wastes were used as a carbon source for the production of superabsorbent BC spheres by agitated culture, applied for fertilizer retention, which was possible due to their spherical shape and high porosity. Also, rice noodle washing wastes have been used, containing 11.11% carbohydrate and 0.1% protein, which produced cellulose with higher tensile strength and elongation at break than that obtained with HS medium, which is due to the close packing of the microfibers (Sutthiphatkul et al. 2020).

Crop Residues

Residues linked to crop production are some of the most used by acetic acid bacteria for the production of BC due to their composition and acidity level. Such is the case of *K. xylinus* ATCC 53582, which used rotten fruits, including plums, green grapes, pineapples and apples, for the production of 6 g/L of cellulose, higher than that obtained with conventional HS medium and with a similar compact nanostructure (Jozala et al. 2014). Also, pineapple waste has been used as a base medium, adding other carbon and nitrogen sources with which up to 176.47 g of hydrogel was produced by agitated fermentation (Ch'ng et al. 2020).

Andritsou et al. (2018) identified that grape and orange fruit juices had better effect on production than their peels using *K. sucrofermentans* DSM 15973, due to their high sucrose, fructose and glucose ratios with which it obtained 6.70 g/L of cellulose. Likewise, pineapple juice is a rich source of glucose, fructose and sucrose, in addition to organic acids and vitamins, which are consumed by *K. xylinus* 893 at a higher rate in the first 6 days of fermentation, after which it decreases (Ivanova et al. 2019, Kongruang 2008). Pineapple processing generates residues such as peels, which together with cane molasses have been used for the production of BC (2.80 g/L), showing evidence of ribbon-like structures and alteration in the assembly of microfibrils compared to the conventional medium, which is due to the presence of polysaccharides in the pineapple skin (Castro et al. 2011). Thus, pineapple, melon and papaya juices have been used as a carbon source by acetic acid bacteria, the latter two being the best substrates. A greater decrease in pH was evidenced with pineapple juice during cultivation compared to the other substrates, due to the production of gluconic acid, which would have an inhibitory effect on cellulose production (Adebayo-Tayo et al. 2017).

Mango pulp contains between 1.69 to 1.89 g/L reducing sugars, which varies according to fruit type, degree of ripening, and growing microclimates (Calderón-Toledo et al. 2021, Molina-Ramírez et al. 2018). Pulp waste has been evaluated as an alternative carbon source for cellulose production (6.32 g/L) by *K. xylinus* DSMZ2004; however, *K. rhaeticus* SU12 would have high affinity for this substrate due to the high levels of cellulose produced (25.24 g/L). In both cases, decreased crystallinity and increased stability at high temperatures have been observed due to the mango pectin added to the nanofibers (Calderón-Toledo et al. 2021, García-Sánchez et al. 2020). Likewise, hydrolyzed mango peels as the sole nutrient source induced the production of 1.22 g/L BC using *Achromobacter* S3, which obtained higher crystallinity compared to HS medium (Hasanin et al. 2023).

Jasmine rice hydrolysate produced up to 401 g/L of cellulose hydrogel using *K. nataicola* TISTR 975 due to the released glucose, whereby these sugars would act as energy source and precursors of the polymeric composition (Photphisutthiphong and Vatanyoopaisarn 2020). Low-solids effluents from potatoes have been highlighted as a better source of BC production compared to cheese whey permeate or sugar refining from beet, due to their composition rich in sugars (0.4% glucose and 15% starch) (Thompson and Hamilton 2001). Likewise, discarded potato peels release 12 to 20 g/L of total sugars, through acid hydrolysis. These produce 4.70 g/L BC, with fewer amorphous components within their structure, leading to an increase in surface area and a decrease in pore volume compared to that produced in HS medium (Abdelraof et al. 2019).

The processing of sisal leaf fibers (*Agave sisalana*) generates as waste a juice or liquid, which has 15 g/L of sugars, including glucose, sucrose and fructose, and 0.62 g/L of nitrogen. This medium supplemented with 7.50 g/L of yeast extract induced 2.60 g/L of BC at 5 days of incubation, three times higher than that obtained in HS medium (0.90 g/L), with similar three-dimensional structure (Lima et al. 2017). Cashew juice and sisal liquid were also used as the sole source of carbon and nitrogen for the production of biocomposites associated with hydroxyapatite, which showed higher deposition of apatite and the latter substrate, lower degree of desorption compared to that of the standard medium (Duarte et al. 2015).

Matured coconut water was used as the sole source of carbon and nitrogen, without addition of salts, for the production of 1.55 g/L of cellulose at 5 days of incubation (Esa et al. 2019). Likewise, coconut residues, discarded from coconut milk processing, were applied as part of the medium with glucose and ammonium sulfate for the production of cellulose acetate, which evidenced higher elongation at break and tensile strength compared to that produced chemically. These membranes had the ability to selectively remove CO_2 from the CO_2/CH_4 mixture, which would represent a reduction in environmental pollution from its chemical production (Khamwichit et al. 2021).

Apple waste, comprising peels, seeds, core, stems and discarded apples, enriched with sugar cane in the ratio 1.5/2.3 were identified as substrate for the production of 2.50 g/L BC, superior to conventional HS medium (1.80 g/L), while the use of only apple waste produced 0.36 g/L BC. The crystallinity of the films using both substrates exceeded that of cellulose using each separately, but also evidenced thinner nanofibers that cross-linked with each other, which caused slight increase in their water holding capacity (Urbina et al. 2017).

Coffee fruit hulls are one of the major wastes in coffee processing, which after extraction are discharged for putrefaction. However, it has up to 7% of total sugars and its supplementation improves BC production despite its high levels of sugars, which would be converted into glucose and subsequently polymerized to cellulose. Likewise, polyphenols (0.8%) would act as stimulators of cellulose production in *K. hansenii* UAC09 by preventing the destruction of c-di-GMP from phosphodiesterases, despite having been previously described as growth inhibitors, so the metabolic system of the species and the concentration in the medium would be of particular importance (Rani and Appaiah 2013).

Banana and passion fruit peels dried and hydrolyzed with amyloglucosidase have also been used as the sole carbon source, obtaining 0.89 and 0.31 g/L of BC, respectively, with a higher water retention capacity compared to the conventional medium due to the larger space in the structure, which allows trapping more water molecules (Moukamnerd et al. 2020). Decomposing banana juice has about 36 g/L of glucose and produced 4.81 g/L of BC, being the only nutrient source of nutrients. In addition, it presented higher tensile strength and breaking strength, with decrease in crystallinity compared to HS medium due to the presence of other polymers from the medium (Molina-Ramirez et al. 2018).

Carob (*Ceratonia siliqua* L.) and white haricot bean (*Phaseolus vulgaris*) hydrolysates were used as a source of carbon and nitrogen, due to their high content of sugars (55%, w/v) and proteins (19.3%, w/v), in addition to their pH stabilizing effect. Thus, the cellulose produced evidenced higher crystallinity (78%) than the conventional medium (61%) due to the higher presence of hydroxyl groups, which allows the generation of intra- and intermolecular bonds (Bilgi et al. 2016). Additionally, the use of tomato juice produced 7.8 g/L BC, which presented an arrangement of fine cellulose ribbons (50–60 nm) of high purity (Kumar et al. 2019).

Horticultural Waste

The rhizomes of lotus (*Nelumbo nucifera*), a co-product of lotus cultivation widely spread in East and Southeast Asia, contain per 100 g: starch (9.25 g), crude protein (1.97 g), sugars (1.97 g) and fiber (0.8 g) (Chen et al. 2019). This substrate when hydrolyzed with amylases and glucoamylases generated the production of 2.1 g/L BC, lower than that obtained with HS (6.6 g/L), but exhibited higher fiber diameter and crystallinity index, stability at high temperatures and higher tensile strength compared to the conventional medium (Nie et al. 2022). Also, hydrolyzed elephant grass (*Pennisetum purpureum*), which is a rich source of xylose, allowed the production of 6.4 g/L BC, which presented higher crystallinity as the culture time increased, which would be related to the formation of hydrogen bridges oriented in the same plane (Yang et al. 2013).

Agricultural Industrial Wastes

The residues in this category do not have a food application, such as undyed cotton-based textile residues, which by chemical pretreatment and hydrolysis with cellulases release reducing sugars (81.6%, w/w) that were transformed into 10.8 g/L of BC. The increase in chemical concentrations for pretreatment is inversely proportional to production costs, so it requires complete chemical removal prior to fermentation (Hong et al. 2012).

CONCLUSIONS AND PERSPECTIVES

Bacterial cellulose is a biopolymer of great industrial value due to the physicochemical properties it possesses, which gives it applicability in the biomedical, food, environmental, electronic and electrical areas, among others. It can be produced by several bacteria, acetic acid bacteria being the most described due to their high production levels. However, the bioprocessing of this polymer requires the application of high-cost substrates, which has led to an increase in research related to the use of agricultural waste. The utilization of these wastes requires a characterization of their physicochemical composition, which mostly contain polymers such as cellulose, hemicellulose, starch, lignin, polyphenols and proteins. This allows the choice of pretreatment and treatment, in order to improve the lignocellulosic waste yield on sugars. Thus, different agricultural residues have been used for the production of BC, obtaining different levels of production with particular morphological characteristics, which have greater applicability depending on the industrial sector. Currently, some of these residues are used in the production of functional BC, which has had promising results of similar or superior quality compared to that produced using conventional medium. However, one of the biggest challenges is the complexity of these wastes, which makes it necessary to process them prior to production, generating additional cost and the release of co-products that can affect cell metabolism. These disadvantages limit scaling up and should be previously studied in order to improve their profitability.

ACKNOWLEDGEMENTS

This research was partially supported by the Universidad Nacional Mayor de San Marcos (project number A2004084) of Peru.

REFERENCES

Abdelraof, M., Hasanin, M.S. and El-Saied, H. 2019. Ecofriendly green conversion of potato peel wastes to high productivity bacterial cellulose. Carbohydrate Polymers. 211: 75–83.

Abol-Fotouh, D., Hassan, M.A., Shokry, H., Roig, A., Azab, M.S. and Kashyout, A.E.H.B. 2020. Bacterial nanocellulose from agro-industrial wastes: low-cost and enhanced production by *Komagataeibacter saccharivorans* MD1. Scientific Reports. 10(1): 1–14.

Adebayo-Tayo, B., Akintunde, M. and Sanusi, J. 2017. Effect of different fruit juice media on bacterial cellulose production by *Acinetobacter* sp. BAN1 and *Acetobacter pasteurianus* PW1. Journal of Advances in Biology and Biotechnology. 14(3): 1–9.

Aditya, T., Allain, J.P., Jaramillo, C. and Restrepo, A.M. 2022. Surface modification of bacterial cellulose for biomedical applications. International Journal of Molecular Sciences. 23(2).

Aguilar-Reynosa, A., Romaní, A., Rodríguez-Jasso, R. Ma., Aguilar, C.N., Garrote, G. and Ruiz, H.A. 2017. Microwave heating processing as an alternative pretreatment in second-generation biorefinery: An overview. Energy Conversion and Management. 136: 50–65.

Agustin, Y.E. and Padmawijaya, K.S. 2018. Effect of acetic acid and ethanol as additives on bacterial cellulose production by *Acetobacter xylinum*. IOP Conference Series. Earth and Environmental Science. 209(1): 1–5.

Akintunde, M.O., Adebayo-Tayo, B.C., Ishola, M.M., Zamani, A. and Horváth, I.S. 2022. Bacterial cellulose production from agricultural residues by two *Komagataeibacter* sp. strains. Bioengineered. 13(4): 10010–10025.

Amorim, L.F.A., Mouro, C., Riool, M. and Gouveia, I.C. 2022. antimicrobial food packaging based on prodigiosin-incorporated double-layered bacterial cellulose and chitosan composites. Polymers. 14(2).

Andritsou, V., De Melo, E.M., Tsouko, E., Ladakis, D., Maragkoudaki, S., Koutinas, A.A., et al. 2018. Synthesis and characterization of bacterial cellulose from citrus-based sustainable resources. ACS Omega. 3(8): 10365–10373.

Ashrafi, Z., Lucia, L. and Krause, W. 2019. Bioengineering tunable porosity in bacterial nanocellulose matrices. Soft Matter. 15(45): 9359–9367.

Aswini, K., Gopal, N.O. and Uthandi, S. 2020. Optimized culture conditions for bacterial cellulose production by *Acetobacter senegalensis* MA1. BMC Biotechnology. 20(1): 1–8.

Augimeri, R.V., Varley, A.J. and Strap, J.L. 2015. Establishing a role for bacterial cellulose in environmental interactions: Lessons learned from diverse biofilm-producing *Proteobacteria*. Frontiers in Microbiology. 6(11): 1410.

Azeredo, H.M.C., Barud, H., Farinas, C.S., Vasconcellos, V.M. and Claro, A.M. 2019. Bacterial cellulose as a raw material for food and food packaging applications. Frontiers in Sustainable Food Systems. 3: 26.

Bagewadi, Z.K., Bhavikatti, J.S., Muddapur, U.M., Yaraguppi, D.A. and MullaI, S.I. 2020. Statistical optimization and characterization of bacterial cellulose produced by isolated thermophilic *Bacillus licheniformis* strain ZBT2. Carbohydrate Research. 491: 107979.

Ballesteros, M. 2010. Enzymatic hydrolysis of lignocellulosic biomass. pp. 159–177. *In*: Waldron, K.W. (ed). Bioalcohol Production: Biochemical Conversion of lignocellulosic Biomass. CRC Press, Cornwall, UK.

Barjasteh, M., Dehnavi, S.M., Ahmadi Seyedkhani, S., Rahnamaee, S.Y. and Golizadeh, M. 2023. Improved biological activities of dual nanofibrous chitosan/bacterial cellulose wound dressing by a novel silver-based metal-organic framework. Surfaces and Interfaces. 36: 102631.

Baruah, J., Nath, B.K., Sharma, R., Kumar, S., Deka, R.C., Baruah, D.C., et al. 2018. Recent trends in the pretreatment of lignocellulosic biomass for value-added products. Frontiers in Energy Research. 6(12): 138.

Benziman, M. and Burger-Rachamimov, H. 1962. Synthesis of cellulose from pyruvate by succinate-grown cells of *Acetobacter xylinum*. Journal of Bacteriology. 84(4): 625–630.

Bilgi, E., Bayir, E., Sendemir-Urkmez, A. and Hames, E.E. 2016. Optimization of bacterial cellulose production by *Gluconacetobacter xylinus* using carob and haricot bean. International Journal of Biological Macromolecules. 90: 2–10.

Biyik, H. and Coban, E.P. 2017. Evaluation of different carbon, nitrogen sources and industrial wastes for bacterial cellulose production. European Journal of Biotechnology and Bioscience. 5(1): 74–80.

Bodea, I.M., Beteg, F.I., Pop, C.R., David, A.P., Dudescu, M.C., Vilău, C., et al. 2021. Optimization of moist and oven-dried bacterial cellulose production for functional properties. Polymers. 13(13): 2034.

Bricotte, L., Chougrani, K., Alard, V., Ladmiral, V. and Caillol, S. 2023. Dihydroxyacetone: A User Guide for a Challenging Bio-Based Synthon. Molecules. 28(6): 2724.

Buldum, G., Bismarck, A. and Mantalaris, A. 2018. Recombinant biosynthesis of bacterial cellulose in genetically modified *Escherichia coli*. Bioprocess and Biosystems Engineering. 41(2): 265–279.

Calderón-Toledo, S., Horue, M., Alvarez, V.A., Castro, G.R. and Zavaleta, A.I. 2021. Isolation and partial characterization of *Komagataeibacter* sp. SU12 and optimization of bacterial cellulose production using *Mangifera indica* extracts. Journal of Chemical Technology and Biotechnology, 06.

Cao, S., Li, Q., Zhang, S., Liu, K., Yang, Y. and Chen, J. 2022. Oxidized bacterial cellulose reinforced nanocomposite scaffolds for bone repair. Colloids and Surfaces B. Biointerfaces. 211.

Carreira, P., Mendes, J.A.S., Trovatti, E., Serafim, L.S., Freire, C.S.R., Silvestre, A.J.D., et al. 2011. Utilization of residues from agro-forest industries in the production of high-value bacterial cellulose. Bioresource Technology. 102: 7354–7360.

Castro, C., Zuluaga, R., Putaux, J.L., Caro, G., Mondragon, I. and Gañán, P. 2011. Structural characterization of bacterial cellulose produced by *Gluconacetobacter swingsii* sp. from Colombian agroindustrial wastes. Carbohydrate Polymers. 84: 96–102.

Cazón, P., Velazquez, G. and Vázquez, M. 2020. UV-protecting films based on bacterial cellulose, glycerol, and polyvinyl alcohol: effect of water activity on barrier, mechanical, and optical properties. Cellulose. 27(14): 8199–8213.

Cazón, P. and Vázquez, M. 2021. Bacterial cellulose as a biodegradable food packaging material: a review. Food Hydrocolloids. 113: 106530.

Ch'ng, C.H., Rahman, M.R.A., Muhamad, I.I., Páe, N. and Zaidel, D.N.A. 2020. Optimization of bacterial cellulose production from pineapple waste using different fermentation method. Chemical Engineering Transactions. 78: 559–564.

Chand Singh, O., Panesar, P.S. and Chopra, H.K. 2016. Production and optimization of bacterial cellulose with different carbon and nitrogen sources using *Gluconacetobacter xylinus* C18 strain. Asian Journal of Chemistry. 28(6): 1397–1398.

Chandel, A.K., da Silva, S.S., Carvalho, W. and Singh, O.V. 2012. Sugarcane bagasse and leaves: Foreseeable biomass of biofuel and bio-products. Journal of Chemical Technology and Biotechnology. 87(1): 11–20.

Chen, L., Hong, F., Yang, X. and Han, S. 2013. Biotransformation of wheat straw to bacterial cellulose and its mechanism. Bioresource Technology. 135: 464–468.

Chen, G., Zhu, M. and Guo, M. 2019. Research advances in traditional and modern use of Nelumbo nucifera: phytochemicals, health promoting activities and beyond. Critical Reviews in Food Science and Nutrition. 59: S189–S209.

Choi, S.M. and Shin, E.J. 2020. The nanofication and functionalization of bacterial cellulose and its applications. Nanomaterials. 10(3): 551.

Çoban, E.P. and Biyik, H. 2011. Effect of various carbon and nitrogen sources on cellulose synthesis by *Acetobacter lovaniensis* HBB5. African Journal of Biotechnology. 10(27): 5346–5354.

Costa, A.F.S., Almeida, F.C.G., Vinhas, G.M. and Sarubbo, L.A. 2017. Production of bacterial cellulose by *Gluconacetobacter hansenii* using corn steep liquor as nutrient sources. Frontiers in Microbiology. 8(10): 1973.

Dahman, Y., Jayasuriya, K.E. and Kalis, M. 2010. Potential of biocellulose nanofibers production from agricultural renewable resources: Preliminary study. Applied Biochemistry and Biotechnology. 162(6): 1647–1659.

de Costa, A.F.S., Do Nascimento, V.R., de Amorim, J.D.P., de Gomes, E.A.S., de Araújo, L. M. and Sarubbo, L.A. 2018. Residue from the production of sugar cane: an alternative nutrient used in biocellulose production by *Gluconacetobacter hansenii*. Chemical Engineering Transactions. 64: 7–12.

de Olyveira, G.M., Filho, L.X., Basmaji, P. and Costa, L.M.M. 2011. Bacterial nanocellulose for medicine regenerative. Journal of Nanotechnology in Engineering and Medicine. 2(3): 1–8.

Dhar, P., Pratto, B., Gonçalves Cruz, A.J. and Bankar, S. 2019. Valorization of sugarcane straw to produce highly conductive bacterial cellulose/graphene nanocomposite films through in situ fermentation: Kinetic analysis and property evaluation. Journal of Cleaner Production. 238: 117859.

Diaz-Ramirez, J., Urbina, L., Eceiza, A., Retegi, A. and Gabilondo, N. 2021. Superabsorbent bacterial cellulose spheres biosynthesized from winery by-products as natural carriers for fertilizers. International Journal of Biological Macromolecules. 191: 1212–1220.

Duarte, E.B., das Chagas, B.S., Andrade, F.K., Brígida, A.I.S., Borges, M.F., Muniz, C.R., et al. 2015. Production of hydroxyapatite–bacterial cellulose nanocomposites from agroindustrial wastes. Cellulose. 22(5): 3177–3187.

Duque-Acevedo, M., Belmonte-Ureña, L.J., Cortés-García, F.J. and Camacho-Ferre, F., 2020. Agricultural waste: Review of the evolution, approaches and perspectives on alternative uses. Global Ecology and Conservation. 22: e00902.

El-Gendi, H., Taha, T.H., Ray, J.B. and Saleh, A.K. 2022. Recent advances in bacterial cellulose: a low-cost effective production media, optimization strategies and applications. Cellulose. 29(14): 7495–7533.

Esa, F., Rahman, N.A., Kalil, M.S. and Tasirin, S.M. 2019. Fermentation optimization on bacterial cellulose production by *Acetobacter xylinum* 0416. International Journal of Engineering and Technology. 8(1.2): 262–265.

Eskilson, O., Zattarin, E., Berglund, L., Oksman, K., Hanna, K., Rakar, J., et al. 2023. Nano-cellulose composite wound dressings for real-time pH wound monitoring. Materials Today Bio. 19.

Fijałkowski, K., Peitler, D., Rakoczy, R. and Zywicka, A. 2016. Survival of probiotic lactic acid bacteria immobilized in different forms of bacterial cellulose in simulated gastric juices and bile salt solution. LWT - Food Science and Technology. 68: 322–328.

García-Sánchez, M.E., Robledo-Ortíz, J.R., Jiménez-Palomar, I., González-Reynoso, O. and González-García, Y. 2020. Production of bacterial cellulose by *Komagataeibacter xylinus* using mango waste as alternative culture medium. Revista Mexicana de Ingeniería Química. 19(1): 851–865.

Gomes Silva, M., da Silva Santana, I.L., Alves Henrique, M., da Rocha Santana, R.M., Vinhas, G.M. and Napoleão, D.C. 2023. Production and application of bacterial Cellulose/Fe nanocomposite for degradation of aqueous mixture of textile dye. Environmental Nanotechnology, Monitoring & Management. 19: 100770.

Gomes, R.J., Hernan, R.J. and Spinosa, A. 2018. Acetic acid bacteria in the food industry: systematics, characteristics and applications. Food Technology and Biotechnology. 0262: 0–2.

Gopu, G. and Govindan, S. 2018. Production of bacterial cellulose from *Komagataeibacter saccharivorans* strain BC1 isolated from rotten green grapes. Preparative Biochemistry and Biotechnology. 48(9): 842–852.

Gorgieva, S. and Trček, J. 2019. Bacterial cellulose: Production, modification, and perspectives in biomedical applications. Nanomaterials: Nanomaterials for Environmental and Energy Technologies. 9(10): 1352.

Guan, F. and Guo, C.F. 2022. Flexible, high-strength and porous nano-nano composites based on bacterial cellulose for wearable electronics: a review. Soft Science: Soft Materials for Emerging Technologies. 2(3): 16.

Gwon, H., Park, K., Chung, S.C., Kim, R.H., Kang, J.K., Ji, S.M., et al. 2019. A safe and sustainable bacterial cellulose nanofiber separator for lithium rechargeable batteries. Proceedings of the National Academy of Sciences of the United States of America. 116(39): 19288–19293.

Hasanin, M.S., Abdelraof, M., Hashem, A.H. and El-Saied, H. 2023. Sustainable bacterial cellulose production by *Achromobacter* using mango peel waste. Microbial Cell Factories: Microbial Biotechnology for Sustainable Development. 22(1).

Hestrin, S. and Schramm, M. 1954. Synthesis of cellulose by *Acetobacter xylinum*. Biochemical Journal. 58: 345–352.

Ho, A.L., Carvalheiro, F., Duarte, L.C., Roseiro, L.B., Charalampopoulos, D. and Rastall, R.A. 2014. Production and purification of xylooligosaccharides from oil palm empty fruit bunch fibre by a non-isothermal process. Bioresource Technology. 152: 526–529.

Hong, F., Guo, X., Zhang, S., Han, S., Yang, G. and Jönsson, L.J. 2012. Bacterial cellulose production from cotton-based waste textiles: Enzymatic saccharification enhanced by ionic liquid pretreatment. Bioresource Technology: Bioresource Utilization and Sustainable Production. 104: 503–508.

Huang, C., Yang, X.Y., Xiong, L., Guo, H.J., Luo, J., Wang, B., et al. 2014. Utilization of corncob acid hydrolysate for bacterial cellulose production by *Gluconacetobacter xylinus*. Applied Biochemistry and Biotechnology. 175(3): 1678–1688.

Huang, C., Yang, X.Y., Xiong, L., Guo, H.J., Luo, J., Wang, B., et al. 2015. Evaluating the possibility of using acetone-butanol-ethanol (ABE) fermentation wastewater for bacterial cellulose production by *Gluconacetobacter xylinus*. Letters in Applied Microbiology. 60(5): 491–496.

Huang, C., Guo, H.J., Xiong, L., Wang, B., Shi, S.L., Chen, X.F., et al. 2016. Using wastewater after lipid fermentation as a substrate for bacterial cellulose production by *Gluconacetobacter xylinus*. Carbohydrate Polymers. 136: 198–202.

Ivanova, N.N., Khomich, L.M., Perova, I.B. and Eller, K.I. 2019. Pineapple juice nutritional profile. *Voprosy Pitaniia*. 88(2): 73–82.

Jacek, P., Dourado, F., Gama, M. and Bielecki, S. 2019. Molecular aspects of bacterial nanocellulose biosynthesis. Microbial Biotechnology. 12(4): 633–649.

John, I., Yaragarla, P., Muthaiah, P., Ponnusamy, K. and Appusamy, A. 2017. Statistical optimization of acid-catalyzed steam pretreatment of citrus peel waste for bioethanol production. Resource-Efficient Technologies. 3(4): 429–433.

Jozala, A.F., Pértile, R.A.N., dos Santos, C.A., Santos-Ebinuma, V. de C., Seckler, M.M., Gama, F.M., et al. 2014. Bacterial cellulose production by *Gluconacetobacter xylinus* by employing alternative culture media. Applied Microbiology and Biotechnology. 99(3): 1181–1190.

Kamal, T., Khalil, A., Bakhsh, E.M., Khan, S.B., Chani, M.T.S. and Ul-Islam, M. 2022. Efficient fabrication, antibacterial, and catalytic performance of Ag-NiO loaded bacterial cellulose paper. International Journal of Biological Macromolecules. 206: 917–926.

Keshk, S.M. and Nada, A.M. 2003. Heterogeneous derivatization of bacterial and plant cellulose. Biosciences, Biotechnology Research Asia. 1(1): 39–42.

Keshk, S.M.A.S. 2014. Vitamin C enhances bacterial cellulose production in *Gluconacetobacter xylinus*. Carbohydrate Polymers. 99: 98–100.

Khamwichit, A., Wattanasit, S. and Dechapanya, W. 2021. Synthesis of bio-cellulose acetate membrane from coconut juice residues for carbon dioxide removal from biogas in membrane unit. Frontiers in Energy Research. 9: 670904.

Kongruang, S. 2008. Bacterial cellulose production by *Acetobacter xylinum* strains from agricultural waste products. Applied Biochemistry and Biotechnology. 148(1–3): 245–256.

Kouda, T., Yano, H. and Yoshinaga, F. 1997. Effect of agitator configuration on bacterial cellulose productivity in aerated and agitated culture. Journal of Fermentation and Bioengineering. 83(4): 371–376.

Kreydie, S.H. and Al-Abdaly, B.I. 2021. Synthesis, characterization, and evaluation of inhibition corrosion of bacterial cellulose/metal oxides nanocomposites. Eurasian Chemical Communications. 3(10): 706–714.

Kumar, V., Sharma, D.K., Bansal, V., Mehta, D., Sangwan, R.S. and Yadav, S.K. 2019. Efficient and economic process for the production of bacterial cellulose from an isolated strain of *Acetobacter pasteurianus* of RSV-4 bacterium. Bioresource Technology. 275: 430–433.

Kumari, M., Padhi, S., Sharmas S., Phukon, L.C., Singh, S.P. and Rai, A.K. 2021. Biotechnological potential of psychrophilic microorganisms as the source of cold-active enzymes in food processing applications. 3 Biotech. 11(11): 479.

Kuo, C.H., Chen, J.H., Liou, B.K. and Lee, C.K. 2016. Utilization of acetate buffer to improve bacterial cellulose production by *Gluconacetobacter xylinus*. Food Hydrocolloids. 53: 98–103.

Lahiri, D., Nag, M., Dutta, B., Dey, A., Sarkar, T., Pati, S., et al. 2021. Bacterial cellulose: Production, characterization, and application as an antimicrobial agent. International Journal of Molecular Sciences. 22(23): 12984.

Li, J., Chen, G., Zhang, R., Wu, H., Zeng, W. and Liang, Z. 2019. Production of high crystallinity type-I cellulose from *Komagataeibacter hansenii* JR-02 isolated from Kombucha tea. Biotechnology and Applied Biochemistry. 66(1): 108–118.

Li, Y., Sun, Z., Liu, D., Lu, S., Li, F., Gao, G., et al. 2021. Bacterial cellulose composite solid polymer electrolyte with high tensile strength and lithium dendrite inhibition for long life battery. Energy and Environmental Materials. 4(3): 434–443.

Lima, H.L.S., Nascimento, E.S., Andrade, F.K., Brígida, A.I.S., Borges, M.F., Cassales, A.R., et al. 2017. Bacterial cellulose production by *Komagataeibacter hansenii* ATCC 23769 using sisal juice—an agroindustry waste. Brazilian Journal of Chemical Engineering. 34(3): 671–680.

Lin, D., Lopez-Sanchez, P., Li, R. and Li, Z. 2014. Production of bacterial cellulose by *Gluconacetobacter hansenii* CGMCC 3917 using only waste beer yeast as a nutrient source. Bioresource Technology. 151: 113–119.

Lin, S., Huang, S., Ting, Y., Hsu, S., Sam, H. and Cheng, K. 2022. Evaluation of detoxified sugarcane bagasse hydrolysate by atmospheric cold plasma for bacterial cellulose production. International Journal of Biological Macromolecules. 204.

Luo, H., Xie, J., Xiong, L., Zhu, Y., Yang, Z. and Wan, Y. 2019a. Fabrication of flexible, ultra-strong, and highly conductive bacterial cellulose-based paper by engineering dispersion of graphene nanosheets. Composites Part B: Engineering. 162: 484–490.

Luo, X., Liu, J., Zheng, P., Li, M., Zhou, Y., Huang, L., et al. 2019b. Promoting enzymatic hydrolysis of lignocellulosic biomass by inexpensive soy protein. Biotechnology for Biofuels. 12(1).

Lustri, W.R., Barud, H.G., da S. Barud, H., Peres, M.F.S., Gutierrez, J., Tercjak, A., et al. 2015. Microbial cellulose—Biosynthesis mechanisms and medical applications. pp. 133–137. *In:* M. Poletto and H.L. Ornagui Junior (eds). Cellulose—Fundamental Aspects and Current Trends. InTech, Croatia.

Mallick, S., Nag, M., Lahiri, D., Pandit, S., Sarkar, T., Pati, S., et al. 2022. Engineered nanotechnology: an effective therapeutic platform for chronic cutaneous wound. Nanomaterials. 12(5): 778.

Mariadoss, A.V.A., Saravanakumar, K., Sathiyaseelan, A., Sivakumar, A.S., Zhang, X., Bin Choi, H., et al. 2023. Cellulose-graphene oxide nanocomposites encapsulated with green-synthesized silver nanoparticles as an effective antibacterial agent. Materials Today Communications. 35: 105652.

Maruthupandy, M., Riquelme, D., Rajivgandhi, G., Muneeswaran, T., Cho, W.S., Anand, M., et al. 2021. Dual-role of graphene/bacterial cellulose/magnetite nanocomposites as highly effective antibacterial agent and visible-light-driven photocatalyst. Journal of Environmental Chemical Engineering. 9(5): 106014.

Mazeau, K. 2015. The hygroscopic power of amorphous cellulose: a modeling study. Carbohydrate Polymers. 117: 585–591.

Meneguin, A., Pacheco, G., Silva, J.M., de Araujo, F.P., Silva-Filho, E.C., Bertolino, L.C., et al. 2021. Nanocellulose/palygorskite biocomposite membranes for controlled release of metronidazole. International Journal of Biological Macromolecules. 188: 689–695.

Molina-Ramírez, C., Castro, M., Osorio, M., Torres-Taborda, M., Gómez, B., Zuluaga, R., et al. 2017. Effect of different carbon sources on bacterial nanocellulose production and structure using the low pH-resistant strain *Komagataeibacter medellinensis*. Materials. 10(6).

Molina-Ramírez, C., Castro, C., Zuluaga, R. and Gañán, P. 2018. Physical characterization of bacterial cellulose produced by *Komagataeibacter medellinensis* using food supply chain waste and agricultural by-products as alternative low-cost feedstocks. Journal of Polymers and the Environment. 26(2): 830–837.

Moukamnerd, C., Ounmuang, K., Konboa, N. and Insomphun. C. 2020. Bacterial cellulose production by *Komagataeibacter nataicola* TISTR 2661 by agrowaste as a carbon source. Chiang Mai Journal of Science. 47(1): 16–27.

Naik, M.L., Sajjan, A.M., Ashwini, M., Achappa, S., Khan, T.M.Y., Banapurmath, N.R., et al. 2022. Nanobacterial cellulose production and its antibacterial activity in biodegradable poly(vinyl alcohol) membranes for food packaging applications. ACS Omega. 7(48): 43559–43573.

Naomi, R., Idrus, R.B.H. and Fauzi, M.B. 2020. Plant-vs. Bacterial-derived cellulose for wound healing: A review. International Journal of Environmental Research and Public Health. 17(18): 1–25.

Naritomi, T., Kouda, T., Yano, H. and Yoshinaga, F. 1998. Effect of ethanol on bacterial cellulose production in continuous culture from fructose. Journal of Fermentation and Bioengineering. 85(6): 598–603.

Nashiruddin, N.I., Manas, N.H.A., Rahman, R.A., Azelee, N.I.W., Dailin, D.J. and Shaarani, S.M. 2020. Pretreatment and enzymatic hydrolysis of lignocellulosic biomass for reducing sugar production. pp. 1–27. *In*: Z.A. Zakaria, C.N. Aguilar, R.D. Kusumaningtyas and P. Binod (eds). Valorisation of Agro-industrial Residues—Volume II: Non-Biological Approaches. Springer, Switzerland.

Nath, P.C., Debnath, S., Sharma, M., Sridhar, K., Nayak, P.K. and Inbaraj, B.S. 2023. Recent advances in cellulose-based hydrogels: food applications. Foods. 12(2): 350.

Ng, H.M., Sin, L.T., Tee, T.T., Bee, S.T., Hui, D., Low, C.Y., et al. 2015. Extraction of cellulose nanocrystals from plant sources for application as reinforcing agent in polymers. Composites Part B: Engineering. 75: 176–200.

Nie, W., Zheng, X., Feng, W., Liu, Y., Li, Y. and Liang, X. 2022. Characterization of bacterial cellulose produced by *Acetobacter pasteurianus* MGC-N8819 utilizing lotus rhizome. LWT — Food Science and Technology. 165: 113763.

Padhi, S., Sarkar, P., Abedin, M.M., Sirohi, R., Singh, S.P. and Rai, A.K. 2023. Microbial enzymes and metabolites in health care: an overview. pp. 3–23. *In*: Sirohi, R., Rai, A.K., de Souza Vandenberghe, L.P. and Parameswaran, B. (eds). Microbial Enzymes and Metabolites for Health and Well-Being, CRC Press.

Palacios, S., Ruiz, H.A., Ramos-Gonzalez, R., Martínez, J., Segura, E., Aguilar, M., et al. 2017. Comparison of physicochemical pretreatments of banana peels for bioethanol production. Food Science and Biotechnology. 26(4): 993–1001.

Palamae, S., Dechatiwongse, P., Choorit, W., Chisti, Y. and Prasertsan, P. 2017. Cellulose and hemicellulose recovery from oil palm empty fruit bunch (EFB) fibers and production of sugars from the fibers. Carbohydrate Polymers. 155: 491–497.

Palmonari, A., Cavallini, D., Sniffen, C.J., Fernandes, L., Holder, P., Fagioli, L., et al. 2020. Short communication: Characterization of molasses chemical composition. Journal of Dairy Science. 103(7): 6244–6249.

Peiravi-Rivash, O., Mashreghi, M., Baigenzhenov, O. and Hosseini-Bandegharaei, A. 2023. Producing bacterial nano-cellulose and keratin from wastes to synthesize keratin/cellulose nanobiocomposite for removal of dyes and heavy metal ions from waters and wastewaters. Colloids and Surfaces A: Physicochemical and Engineering Aspects. 656(Part A): 130355.

Photphisutthiphong, Y. and Vatanyoopaisarn, S. 2020. The production of bacterial cellulose from organic low-grade rice. Current Research in Nutrition and Food Science. 8(1): 206–216.

Phruksaphithak, N., Kaewnun, C. and O-Thong, S. 2017. Productions and properties of bacterial cellulose from oil palm shoot juices felled medium and coconut medium. Key Engineering Materials. 728: 271–276.

Phukon, L.C., Abedin, M.M., Chourasia, R., Sahoo, D., Parameswaran, B. and Rai, A.K. 2020. Production of xylanase under submerged fermentation from *Bacillus firmus* HS11 isolated from Sikkim Himalayan region. Indian Journal of Experimental Biology (IJEB). 58(08): 563–570.

Phukon, L.C., Chourasia, R., Abedin, M.M., Sharma, S., Singh, S.P. and Rai, A.K. 2024. Exopolysaccharides from extremophiles and their potential applications. pp. 139–163. *In*: Bhatia, S.K. Panesar, P.S. and Mehariya, S. (eds). Microbial Exopolysaccharides. CRC Press.

Picheth, G.F., Pirich, C.L., Sierakowski, M.R., Woehl, M.A., Sakakibara, C.N., de Souza, C.F., et al. 2017. Bacterial cellulose in biomedical applications: a review. International Journal of Biological Macromolecules. 104: 97–106.

Pogorelova, N., Rogachev, E., Digel, I., Chernigova, S. and Nardin, D. 2020. Bacterial cellulose nanocomposites: Morphology and mechanical properties. Materials. 13(12): 1–16.

Raiszadeh-Jahromi, Y., Rezazadeh-Bari, M., Almasi, H. and Amiri, S. 2020. Optimization of bacterial cellulose production by *Komagataeibacter xylinus* PTCC 1734 in a low-cost medium using optimal combined design. Journal of Food Science and Technology. 57(7): 2524–2533.

Rangaswamy, B.E., Vanitha, K.P. and Hungund, B.S. 2015. Microbial cellulose production from bacteria isolated from rotten fruit. International Journal of Polymer Science. 2015: 1–8.

Rani, M.U. and Appaiah, A. 2011. Optimization of culture conditions for bacterial cellulose production from *Gluconacetobacter hansenii* UAC09. Annals of Microbiology. 61(4): 781–787.

Rani, M.U. and Appaiah, K.A.A. 2013. Production of bacterial cellulose by *Gluconacetobacter hansenii* UAC09 using coffee cherry husk. Journal of Food Science and Technology. 50(4): 755–762.

Ravindran, R. and Jaiswal, A.K. 2016. A comprehensive review on pre-treatment strategy for lignocellulosic food industry waste: challenges and opportunities. Bioresource Technology. 199: 92–102.

Ravindran, R., Hassan, S.S., Williams, G.A. and Jaiswal, A.K. 2018. A review on bioconversion of agro-industrial wastes to industrially important enzymes. Bioengineering. 5(4): 93.

Revin, V., Liyaskina, E., Nazarkina, M., Bogatyreva, A. and Shchankin, M. 2018. Cost-effective production of bacterial cellulose using acidic food industry by-products. Brazilian Journal of Microbiology. 49: 151–159.

Rodriguez da Silva, I.G., Dos Santos Pantoja, B.T., Rodriguez Almeida, G.H.D., Oliveira Carreira, A.C. and Miglino, M.A. 2022. Bacterial cellulose and ECM hydrogels: an innovative approach for cardiovascular regenerative medicine. International Journal of Molecular Sciences. 23(7).

Römling, U. and Galperin, M.Y. 2015. Bacterial cellulose biosynthesis: diversity of operons, subunits, products, and functions. Trends in Microbiology. 23(9): 545–557.

Ruka, D.R., Simon, G.P. and Dean, K.M. 2012. Altering the growth conditions of *Gluconacetobacter xylinus* to maximize the yield of bacterial cellulose. Carbohydrate Polymers. 89(2): 613–622.

Ruka, D.R., Simon, G.P. and Dean, K.M. 2015. Bacterial cellulose and its use in renewable composites. pp. 89–130. *In*: Thakur, V.K. (ed.). Nanocellulose Polymer Nanocomposites: Fundamentals and Applications. John Wiley and Sons, New Jersey. USA.

Sajadi, E., Babaipour, V., Deldar, A.A., Yakhchali, B. and Fatemi, S.S.A. 2017. Enhancement of crystallinity of cellulose produced by *Escherichia coli* through heterologous expression of bcsD gene from *Gluconacetobacter xylinus*. Biotechnology Letters. 39(9): 1395–1401.

Saleh, A.K., El-Gendi, H., Soliman, N.A., El-Zawawy, W.K. and Abdel-Fattah, Y.R. 2022. Bioprocess development for bacterial cellulose biosynthesis by novel *Lactiplantibacillus plantarum* isolate along with characterization and antimicrobial assessment of fabricated membrane. Scientific Reports. 12(1): 2181.

Sarkar, P., Abedin, M.M., Singh, S.P., Pandey, A. and Rai, A.K. 2022. Microbial production and transformation of polyphenols. pp. 189–208. *In*: Rai, A.K., Singh, S.P., Pandey, A., Larroche, C. and Soccol, C.R. (eds). Current Developments in Biotechnology and Bioengineering.

Semjonovs, P., Ruklisha, M., Paegle, L., Saka, M., Treimane, R., Skute, M., et al. 2017. Cellulose synthesis by *Komagataeibacter rhaeticus* strain P 1463 isolated from Kombucha. Applied Microbiology and Biotechnology. 101(3): 1003–1012.

Sezali, N.A.A., Ong, H.L., Jullok, N., Villagracia, A.R. and Doong, R.A. 2021. A review on nanocellulose and its application in supercapacitors. Macromolecular Materials and Engineering. 306(12).

Shaheen, T.I., El-Gamal, M.S., Desouky, S.E., Hassan, S.E.D. and Alemam, A.M. 2022. Benign production of AgNPs/bacterial nanocellulose for wound healing dress: antioxidant, cytotoxicity and *in vitro* studies. Journal of Cluster Science. 33(6): 2735–2751.

Shoda, M. and Sugano, Y. 2015. Recent advances in bacterial cellulose production. Biotechnology and Bioprocess Engineering. 10: 1–9.

Singh, O., Panesar, P.S. and Chopra, H.K. 2017. Response surface optimization for cellulose production from agro industrial waste by using new bacterial isolate *Gluconacetobacter xylinus* C18. Food Science and Biotechnology. 26(4): 1019–1028.

Skiba, E.A., Budaeva, V.V., Ovchinnikova, E.V., Gladysheva, E.K., Kashcheyeva, E.I., Pavlov, I.N., et al. 2020. A technology for pilot production of bacterial cellulose from oat hulls. Chemical Engineering Journal. 383: 123128.

Škraban, J., Cleenwerck, I., Vandamme, P., Fanedl, L. and Trček, J. 2018. Genome sequences and description of novel exopolysaccharides producing species *Komagataeibacter pomaceti* sp. nov. and reclassification of *Komagataeibacter kombuchae* (Dutta and Gachhui 2007) Yamada et al. 2013 as a later heterotypic synonym of *Komagataeibacter hansenii* (Gosselé et al. 1983) Yamada et al. 2013. Systematic and Applied Microbiology. 41(6): 581–592.

Spiers, A.J., Bohannon, J., Gehrig, S.M. and Rainey, P.B. 2003. Biofilm formation at the air-liquid interface by the *Pseudomonas fluorescens* SBW25 wrinkly spreader requires an acetylated form of cellulose. Molecular Microbiology. 50(1): 15–27.

Sriplai, N., Mongkolthanaruk, W., Eichhorn, S.J. and Pinitsoontorn, S. 2018. Magnetically responsive and flexible bacterial cellulose membranes. Carbohydrate Polymers. 192: 251–262.

Stickel, J.J., Elander, R.T., Mcmillan, J.D. and Brunecky, R. 2014. Enzymatic hydrolysis of lignocellulosic biomass. pp. 77–103. *In*: V.S. Bisaria and A. Condo (eds). Bioprocessing of Renewable Resources to Commodity Bioproduct. John Wiley and Sons, New Jerser, USA.

Sudying, P., Laingaumnuay, N. and Jaturapiree, P. 2019. Production and characterization of bacterial cellulose from rice washing drainage (RWD) by *Komagataeibacter nataicola* Li1. Key Engineering Materials. 824: 30–37.

Sutthiphatkul, T., Suwanposri, A. and Ochaikul, D. 2020. Optimization of bacterial cellulose production from wastewater of noodle processing by *Komagataeibacter* sp. PAP1 and bio-cellulose paper production. Walailak Journal of Science and Technology. 17(11): 1241–1251.

Suwanposri, A., Yukphan, P., Yamada, Y. and Ochaikul, D. 2014. Statistical optimisation of culture conditions for biocellulose production by *Komagataeibacter sp.* PAP1 using soya bean whey. Maejo International Journal of Science and Technology. 8(01): 1–14.

Thompson, D.N. and Hamilton, M.A. 2001. Production of bacterial cellulose from alternate feedstocks. Applied Biochemistry and Biotechnology. 91–93: 503–513.

Thongwai, N., Futui, W., Ladpala, N., Sirichai, B., Weechan, A., Kanklai, J., et al. 2022. Characterization of bacterial cellulose produced by *Komagataeibacter maltaceti* P285 isolated from contaminated honey wine. Microorganisms. 10(3): 528.

Tiwari, G., Sharma, A., Dalela, M., Gupta, R., Sharma, S. and Kuhad, R.C. 2017. Microwave assisted alkali pretreatment of fruit peel wastes for enzymatic hydrolysis. Indian Journal of Agricultural Sciences. 87(4): 496–499.

Tsouko, E., Kourmentza, C., Ladakis, D., Kopsahelis, N., Mandala, I., Papanikolaou, S., et al. 2015. Bacterial cellulose production from industrial waste and by-product streams. International Journal of Molecular Sciences, 16(7): 14832–14849.

Urbina, L., Hernández-Arriaga, A.M., Eceiza, A., Gabilondo, N., Corcuera, M.A., Prieto, M.A., et al. 2017. By-products of the cider production: an alternative source of nutrients to produce bacterial cellulose. Cellulose. 24(5): 2071–2082.

Vigentini, I., Fabrizio, V., Dellacà, F., Rossi, S., Azario, I., Mondin, C., et al. 2019. Set-up of bacterial cellulose production from the genus *Komagataeibacter* and its use in a gluten-free bakery product as a case study. Frontiers in Microbiology. 10(1953): 1–13.

Yang, X.Y., Huang, C., Guo, H.J., Xiong, L., Li, Y.Y., Zhang, H.R., et al. 2013. Bioconversion of elephant grass (*Pennisetum purpureum*) acid hydrolysate to bacterial cellulose by *Gluconacetobacter xylinus*. Journal of Applied Microbiology. 115(4): 995–1002.

Yang, H., Yu, Y., Fu, C. and Chen, F. 2019. Bacterial acid resistance toward organic weak acid revealed by RNA-seq transcriptomic analysis in *Acetobacter pasteurianus*. Frontiers in Microbiology. 10: 1–14.

Yanti, N., Ahmad, S. and Muhiddin, N. 2018. Evaluation of inoculum size and fermentation period for bacterial cellulose production from sago liquid waste. Journal of Physics: Conference Series. 1116: 052076.

Yassine, F., Bassil, N., Flouty, R., Chokr, A., El Samrani, A., Boiteux, G., et al. 2016. Culture medium pH influence on *Gluconacetobacter* physiology: Cellulose production rate and yield enhancement in presence of multiple carbon sources. Carbohydrate Polymers. 146: 282–291.

Yi, X., Wang, F., Wu, Y., He, J. and Huang, Y. 2021. Aramid nanofibers/Bacterial cellulose nanocomposite aerogels for high-efficient cationic dye removal. Materials Chemistry and Physics. 272: 124985.

Zahan, K.A., Páe, N. and Muhamad, I.I. 2015. Monitoring the effect of pH on bacterial cellulose production and *Acetobacter xylinum* 0416 growth in a rotary discs reactor. Arabian Journal for Science and Engineering. 40(7): 1881–1885.

Zeng, X., Small, D.P. and Wan, W. 2011. Statistical optimization of culture conditions for bacterial cellulose production by *Acetobacter xylinum* BPR 2001 from maple syrup. Carbohydrate Polymers. 85(3): 506–513.

Zhang, H., Yan, X., Jiang, Y. and Cong. 2011. Development and characteristic of bacterial cellulose for antimicrobial wound dressing. Advanced Materials Research. 152–153: 978–987.

Zhang, K., Pei, Z. and Wang, D. 2016. Organic solvent pretreatment of lignocellulosic biomass for biofuels and biochemicals: a review. Bioresource Technology. 199: 21–33.

Zhang, H., Xu, X., Chen, X., Yuan, F., Sun, B., Xu, Y., et al. 2017. Complete genome sequence of the cellulose-producing strain *Komagataeibacter nataicola* RZS01. Scientific Reports. 7(1): 1–8.

Zhang, W., Zhang, Q.Y., Wang, J.J., Zhang, L. and Dong, Z.Z. 2022. Efficiency assessment of bacterial cellulose on lowering lipid levels in vitro and improving lipid metabolism in vivo. Molecules. 27(11): 3495.

Zhao, H., Xia, J., Wang, J., Yan, X., Wang, C., Lei, T., et al. 2018. Production of bacterial cellulose using polysaccharide fermentation wastewater as inexpensive nutrient sources. Biotechnology and Biotechnological Equipment. 32(2): 350–356.

Zhao, X.B., Wang, L. and Liu, D.H. 2007. Effect of several factors on peracetic acid pretreatment of sugarcane bagasse for enzymatic hydrolysis. Journal of Chemical Technology and Biotechnology. 82(12): 1115–1121.

Chapter 15

Management of Agri-Food Wastes via Novel Biotechnological Methods: Emerging Trends and Technological Advances

Soubhagya Tripathy*[1], Deepak Kumar Verma*[1], Ami R. Patel[2], Smita Singh[3], Gemilang Lara Utama[4,5], Alaa Kareem Niamah*[6], Mónica L. Chávez González[7], Prem Prakash Srivastav[1] and Cristobal Noe Aguilar*[7]

[1]Agricultural and Food Engineering Department, Indian Institute of Technology Kharagpur, Kharagpur 721302, West Bengal, India.

[2]Division of Dairy Microbiology, Mansinhbhai Institute of Dairy and Food Technology-MIDFT, Dudhsagar Dairy Campus, Mehsana-384002, Gujarat, India.

[3]Department of Allied Health Sciences, Chitkara School of Health Sciences, Chitkara University, Punjab 140401, India.

[4]Faculty of Agro-Industrial Technology, Universitas Padjadjaran, Sumedang 45363, Indonesia.

[5]Center for Environment and Sustainability Science, Universitas Padjadjaran, Bandung 40132, Indonesia.

[6]Department of Food Science, College of Agriculture, University of Basrah, Basra City, Iraq.

[7]Bioprocesses and Bioproducts Group, Food Research Department, School of Chemistry. Autonomous University of Coahuila, Saltillo Campus. 25280. Coahuila, México.

* For Correspondence: rajadkv@rediffmail.com; cristobal.aguilar@uadec.edu.mx; mr.soubhagyatripathy@gmail.com

INTRODUCTION

A common correlation exists between the persistent rise in the human population around the globe and the concurrent rise in the demand for food and natural resources on a global scale (Zhang et al. 2018, Torres-León et al. 2019, Sharma et al. 2020). Because of strong industrialization and poor procedures for managing agri-food wastes (AFWs), large volumes of AFWs are produced and accumulated in a variety of sectors, including food supply chains, our households, and the food processing industry (Paritosh et al. 2017, Hegde et al. 2018, Xiong et al. 2019, Halog and Anieke 2021, Barrett et al. 2022a, b, Abbate et al. 2023, Bhat et al. 2023). This is a problem in many parts of the world. According to the data that were published by FAO in the year 2012, roughly 1.3 billion tons of AFWs are produced each year. It is expected that this quantity would increase over the next quarter century as a result of the expansion of the economy and the human population (Paritosh et al. 2017). In light of this fact, scientific research has made a significant number of efforts in the past few years to either enhance the current system for managing AFWs or, more ideally, to lessen the accumulation of waste from food processing and agricultural practices by reusing AFWs as a feedstock for the production of high-value products (Tsui and Wong 2019, Wang et al. 2021, Capanoglu et al. 2022, Kover et al. 2022, Hoehn et al. 2023, Kour et al. 2023).

In general, the wastes from agriculture and food processing are high in carbon and nitrogen sources including carbohydrates, proteins, and lipids (Fan et al. 2000, Ezejiofor et al. 2014, Hegde et al. 2018, Mazlan et al. 2019, Capanoglu et al. 2022, Kover et al. 2022). For the production of biofuel, enzymes, probiotics, and other high-value VBPs, they make ideal industrial feedstocks (Bilal and Iqbal 2019, Ng et al. 2020b, Torres-León et al. 2021, Kumar et al. 2022a, b). In addition, they may be converted into biodegradable polymers by a variety of highly efficient bioprocesses (Vijayaraghavan et al. 2016, Bilal and Iqbal 2019, Ng et al. 2020b, Chavan et al. 2022). The materials that are considered to be bio-waste have been considered to be viable feedstock for the biotechnological manufacture of high-value bio-based value-added bioproducts/byproducts (VBPs), particularly for the production of biofuels (Sharma et al. 2021, Sufficiency et al. 2022, David et al. 2023, Chatterjee and Mohan 2023). The ongoing increase in the costs of fossil fuels and natural biomaterials on the market is a direct result of the decreasing availability of natural resources. In addition, there is a demand for bioenergy, biofuels, natural bioactive compounds (BACs), and a variety of other types of biomaterials to satisfy the expanding market (Gaurav et al. 2017, Hiloidhari et al. 2020, Ng et al. 2020b, Banwo et al. 2021, Mehmood et al. 2021). Therefore, they are all together stressing the case for research into greener, cheaper bioprocesses and the hunt for more sustainable feedstocks to maximize profits. The development of the circular economy was spurred on by the adoption of environmentally responsible methods for the repurposing of waste products from food processing and agricultural operations. These methods convert AFWs into a variety of useful VBPs derived from biomaterials (Maina et al. 2017, Paritosh et al. 2017, Dahiya et al. 2018, Bilal and Iqbal 2019, Ng et al. 2020b, Sharma et al. 2021, Bhat et al. 2023). In such circumstances, positive economic growth combined with a rapid rise in the number of employment prospects is anticipated as a result of the growth and advancement of bio-based end products that are marketable to consumers. The principles of a circular economy provide the framework for the use and recycling of AFWs (Maina et al. 2017, Dahiya et al. 2018, Halog and Anieke 2021, Sharma et al. 2021, Abbate et al. 2023, Bhat et al. 2023). This orientation provides the stakeholders and manufacturers with the motivation they need to embrace bioprocesses that are more sustainable and have fewer negative effects on the environment. The utilization of AFWs as feedstocks in biotechnological processes for the production of VBPs has the potential to address environmental pollution caused by the accumulation of these wastes. However, before scaling-up and implementing these processes on an industrial scale, it is necessary to conduct thorough and specific experimental trials, taking into account the economic considerations (Hegde et al. 2018, Kumar et al. 2022a, b). Therefore, it is imperative to engage in a discourse on the present state of the AFWs management system and the growing biotechnological advancements in the repurposing and use of AFWs as raw materials for the development of VBPs with significant economic worth.

As has been pointed out in the previous paragraph about the management of AFWs, the role of biotechnology has emerged as a potentially useful tool. Because of this, the applications of biotechnology in this field satisfy the present requirements of society in terms of human health, food, and the reduction of pollution in the environment. Therefore, the goals of the current chapter focus on AFWs in the agricultural and food industries. In addition to this, this chapter emphasizes a variety of cutting-edge biotechnological methods, such as microbial fermentation, enzymatic processing, and extraction technology. This is due to the fact that these cutting-edge biotechnological methods are very significant and have been regarded as the most effective for the management of AFWs. Within this context, the chapter also expresses worry over the existing state of AFWs and the potential outcomes for it in the future. Finally, the chapter concludes with some recommendations and future possibilities for researchers, scientists, and industrial people who are associated with the subject area that has been discussed. These individuals are working toward innovation and searching for new opportunities in the management of AFWs due to various benefits, such as the production of VBPs, the conversion of AFWs into bioenergy, and the reduction of environmental pollution. The recommendations and future possibilities are presented in this chapter. Therefore, this chapter would appeal to researchers, scientists, and industry people not only to help decrease AFWs but also to provide economic possibilities and promote sustainable development.

AGRI-FOOD WASTES AT A GLANCE

The prevailing approach to managing AFWs, which remains prevalent globally, encompasses various methods such as incineration, landfills, composting, and utilization as animal feed (Bilal and Iqbal 2019, Awasthi et al. 2020, Esparza et al. 2020, Kassim et al. 2022, Kour et al. 2023). The incineration of AFWs containing a significant amount of water is frequently associated with energy dissipation resulting from the evaporation of water during the incineration process. Additionally, the incineration process may lead to the emission of dioxins, which can potentially cause substantial harm to the environment. Furthermore, the incineration of substrates leads to a decrease in the potential for the retrieval of valuable BACs and nutrients due to the modification of their chemical makeup, resulting in diminished commercial worth (Xiong et al. 2019, Caldeira et al. 2020, Ng et al. 2020b, Wang et al. 2021, Sufficiency et al. 2022). Currently, a significant portion of AFWs continues to be disposed of in landfills, which has raised significant environmental concerns. This is primarily due to the favorable conditions provided by the high water content and nutritional composition of these wastes, which serve as ideal breeding grounds for diverse microorganisms. Consequently, this situation contributes to the proliferation of harmful bacteria, leading to the potential spread of infectious diseases. Improper and excessive disposal of AFWs in landfills leads to the release of greenhouse gases, including methane and carbon dioxide. This emission has been identified as a major source of air pollution (Karthikeyan et al. 2017, Usmani et al. 2020, Mariappan et al. 2023, Xu et al. 2023). Furthermore, the escalation in the production of greenhouse gases is resulting in alterations to the Earth's climate system and the phenomenon of global warming, both of which have adverse effects on various forms of life (Usmani et al. 2020, Mariappan et al. 2023). The utilization of AFWs as animal feeds is considered the most economically efficient method for managing AFWs over an extended period (Bilal and Iqbal 2019, Mehmood et al. 2021, Kassim et al. 2022, Bala et al. 2023, Jasrotia et al. 2023). However, the application of this approach is subject to strict regulations due to the presence of diverse and unidentified BACs in the AFWs. Furthermore, the process of AFWs decomposition during conversion can sometimes be unpredictable and lead to the formation of undesired by-products, thereby adding complexity to the AFWs management system as a whole (Mehmood et al. 2021, Chavan et al. 2022, Kassim et al. 2022). Land spreading, also referred to as composting in scientific literature, is a highly efficient method for mitigating AFWs accumulation in landfills and obviating the necessity for farmers to rely on biofertilizers (Awasthi et al. 2020, Chojnacka et al. 2020, Ashokkumar et al. 2022). However, composting exhibits a higher cost in comparison to the aforementioned techniques, thus limiting its widespread implementation in global AFWs management strategies.

BIOTECHNOLOGICAL APPROACHES FOR AGRI-FOOD WASTES MANAGEMENT

AFWs encompass both preprocessed agricultural waste and biodegradable waste from kitchen leftovers, which are discarded by a range of industries, notably the food-processing sector and households (Ezejiofor et al. 2014, Paritosh et al. 2017, Kover et al. 2022, Kumar et al. 2022a, b). AFWs, derived from various sources and possessing diverse chemical compositions, exhibit great potential as sustainable feedstocks for various bioprocesses aimed at producing bioenergy and other commercially VBPs (Paritosh et al. 2017, Tsui and Wong 2019, Ng et al. 2020b,

Torres-León et al. 2021, Ashokkumar et al. 2022, Kumar et al. 2022b, Bala et al. 2023). In recent years, a multitude of scientific investigations have been conducted to explore novel and environmentally friendly bioprocesses for the conversion of AFWs into VBPs. This research is driven by the recognition of the economic opportunities associated with transforming VBPs derived from food processing and agricultural waste, as opposed to the detrimental consequences of landfill-based food decomposition (Maina et al. 2017, Dahiya et al. 2018, Ng et al. 2020b, Sharma et al. 2021, Tsui and Wong 2019, Ashokkumar et al. 2022, Kumar et al. 2022b, Kassim et al. 2022, Xu et al. 2023). One of the primary obstacles faced by researchers in utilizing AFWs as raw materials is the lack of precise knowledge regarding the specific chemical composition percentages of different AFWs. This uncertainty hampers the ability to accurately predict the synthesis of BPs during the bioconversion processes (Bilal and Iqbal 2019, Awasthi et al. 2020, Kumar et al. 2022a, Bhat et al. 2023). Given these considerations, additional research and refinement of innovative biotechnological methods have been conducted to enhance the efficiency of producing specific VBPs and sustainable bioenergy. The aim is to maximize the reuse and recycling of waste generated from food processing and agricultural operations. The subsequent subsections present a comprehensive analysis of various emerging and innovative methodologies aimed at achieving sustainable production of BPs derived from AFWs.

Microbial Fermentation

Microbial fermentation is considered a promising method for the conversion of AFWs into valuable final BPs (Sadh et al. 2018a, b, Bautista-Hernández et al. 2022). The choice of fermentation strategies for the bioconversion process is heavily influenced by the specific AFWs utilized as the raw material for the fermentation (Vijayaraghavan et al. 2016, Sadh et al. 2018a, Bautista-Hernández et al. 2022, Chavan et al. 2022). Solid-state fermentation is typically employed for solid substrates to optimize nutrient assimilation (de Castro et al. 2015, Sadh et al. 2018a, Naik et al. 2019, Leite et al. 2019, Gmoser et al. 2019, Torres-León et al. 2019, Bautista-Hernández et al. 2022), while submerged fermentation is predominantly utilized for liquefied substrates (Sadh et al. 2018b, Gmoser et al. 2019, Sepúlveda et al. 2020, Sarkar et al. 2022). In the realm of AFWs fermentation, submerged fermentation is widely employed in industrial-scale applications due to its advantageous characteristics, such as its economically efficient processing cost and remarkable productivity levels. Furthermore, the utilization of submerged fermentation has been widely adopted in the field of enzyme production owing to its advantageous features such as streamlined process control and efficient sterilization protocols. Figure 15.1 illustrates the microbial fermentation procedure employed to convert AFWs into BPs.

AFWs are commonly regarded as highly suitable and sustainable raw materials for microbial fermentation processes aimed at the production of diverse BACs and BPs. This is primarily due to their abundant presence of fatty acids, proteins, minerals, and phenolic BACs (Paritosh et al. 2017, Torres-León et al. 2021, Mehmood et al. 2021, Kumar et al. 2022a, Sufficiency et al. 2022).

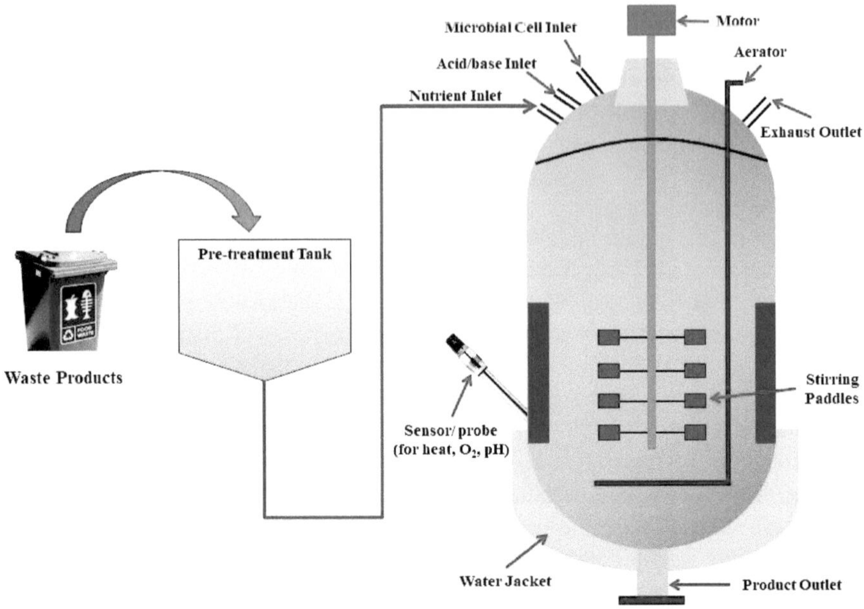

Figure 15.1 A schematic diagram illustrating the microbial fermentation process for the conversion of food waste into bioproducts. This process involves the enzymatic breakdown of nutrients, primarily sugars and carbohydrates, by microbial metabolic enzymes in an anaerobic environment. In the fermentation process, a diverse array of microorganisms is typically present, each possessing a unique complement of enzymes. These enzymes catalyze various biochemical reactions, resulting in the production of intricate mixtures of bioproducts. These bioproducts contribute to the distinctive olfactory and gustatory qualities observed in fermented foods (Maicas 2020, Sharma et al. 2020).

The process of converting agricultural and food industry waste into BPs that have increased functionality can effectively decrease production expenses and mitigate environmental contamination by obviating the necessity for food waste management (Bilal and Iqbal 2019, Xiong et al. 2019, Caldeira et al. 2020, Mehmood et al. 2021, Wang et al. 2021, Barrett et al. 2022a, Sufficiency et al. 2022, Bala et al. 2023, Kour et al. 2023). AFWs frequently function as cost-effective carbon and nitrogen substrates for the proliferation of microorganisms, which can be harnessed for the process of fermenting BACs (de Castro et al. 2015, Vijayaraghavan et al. 2016, Javed et al. 2019, Naik et al. 2019, Capanoglu et al. 2022, Kover et al. 2022, Kumar et al. 2022a). Alternatively, a wide range of BACs such as proteins, enzymes, antioxidants, and pigments can be synthesized via microbial fermentation utilizing AFWs as raw material in an unrefined state (Vijayaraghavan et al. 2016, Mathias et al. 2017, Moreira et al. 2018, Sadh et al. 2018a, b, Catalkaya and Kahveci 2019, Gmoser et al. 2019, Leite et al. 2019, Deng et al. 2020). Table 15.1 presents a comprehensive overview of multiple BACs that are synthesized through diverse microbial fermentation techniques employing AFWs as environmentally friendly raw materials. Lactic acid and ethanol are prominent BPs that can be generated via the process of fermentation using AFWs (Jawad et al. 2013, Tang et al. 2016, Li et al. 2018, Waqas et al. 2019, Liu et al. 2020, Zain et al. 2021).

Lactic acid is generated through the hydrolysis and acidogenesis mechanisms, which represent the initial stages of anaerobic digestion (Wu et al. 2015, Gong et al. 2021, Yadav et al. 2022, Zhang et al. 2023b). The optimization of operating conditions, such as inoculum pH, temperature, and organic loading rate, can lead to a significant increase in lactic acid yield through microbial fermentation of AFWs (Tang et al. 2016, Tang et al. 2017, Yadav et al. 2022). The hydrolysis and acidification processes are commonly carried out under pH conditions of 4 and 5 to optimize lactic acid fermentation when utilizing AFWs as substrates (Wu et al. 2015). In addition, variations in operating temperature have a direct impact on the rate at which the substrate is converted and the level of microbial activity, consequently influencing the production of lactic acid (Liang et al. 2014). Previous studies have utilized AFWs, including potato peel and fruit and vegetable wastes, as substrates for the microbial fermentation process to produce lactic acid (Liang et al. 2014, Wu et al. 2015, Gupta et al. 2022, Chatterjee and Mohan 2023). In general, the utilization of mango peel waste has been scientifically demonstrated as a promising and economically feasible substrate for the process of lactic acid fermentation (Jawad et al. 2013). The duration of fermentation required for achieving the highest lactic acid production is significantly decreased when AFWs are utilized as the exclusive substrate for fermentation, along with the implementation of thermal pre-treatment (Li et al. 2018).

Bioethanol is a sustainable biofuel with the potential to serve as a viable alternative to petroleum-derived energy sources, thereby mitigating the detrimental environmental impacts associated with the use of such sources (Han et al. 2020, Singh et al. 2021, David et al. 2023, Reshmy et al. 2023). The AFWs, which are abundant in carbohydrates, serve as a favorable substrate for the production of bioethanol via the process of fermentation (Sharma et al. 2023). Bioethanol is commonly produced using a two-step process, consisting of enzymatic hydrolysis of carbohydrates and subsequent fermentation of the resulting simple sugars by the yeast *Saccharomyces cerevisiae* (Parmar and Rupasinghe 2013, Singh et al. 2021 Sharma et al. 2023). Pre-processing of AFWs, such as thermal treatment or hydrolysis, is necessary before initiating the fermentation procedure owing to the intricate chemical constituents present in these wastes. These constituents pose challenges in terms of digestion and utilization in the fermentation process (Kim et al. 2011, Taheri et al. 2021, Chatterjee and Mohan 2022, David et al. 2023). The AFW materials utilized for ethanol production primarily consist of energy-dense substances, including sugarcane bagasse, corn residue, apple pomace, potato, rice, sugarcane, and various others (Amores et al. 2013, Kiran et al. 2014, Hernández et al. 2019, Aghaei et al. 2022, de Almeida and Colombo 2023, Sharma et al. 2023).

Enzymatic Processing

Based on the preceding sections, it is evident that AFWs retain a significant concentration of diverse chemicals even after undergoing various disposal

Table 15.1 Some important studies on various bioactive compounds derived from agri-food wastes through the process of microbial fermentation, employing diverse microorganisms

Agri-food wastes-based feedstock	Employed microorganisms	Type of microbial fermentation	Produced bioactive compounds	References
Apple, banana and grape waste	*Trichoderma reesei RUT-C30 and T. atroviride Tal3*	Solid state fermentation	Cellulase, esterase, citric, malic acids, gibberellins and oxylipins	Bulgari et al. (2023)
Sugarcane bagasse	*Lactobacillus acidophilus* MTCC 10307	—	Ferulic acid, vanillic acid and vanillin	Pattnaik et al. (2022)
Waste Fish Oil	*Yarrowia lipolytica*	Batch fermentation	Docosahexaenoic (DHA), Eicosapentaenoic acid (EPA), Eicosenic and Erucic acids	Fabiszewska et al. (2021)
Biofuel waste	*Y. lipolytica*	—	Vitamin B	Jach et al. (2021)
Sorghum Silage	*L. farciminis (LF18) and L. plantarum (LP23)*	—	Lactic acid	Xie et al. (2021)
Sugarcane bagasse	*Planococcus* sp. TRC1	Solid state fermentation	β-carotene	Majumdar et al. (2020)
Orange peel waste	*Aspergillus fumigatus*	Submerged fermentation	Ellagic acid	Sepulveda et al. (2020)
Apple peel	*Humicola insolens*	–	Pectin oligosaccharides	Wang et al. (2020)
Wheat bread waste	*Neurospora intermedia*	Submerged fermentation and solid-state fermentation	Carotenoids	Gmoser et al. (2019)
Corncob, wheat bran, rice husk, orange peel, pomegranate peel	*A. niger*	Submerged fermentation	Xylanase	Javed et al. (2019)
Kitchen waste (rice, noodles, meat, chicken, bones, vegetables)	*Starmerella bombicola*	Fed-batch fermentation	Sophorolipids	Kaur et al. (2019)
Brewer's spent grain	*A. niger*	–	β-glucosidase	Leite et al. (2019)
Wheat bran	*Aspergillus* spp.	Solid-state fermentation	Pullulanase	Naik et al. (2019)

Citrus peel waste	*Pichia pastoris*	Shake-flask fermentation	Oligogalacturonides	Yang et al. (2020)
Coffee waste	*Rhodotorula mucilaginosa*	Batch fermentation	Carotenoids	Moreira et al. (2018)
Turbot skin waste	*A. oryzae*	Eco-friendly fermentation	Antioxidant peptides (preservatives)	Fang et al. (2017)
Brewery waste	*L. delbrueckii*	Lactic acid bacteria fermentation	Protease	Mathias et al. (2017)
Dairy waste	*Saccharomyces cerevisiae*	Shake flask fermentation	Single-cell protein	Suman et al. (2017)
Dairy waste	*L. acidophilus*	Shake flask fermentation	Single-cell protein	Suman et al. (2017)
Wheat bran	*Bacillus halodurans*	Solid-state fermentation	Fibrinolytic enzyme	Vijayaraghavan et al. (2016)
Wheat bran, soybean meal, cottonseed meal, orange peel	*A. niger*	Solid state fermentation	Protease	de Castro et al. (2015)
Wheat waste	*Aporidiobolux salmonicolor, Xanthophyllomyces dendrorhous, Yamadazyma guilliermondii, Yarrowia lipolytica*	Solid state fermentation	Astaxanthin	Dursun and Dalgic (2016)
Pineapple peel	*S. cerevisiae*	Submerged yeast fermentation	Single-cell protein	Mensah and Twumasi (2017)

– Not reported

methods. These chemicals can be extracted through a range of approaches (Radenkovs et al. 2018, Bilal and Iqbal 2019, Taheri et al. 2021). Extraction methodologies employing chemical solvents are widely employed techniques for extracting industrially valuable BACs from agro-industrial waste. However, the impact on substrates can be detrimental due to the vigorous treatments employed in the extraction conditions; thus, these techniques may not always be the optimal choice (Zhang et al. 2018, Catalkaya and Kahveci 2019, Verma and Srivastav 2020a, b, 2022, Tripathy et al. 2021, 2022). Additional alternative approaches, including ultrasound, microwaves, electrical pulses, and supercritical fluids (SCFs), among various others, have been suggested as potential methods (Verma et al. 2018, Chávez-González et al. 2020, Singh et al. 2021, Gupta et al. 2022). Nevertheless, notwithstanding the superior productivity and reduced ecological repercussions associated with these techniques, their precision in the hydrolysis of plant materials for ensuring the comprehensive liberation of target chemicals is comparatively inferior. As a result of this trend, the application of enzymatic systems on agricultural and food industry residues has facilitated enhanced utilization of this particular type of material. This trend can be attributed to the inherent advantages of enzymatic catalysis, including the remarkable substrate specificity exhibited by enzymatic systems. This specificity greatly enhances the efficiency of recovering specific BACs of interest (Radenkovs et al. 2018, Mazlan et al. 2019, Torres-León et al. 2021, Taheri et al. 2021). The utilization of enzymatic catalysis has facilitated the enhanced application of this particular type of substance. Figure 15.2 illustrates the standard procedure employed for the synthesis of nanocellulose from lignocellulosic biomass through enzymatic hydrolysis. This mechanism encompasses all stages of enzymatic hydrolysis.

The utilization of enzyme complexes is to catalyze the hydrolysis of various AFW materials (Vijayaraghavan et al. 2016, Bilal and Iqbal 2019, Salimi et al. 2019). Consequently, the potential for product release exists due to significant interest in improving the techniques for extracting and recovering a specific metabolite (Andler and Goddard 2018, Salimi et al. 2019, Padhi et al. 2023). A wide range of enzymes have been examined to fulfill this objective, taking into consideration the precise composition of the substance under investigation. The enzymes that are commonly utilized include α-amylase, cellulases, xylanases, pectinases, proteases, chitinase, and various others (as shown in Table 15.2). The majority of the existing literature pertaining to the utilization of hydrolytic enzyme systems in AFWs primarily emphasizes the assessment of catalytic conditions (as presented in Table 15.2). These conditions encompass a range of factors including temperature variations, pH levels, agitation intensity, enzyme concentration, reaction duration, and other relevant parameters (Andler and Goddard 2018, Mazlan et al. 2019, Esparza et al. 2020, Phukon et al. 2020, 2022, Kumari et al. 2021, Rojas et al. 2022). The products obtained through catalysis exhibit a high degree of complexity, encompassing a wide range of substances such as antioxidants (Leite et al. 2019), pigments (Catalkaya and Kahveci 2019), protein hydrolysates with proven bioactivity (Murphy et al. 2018), oligosaccharides (Mazlan et al. 2019), growth-promoting substrates (Kumar et al. 2019b), and more. The products exhibit substantial utility across various industrial sectors. One notable development over the past decade has been the advancement in monosaccharide production for bioethanol synthesis (Ebikade et al. 2018, Hernández et al. 2019, Han et al. 2020, Sharma et al. 2023).

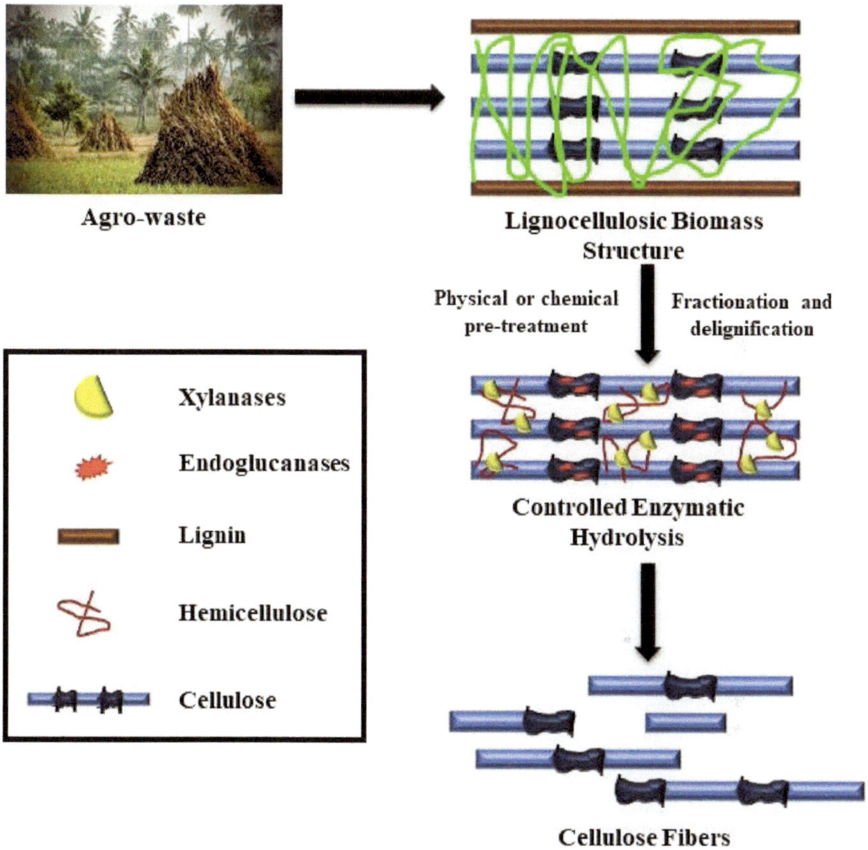

Figure 15.2 A schematic diagram illustrating the process of enzymatic hydrolysis of lignocellulosic biomass for the production of cellulose (Pradhan et al. 2022). The enzymatic processing technology is a nascent pathway for the acquisition of nanocellulose from lignocellulosic biomass, encompassing multiple sequential stages. The initial stage involves the pretreatment of lignocellulosic biomass to separate and extract the primary constituents, namely cellulose, hemicellulose, and lignin, typically employing physical or chemical pretreatment techniques. The enzymatic hydrolysis of fiber samples is subjected to regulated conditions in the subsequent step. An enzymatic mixture is employed to dissolve the pre-treated substance in the buffered solution. The primary function of the enzymatic mixture is to hydrolyze the cellulose polymer, leading to the depolymerization of cellulose into smaller oligosaccharides.

The utilization of global fast-food waste has been employed for the production of bioethanol on a global scale (Yang et al. 2014, Han et al. 2020, Liu et al. 2020). For instance, the analysis of pizza remnants has recently been conducted (Liu et al. 2020). This investigation was prompted by the fact that pizza is known to

Table 15.2 Detail information of some important investigations concentrated on various bioactive compounds/bioproducts derived from agri-food wastes through enzymatic processing employing diverse enzyme types

Agri-food wastes	Employed enzyme	Type of enzymatic processing	Conditions for enzymatic processing	Produced bioproducts/ bioactive compounds	References
Indian black rice straw	Cellic® CTec3	Enzymatic hydrolysis	A solution was prepared by adding 5 mL of Cellic® CTec3 to 1 L of a 50 mM sodium citrate buffer with a pH of 3.76.	Fermentable sugar	Moirangthem et al. (2022)
Fish waste	Protease	Enzymatic Hydrolysis	The substrate concentration ranged from 5 to 45 g/L. The pH was set at 8, the temperature was maintained at 50°C, and the reaction time was 180 minutes. The stirring speed was set to 300 rpm.	Protein hydrolysate Collagen Oil	Araujo et al. (2021)
Citrus by-products	Commercial β-glucosidase, tannase, and cellulase	Enzyme-assisted extraction	The temperature was 40°C, and the time was measured using either a 12-hour or 24-hour format, while the agitation level was set at 120 rpm.	Flavanone aglycones	Barbosa et al. (2021)
Tomato pomace	Cellulase	Enzymatic Hydrolysis	The concentration of the enzyme ranged from 1% to 4%, while the duration of hydrolysis varied between 2 and 6 hours.	Monomeric sugars, Alcohols Esters	Kılmanoğlu et al. (2021)
Brewers' spent grain	Xylanase	Enzymatic Hydrolysis	The pH level was adjusted to a range of 4.6–5.0, and the hydrolysis process was allowed to proceed for a duration of 48 h.	Reducing sugars, Polyhydroxyalkanoates	Martínez-Avila et al. (2021)
Jujube peels	Cocktail enzyme	Enzyme-assisted extraction	The pH of the solution was adjusted to a value of 7.0, the temperature was maintained at 43°C, and the enzyme period lasted for 97 min.	Pigments	Shen et al. (2021)

Agri-food wastes	Employed enzyme	Type of enzymatic processing	Conditions for enzymatic processing	Produced bioproducts/ bioactive compounds	References
Raspberry wine residues	Pectinase	Enzyme-assisted extraction	The enzyme dosage used in the experiment was 0.16%. The temperature was carefully controlled and maintained at 44°C throughout the extraction process. The extraction time lasted for 30 min.	Anthocyanins	Xue et al. (2021)
Bovine bone	Mix proteases	Enzyme hydrolysis	The enzyme dosage used in the experiment was 76.25×10^4 U/g. The pH was set at 9.0 while the incubation time was 2.01 h throughout the extraction process.	Low molecular weight peptides	Yang et al. (2021)
Grape pomace	Cocktail hydrolytic enzymes	Enzyme-assisted extraction	—	Phenolic compounds	Teles et al. (2021)
Sunflower wastes	Multienzyme complex viscozyme	Enzyme-assisted extraction	The concentration of the enzyme was 0.25%. The system was subjected to agitation at a rotational speed of 200 rpm for a duration of 2 h while maintaining a temperature of 40°C.	Carotenoids and phenolic compounds	Ricarte et al. (2020)
Shrimp shell	Recombinant aspartic proteases and recombinant chitinase	Enzyme-based biorefinement process	Various quantities of enzymes, durations of reactions, and concentrations of substrates were utilized and assessed.	Protein and chitin	Deng et al. (2020)
Tomato waste	Cellulase	Enzyme-assisted extraction	The temperature was maintained at 40°C, the reaction duration was 5 h, and the ratio of enzyme to substrate was 0.2 mL/g.	Lycopene	Catalkaya and Kahveci (2019)
Tomato waste	Pectinase	Enzyme-assisted extraction	The temperature was maintained at 40°C, the reaction duration was 5 h, and the ratio of enzyme to substrate was 0.2 mL/g.	Lycopene	Catalkaya and Kahveci (2019)

Table 15.2 (*Contd.*)

Agri-food wastes	Employed enzyme	Type of enzymatic processing	Conditions for enzymatic processing	Produced bioproducts/ bioactive compounds	References
Sago bagasse	Amylase	Enzymatic hydrolysis	The substrate loading was set at 1% (w/v). The enzyme loading was 0.08 g, corresponding to 2000 units of enzyme activity. The temperature during the hydrolysis process was maintained at 42°C. The hydrolysis reaction was allowed to proceed for a duration of 3 h. The pH of the reaction mixture was adjusted to 7.0.	Plant growth-promoting	Kumar et al. (2019b)
Oil palm frond bagasse	Xylanase (Cellic Htec2)	Enzymatic hydrolysis	The substrate loading was set at 1% (w/v), while the enzyme loading was 50 U/mL. The temperature was maintained at 55°C and the reaction time was set to 4 h. The agitation speed was set at 50 rpm.	Xylooligosaccharides	Mazlan et al. (2019)
Defatted soy flakes	Corolase PP (chymotrypsin, elastase, dipeptidase, aminopeptidase activities)	Enzymatic hydrolysis	The temperature was maintained at 50°C, the pH was adjusted to 7.5, and the ratio of enzyme to soy protein isolates was 3:1000.	Bioactive peptides	Guan et al. (2018)
Mussel shell	Multifect PR 6 L and multifect PR 7 L peptidases	Enzymatic cleaning process by the utilization of either thermogravimetric analysis or manual recovery methods.	PR 6 L material immersed in seawater or tap water at a temperature of 55°C for a duration of 4 h, with a dosage of 1.0–2.0 mL/g of material. On the other hand, the PR 7 L material was exposed to a concentration of 6 mg/L of water at a temperature of 25°C for a duration of 10 h.	Protein hydrolysate	Murphy et al. (2018)

– Not reported

contain significant amounts of starch, proteins, and fats. In the study conducted by Liu et al. (2020), waste pizza was employed as a substrate for the production of bioethanol. The experiment was carried out at a temperature of 90°C, with a stirring speed of 300 rpm, and 150 min. The enzyme utilized in the process was α-amylase, which exhibited a specific activity of 7×10^4 U/mL. The effects of different volumes of the enzyme (0.02 mL/L, 0.04 mL/L, and 0.08 mL/L) and various ratios of waste pizza mass (50 g/L, 100 g/L, and 160 g/L) were investigated. In the study conducted by Liu et al. (2020), it was observed that the maximum ethanol production and ethanol yield reached values of 27.58 g/L and 0.292 g/g waste pizza, respectively. In contrast, Han et al. (2020) conducted an assessment on the utilization of hamburger waste for the production of ethanol. They employed a-amylase enzyme with a specific activity of 7×10^4 U/mL, operating at temperatures ranging from 90 to 95 °C. The enzyme volumes used were 0.02, 0.08, and 0.14 mL/L. The maximum recorded ethanol production reached a concentration of 27.4 g/L, with an ethanol yield of 0.271 g/g of waste hamburger. Contrary to this, Salimi et al. (2019) employed household food waste as a feedstock for the production of bioethanol. The dehydrated food waste (20% w/w) underwent treatment with the amylolytic enzyme (NS22109) at a temperature of 65 °C for a duration of 1 h. The enzyme loadings used were 15, 30, 45, and 60 μL/g. Subsequently, cellulolytic enzymes (NS22177) were employed at concentrations of 90, 130, 175, 280, 560, 830, and 1120 mg/g for a duration of 5 h at a temperature of 50 °C. The application of enzymatic hydrolysis on AFWs has been explored to enhance enzymatic catalysis and achieve increased yields. One approach involves employing enzymes in conjunction with techniques like ultrasound, resulting in yields exceeding 10% (Li et al. 2019). Alternative methodologies, such as the utilization of microwaves, have also exhibited enhancements in the extraction process.

Extraction Technology

Extraction is a commonly employed method for the retrieval of BACs from AFW sources. This process aims to produce high quantities of valuable and pure products that can be utilized in the pharmaceutical, cosmetics, and food industries (Kumar et al. 2017, Sagar et al. 2018, Zhang et al. 2018, Chávez-González et al. 2020, Ng et al. 2020b, Capanoglu et al. 2022, Bala et al. 2023). Table 15.3 presents a comprehensive overview of various extraction techniques employed in the retrieval of valuable BACs and BPs from AFWs, as reported in multiple published scientific literature sources. Antioxidative BACs represent significant constituents that can be efficiently extracted from the BPs of fruits and vegetables owing to their abundant presence in seeds, peels, and kernels, which are rich in antioxidants (Sagar et al. 2018, Torres-León et al. 2019, Xiong et al. 2019, Sepúlveda et al. 2020, Gupta et al. 2022). In addition, functional lipids, proteins, and starch are also prominent functional BACs that can be extracted from diverse sources of AFWs, including dairy, marine, and meat industries (Phongthai et al. 2016, Santana et al. 2017, Guo et al. 2018, Ghosh et al. 2019, Kaur et al. 2019, Kurup et al. 2019). The extraction of BACs from AFWs is contingent upon the

Table 15.3 A comprehensive overview of significant research studies conducted on various bioactive compounds/bioproducts derived from agri-food waste through extraction technology employing diverse extraction methods

Agri-food wastes	Conditions for extraction	Targeted bioactive compound(s)/ bioproduct(s)	Production yield	References
(A) Solvent extraction				
Kraft black liquor	Extraction was performed using ethyl acetate in a 1:1 volume-to-volume ratio.	Total phenols	3.15 g/L (5% of initial lignin)	Curmi et al. (2022)
Grape seeds	An extraction process was conducted using an 80% (volume/volume) solution of acetone.	Total phenols	24.6 mg/GAE g	Di Stefano et al. (2022)
Sugar cane bagasse	A volume of 1 g of the specimen was measured and placed into a volumetric flask. The flask was then filled with 25 mL of methanol and subsequently subjected to agitation for a duration of 48 h.	Phenolic compounds	35.19 mg GAE/100 g	Srihanam et al. (2021)
Centrolophus niger skin	The skin samples were immersed in a solution containing a mixture of acetic, citric, hydrochloric, lactic, tartaric, formic, and sulfuric acid at a concentration of 0.5 M (1:25 w/v). The immersion was carried out at a temperature of 4°C for durations of 24, 48, and 72 h.	Acid soluble collagen	45%	Bhuimbar et al. (2019)
Pineapple peel	An 80% concentration of acetone was employed for the extraction process.	Phenolic compounds	1.62 g/GAE	Nurdalilah et al. (2018)
(B) Aqueous biphasic system				
Cruciferous vegetable (kale) waste	Extraction was performed using a molar ratio of 3:1 of glycerol to betaine.	Polyphenols	91.7%	Lee et al. (2023)
Red beet waste	The extraction process involved the utilization of quaternary ammonium-based ionic liquids and polypropyleneglycol 400 g mol^{-1} at a temperature of 20°C, a duration of 70 minutes, and a solid-to-liquid ratio of 0.12.	Betalains and chlorophyll	6.67 wt% for betalains and 1.82 wt% for chlorophylls (per weight of biomass)	Rosa et al. (2023)

Agri-food wastes	Conditions for extraction	Targeted bioactive compound(s)/ bioproduct(s)	Production yield	References
Evodia lepta residue	A mixture of levulinic acid and glycerol in a molar ratio of 1.5:1, along with a 30% solution of K_2HPO_4 (0.4 g/mL), was employed for extraction. The solid-to-DES ratio used was 1:20 g/mL, and the extraction process was carried out for 40 minutes at a temperature of 75°C.	Skimmianine, dictamnine, and evodiamine	72.03, 39.16, and 80.67%, respectively	Yu et al. (2023)
Ginkgo biloba Exocarp	The hydrophobic DES MHL, consisting of l-menthol, 1-hexanol, and lauric acid in a ratio of 1:1.6:0.4, and the hydrophilic DES CCE, composed of choline chloride, citric acid, and ethylene glycol in a ratio of 3:2:1, when combined in a volume ratio of 1:1, exhibited the most favorable extraction yields.	Procyanidine, flavonoids, and terpene trilactones	60.97, 75.97, and 83.79%, respectively	Wu et al. (2022)
Ipomoea batatas slurry	Polyethylene glycol and sodium citrate were employed for the extraction process.	β-amylase	78.24%	Yusree et al. (2022)
(C) Enzyme-assisted extraction				
Berries residues	The sample (2.31 g) was diluted in potassium phosphate buffer (100 μM at pH 6) and the enzyme (20 mL), and then incubated for 2 h at 100 rpm and 50°C in a shaker (SL 222/Solab). After incubation, the extracts were placed in an ice bath to stop the enzymes' activity and speed up the chilling process. Then, after cooling to –18°C, they were centrifuged at 12,857 × g (10,000 rpm) for 10 minutes, and the resulting supernatant was filtered using Whatman No 2 filter paper.	Rutin, and naringenin	15737.13 μg/g DW and 1347.50 μg/g DW, respectively	Silva et al. (2022)

Table 15.3 (*Contd.*)

Agri-food wastes	Conditions for extraction	Targeted bioactive compound(s)/ bioproduct(s)	Production yield	References
Coffee coproducts	For 30 min, 8 g of materials were mixed into a suspension in 200 mL of phosphate buffer. The reaction was continued at 50°C for 30, 60, or 120 min after 5 or 15 U of cellulase were introduced. After being rinsed twice with distilled water and centrifuged at 3500 g for 10 min, the solution was neutralized with 0.5 M NaOH. Freeze drying was used on the frozen supernatant.	Chlorogenic acid, Caffeic acid, and Caffeine	0.18, 0.02, and 0.60 mg GAE/g	Belmiro et al. (2021)
Mango peel	Mango peel extracts were subjected to pretreatment using an alcalase enzyme solution with a concentration of 3.3%. The pretreatment process was carried out at a pH of 5.5 and a temperature of 63°C for 110 min.	Polyphenolic compounds	25.64 ± 0.98 mg GAE/g	Sharif et al. (2021)
Tomato peel	The peels underwent a 5% (w/v) Viscozyme L solution modification process at pH 6, with a solid-to-liquid ratio of 1:10 (w/v), for 35 min.	Lycopene	170.67 mg/100 g	Gu et al. (2020)
Tomato waste	A blend of cellulolytic and pectinolytic enzymes was employed. The experimental parameters were set as follows: the temperature of the enzymatic reaction was maintained at 40°C, the duration of the enzymatic reaction was 5 h, the enzyme-to-substrate ratio was 0.2 mL/g, the solvent-to-substrate ratio was 5 mL/g, the extraction process lasted for 1 h, and the enzyme to enzyme ratio was 1.	Lycopene	11.5 mg/g	Catalkaya and Kahveci (2019)

Agri-food wastes	Conditions for extraction	Targeted bioactive compound(s)/ bioproduct(s)	Production yield	References
(D) Microwave-assisted extraction				
Pineapple peel	The optimal conditions determined from the BBD experiment were a solvent-to-substrate ratio of 20 mL/g, microwave power of 600 W, and an extraction time of 40 min.	Bromelain	1647.612 GDU $g_{concentrate}^{-1}$	Bansod et al. (2023)
Musa sapientum Waste	Concisely, 300 g of powdered material from each component was subjected to extraction using a glycerol-ethanol mixture in a 1:1 volume ratio (1.5 L). This extraction process was carried out using microwave-assisted extraction, employing a power of 900 W and consisting of 15 irradiation cycles. Each cycle involved 1 min of power being applied followed by 1 m of power being turned off.	Vitamin C, vitamin E, phenolic content, and flavonoid content	10.47 mg/100 g, 4.77 mg/100 g, 39 mg GAE/100 g extract, and 26 mg QE/100 g extract.	Bilhman et al. (2023)
Punica granatum waste	The optimized experimental parameters included an ethanol concentration of 50%, a solvent-to-solid ratio of 25:1 mL/g, an extraction time of 4 minutes, and a microwave power of 450 W.	Total ellagitannin content and flavonoid content	279.2 mg TAE/g DM and 25 mg QE/g DM, respectively	Mali and Kumar (2023)
Olive pomace	The conditions that had been optimized for this experiment were as follows: a power input of 800 W, a duration of 180 s, and a concentration of 1 g/50 mL.	Total phenolic content and ascorbic acid	15.30 mg GAE/gdw and 10 mg AAE/gdw	Marđokić et al. (2023)
Walnut Hull	The optimal parameters for the microwave-assisted extraction method were documented as follows: a microwave power of 363.64 W, an extraction time of 3.133 min, and a solvent volume of 39.999 mL/g of the sample.	Total phenol content and flavonoid content	83.535 mgGAE/g and 18.98 mgQAE/g	Singh et al. (2023)

Table 15.3 (*Contd.*)

Agri-food wastes	Conditions for extraction	Targeted bioactive compound(s)/ bioproduct(s)	Production yield	References
Citrullus lanatas fruit rinds	The experimental parameters were determined to be an applied microwave power of 477 W, an irradiation time of 128 s, a pH value of 1.52, and a solid-to-liquid ratio of 1:20.3 g/mL, respectively.	Pectin	25.79%	Maran et al. (2014)
Rice bran	The experimental parameters yielding the best results were a microwave power of 1000 W, an extraction time of 90 s, and a solid-to-liquid ratio of 0.89 g rice bran per 10 mL of distilled water.	Protein hydrolysate	4.37%	Phongthai et al. (2016)
Persian Gulf shrimp waste	The conditions that had been optimized for this experiment were a power input of 720 W and a duration of 20 s.	Chitosan	19.47%	Nouri et al. (2016)
(E) Pressurized liquid extraction				
Granadilla waste	The process of PLE was conducted at a temperature of 70°C and a pressure of 10 MPa, using a solvent mixture consisting of 75% ethanol.	Phenolic compounds	96 ± 8mg GAE/g biomass	Vardanega et al. (2023)
Tahiti lime by-product	The solvent utilized in the experiment consisted of a blend of ethanol and water in a weight ratio of 3:1. The pressure within the system was maintained at a constant value of 10.0±1.0 MPa. The temperatures of 60, 85, and 110°C were subjected to evaluation. The rate of solvent flow was measured to be 4.74 g/min. Extraction procedures were conducted for 40 min, yielding a solvent-to-feed mass ratio (S/F) of 47.4 kg solvent/kg of dry lime pomace.	Phenolic compounds, hesperidin, and narirutin	17.66 mg GAE/g DP, 1.42 mg/g DP, and 0.22 mg/g DP, respectively.	Oliveira et al. (2022)

Agri-food wastes	Conditions for extraction	Targeted bioactive compound(s)/ bioproduct(s)	Production yield	References
Pomegranate by-Products	The extraction process involved subjecting ethanol to pressurization at a temperature of 60°C and a pressure of 40 bar.	α, β punicalagin, and ellagic acid	48, 146, and 25.6 mg/100 g, respectively	Toledo-Merma et al. (2022)
Pomegranate by-Products	The extraction process involved utilizing a temperature of 200°C, an ethanol concentration of 77%, and a pressure of 1500 psi for 20 min.	Punicalagin	17 ± 3.6 mg/g DW	García et al. (2021)
Passion fruit rinds	A total mass of 3.0 g of recently harvested passion fruit rinds was exposed to a pressure of 10 MPa and a temperature of 90°C.	Isoorientin, vicenin, isovitexin, vitexin, and orientin	106, 26, 9, and 2.55 μg/g dr	Pereira et al. (2021)
Beta vulgaris L.	The PLE experiment was conducted at a controlled temperature of 40°C, with varying pressures of 7.5, 10, and 12.5 MPa. The flow rate of the experiment was maintained at a constant value of 3 mL/min.	Phenolic compounds	0.6128 mg/g	Lasta et al. (2019)
(F) Ultrasound-assisted extraction				
Olive pomace	Solvent: Ethanol Time: 28 min Power: 490 W	Hydroxytyrosol and tyrosol	36 and 14 mg g^{-1} of extract	Madureira et al. (2023)
Blueberry leaves	Solvent: NEDS Time: 45 min Temperature: 65°C Solid to solvent: 5:1 (w/v)	Phenolic compounds	195.5 mg GAE/g DBL	Santos-Martín et al. (2023)
Allium sativum leaves	Solvent: 50% Ethanol Time: 13 min Amplitude: 53%	Phenolic compounds and flavonoids	9.9 mg GAE/g and 6.8 mg QE/g, respectively	Shekhar et al. (2023)

Table 15.3 (*Contd.*)

Agri-food wastes	Conditions for extraction	Targeted bioactive compound(s)/ bioproduct(s)	Production yield	References
Benincasa hispida rinds	Solvent: Ethanol Time: 45 min Power: 280 W Solid to solvent: 1:10 (w/v)	Phenolic compounds and flavonoids	23.39 ± 0.22 μg GAE/mg and 11.17 ± 0.17 μg RE/g	Yusoff et al. (2023)
Citrullus lanatus rinds	Solvent: Ethanol Time: 45 min Power: 280 W Solid to solvent: 1:10 (w/v)	Phenolic compounds and flavonoids	15.10 ± 0.58 μg GAE/mg and 6.16 ± 0.09 μg RE/g	Yusoff et al. (2023)
Walnut processing waste	Solvent: Acidified distilled water Time: 10 min Power: 200 W Solid to solvent: 1:15 (w/v)	High-methylated pectin	12.78%	Asgari et al. (2019)
(G) Supercritical fluid extraction				
Orange, tangerine, and lemon peels	Pressure: 30 MPa Temperature: 60°C Time: 5 h Flow rate: 10 mL min^{-1}	Naringin	35.26, 44.05 and 19.86 mg g^{-1}	Romano et al. (2022)
Vitis vinifera waste	Pressure: 250 bar Temperature: 60°C Time: 15 min Flow rate: 2 mL min^{-1}	Trans-resveratrol, β-sitosterol and α-tocopherol	140.7 2942.5, and 333.6 μg/g dm	Aresta et al. (2020)
Avocado waste	Pressure: 25 MPa Temperature: 80°C Time: 38 min Flow rate: 2.0 mL/min	Phenolic content	5136 mg GAE/100 g	Paramos et al. (2020)
Tomato pomace	Pressure: 38 MPa Temperature: 80°C	Lycopene	619.3 mg kg^{-1} Trans 53.82:C is 46.17	Scaglia et al. (2020)

Agri-food wastes	Conditions for extraction	Targeted bioactive compound(s)/ bioproduct(s)	Production yield	References
(H) Supercritical fluid and pressurized fluid extraction				
Turmeric waste	SCF: Pressure: 24 MPa Temperature: 50 °C Time: 20 min Flow rate: 4.5×10^{-4} kg/s PFE: Pressure: 10 MPa Temperature: 50 °C Time: 60 min Solid to solvent: 1:9.5	Starch	80%	Santana et al. (2017)
(I) Pulsed-electric field and mechanical pressing				
Waste chicken meat	A series of high-voltage pulses, followed by low-voltage pulses of longer duration, were applied. The total amount of energy invested was measured to be 38.4 ± 1.2 J/g, with an additional load of 2 kg.	Water soluble proteins	12.00%	Ghosh et al. (2019)
(J) Mechanical extraction				
Pineapple stem	—	Amylose	34.4%	Nakthong et al. (2017)

Table 15.3 (*Contd.*)

Agri-food wastes	Conditions for extraction	Targeted bioactive compound(s)/ bioproduct(s)	Production yield	References
(K) Food-grade sodium lignosulphonate				
Dairy wastewater	The experiments were conducted within a pH range of 1.0–3.5, with varying concentrations of Na-lignosulphonate (0.002%, 0.004%, 0.008%, 0.016%, and 0.032% w/v), at a temperature of 22°C. The experiments were additionally performed at a temperature of 40°C under acidic conditions with pH values of 1.0, 2.0, and 3.5.	Lipids, and Caseins	96%, and 46%	Kurup et al. (2019)
(L) Low-voltage direct-current electric field				
Squid viscera	Electric field intensities of 5, 10, and 15 V/cm were administered to the squid visceral samples for durations of 1, 3, 6, 9, and 12 h at an intensity of 5 V/cm; 0.5, 1, 1.5, 2, and 2.5 h at an intensity of 10 V/cm; and 10, 20, 30, 40, and 50 min at an intensity of 15 V/cm.	Lipid	72.1%	Guo et al. (2018)

specific waste type and the methodological approach employed for extraction. These aspects have been thoroughly discussed in the subsequent sections.

Solvent Extraction

The process of revalorizing AFWs to extract BACs with increased value offers both economic and environmental advantages. This approach facilitates the establishment of a circular economy for the waste generation process. Depending on the origin, diverse BACs can be extracted, including pigments, polyphenols, bioactive peptides, fermentable sugars, and other substances. The utilization of solvents, such as n-hexane, chloroform, dichloromethane, petroleum ether, benzene, and toluene, to extract BACs is widely employed due to its ability to yield high extraction efficiencies (Kitrytė et al. 2020, Teles et al. 2021). Table 15.3A presents a concise overview of recent scientific investigations on various BACs and BPs derived from AFWs through the utilization of solvent extraction techniques. Nevertheless, the development of an ecologically sustainable and secure procedure necessitates the complete elimination of the utilization of these solvents to the maximum extent feasible, or alternatively, employing solvents that are widely acknowledged as safe (Santana and Macedo 2019). One additional detrimental consequence of solvent usage is the production of cell wall compounds that engage in interactions with phenolic BACs. Consequently, when these specific molecules are the focus, their extraction efficiency diminishes, as does their biological functionality, including antimicrobial, antioxidant, antiproliferative, and anti-inflammatory activities.

Aqueous Biphasic System

Aqueous biphasic system (ABS) represents a category of liquid-liquid extraction methods that exhibit enhanced biocompatibility compared to other techniques (Khoo et al. 2020, Kee et al. 2020). The combination of two water-soluble compounds, such as polymers, salts, ionic liquids, alcohols, polysaccharides, and deep eutectic solvents, at a critical concentration, leads to the formation of two immiscible aqueous phases in ABS. This establishes a moderate milieu that is conducive to a diverse array of biomolecules, including proteins, antibiotics, and colorants, among other substances. In general, the ABS is primarily employed in two primary domains: (1) downstream processing and (2) sample pretreatment, primarily for analyte concentration purposes. The latter is essential for the analytical quantification of diverse sample categories, particularly when the analytes are measured in parts per million (ppm), as is the situation with the identification of various elements in agro-food waste. Consequently, ABS has been extensively suggested for use in downstream processing due to its capability to facilitate the separation, concentration, and purification of BACs. This process is typically accomplished in a single step, making it a well-regarded method that is both straightforward and cost-effective. Additionally, the ABS offers the advantage of easy scalability, as depicted in Figure 15.3 (Khoo et al. 2020, Kee et al. 2020, Pereira et al. 2020, Saha et al. 2022). Table 15.3B provides a concise overview of the latest research conducted on ABS, specifically their application in the extraction of various BACs and BPs derived from AFW materials.

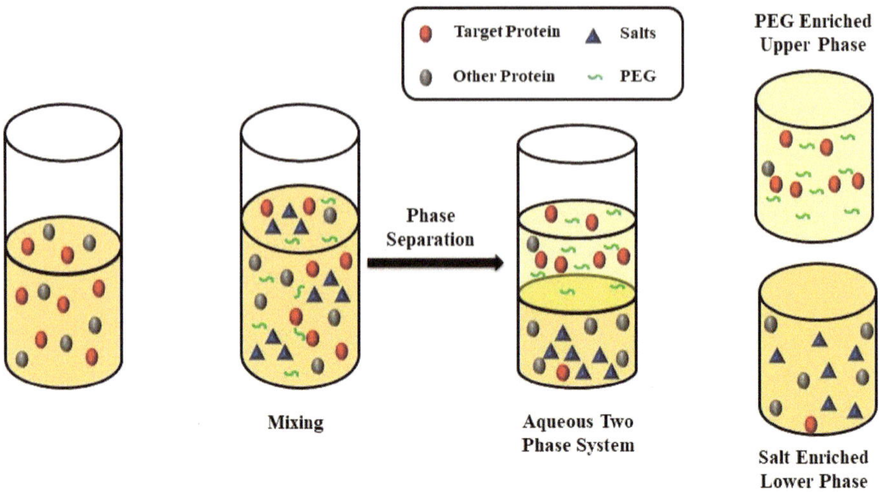

Figure 15.3 Schematic presentation on the separation of proteins using an aqueous biphasic system (Saha et al. 2022).

In recent scientific investigations, ABS with a significant amount of water and a reduced interfacial tension has been extensively utilized for the extraction of diverse BACs from AFW materials (Asenjo and Andrews 2011, Goja 2013). The single-step operation of ABS is expected to yield a high recovery rate and increase the purification fold of target BACs. This makes ABS a commonly preferred and cost-effective method for the industrial-scale production of diverse BACs (Selvaraj et al. 2011, Wu et al. 2017). In recent studies, ABS has demonstrated its efficacy in extracting various natural BACs, including enzymes (Patchimpet et al. 2021), proteins (Saravana Pandian et al. 2022), and food additives (Yu et al. 2021). These valuable BACs have already been extracted from various sources, which can be broadly categorized as expired food products and by-products of the agricultural industry, such as peels, seedcakes, fish residues, and similar sources (Dimitrijević et al. 2020, Gao et al. 2020, Jiménez-Velázquez et al. 2020, Ng et al. 2020a, Veloso et al. 2020, Patchimpet et al. 2021, Yu et al. 2021, Saravana Pandian et al. 2022, Zhu et al. 2022).

Due to the persistent hesitance of the industry to incorporate ABS in the production process and the critique of limited optimization of fundamental partition parameters (specifically pH, temperature, and phase-forming components) solely at the laboratory scale, ongoing research is gradually introducing additional factors to enhance the appeal of this technology for commercial applications. These strategies encompass the utilization of thermoseparating polymers to mitigate recycling expenses (Gao et al. 2020), pH-driven ABS to optimize energy efficiency (Yu et al. 2021), enhanced computational models (Saravana Pandian et al. 2022), and the adoption of continuous or semicontinuous operations (Veloso et al. 2020). In their study, Gao et al. (2020) introduced a two-step ABS process employing a straightforward methodology for the extraction of polysaccharides from seed cake. Following a solid-liquid extraction utilizing deep eutectic solvents, a thermosensitive polymer was introduced to facilitate the formation of an ABS. The product subsequently underwent migration towards the phase rich in polymer, which was then separated from the phase rich in the deep eutectic solvent. The separated polymer-rich phase was then subjected to heating to induce phase separation, resulting in the formation of a second product of ABS. The product and polymer were recovered in separate phases during this process. In this manner, a separation process with enhanced efficiency was conducted, surpassing the performance of the traditional system, while also being more cost-effective than enzymatic extraction. The valorization of AFWs has also been demonstrated, wherein fructose conversion was carried out using an ABS (Yu et al. 2021). The scientists successfully substituted crude fructose with apple peel and conducted concurrent extraction and conversion within the ABS (Li et al. 2020, Yu et al. 2021).

Enzyme-assisted Extraction

The utilization of enzymes to facilitate the extraction of BACs is increasingly being recognized as a feasible substitute for conventional solvent extraction techniques. Enzyme-assisted extraction (EAE) is an economically viable and ecologically sustainable extraction method that relies on the enzymatic property of catalyzing reactions with precise specificity and regioselectivity (Nadar et al. 2018, Gligor et al. 2019). Additionally, enzymes exhibit the capability to facilitate reactions under gentle conditions, characterized by low temperatures and short durations. In this particular method of extraction, enzymes are employed to perturb the intricate structure of the cell wall and catalyze the hydrolysis of its constituent components (Marathe et al. 2017). Consequently, the cellular membrane undergoes disintegration, leading to an increase in cell permeability (Nadar et al. 2018, Gligor et al. 2019). This results in the release of intracellular components, as depicted in Figure 15.4A. This disintegration facilitates the infiltration of the solvent, which may consist of water or organic compounds, to enhance the process of metabolism (Das et al. 2021). Targeted BACs are frequently sequestered within the cellular compartments of plants, necessitating the traversal of additional barriers for their access. The impediments encountered are primarily composed of extracellular cell walls, cell walls, and oleosomes (oil bodies). However, these barriers can undergo natural hydrolysis facilitated by specific enzymes. Hence, the enzymes predominantly employed for the hydrolysis of plant cell wall compounds encompass four distinct categories, which are as follow: (1) hemicellulases, (2) cellulase, (3) proteases, and (4) pectinases. Each of these enzymes possesses the capability to execute a targeted extraction, either individually or in conjunction, of the BACs existing within the sample (Santana and Macedo 2019, Bilal and Iqbal 2019, Kumar et al. 2020, Barbosa et al. 2021, Jha and Sit 2022). In contrast to conventional techniques, EAE methods have demonstrated reduced energy consumption, accelerated extraction rates, enhanced extraction yields, and simplified recovery processes with decreased solvent utilization (Nadar et al. 2018). However, EAE does present certain drawbacks, such as the elevated cost associated with the enzymes employed, thereby restricting the application of this technique solely to BACs of significant value. Simultaneously, the employed enzymes are incapable of degrading the cellular matrix. Therefore, an additional stage of sample separation and purification is necessary. Furthermore, the enzymes exhibit suboptimal recyclability and reusability, experiencing a decline in their catalytic efficacy following multiple applications. Hence, the aforementioned drawbacks of EAE contribute to its elevated cost and consequently hinder its viability for implementation on a large-scale industrial level (Das et al. 2021). The factors that can be adjusted to enhance the efficiency of EAE include pH and system temperature, enzyme concentration and composition, extraction duration, liquid-solid ratio, and substrate particle size (Marathe et al. 2017). Therefore, through the manipulation of these operational parameters, this approach can attain superior efficiency in extracting diverse BACs, including phenols, pigments, and anthocyanins, from various natural sources when compared to traditional methodologies. Hence, a primary benefit of this approach lies in the ability to tailor an enzymatic blend based on the specific matrix and target BACs, thereby maximizing the potential yield (Marathe et al. 2017, Nadar et al. 2018, Gligor et al. 2019, Das et al. 2021, Rojas et al. 2022). Table 15.3C provides a comprehensive summary of the most recent scientific investigations on EAE techniques employed in the retrieval of various BACs and BPs derived from AFW materials. The utilization of EAE enables the preservation of molecular functionality through the application of gentle conditions (Yang et al. 2021). Furthermore, the utilization of enzymes for the extraction of value-added BACs facilitates environmentally friendly procedures, eliminating the need for solvents and mechanical pressing methods (Cheng et al. 2019). Phenolic BACs have been isolated from berry pomaces (Saad et al. 2019), grape (Vitis spp.) pomace (Montibeller et al. 2019, Teles et al. 2021), pomegranate (Punica granatum) peels (Nag and Sit 2018, Kumar et al. 2020), citric peels (Kantifedaki et al. 2018), and soybeans (Glycine max) (Kumar et al. 2019a) through the utilization of pectinases, cellulases, and β-glucosidase enzymes. The extraction of anthocyanins and pigments is primarily documented in literature using grape pomace (Montibeller et al. 2019), raspberry wine residues (Xue et al. 2021), tomato waste (Catalkaya and Kahveci 2019, Gu et al. 2020), black soybean seed coat (Kumar et al. 2019a), and sunflower waste (Ricarte et al. 2020) as sources. In these studies, the pectinase enzyme is commonly employed as the primary catalyst for the recovery process. In the most recent experiment, peptides were extracted from the bovine bone using a commercially available blend of proteolytic enzymes (Yang et al. 2021). The majority of these isolated BACs are designed for utilization as functional food ingredients, nutraceuticals, or food supplements; thus, it is crucial to assess their resistance to processing (Ricarte et al. 2020). Nevertheless, a limited number of studies have assessed the stability of the BACs under investigation and their corresponding biological characteristics. In this context, it has been demonstrated by Rivero-Pino et al. (2020) that the bioactive peptides derived from wasted sardines, which underwent hydrolysis using a commercially available protease (Alcalase), retained their antihypertensive and antioxidant properties following various thermal treatments, high-pressure homogenization, ultrasound processing, and spray drying. While EAE exhibits selectivity, enhanced recovery of target BACs has been documented when it is synergistically employed in conjunction with another extraction technique

(Kitrytė et al. 2020). In this context, enhanced processing efficiency can be attained through the integration of various methodologies, thereby mitigating the potential for degradation during the extraction process. For example, the study showcased the synergistic impact of utilizing a blend of commercially

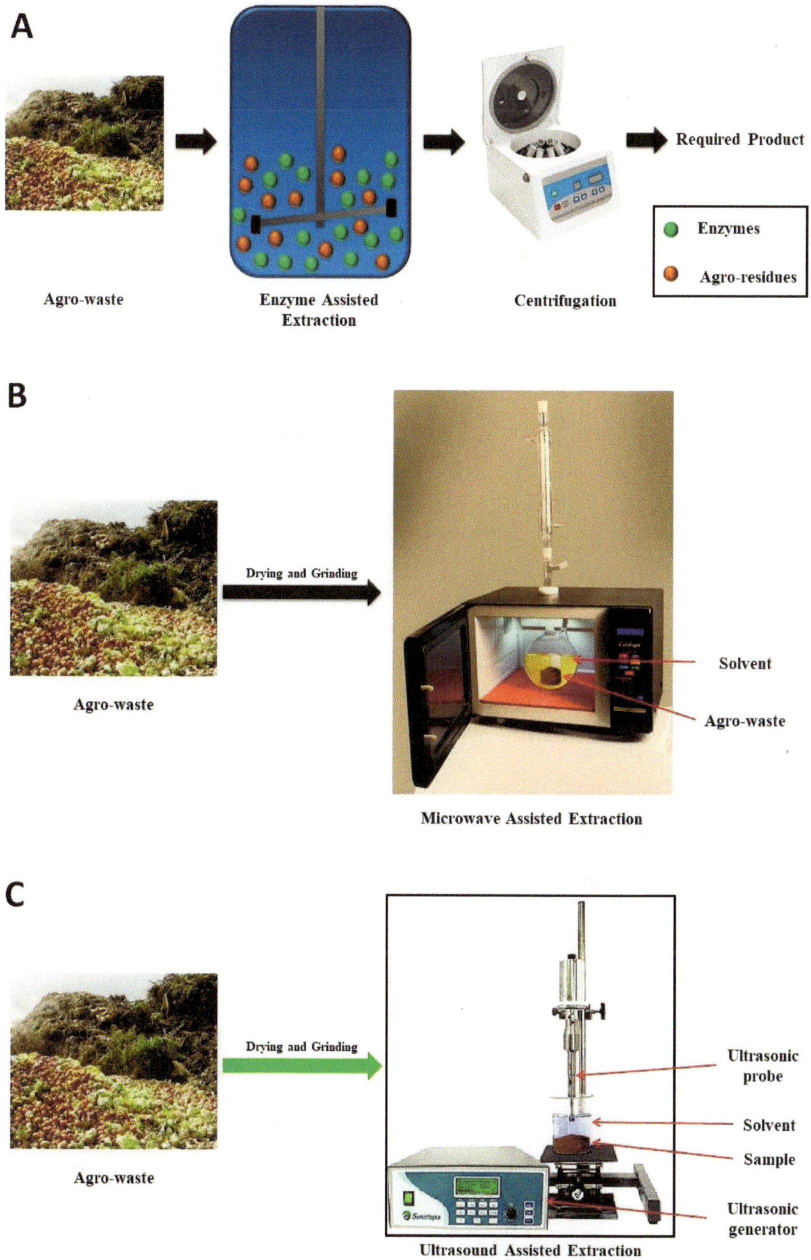

Figure 15.4 *(Contd.)*

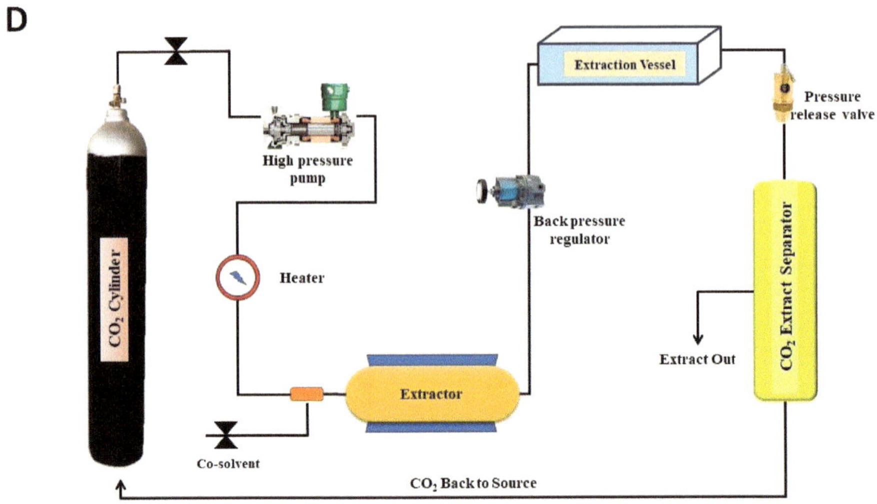

Figure 15.4 Schematic representation illustrating the utilization of advanced extraction methodologies employed for the isolation of multiple valuable bioactive compounds derived from diverse value-added bioproducts generated from agri-food waste sources. These methodologies include (A) Enzyme-assisted extraction (Saad et al. 2019), (B) Microwave-assisted extraction (del Pilar Sánchez-Camargo et al. 2021), (C) Ultrasound-assisted extraction (Poyraz et al. 2021), and (D) Supercritical fluid extraction (Gallego et al. 2019).

available cellulases and microwave-assisted extraction (MAE) (Kumar et al. 2020). This combination proved to be more effective in enhancing the extraction of polyphenolic BACs from pomegranate peels compared to using each technology individually. Furthermore, a study has documented the utilization of enzyme-assisted and microwave-assisted methodologies for the extraction of anthocyanins from black soybean seed coats (Kumar et al. 2019a).

The utilization of ultrasound and enzymes has facilitated the extraction of proteins (Yan et al. 2020), phenolic BACs (Shahram et al. 2019), vegetable oil (Amigh and Taghian Dinani 2020), and lycopene (Rahimpour and Taghian Dinani 2018) from diverse AFW sources. The integration of these methodologies yields optimal extraction efficiency. Cavitation waves enhance the permeability of plant cells, thereby facilitating the entry of enzymes (Shahram et al. 2019). At low frequencies and intensities, these waves can induce a state of hyperactivity in the enzymes, leading to an increased rate of catalyzed reactions. Furthermore, the application of these techniques can vary depending on the specific technology, processing conditions, and characteristics of the enzymes. They can be implemented concurrently or sequentially as deemed appropriate. The latter process occurs when the extractive technology induces enzyme inactivation by altering its molecular structure, thereby impacting its ability to catalyze reactions (Mohamad et al. 2019). The extraction of α-linolenic and linoleic fatty acids from lingonberry pomace (Kitrytė et al. 2020) and phenolic BACs from grape pomace (Teles et al. 2021) has been achieved through a combination of EAE, followed by SCF-CO_2 extraction and high hydrostatic pressure treatment.

Microwave-assisted Extraction

Microwave-assisted extraction (MAE) employs electromagnetic radiation within the microwave spectrum to harness energy for the extraction process. The phenomenon of microwave absorption involves the stimulation of BACs, leading to their transition to elevated rotational states (Verma et al. 2020, Prabhakar et al. 2023a, b). Consequently, the surplus energy is subsequently dispersed within the surrounding medium, manifesting as thermal energy (Verma et al. 2020). This technique is predicated on the expedited application of microwave radiation, which induces the disintegration of plant cell walls as the cellular framework undergoes expansion (Figure 15.4B) (Chávez-González et al. 2020, Carmen et al. 2023, Zhang et al. 2023a). Consequently, the methodology relies on targeted or non-targeted implementations of non-ionizing electromagnetic waves within the frequency range of 300 MHz to 300 GHz on a given sample (Leonelli et al. 2012, Barba and d'Amore 2012, Chávez-González et al. 2020, Prabhakar et al. 2023a, b). There are three sequential stages involved in this particular extraction process: the initial phase entails the segregation of dissolved substances from the sample matrix, followed by an elevation in temperature and pressure, and ultimately, the solvent permeates into the sample matrix (Jara-Palacios et al. 2019, Chávez-González et al. 2020, Prabhakar et al. 2023a, b). Table 15.3 D provides a comprehensive overview of the recent scientific investigations conducted on the application of MAE techniques for the efficient retrieval of various valuable BACs and BPs derived from AFW sources.

The utilization of MAE has emerged as a promising technique for the extraction of BACs. This method offers several advantages over traditional extraction methods, including enhanced extraction efficiency and improved recovery results. By applying microwave energy, the extraction process is accelerated, leading to reduced extraction times compared to conventional techniques (Chávez-González et al. 2020, Prabhakar et al. 2023a, b). Additionally, MAE has been found to enhance the recovery of BACs, ensuring a higher yield of target BACs. These outcomes highlight the potential of MAE as a valuable tool in the field of BACs (Prabhakar et al. 2023a, b). Similar to other techniques, the extraction efficiency can be influenced by specific parameters. In this particular scenario, the parameters of interest encompass temperature, frequency, irradiation duration, solid-to-liquid ratio, microwave power, extraction pressure, solvent characteristics and composition, as well as matrix particle size (Chávez-González et al. 2020). The selection of an appropriate solvent that effectively absorbs microwave energy is a crucial factor in MAE. The crucial factor for the recovery of the desired BACs lies in their capacity to solubilize the target BACs and effectively absorb microwave energy (Prabhakar et al. 2023a, b). Hence, solvents with a significantly elevated dielectric constant are predominantly employed for this particular extraction method. Therefore, polar solvents, such as water and ethanol, have the ability to absorb electromagnetic energy and subsequently convert it into heat (Chávez-González et al. 2020, Verma et al. 2020, Prabhakar et al. 2023a, b). This phenomenon promotes the extraction process by causing internal warming, which in turn leads to cell disruption (Arruda et al. 2021).

The application of MAE technology enhances the efficiency of anthocyanin extraction by promoting accelerated mass transfer rates. Consequently, the concentration of anthocyanins released into the extraction medium is elevated (Arruda et al. 2021). The extraction method discussed in previous research has been found to offer several notable advantages. These include a decreased time requirement, a reduced solvent volume, a higher yield in comparison to traditional extraction methods, and improved energy efficiency resulting from its rapid processing time (Arruda et al. 2021). Nevertheless, a drawback associated with this approach is the detrimental impact of microwave energy on anthocyanins. Specifically, the escalation of microwaves results in a reduction in the efficacy of anthocyanin extraction due to their degradation (Kaur et al. 2018). In a study conducted by Kulkarni and Rathod (2016), MAE was utilized as a method for extracting mangiferin from the leaves of Mangifera indica. The extraction process involved the use of water as a solvent. In the conducted study, the researchers were able to achieve a maximum yield of 55 mg/g using a solvent-to-solid ratio of 20:1 and applying a power of 272 W for 5 min. In comparison, the traditional Soxhlet extraction method resulted in a yield of 57 mg/g but required a significantly longer extraction time of 5 h (Kulkarni and Rathod 2016). The study concluded that the primary factors that exhibit variation based on the source of raw materials, matrix permeability, and targeted BACs are the optimal microwave circumstances and the concentration of the solvent.

Pressurized Liquid Extraction

Pressurized liquid extraction (PLE) is a contemporary method that has gained recognition as an effective technique for extracting BACs. PLE involves the utilization of water and/or organic solvents, along with elevated temperature and pressure, to obtain these BACs (Sumere et al. 2018). This technique has emerged as a promising alternative for extraction due to its ability to enhance the extraction efficiency and yield of desired BACs. By subjecting the sample to increased pressure and temperature, PLE facilitates the extraction process by promoting the dissolution of target BACs into the solvent. Consequently, PLE offers a viable approach for the extraction of BACs, providing researchers with a valuable tool for obtaining these substances. PLE, alternatively referred to as accelerated solvent extraction (Mottaleb and Sarker 2012), pressurized fluid extraction (Vardanega et al. 2023), enhanced solvent extraction (Seabra et al. 2010), or high-pressure solvent extraction (Khan et al. 2019), is a method utilized in scientific research and industrial applications (Jha and Sit 2022). PLE is a highly effective method that enables rapid and efficient extraction of BACs with varying polarities (Fierascu et al. 2019). PLE offers several advantages over traditional solid-liquid extraction, including reduced extraction time and solvent usage.

The underlying principle of this methodology is predicated upon the utilization of elevated pressure to maintain the solvent in a liquid state at exceedingly high temperatures that surpass its boiling point. This approach serves to enhance the solubility of the compounds, facilitate the wetting of the sample, and enable the penetration of the matrix (Silva et al. 2017). The aforementioned conditions give rise to diminished surface tension and viscosity, thereby facilitating the infiltration of the solvent into the solid matrix and resulting in an augmented rate of mass transfer and diffusion. Simultaneously, it is noteworthy that the existence of air within the vacuoles of plant cells gives rise to a phenomenon known as high-pressure leakage. This occurrence leads to the denaturation of proteins within the cell membrane, consequently enhancing the accessibility of desired BACs during the extraction process (Picot-Allain et al. 2021). The pressure range employed in this particular extraction method spans from 100 MPa to values exceeding 1000 MPa. This wide pressure range is crucial for enhancing the solvent's ability to penetrate the cell membrane, thereby increasing the bioaccessibility of the desired BACs. According to the existing literature, there is a positive correlation between hydrostatic pressure and the release of cellular components. As the pressure increases, the extraction efficiency also increases (Lefebvre et al. 2021, Tena and Asuero 2022). However, it is important to note that higher pressure can result in a decrease in selectivity. One notable benefit of employing PLE is the potential to utilize a diverse range of solvents, thereby facilitating the retrieval of both polar and non-polar BACs. Therefore, environmentally friendly solvents, such as ethanol, ethyl lactate, ethyl acetate, and aqueous solutions, can be utilized (Alvarez-Rivera et al. 2020). Additional benefits of this technique include the utilization of a reduced quantity of solvent, a decreased duration of extraction, and an enhanced bioaccessibility of the BACs found within the sample. Nevertheless, the primary drawback of these extraction techniques is the economic burden associated with the energy expenditure necessary to generate elevated pressures (Tena and Asuero 2022). By manipulating the extraction conditions, it is possible to retrieve distinct groups of polyphenols, including phenolic acids, flavonols, anthocyanins, and stilbenes (Allcca-Alca et al. 2021, Rifna et al. 2021). Recently, several studies have investigated various parameters that can be potentially modified to optimize the extraction process (Tena and Asuero 2022, Vardanega et al. 2023). These parameters include pressure, solvent composition, particle size, temperature, extraction time, liquid-solid ratio, and extraction duration. This ensures the overall efficacy of the extraction process for BACs from agricultural wastes, specifically vegetable and fruit peels (Rifna et al. 2021).

Table 15.3 E presents a comprehensive summary of recent studies conducted on PLE techniques employed for the extraction of various BACs and BPs derived from AFWs. These studies have explored the potential of PLE as an efficient and sustainable method for extracting valuable BACs from waste materials in the agricultural and food industry. By utilizing PLE, researchers have sought to maximize the extraction efficiency and yield of BACs and BPs, thereby contributing to the development of a more sustainable and economically viable waste management strategy. The studies included in Table 15.3 E highlight the diverse range of BACs and BPs that can be obtained through PLE, including but not limited to BACs, antioxidants, essential oils, and pigments. The findings presented in these studies underscore the potential of PLE as a promising extraction technique for the valorization of AFWs, offering a sustainable solution for the utilization of these waste materials and the production of BPs.

Limited data regarding the extraction of BACs from pomegranate peel through the utilization of PLE has been documented in the existing literature. In an initial investigation, it was demonstrated that the total phenolic content obtained through PLE using water was three times greater than that obtained through solid-liquid extraction using water under atmospheric pressure conditions (Çam and Hışıl 2010). However, another study investigated the utilization of a novel approach involving the combination of ultrasound and PLE as a viable and sustainable method for the extraction of phenolic BACs from pomegranate peel. This research aimed to explore the potential of this approach as a clean and environmentally friendly alternative to conventional extraction methods (Sumere et al. 2018). A recent study focused on the utilization of pressurized water and ethanol for the extraction of BACs from pomegranate peel (García et al. 2021). The main aim of this study was to determine the optimal conditions for PLE to obtain a pomegranate peel extract with the highest total phenolic content, punicalagin concentration, and antibacterial activity. Specifically, the focus was on investigating the effects of ethanol percentage and process temperature on these desired outcomes. According to the outcomes of the multiple response optimization, the optimal operational parameters for the process were determined to be a temperature of 200°C and an ethanol concentration of 77%. In optimal conditions, the pomegranate peel extract using PLE exhibited a total phenolic content of 164.3 ± 10.7 mg GAE/g DW and a punicalagin content of 17 ± 3.6 mg/g DW (García et al. 2021).

Ultrasound-assisted Extraction

The utilization of oscillating sound pressure waves with a frequency exceeding 20 kHz has been observed to facilitate the acceleration of mass transfer between the solvent and the sample matrix (Barba et al. 2015, Al-Hilphy et al. 2016, Zhu et al. 2022, Tripathy and Srivastav 2023a, b). Ultrasonic-assisted extraction (UAE) is a technique that relies on the phenomenon of cavitation, which is a mechanical process induced by ultrasounds. This process facilitates mass transfer and BACs extraction, all the while operating at low temperatures and achieving short extraction durations (Rahimpour and Taghian Dinani 2018, Sumere et al. 2018, Chávez-González et al. 2020, Zhu et al. 2022, Martínez-Olivo et al. 2023, Tripathy and Srivastav 2023a, b). Cavitation induces mechanical phenomena within the matrix, enhancing the process of mass transfer and disrupting the matrix structure (Al-Hilphy et al. 2016, Low et al. 2022, Bucur et al. 2023). As a consequence, the recovery of the desired BACs is heightened (Figure 15.4C). Furthermore, the phenomenon of cavitation gives rise to the formation of bubbles which undergo growth until reaching a critical size, subsequently undergoing collapse. This collapse event results in the liberation of significant amounts of energy, leading to an elevation in both the temperature and pressure of the surrounding solvent (Al-Hilphy et al. 2016, Low et al. 2022, Bucur et al. 2023, Martínez-Olivo et al. 2023). UAE can be conducted through two methods: direct probe irradiation or utilization of an ultrasonic bath. The latter option exhibits greater cost-effectiveness and ease of use, albeit with certain drawbacks. Specifically, it typically operates at a singular frequency of either 20 or 40 kHz, resulting in uneven distribution of energy within the system (Chávez-González et al. 2020, Martínez-Olivo et al. 2023, Tripathy and Srivastav 2023a, b). Consequently, this uneven distribution restricts the overall efficiency of extraction. The process of direct UAE involves utilizing a probe containing the sample, which is placed within a transducer that is submerged in the extraction vessel. In this setup, the ultrasound waves disperse throughout the surrounding environment with minimal energy loss. In this manner, greater ultrasonic intensity is achieved compared to the system utilizing an ultrasonic bath, as the ultrasound energy is focused on a particular region of the sample, thereby increasing the efficiency of extraction (Chávez-González et al. 2020, Low et al. 2022, Martínez-Olivo et al. 2023, Tripathy and Srivastav 2023a, b). This extraction technique has the potential for large-scale implementation. However, it is imperative to possess distinct attributes such as sustainability, profitability, safety, and environmental compatibility. These objectives can be attained through the augmentation of the process and the mitigation of energy consumption. The factors that impact the efficacy of UAE include the intensity, amplitude, power, and frequency of the ultrasonic waves (Chávez-González et al. 2020, Gavahian et al. 2022, Martínez-Olivo et al. 2023, Tripathy and Srivastav 2023a, b). UAE is a widely employed technique for the extraction of BACs from diverse natural matrices. However, it is imperative to optimize the extraction parameters to achieve maximum BACs yield through the utilization of UAE (Chávez-González et al. 2020, Tripathy and Srivastav 2023a, b). The primary matrix-specific factors that can be optimized to enhance the extraction efficiency include the duration of extraction, composition of the solvent, temperature, ratio of liquid to solid, particle size, moisture content in the sample, and power of ultrasound (Nag and Sit 2018, Chávez-González et al. 2020, Martínez-Olivo et al. 2023, Tripathy and Srivastav 2023a, b).

Scientific communities are currently utilizing UAE to extract and recover BACs from AFWs. This information can be found in Table 15.3 F of relevant research. Several of them have been subject to critical discussion in this context. A study investigated the extraction of BACs from purple eggplant peels using the UAE (Ferarsa et al. 2018). Specifically, the focus was on the recovery of anthocyanins, which are known to possess various health benefits. The experimental results indicated that the quantity of anthocyanins extracted from the purple eggplant peels was directly proportional to the duration of ultrasonication (Ferarsa et al. 2018). In the context of anthocyanin extraction, it had been observed that the highest concentration of anthocyanins was achieved following a sonication period of 30 min. This peak concentration was measured to be approximately 29.011 mg GAE/g DM in peel extracts. It was worth noting that acidified water was utilized as the solvent in this particular extraction process. Furthermore, in this groundbreaking study conducted by Ferarsa et al. (2018), the presence of two distinct anthocyanins, specifically malvidin-3-rutinoside-5-glucoside and cyanidin-3-rutinoside, was successfully identified in the extract derived from the peel of eggplants. This discovery marked the first instance in which these particular anthocyanins had been detected in eggplant peel extract. This study also investigated the effects of utilizing an ultrasonic probe on the extraction of anthocyanins and demonstrated that the ultrasonic probe was found to enhance the extraction process, while microscopic analysis conducted subsequently revealed the occurrence of cell denaturation and damage (Ferarsa et al. 2018). Kaderides et al. (2015) conducted an experiment where they investigated the impact of varying the amplitude level on the extraction yield of phenolic BACs from pomegranate peel. They found that increasing the amplitude level up to 40% resulted in a higher extraction yield. This can be attributed to the increased contact surface area between the solid matrix (pomegranate peel) and the solvent surface, which in turn enhanced the mechanical and cavitation effects of ultrasounds (Kaderides et al. 2015). The observed phenomenon of a more forceful bubble collapse occurring within a shorter duration was attributed to the heightened amplitude of ultrasound waves. It was well established that the resonant size of a bubble was directly influenced by the amplitude of these waves (Kaderides et al. 2015). Moreover, the application of a high-speed jet had been observed in the study to enhance the rate of solvent penetration into the matrix and facilitate the release of phenolic BACs through the disruption of cell walls. Adequate extraction of phenolic BACs and flavonols from grape pomace was achieved using ultrasonic treatment at a frequency of 40 kHz, power density of 150 W/L, and an extraction time of 25 min (González-Centeno et al. 2014). In a study conducted by González-Centeno et al. (2014), it was observed that the phenomenon of cavitation during extraction processes was subject to the influence of extraction temperature. The impact of elevated temperatures on the rate of solvent diffusion can be attributed to the disruption of intermolecular bonds between the solvent and matrix. Extended extraction durations resulted in the emergence of unfavorable alterations in the extracted solution. Conversely, the reduced size of the sample matrix facilitated greater interaction between the exposed surface and the solvent, thereby promoting the destruction of cell pores and subsequently augmenting internal diffusion of the solute into the solvent (González-Centeno et al. 2014). In a study conducted by Jadhav et al. (2016), it was reported that the oil yield obtained from date seeds exhibited a significant correlation with the particle size of the samples. The researchers discovered that samples with particle sizes ranging from 0.54 to 1.5 mm yielded the highest amount of oil (Jadhav et al. 2016).

In this research effort, multiple groups of esteemed researchers undertook a comparative investigation (Drosou et al. 2015, Safdar et al. 2017). The primary objective of their study was to evaluate and compare the efficacy of two distinct extraction methods, namely soxhlet extraction and UAE. The researchers sought to determine the composition and concentration of phenolic BACs in the extracted samples. Upon conducting their experiments, the researchers made use of UAE and soxhlet extraction techniques. Subsequently, they meticulously analyzed the obtained extracts to ascertain the presence and quantity of phenolic BACs. It was worth noting that the researchers specifically focused on the ethanol:water (1:1) extracts derived from the UAE method. After careful analysis and thorough examination, the researchers arrived at a noteworthy finding. They observed that the ethanol:water (1:1) extracts obtained through UAE exhibited the highest concentration of phenolic BACs (Drosou et al. 2015, Safdar et al. 2017). This discovery underscores the potential superiority of the UAE method in extracting and preserving phenolic BACs. In conclusion, this comparative study conducted by various groups of researchers sheds light on the differential efficacy of soxhlet extraction and UAE methods. The researchers' findings indicated that the UAE technique, specifically utilizing ethanol:water (1:1) extracts, yields the highest concentration of phenolic BACs.

Therefore, based on the scientific evidence presented in the published literature and critically analyzed in the preceding paragraphs of this section, it can be inferred that UAE has garnered significant recognition for its multitude of advantages when compared to traditional extraction methods. UAE has the ability to reduce extraction time, enhance extraction rate, and improve yield. By utilizing ultrasonic waves, the accelerated extraction process of UAE allows for more efficient extraction of target BACs. This reduction in time not only increases productivity but also minimizes the risk of degradation or alteration of the extracted BACs. UAE enhances the extraction rate, leading to more effective extraction of desired BACs. The application of ultrasonic waves promotes the disruption of cell walls and membranes, facilitating the release of intracellular components. This enhanced extraction rate ensures a more comprehensive extraction of target BACs, thereby maximizing the efficiency of the extraction process. In addition to time and extraction rate, UAE demonstrates a higher yield.

Supercritical Fluid Extraction

Supercritical fluid extraction (SCFE) has emerged as a promising environmentally sustainable alternative to traditional organic solvents. This method utilizes supercritical carbon dioxide (SC-CO_2) as the extraction medium, which offers several advantages over conventional solvents. SC-CO_2 is generally recognized as non-toxic, cost-effective, and non-explosive, making it an attractive choice for various extraction processes (Kumar et al. 2017, Santana et al. 2017, Verma et al. 2018, Singh et al. 2022). The selection of CO_2 as the preferred solvent in extraction processes is often attributed to its critical point, which occurs at a temperature of 304 K and a pressure of 7.4 MPa. This critical point allows for the extraction process to be carried out at low temperatures, effectively preventing the thermal degradation of biomolecules (Singh et al. 2022). The diminished polarity of CO_2 amplifies the efficacy of extracting biomolecules with non-polar or slight polar characteristics, such as lipids and volatile oils, under supercritical conditions (Zhang et al. 2018, Singh et al. 2022). During the extraction process, the unprocessed material is first introduced into a specialized container equipped with controllers for temperature and pressure. Subsequently, the material is subjected to controlled pressurization using a pump, which ensures precise regulation of temperature conditions, while fluid is used as the pressurizing medium. The solutes present in the fluid are transported to the separators, where the solutes accumulate at the lower portion while the fluid is either recycled or discharged into the surrounding environment (Figure 15.4D) (Da Silva et al. 2016, Singh et al. 2022). The critical point of a fluid is characterized by its transition from gas-like to liquid-like behavior when subjected to specific conditions of temperature (critical temperature) and pressure (critical pressure). At this critical point, the fluid exhibits unique properties, allowing it to diffuse into a solid matrix similar to a gas, while also possessing a significant capacity to dissolve compounds akin to a liquid. SCFs possess several notable advantages in terms of diffusivity, solute capacity, and low viscosity compared to conventional solvents (Soquetta et al. 2018, Singh et al. 2022). These attributes are accountable for enhanced extraction efficiencies and reduced extraction durations. CO_2 with a critical temperature of 31°C and a critical pressure of 74 bar is the prevailing SCF employed in various food-related applications. Nevertheless, the solvation capacity of SC-CO_2 to dissolve BACs within a solid matrix is diminished as a result of its low polarity. Hence, it is frequently employed in combination with a co-solvent or a modifier (Singh et al. 2022).

The utilization of SCFE as a method for the isolation and purification of BACs from diverse biomass sources within the agricultural and food sector has garnered significant attention in recent scientific investigations. Table 15.3G presents a comprehensive overview of recent scientific investigations on the utilization of SCFE techniques for the extraction of various BACs and BPs derived from AFW materials. The extraction method for BACs from AFWs, which is guided by activity, efficiency, and optimization, plays a crucial role in the search for lead-potent molecules or BACs. The SCFE process can be utilized to obtain extracts from these AFW materials. This can be accomplished by incorporating modifiers such as ethanol, methanol, and ethyl acetate, which serve to enhance the polarity of the biomass. Consequently, this polarity enhancement leads to an increased yield of nutraceutical compounds. The AFW biomass is primarily composed of fruit and seeds, which constitute the second largest proportion. BACs obtained through extraction processes have various applications in the fields of nutrition and food science (Huang et al. 2018, Khanpit et al. 2021). These BACs can be utilized as dietary fibers, prebiotics, and natural gelling agents. The process of extracting BACs from AFWs biomass primarily involves the use of conventional solvents. These solvents are capable of extracting a diverse array of components, including fat, cholesterol, pigments, protein isolates, dietary fibers, complex polysaccharides, and polyphenols (Gerardi et al. 2020, Cruz-Casillas et al. 2021, Farag et al. 2022). The polyphenols themselves encompass various subclasses such as flavonoids, terpenoids, and carotenoids. The SCFE approach is a cost-effective and efficient extraction technique utilized for the isolation of biomolecules possessing pharmaceutical and nutraceutical properties from agricultural and food industry by-products. The AFW biomass produced by the beverage and frozen product industries is predominantly composed of a lipid-rich fraction (Castro-Vargas et al. 2019, Romano et al. 2020, Idris et al. 2022). This fraction consists of seeds, kernels, and pomace derived from various fruits, accounting for approximately 50% of the total waste biomass.

The extraction of commercially significant fatty acids and dietary fats from fruit seeds is commonly achieved through the utilization of SCFE techniques. SCFE is a well-established method that offers numerous advantages in terms of efficiency and selectivity. This process involves the use of supercritical fluids, typically CO_2, as the solvent to extract the desired BACs from the fruit seeds. Fatty acids and dietary fats play a crucial role in various industries, including food, pharmaceuticals, and cosmetics. These BACs are known for their nutritional value, as well as their functional properties, such as flavor enhancement, and texture improvement. Lipid intermediates, such as fat-soluble vitamins, pigments, sterols, phytosterols, and waxes, along with polar fractions comprising polyphenols like caffeine, chlorogenic acid, and polysaccharides, can be derived from AFW sources (Al Bulushi et al. 2018, Milea et al. 2020, Páramos et al. 2020). The extraction and isolation of BACs derived from AFW materials can be categorized into two main fractions: polar and non-polar. The polar fractions obtained from the peels or bagasse consist primarily of polyphenols such as chlorogenic acid, tocopherols, tocotrienols, caffeine, limonin, quercetin, flavonoids, catechins, ascorbic acid, resveratrol, and water-soluble pigments such as anthocyanins. On the other hand, non-polar fractions can be extracted from the seeds of fruits. Seeds predominantly serve as reservoirs for lipids and lipid intermediates, thereby making them a valuable source for lipid extraction. SCFE has emerged as an effective method for the efficient retrieval of these lipid compounds from seeds. The composition of major seed lipids primarily consists of saturated and polyunsaturated fatty acids. The utilization of valorized AFWs as a substitute for animal fats offers a convenient solution for incorporating saturated fats into various applications.

One of the primary constraints associated with the process of SCFE is the relatively sluggish extraction kinetics (Singh et al. 2022). In the pursuit of enhancing the efficiency of extraction, it is recommended to incorporate supplementary techniques such as ultrasound or enzyme treatment in conjunction with SCFE. This integration aims to amplify the mass transfer mechanism by disrupting the structural integrity of plant matrices.

Pulsed-electric Field Extraction

Pulsed electric field extraction (PEFE) has emerged as a promising technology for the extraction of BACs. This innovative method involves the application of short, high-voltage electric pulses to disrupt the cell membranes of biological materials, facilitating the release of desired BACs. PEFE offers several advantages over traditional extraction techniques, including increased extraction efficiency, reduced processing time, and improved product quality. As a result, PEFE has gained significant attention in various industries, such as food, pharmaceuticals, and cosmetics, for the extraction of valuable BACs. This paper aims to provide an overview of PEFE technology, its principles, and its applications in the extraction of BACs.

PEFE technique employs electric pulses to induce cellular destruction, without relying on thermal energy. The application of electric pulses is a common practice in various fields of research. These pulses are typically characterized by their short duration, which can range from milliseconds to nanoseconds (Kolb et al. 2006, Luengo et al. 2016). Additionally, they are administered at a moderate electric field strength, as reported by Kolb et al. (2006) and Luengo et al. (2016). The accumulation of charges on either side of the membrane surface, induced by exposure to electric fields, results in the generation of transmembrane potential on the cell surface. According to Barbosa-Pereira et al. (2018), cell electroporation occurs when the transmembrane potential surpasses a specific critical threshold, leading to the formation of pores in the weaker regions of the cell membrane. It facilitates a significant enhancement in cellular membrane permeability, thereby promoting the efflux of intracellular compounds. Hence, it is well-established that this phenomenon leads to enhanced extraction efficiency and kinetics, accompanied by decreased energy requirements and minimal ecological footprint (Toepfl et al. 2006, Arnal et al. 2018).

Moreover, PEFE demonstrates utility in the efficient extraction of thermally fragile BACs from the sample matrix (Puértolas and Barba 2016). In the study conducted by Luengo et al. (2013), it was demonstrated that orange peels treated with PEF exhibited an elevation in total phenolic content. The total phenols increased from an initial value of 11.76 mg GAE/100 g to 14.14, 26.92, 29.81, and 34.80 mg GAE/100 g when subjected to PEF intensities of 1, 3, 5, and 7 kV/cm, respectively. This finding suggests that PEF treatment has the potential to enhance the phenolic content of orange peels, which could have implications for their utilization in various applications.

In the realm of grape pomace processing, the extraction of anthocyanins is a topic of great interest (Barba et al. 2015, Montibeller et al. 2019, Tena and Asuero 2022). Anthocyanins are a class of pigments responsible for the vibrant colors observed in various fruits, including grapes (Arruda et al. 2021, Xue et al. 2021, Tena and Asuero 2022). These BACs possess numerous health benefits and have been linked to potential therapeutic applications (Silva et al. 2017, Arruda et al. 2021). Therefore, the development of efficient extraction methods for anthocyanins from grape pomace is of utmost importance. In recent studies, two extraction techniques, namely PEFE and high-voltage electric discharge, have been investigated for their efficacy in anthocyanin extraction from grape pomace (Barba et al. 2015). This study explored the potential application of a non-thermal treatment method for selective extraction, wherein the temperature was maintained below 5 °C. This technique aimed to preserve sensitive BACs during the extraction process. The application of PEFE was effective in reducing the extraction effort when used in conjunction with a classical extraction process (Barba et al. 2015, Barbosa-Pereira et al. 2018, Ghosh et al. 2019).

WHERE DOES AGRI-FOOD WASTE STAND CURRENTLY AND WHERE MIGHT IT GO IN THE FUTURE?

The rapid increase in the worldwide population leads to a surge in the need for food production and processing, resulting in the generation of substantial quantities of AFWs. The food processing activities along the supply chain result in the production of predominantly biodegradable organic waste (Sharma et al. 2020). The Food and Agriculture Organization (FAO) defines AFWs as the deterioration in both the quality and quantity of food that occurs throughout the entire supply chain system, specifically at the stages of production, post-harvest, and processing (Tsang et al. 2019). The minimization of AFWs is a pressing concern in various sectors, encompassing a wide range of perishable items. This waste category comprises fresh produce, including fruits and vegetables, as well as dairy products, bakery items, and meat. The sources of such waste are diverse, originating from households, restaurants, hotels, food processing facilities, industrial kitchens, and agricultural practices (Sharma et al. 2020, Arun et al. 2020, Gupta et al. 2022, Chourasia et al. 2022). Despite efforts to mitigate this issue, it is important to acknowledge that the complete elimination of AFWs remains an elusive goal. The FAO has reported that a significant proportion, approximately 33%, of the overall food supply allocated for human consumption is squandered each year (Sharma et al. 2020, Kannah et al. 2020). On a global scale, the annual production of wasted food amounts to 1.3 billion tons, which corresponds to approximately 28% of the total agricultural area utilized (Sharma et al. 2020). The management of these wastes typically involves two primary methods: incineration and landfilling. Incineration is a process in which the waste materials are subjected to high temperatures, resulting in their combustion. This method is commonly employed to reduce the volume of waste and eliminate potentially harmful substances. On the other hand, landfilling involves the disposal of waste materials in designated areas. The aforementioned alternatives, however, are accompanied by elevated costs and possess their respective drawbacks. Incineration, for instance, leads to the release of flue gas, while landfills contribute to water source contamination (Kannah et al. 2020). In light of this, scientists and researchers have undertaken various endeavors in recent years to mitigate the quantity of AFWs produced and to repurpose said waste (Ng et al. 2020b). Consequently, there has been a notable surge of 90% in research on the recuperation of AFWs during the past decade (Sharma et al. 2020). The utilization of AFWs, which are abundant in organic matter and nutrients, presents a promising opportunity for the production of BPs (Kannah et al. 2020). AFWs are known to contain a significant amount of valuable BACs, including sugars, proteins, starch, micronutrients, BACs, and dietary fibers. These BACs can serve as raw materials for the biotechnological process of converting them into high-value VBPs. Examples of such VBPs include biofuel, enzymes, probiotics, BACs, and biodegradable plastics (Torres-León et al. 2019, Ng et al. 2020b). The implementation of sustainable methodologies for the utilization of AFWs in the production of valuable VBPs has played a significant role in the establishment of a circular bioeconomy (Ng et al. 2020b). The development of novel bioprocesses, including the utilization of AFWs through enzymatic and microbiological means, plays a crucial role in establishing a secure and sustainable food system. This imperative aligns with the objectives outlined in the Global Sustainability Goals, thereby highlighting the significance of such advancements in the field of food science and technology (Lillford et al. 2020).

In brief, the escalating global human population has led to significant AFWs generation, necessitating the exploration of sustainable methodologies for managing this waste. This is crucial to address the environmental issues arising from the accumulation of AFWs. The conversion of AFWs into valuable VBPs has demonstrated feasibility for industrial implementation, because of the significant economic value and sustainability potential exhibited by optimized bioprocesses. Nevertheless, it is imperative to consider the financial implications associated with the investment cost of collecting and categorizing AFWs, with a specific focus on their chemical composition, to align with the objectives of the circular bioeconomy. Furthermore, the significant obstacles in the industrial implementation of the biotransformation of AFWs into valuable VBPs include the considerable initial investment cost required for conducting extensive biorefinery experimental trials to achieve optimal operation efficiency, as well as the infrastructure requirements. However, innovative and emerging sustainable methodologies for the utilization of AFWs and their transformation into valuable VBPs have resulted in the elevated commercial worth of VBPs and the emergence of new employment prospects. Moreover, the accumulation of AFWs can be effectively addressed through the effective management of natural resources.

CONCLUSIONS, RECOMMENDATIONS, AND FUTURE POSSIBILITIES

AFWs are produced whenever agricultural or food processing operations are carried out, and the food sector has a significant obstacle when it comes to the management of these AFWs. For the management of AFWs, biotechnological techniques have emerged as a potentially useful tool, and in comparison to more conventional AFWs management practices, these approaches provide several distinct benefits. Utilizing microbes and enzymes to turn AFWs into VBPs is one of the most important biotechnological techniques for the management of AFWs. For instance, agricultural leftovers have the potential to be utilized as a substrate in the manufacturing of enzymes, biofuels, and bioplastics. This method not only helps to cut down on waste but also generates new commercial prospects by making novel VBPs out of resources that would otherwise be thrown away. The production of biogas from organic waste materials such as food leftovers and animal manure can also be accomplished by the application of another biotechnological method known as anaerobic digestion. To lessen one's reliance on fossil fuels, one can turn to biogas, which is a form of renewable energy that can be used to produce both power and heat, as well as fuel for automobiles. The negative effects that wastes from agriculture and food production have on the surrounding environment can also be mitigated via the use of biotechnological methods. For instance, microbes may be utilized to breakdown organic contaminants in wastewater, lessening the detrimental effects that waste discharge has on the environment. In addition, bioremediation techniques may be utilized to decontaminate polluted soil and water by capitalizing on the inherent capacity of microorganisms to degrade pollutants. This can be accomplished through the use of natural processes. When one looks ahead, one sees that biotechnological methods for the management of AFWs have a bright future. It is anticipated that the development of new biotechnological tools and techniques, such as synthetic biology and genome editing, would significantly improve the efficiency and efficacy of existing methods of AFWs management. In addition, there is an ever-increasing demand for VBPs that are sustainable and favorable to the environment, which is expected to boost the use of biotechnological techniques for AFWs management in the agricultural and food sector. In a nutshell, biotechnological techniques have recently emerged as a potentially useful tool for the management of AFWs. Compared to more conventional methods of AFWs management, these approaches offer several benefits. The application of biotechnological strategies to the problem of AFWs management in the agricultural and food sector appears to have a promising future, owing to the development of new instruments and methods.

REFERENCES

Abbate, S., Centobelli, P., Cerchione, R., Giardino, G. and Passaro, R. 2023. Coming out the egg: Assessing the benefits of circular economy strategies in agri-food industry. Journal of Cleaner Production. 385: 135665.

Aghaei, S., Alavijeh, M.K., Shafiei, M. and Karimi, K. 2022. A comprehensive review on bioethanol production from corn stover: Worldwide potential, environmental importance, and perspectives. Biomass and Bioenergy. 161: 106447.

Al-Bulushi, K., Attard, T.M., North, M. and Hunt, A.J. 2018. Optimisation and economic evaluation of the supercritical carbon dioxide extraction of waxes from waste date palm (Phoenix dactylifera) leaves. Journal of Cleaner Production. 186: 988–996.

Al-Hilphy, A.R.S., Verma, D.K., Niamah, A.K., Billoria, S. and Srivastar, P. 2016. Principles of ultrasonic technology for treatment of milk and milk products. pp. 178–202. *In*: Meghwal, M. and Goyal, M.R. (eds). Food process engineering: Emerging trends in research and their applications. Apple Academic Press.

Allcca-Alca, E.E., León-Calvo, N.C., Luque-Vilca, O.M., Martínez-Cifuentes, M., Pérez-Correa, J.R., Mariotti-Celis, M.S., et al. 2021. Hot pressurized liquid extraction of polyphenols from the skin and seeds of Vitis vinifera L. cv. Negra Criolla pomace a Peruvian native pisco industry waste. Agronomy. 11(5): 866.

Alvarez-Rivera, G., Bueno, M., Ballesteros-Vivas, D., Mendiola, J.A. and Ibanez, E. 2020. Pressurized liquid extraction. pp. 375–398. *In*: Poole, C.F. (ed.). Liquid-phase Extraction. Elsevier.

Amigh, S. and Taghian Dinani, S. 2020. Combination of ultrasound-assisted aqueous enzymatic extraction and cooking pretreatment for date seed oil recovery. Heat and Mass Transfer. 56(8): 2345–2354.

Amores, I., Ballesteros, I., Manzanares, P., Sáez, F., Michelena, G. and Ballesteros, M. 2013. Ethanol production from sugarcane bagasse pretreated by steam explosion.

Andler, S.M. and Goddard, J.M. 2018. Transforming food waste: how immobilized enzymes can valorize waste streams into revenue streams. npj Science of Food. 2(1): 19.

Araujo, J., Sica, P., Costa, C. and Márquez, M.C. 2021. Enzymatic hydrolysis of fish waste as an alternative to produce high value-added products. Waste and Biomass Valorization. 12: 847–855.

Aresta, A., Cotugno, P., De Vietro, N., Massari, F. and Zambonin, C. 2020. Determination of polyphenols and vitamins in wine-making by-products by supercritical fluid extraction (SFE). Analytical Letters. 53(16): 2585–2595.

Arnal, Á.J., Royo, P., Pataro, G., Ferrari, G., Ferreira, V.J., López-Sabirón, A.M., et al. 2018. Implementation of PEF treatment at real-scale tomatoes processing considering LCA methodology as an innovation strategy in the agri-food sector. Sustainability. 10(4): 979.

Arruda, H.S., Silva, E.K., Peixoto Araujo, N.M., Pereira, G.A., Pastore, G.M. and Marostica Junior, M.R. 2021. Anthocyanins recovered from agri-food by-products using innovative processes: Trends, challenges, and perspectives for their application in food systems. Molecules. 26(9): 2632.

Arun, K.B., Madhavan, A., Sindhu, R., Binod, P., Pandey, A., Reshmy, R., et al. 2020. Remodeling agro-industrial and food wastes into value-added bioactives and biopolymers. Industrial Crops and Products. 154: 112621.

Asenjo, J.A. and Andrews, B.A. 2011. Aqueous two-phase systems for protein separation: a perspective. Journal of Chromatography A. 1218(49): 8826–8835.

Asgari, K., Labbafi, M., Khodaiyan, F., Kazemi, M. and Hosseini, S.S. 2020. High-methylated pectin from walnut processing wastes as a potential resource: Ultrasound assisted extraction and physicochemical, structural and functional analysis. International Journal of Biological Macromolecules. 152: 1274–1282.

Ashokkumar, V., Flora, G., Venkatkarthick, R., SenthilKannan, K., Kuppam, C., Stephy, G.M., et al. 2022. Advanced technologies on the sustainable approaches for conversion of organic waste to valuable bioproducts: Emerging circular bioeconomy perspective. Fuel. 324: 124313.

Awasthi, S.K., Sarsaiya, S., Awasthi, M.K., Liu, T., Zhao, J., Kumar, S., et al. 2020. Changes in global trends in food waste composting: Research challenges and opportunities. Bioresource Technology: 299: 122555.

Bala, S., Garg, D., Sridhar, K., Inbaraj, B.S., Singh, R., Kamma, S., et al. 2023. Transformation of agro-waste into value-added bioproducts and bioactive compounds: Micro/nano formulations and application in the agri-food-pharma sector. Bioengineering. 10(2): 152.

Bansod, S.P., Parikh, J.K. and Sarangi, P.K. 2023. Pineapple peel waste valorization for extraction of bio-active compounds and protein: Microwave assisted method and Box Behnken design optimization. Environmental Research. 221: 115237.

Banwo, K., Olojede, A.O., Adesulu-Dahunsi, A.T., Verma, D.K., Thakur, M., Tripathy, S., et al. 2021. Functional importance of bioactive compounds of foods with Potential Health Benefits: A review on recent trends. Food Bioscience. 43: 101320.

Barba, A.A. and d'Amore, M. 2012. Relevance of dielectric properties in microwave assisted processes. Microwave Materials Characterization. 6: 91–118.

Barba, F.J., Brianceau, S., Turk, M., Boussetta, N. and Vorobiev, E. 2015. Effect of alternative physical treatments (ultrasounds, pulsed electric fields, and high-voltage electrical discharges) on selective recovery of bio-compounds from fermented grape pomace. Food and Bioprocess Technology. 8: 1139–1148.

Barbosa, P.D.P.M., Ruviaro, A.R. and Macedo, G.A. 2021. Conditions of enzyme-assisted extraction to increase the recovery of flavanone aglycones from pectin waste. Journal of Food Science and Technology. 58(11): 4303–4312.

Barbosa-Pereira, L., Guglielmetti, A. and Zeppa, G. 2018. Pulsed electric field assisted extraction of bioactive compounds from cocoa bean shell and coffee silverskin. Food and bioprocess technology. 11: 818–835.

Barrett, C.B., Benton, T., Fanzo, J., Herrero, M., Nelson, R.J., Bageant, E., et al. 2022a. Socio-technical innovation bundles for agri-food systems transformation. 195. Springer Nature.

Barrett, C.B., Benton, T., Fanzo, J., Herrero, M., Nelson, R.J., Bageant, E., et al. 2022b. The state of agri-food systems and agri-food value chains in 2020. pp. 21–45. *In*: Barrett, C.B., Benton, T., Fanzo, J., Herrero, M., Nelson, R.J., Bageant, E., et al. (eds). Socio-Technical Innovation Bundles for Agri-Food Systems Transformation. Cham: Springer International Publishing.

Belmiro, R.H., de Carvalho Oliveira, L., Geraldi, M.V., Junior, M.R.M. and Cristianini, M. 2021. Modification of coffee coproducts by-products by dynamic high pressure, acetylation and hydrolysis by cellulase: A potential functional and sustainable food ingredient. Innovative Food Science and Emerging Technologies. 68: 102608.

Bhat, R., Sharma, M., Rätsep, R., Malenica, D. and Jõgi, K. 2023. Challenges and prospects of tackling food loss and wastes in the circular economy context. Sustainable Food Value Chain Development: Perspectives from Developing and Emerging Economies. 15–36.

Bhuimbar, M.V., Bhagwat, P.K. and Dandge, P.B. 2019. Extraction and characterization of acid soluble collagen from fish waste: development of collagen-chitosan blend as food packaging film. Journal of Environmental Chemical Engineering. 7(2): 102983.

Bilal, M. and Iqbal, H.M. 2019. Sustainable bioconversion of food waste into high-value products by immobilized enzymes to meet bio-economy challenges and opportunities–a review. Food Research International. 123: 226–240.

Bilhman, S., Ramanathan, S., Dumjun, K., Wunnoo, S., Lethongkam, S., Waen-ngoen, T., et al. 2023. Value-added from microwave-assisted extraction of *Musa sapientum* waste as an alternative safe and effective agent for the treatment of hyperpigmentation. Waste and Biomass Valorization. 14(5): 1477–1488.

Bucur, M.P., Radulescu, M.C., Radu, G.L. and Bucur, B. 2023. Cavitation-effect-based treatments and extractions for superior fruit and milk valorisation. Molecules. 28(12): 4677.

Bulgari, D., Alias, C., Peron, G., Ribaudo, G., Gianoncelli, A., Savino, S., et al. 2023. Solid-state fermentation of trichoderma spp.: a new way to valorize the agricultural digestate and produce value-added bioproducts. Journal of Agricultural and Food Chemistry. 71(9): 3994–4004.

Caldeira, C., Vlysidis, A., Fiore, G., De Laurentiis, V., Vignali, G. and Sala, S. 2020. Sustainability of food waste biorefinery: a review on valorisation pathways, techno-economic constraints, and environmental assessment. Bioresource Technology. 312: 123575.

Çam, M. and Hışıl, Y. 2010. Pressurised water extraction of polyphenols from pomegranate peels. Food Chemistry. 123(3): 878–885.

Capanoglu, E., Nemli, E. and Tomas-Barberan, F. 2022. Novel approaches in the valorization of agricultural wastes and their applications. Journal of Agricultural and Food Chemistry. 70(23): 6787–6804.

Carmen, F., Frances, C. and Barthe, L. 2023. Trends on valorization of pitaya fruit biomass through value-added and green extraction technology–A critical review of advancements and processes. Trends in Food Science and Technology. 138: 339–354.

Castro-Vargas, H.I., Baumann, W., Ferreira, S.R. and Parada-Alfonso, F. 2019. Valorization of papaya (*Carica papaya* L.) agroindustrial waste through the recovery of phenolic antioxidants by supercritical fluid extraction. Journal of Food Science and Technology. 56: 3055–3066.

Catalkaya, G. and Kahveci, D. 2019. Optimization of enzyme assisted extraction of lycopene from industrial tomato waste. Separation and Purification Technology. 219: 55–63.

Chatterjee, S. and Mohan, S.V. 2023. Pretreatment of fruit and vegetable waste for the production of biofuels and high-value products. pp. 285–299. *In*: Mandavgane, S.A., Chakravarty, I. and Jaiswal, A.K. (eds). Fruit and Vegetable Waste Utilization and Sustainability. Academic Press.

Chavan, S., Yadav, B., Atmakuri, A., Tyagi, R.D., Wong, J.W. and Drogui, P. 2022. Bioconversion of organic wastes into value-added products: a review. Bioresource Technology. 344: 126398.

Chávez-González, M.L., Sepúlveda, L., Verma, D.K., Luna-García, H.A., Rodríguez-Durán, L.V., Ilina, A., et al. 2020. Conventional and emerging extraction processes of flavonoids. Processes. 8(4): 434.

Cheng, M.H., Rosentrater, K.A., Sekhon, J., Wang, T., Jung, S. and Johnson, L.A. 2019. Economic feasibility of soybean oil production by enzyme-assisted aqueous extraction processing. Food and Bioprocess Technology. 12: 539–550.

Chojnacka, K., Moustakas, K. and Witek-Krowiak, A. 2020. Bio-based fertilizers: a practical approach towards circular economy. Bioresource Technology. 295: 122223.

Chourasia, R., Phukon, L.C., Abedin, M.M., Padhi, S., Singh, S.P. and Rai, A.K. 2022. Whey valorization by microbial and enzymatic bioprocesses for the production of nutraceuticals and value-added products. Bioresource Technology Reports. 19: 101144.

Cruz-Casillas, F.C., García-Cayuela, T. and Rodriguez-Martinez, V. 2021. Application of conventional and non-conventional extraction methods to obtain functional ingredients from jackfruit (Artocarpus heterophyllus lam.) tissues and by-products. Applied Sciences. 11(16): 7303.

Curmi, H., Chirat, C., Roubaud, A., Peyrot, M., Haarlemmer, G. and Lachenal, D. 2022. Extraction of phenolic compounds from sulfur-free black liquor thanks to hydrothermal treatment before the production of syngas for biofuels. The Journal of Supercritical Fluids. 181: 105489.

Da Silva, R.P., Rocha-Santos, T.A. and Duarte, A.C. 2016. Supercritical fluid extraction of bioactive compounds. TrAC Trends in Analytical Chemistry. 76: 40–51.

Dahiya, S., Kumar, A.N., Sravan, J.S., Chatterjee, S., Sarkar, O. and Mohan, S.V. 2018. Food waste biorefinery: Sustainable strategy for circular bioeconomy. Bioresource Technology. 248: 2–12.

Das, S., Nadar, S.S. and Rathod, V.K. 2021. Integrated strategies for enzyme assisted extraction of bioactive molecules: A review. International Journal of Biological Macromolecules. 191: 899–917.

David, A.J., Abinandan, S., Vaidyanathan, V.K., Xu, C.C. and Krishnamurthi, T. 2023. A critical review on current status and environmental sustainability of pre-treatment methods for bioethanol production from lignocellulose feedstocks. 3 Biotech. 13(7): 233.

de Almeida, M.A. and Colombo, R. 2023. Production chain of first-generation sugarcane bioethanol: characterization and value-added application of wastes. BioEnergy Research. 16(2): 924–939.

de Castro, R.J.S., Ohara, A., Nishide, T.G., Bagagli, M.P., Dias, F.F.G. and Sato, H.H. 2015. A versatile system based on substrate formulation using agroindustrial wastes for protease production by Aspergillus niger under solid state fermentation. Biocatalysis and Agricultural Biotechnology, 4(4): 678–684.

del Pilar Sánchez-Camargo, A., Ballesteros-Vivas, D., Buelvas-Puello, L.M., Martinez-Correa, H.A., Parada-Alfonso, F., Cifuentes, A., et al. 2021. Microwave-assisted extraction of phenolic compounds with antioxidant and anti-proliferative activities from supercritical CO_2 pre-extracted mango peel as valorization strategy. Lwt. 137: 110414.

Deng, J.J., Mao, H.H., Fang, W., Li, Z.Q., Shi, D., Li, Z.W., et al. 2020. Enzymatic conversion and recovery of protein, chitin, and astaxanthin from shrimp shell waste. Journal of Cleaner Production. 271: 122655.

Di Stefano, V., Buzzanca, C., Melilli, M.G., Indelicato, S., Mauro, M., Vazzana, M., et al. 2022. Polyphenol Characterization and Antioxidant Activity of Grape Seeds and Skins from Sicily: A Preliminary Study. Sustainability. 14(11): 6702.

Dimitrijević, A., Tavares, A.P., Almeida, M.R., Vraneš, M., Sousa, AC., Cristóvão, A.C., et al. 2020. Valorization of expired energy drinks by designed and integrated ionic liquid-based aqueous biphasic systems. ACS Sustainable Chemistry and Engineering, 8(14): 5683–5692.

Drosou, C., Kyriakopoulou, K., Bimpilas, A., Tsimogiannis, D. and Krokida, M. 2015. A comparative study on different extraction techniques to recover red grape pomace polyphenols from vinification byproducts. Industrial Crops and Products. 75: 141–149.

Dursun, D. and Dalgıç, A.C. 2016. Optimization of astaxanthin pigment bioprocessing by four different yeast species using wheat wastes. Biocatalysis and Agricultural Biotechnology. 7: 1–6.

Ebikade, E., Lym, J., Wittreich, G., Saha, B. and Vlachos, D.G. 2018. Kinetic studies of acid hydrolysis of food waste-derived saccharides. Industrial and Engineering Chemistry Research. 57(51): 17365–17374.

Esparza, I., Jiménez-Moreno, N., Bimbela, F., Ancín-Azpilicueta, C. and Gandía, L.M. 2020. Fruit and vegetable waste management: Conventional and emerging approaches. Journal of Environmental Management. 265: 110510.

Ezejiofor, T.I.N., Enebaku, U.E. and Ogueke, C. 2014. Waste to wealth-value recovery from agro-food processing wastes using biotechnology: a review. British Biotechnology Journal: 4(4): 418.

Fabiszewska, A.U., Zieniuk, B., Kozłowska, M., Mazurczak-Zieniuk, P.M., Wołoszynowska, M., Misiukiewicz-Stępień, P., et al. 2021. Studies on upgradation of waste fish oil to lipid-rich yeast biomass in Yarrowia lipolytica batch cultures. Foods: 10(2): 436.

Fan, K.W., Chen, F., Jones, E.B.G. and Vrijmoed, L.L.P. 2000. Utilization of food processing waste by Thraustochytrids. pp. 185–194. *In*: Hyde KD, Ho WH, Pointing SB (eds). Aquatic Mycology across the Millenium. Hong Kong. Fungal Diversity Press, New Zealand.

Fang, B., Sun, J., Dong, P., Xue, C. and Mao, X. 2017. Conversion of turbot skin wastes into valuable functional substances with an eco-friendly fermentation technology. Journal of Cleaner Production. 156: 367–377.

Farag, M.A., Eldin, A.B. and Khalifa, I. 2022. Valorization and extraction optimization of Prunus seeds for food and functional food applications: a review with further perspectives. Food Chemistry. 132955.

Ferarsa, S., Zhang, W., Moulai-Mostefa, N., Ding, L., Jaffrin, M.Y. and Grimi, N. 2018. Recovery of anthocyanins and other phenolic compounds from purple eggplant peels and pulps using ultrasonic-assisted extraction. Food and Bioproducts Processing. 109: 19–28.

Fierascu, R.C., Fierascu, I., Avramescu, S.M. and Sieniawska, E. 2019. Recovery of natural antioxidants from agro-industrial side streams through advanced extraction techniques. Molecules. 24(23): 4212.

Gallego, R., Bueno, M. and Herrero, M. 2019. Sub-and supercritical fluid extraction of bioactive compounds from plants, food-by-products, seaweeds and microalgae–An update. TrAC Trends in Analytical Chemistry. 116: 198–213.

Gao, C., Cai, C., Liu, J., Wang, Y., Chen, Y., Wang, L., et al. 2020. Extraction and preliminary purification of polysaccharides from Camellia oleifera Abel. seed cake using a thermoseparating aqueous two-phase system based on EOPO copolymer and deep eutectic solvents. Food Chemistry. 313: 126164.

García, P., Fredes, C., Cea, I., Lozano-Sánchez, J., Leyva-Jiménez, F.J., Robert, P., et al. Jimenez, P. 2021. Recovery of bioactive compounds from pomegranate (Punica granatum L.) peel using pressurized liquid extraction. Foods. 10(2): 203.

Gaurav, N., Sivasankari, S., Kiran, G.S., Ninawe, A. and Selvin, J. 2017. Utilization of bioresources for sustainable biofuels: a review. Renewable and Sustainable Energy Reviews. 73: 205–214.

Gavahian, M., Manyatsi, T.S., Morata, A. and Tiwari, B.K. 2022. Ultrasound-assisted production of alcoholic beverages: From fermentation and sterilization to extraction and aging. Comprehensive Reviews in Food Science and Food Safety. 21(6): 5243–5271.

Gerardi, C., D'amico, L., Migoni, D., Santino, A., Salomone, A., Carluccio, M.A., et al. 2020. Strategies for reuse of skins separated from grape pomace as ingredient of functional beverages. Frontiers in Bioengineering and Biotechnology. 8: 645.

Ghosh, S., Gillis, A., Sheviryov, J., Levkov, K. and Golberg, A. 2019. Towards waste meat biorefinery: Extraction of proteins from waste chicken meat with non-thermal pulsed electric fields and mechanical pressing. Journal of Cleaner Production. 208: 220–231.

Gligor, O., Mocan, A., Moldovan, C., Locatelli, M., Crişan, G. and Ferreira, I.C. 2019. Enzyme-assisted extractions of polyphenols–A comprehensive review. Trends in Food Science and Technology. 88: 302–315.

Gmoser, R., Sintca, C., Taherzadeh, M.J. and Lennartsson, P.R. 2019. Combining submerged and solid state fermentation to convert waste bread into protein and pigment using the edible filamentous fungus N. intermedia. Waste management. 97: 63–70.

Goja, A.M., Yang, H., Cui, M. and Li, C., 2013. Aqueous two-phase extraction advances for bioseparation. Journal of Bioprocess Biotechnique. 4: 140. doi: 10.4172/2155-9821.1000140

Gong, C., Singh, A., Singh, P. and Singh, A. 2021. Anaerobic digestion of agri-food wastes for generating biofuels. Indian Journal of Microbiology. 61(4): 427–440.

González-Centeno, M.R., Knoerzer, K., Sabarez, H., Simal, S., Rosselló, C. and Femenia, A. 2014. Effect of acoustic frequency and power density on the aqueous ultrasonic-assisted extraction of grape pomace (Vitis vinifera L.)—a response surface approach. Ultrasonics Sonochemistry. 21(6): 2176–2184.

Gu, M., Fang, H., Gao, Y., Su, T., Niu, Y. and Yu, L.L. 2020. Characterization of enzymatic modified soluble dietary fiber from tomato peels with high release of lycopene. Food Hydrocolloids. 99: 105321.

Guan, H., Diao, X., Jiang, F., Han, J. and Kong, B. 2018. The enzymatic hydrolysis of soy protein isolate by Corolase PP under high hydrostatic pressure and its effect on bioactivity and characteristics of hydrolysates. Food Chemistry. 245: 89–96.

Guo, Y., Huang, W.C., Wu, Y., Qi, X. and Mao, X. 2018. Application of a low-voltage direct-current electric field for lipid extraction from squid viscera. Journal of Cleaner Production. 205: 610–618.

Gupta, A.K., Gurjar, P.S., Beer, K., Pongener, A., Ravi, S.C., Singh, S., et al. 2022. A review on valorization of different byproducts of mango (Mangifera indica L.) for functional food and human health. Food Bioscience. 48: 101783.

Halog, A. and Anieke, S. 2021. A review of circular economy studies in developed countries and its potential adoption in developing countries. Circular Economy and Sustainability. 1: 209–230.

Han, W., Liu, Y., Xu, X., Huang, J., He, H., Chen, L., et al. 2020. Bioethanol production from waste hamburger by enzymatic hydrolysis and fermentation. Journal of Cleaner Production. 264: 121658.

Hegde, S., Lodge, J.S. and Trabold, T.A. 2018. Characteristics of food processing wastes and their use in sustainable alcohol production. Renewable and Sustainable Energy Reviews. 81: 510–523.

Hernández, C., Escamilla-Alvarado, C., Sánchez, A., Alarcón, E., Ziarelli, F., Musule, R., et al. 2019. Wheat straw, corn stover, sugarcane, and Agave biomasses: chemical properties, availability, and cellulosic-bioethanol production potential in Mexico. Biofuels, Bioproducts and Biorefining. 13(5): 1143–1159.

Hiloidhari, M., Bhuyan, N., Gogoi, N., Seth, D., Garg, A., Singh, A., et al. 2020. Agroindustry wastes: biofuels and biomaterials feedstocks for sustainable rural development. pp. 357–388. *In*: Kumar, R.P., Gnansounou, E., Raman, J.K. and Baskar, G. (eds). Refining Biomass Residues for Sustainable Energy and Bioproducts. Academic Press.

Hoehn, D., Vázquez-Rowe, I., Kahhat, R., Margallo, M., Laso, J., Fernández-Ríos, A., et al. 2023. A critical review on food loss and waste quantification approaches: Is there a need to develop alternatives beyond the currently widespread pathways?. Resources, Conservation and Recycling. 188: 106671.

Huang, L., Zhang, X., Xu, M., An, S., Li, C., Huang, C., et al. 2018. Dietary fibres from cassava residue: Physicochemical and enzymatic improvement, structure and physical properties. AIP Advances. 8(10): 105035.

Idris, S. A., Rosli, N.R. and Aris, R.M.A.R. 2022. Supercritical carbon dioxide extraction of fatty acids compounds from tamarind seeds. Materials Today: Proceedings. 63: S462–S466.

Jach, M.E., Sajnaga, E., Janeczko, M., Juda, M., Kochanowicz, E., Baj, T., et al. 2021. Production of enriched in B vitamins biomass of Yarrowia lipolytica grown in biofuel waste. Saudi Journal of Biological Sciences. 28(5): 2925–2932.

Jadhav, A.J., Holkar, C.R., Goswami, A.D., Pandit, A.B. and Pinjari, D.V. 2016. Acoustic cavitation as a novel approach for extraction of oil from waste date seeds. ACS Sustainable Chemistry and Engineering. 4(8): 4256–4263.

Jara-Palacios, M.J., Santisteban, A., Gordillo, B., Hernanz, D., Heredia, F.J. and Escudero-Gilete, M.L. 2019. Comparative study of red berry pomaces (blueberry, red raspberry, red currant and blackberry) as source of antioxidants and pigments. European Food Research and Technology. 245: 1–9.

Jasrotia, R., Dhar, M. and Langer, S. 2023. Climate change impacts on animal production. pp. 311–333. *In*: Ahmed, M. (ed.). Global Agricultural Production: Resilience to Climate Change. Cham: Springer International Publishing.

Javed, U., Ansari, A., Aman, A. and Qader, S.A.U. 2019. Fermentation and saccharification of agro-industrial wastes: a cost-effective approach for dual use of plant biomass wastes for xylose production. Biocatalysis and Agricultural Biotechnology. 21: 101341.

Jawad, A.H., Alkarkhi, A.F.M., Jason, O.C., Easa, A.M. and Nik Norulaini, N.A., 2013. Production of the lactic acid from mango peel waste—factorial experiment. Journal of King Saud University, Science. 25(1): 39–45.

Jha, A.K. and Sit, N. 2022. Extraction of bioactive compounds from plant materials using combination of various novel methods: a review. Trends in Food Science and Technology. 119: 579–591.

Jiménez-Velázquez, P., Valle-Guadarrama, S., Alia-Tejacal, I., Salinas-Moreno, Y., García-Cruz, L., Pérez-López, A., et al. 2020. Separation of bioactive compounds from epicarp of 'Hass' avocado fruit through aqueous two-phase systems. Food and Bioproducts Processing. 123: 238–250.

Kaderides, K., Goula, A.M. and Adamopoulos, K.G. 2015. A process for turning pomegranate peels into a valuable food ingredient using ultrasound-assisted extraction and encapsulation. Innovative Food Science and Emerging Technologies. 31: 204–215.

Kannah, R.Y., Merrylin, J., Devi, T.P., Kavitha, S., Sivashanmugam, P., Kumar, G., et al. 2020. Food waste valorization: Biofuels and value added product recovery. Bioresource Technology Reports. 11: 100524.

Kantifedaki, A., Kachrimanidou, V., Mallouchos, A., Papanikolaou, S. and Koutinas, A.A. 2018. Orange processing waste valorisation for the production of bio-based pigments using the fungal strains Monascus purpureus and Penicillium purpurogenum. Journal of Cleaner Production. 185: 882–890.

Karthikeyan, O.P., Mehariya, S. and Wong, J.W.C. 2017. Bio-refining of food waste for fuel and value products. Energy Procedia. 136: 14–21.

Kassim, F.O., Thomas, C.P. and Afolabi, O.O. 2022. Integrated conversion technologies for sustainable agri-food waste valorization: A critical review. Biomass and Bioenergy. 156: 106314.

Kaur, A., Singh, D., Kaur, H. and Sogi, D.S. 2018. Drying characteristics and antioxidant properties of Java plum seed and skin waste. Journal of Stored Products and Postharvest Research. 9(4): 36–43.

Kaur, G., Wang, H., To, M.H., Roelants, S.L., Soetaert, W. and Lin, C.S.K. 2019. Efficient sophorolipids production using food waste. Journal of Cleaner Production. 232: 1–11.

Kee, P.E., Ng, T.C., Lan, J.C.W. and Ng, H.S. 2020. Recent development of unconventional aqueous biphasic system: characteristics, mechanisms and applications. Critical Reviews in Biotechnology. 40(4): 555–569.

Khan, S.A., Aslam, R. and Makroo, H.A. 2019. High pressure extraction and its application in the extraction of bio-active compounds: a review. Journal of Food Process Engineering. 42(1): e12896.

Khanpit, V.V., Tajane, S.P. and Mandavgane, S.A. 2021. Dietary fibers from fruit and vegetable waste: methods of extraction and processes of value addition. Biomass Conversion and Biorefinery. 1–20.

Khoo, K.S., Leong, H.Y., Chew, K.W., Lim, J.W., Ling, T.C., Show, P.L., et al. 2020. Liquid biphasic system: a recent bioseparation technology. Processes, 8(2): 149.

Kılmanoğlu, H., Hoşoğlu, M.İ., Güneşer, O. and Yüceer, Y.K. 2021. Optimization of pretreatment and enzymatic hydrolysis conditions of tomato pomace for production of alcohols and esters by Kluyveromyces marxianus. Lwt. 138: 110728.

Kim, J.H., Lee, J.C. and Pak, D. 2011. Feasibility of producing ethanol from food waste. Waste management. 31(9–10): 2121–2125.

Kiran, E.U., Trzcinski, A.P., Ng, W.J. and Liu, Y. 2014. Bioconversion of food waste to energy: a review. Fuel. 134: 389–399.

Kitrytė, V., Kavaliauskaitė, A., Tamkutė, L., Pukalskienė, M., Syrpas, M. and Venskutonis, P.R. 2020. Zero waste biorefining of lingonberry (Vaccinium vitis-idaea L.) pomace into functional ingredients by consecutive high pressure and enzyme assisted extractions with green solvents. Food Chemistry. 322: 126767.

Kolb, J.F., Kono, S. and Schoenbach, K.H. 2006. Nanosecond pulsed electric field generators for the study of subcellular effects. Bioelectromagnetics. 27(3): 172–187.

Kour, R., Singh, S., Sharma, H.B., Naik, T.S.S.K., Shehata, N., Pavithra, N., et al. 2023. Persistence and remote sensing of agri-food wastes in the environment: Current state and perspectives. Chemosphere. 137822.

Kover, A., Kraljić, D., Marinaro, R. and Rene, E.R. 2022. Processes for the valorization of food and agricultural wastes to value-added products: recent practices and perspectives. Systems Microbiology and Biomanufacturing. 2: 50–66.

Kulkarni, V. and Rathod, V. 2016. Green process for extraction of mangiferin from Mangifera indica leaves. Journal of Biologically Active Products from Nature. 6(5–6): 406–411.

Kumar, K., Yadav, A.N., Kumar, V., Vyas, P. and Dhaliwal, H.S. 2017. Food waste: a potential bioresource for extraction of nutraceuticals and bioactive compounds. Bioresources and Bioprocessing. 4: 1–14.

Kumar, M., Dahuja, A., Sachdev, A., Kaur, C., Varghese, E., Saha, S., et al. 2019a. Evaluation of enzyme and microwave-assisted conditions on extraction of anthocyanins and total phenolics from black soybean (Glycine max L.) seed coat. International Journal of Biological Macromolecules. 135: 1070–1081.

Kumar, S., Chinnannan, K., Thamilarasan, S.K., Seralathan, M., Shanmuganathan, R. and Padikasan, I.A. 2019b. Enzymatically hydrolysed sago bagasse improves physiological, biochemical and molecular attributes of Solanum lycopersicum. Biocatalysis and Agricultural Biotechnology. 17: 499–506.

Kumar, M., Tomar, M., Punia, S., Amarowicz, R. and Kaur, C. 2020. Evaluation of cellulolytic enzyme-assisted microwave extraction of Punica granatum peel phenolics and antioxidant activity. Plant Foods for Human Nutrition. 75: 614–620.

Kumar, V., Sharma, N., Umesh, M., Selvaraj, M., Al-Shehri, B.M., Chakraborty, P., et al. 2022a. Emerging challenges for the agro-industrial food waste utilization: A review on food waste biorefinery. Bioresource Technology. 127790.

Kumar, V., Yadav, S.K., Patel, A.K., Mishra, B.B., Ahluwalia, V., Thakur, L.K., et al. 2022b. Bioprocessing of agri-food processing residues into nutraceuticals and bioproducts. pp. 301–322. *In*: Rai, A.K., Singh, S.P., Pandey, A., Larroche, C. and Soccol, C.R. (eds). Current Developments in Biotechnology and Bioengineering. Elsevier.

Kumari, M., Padhi, S., Sharmas S., Phukon, L.C., Singh, S.P. and Rai, A.K. 2021. Biotechnological potential of psychrophilic microorganisms as the source of cold-active enzymes in food processing applications. 3 Biotech. 11(11): 479.

Kurup, G.G., Adhikari, B. and Zisu, B. 2019. Recovery of proteins and lipids from dairy wastewater using food grade sodium lignosulphonate. Water Resources and Industry. 22: 100114.

Lasta, H.F.B., Lentz, L., Rodrigues, L.G.G., Mezzomo, N., Vitali, L. and Ferreira, S.R.S. 2019. Pressurized liquid extraction applied for the recovery of phenolic compounds from beetroot waste. Biocatalysis and Agricultural Biotechnology. 21: 101353.

Lee, S.Y., Liang, Y.N., Stuckey, D.C. and Hu, X. 2023. Single-step extraction of bioactive compounds from cruciferous vegetable (kale) waste using natural deep eutectic solvents. Separation and Purification Technology. 317: 123677.

Lefebvre, T., Destandau, E. and Lesellier, E. 2021. Selective extraction of bioactive compounds from plants using recent extraction techniques: a review. Journal of Chromatography A. 1635: 461770.

Leite, P., Silva, C., Salgado, J.M. and Belo, I. 2019. Simultaneous production of lignocellulolytic enzymes and extraction of antioxidant compounds by solid-state fermentation of agro-industrial wastes. Industrial Crops and Products. 137: 315–322.

Leonelli, C., Veronesi, P. and Cravotto, G. 2012. Microwave-assisted extraction: an introduction to dielectric heating. pp. 1–14. *In*: Chemat, C. and Cravotto, G. (eds). Microwave-assisted Extraction for Bioactive Compounds: Theory and Practice. Boston, MA: Springer US.

Li, J., Zhang, W., Li, X., Ye, T., Gan, Y., Zhang, A., et al. 2018. Production of lactic acid from thermal pretreated food waste through the fermentation of waste activated sludge: effects of substrate and thermal pretreatment temperature. Bioresource Technology. 247: 890–896.

Li, X., Mettu, S., Martin, G.J., Ashokkumar, M. and Lin, C S.K. 2019. Ultrasonic pretreatment of food waste to accelerate enzymatic hydrolysis for glucose production. Ultrasonics Sonochemistry. 53: 77–82.

Li, M., Yu, X., Zhou, C., Yagoub, A.E.A., Ji, Q. and Chen, L. 2020. Construction of an integrated platform for 5-HMF production and separation based on ionic liquid aqueous two-phase system. Journal of Molecular Liquids. 313: 113529.

Liang, S., McDonald, A.G. and Coats, E.R. 2014. Lactic acid production with undefined mixed culture fermentation of potato peel waste. Waste Management. 34(11): 2022–2027.

Lillford, P. and Hermansson, A.M. 2021. Global missions and the critical needs of food science and technology. Trends in Food Science and Technology. 111: 800–811.

Liu, Y., Han, W., Xu, X., Chen, L., Tang, J. and Hou, P. 2020. Ethanol production from waste pizza by enzymatic hydrolysis and fermentation. Biochemical Engineering Journal. 156: 107528.

Low, Z.L., Low, D.Y.S., Tang, S.Y., Manickam, S., Tan, K.W. and Ban, Z.H. 2022. Ultrasonic cavitation: An effective cleaner and greener intensification technology in the extraction and surface modification of nanocellulose. Ultrasonics Sonochemistry. 106176.

Luengo, E., Álvarez, I. and Raso, J. 2013. Improving the pressing extraction of polyphenols of orange peel by pulsed electric fields. Innovative Food Science and Emerging Technologies. 17: 79–84.

Luengo, E., Martínez, J.M., Álvarez, I. and Raso, J. 2016. Effects of millisecond and microsecond pulsed electric fields on red beet cell disintegration and extraction of betanines. Industrial Crops and Products. 84: 28–33.

Madureira, J., Albuquerque, B., Dias, M.I., Pinela, J., Calhelha, R.C., Santos-Buelga, C., et al. 2023. Ultrasound-assisted extraction of hydroxytyrosol and tyrosol from olive pomace treated by gamma radiation: Process optimization and bioactivity assessment. Food and Function. 14(7): 3038–3050.

Maicas, S. 2020. The role of yeasts in fermentation processes. Microorganisms. 8(8): 1142.

Maina, S., Kachrimanidou, V. and Koutinas, A. 2017. A roadmap towards a circular and sustainable bioeconomy through waste valorization. Current Opinion in Green and Sustainable Chemistry. 8: 18–23.

Majumdar, S., Mandal, T. and Mandal, D.D. 2020. Production kinetics of β-carotene from Planococcus sp. TRC1 with concomitant bioconversion of industrial solid waste into crystalline cellulose rich biomass. Process Biochemistry. 92: 202–213.

Mali, P.S. and Kumar, P. 2023. Simulation and experimentation on parameters influencing microwave-assisted extraction of bioactive compounds from Punica granatum waste and its preliminary analysis. Food Chemistry Advances. 100344.

Maran, J.P., Sivakumar, V., Thirugnanasambandham, K. and Sridhar, R. 2014. Microwave assisted extraction of pectin from waste *Citrullus lanatus* fruit rinds. Carbohydrate Polymers. 101: 786–791.

Marathe, S.J., Jadhav, S.B., Bankar, S.B. and Singhal, R.S. 2017. Enzyme-assisted extraction of bioactives. Food bioactives: Extraction and Biotechnology Applications. 171–201.

Marđokić, A., Maldonado, A.E., Klosz, K., Molnár, M.A., Vatai, G. and Bánvölgyi, S. 2023. Optimization of conditions for microwave-assisted extraction of polyphenols from olive pomace of Žutica variety: Waste valorization approach. Antioxidants. 12(6): 1175.

Mariappan, S., David Raj, A., Kumar, S. and Chatterjee, U. 2023. Global Warming Impacts on the Environment in the Last Century. pp. 63–93. *In*: Chatterjee, U., Akanwa, A.O., Kumar, S., Singh, S.K. and Roy, A.D. (eds). Ecological Footprints of Climate Change: Adaptive Approaches and Sustainability. Cham: Springer International Publishing.

Martínez-Avila, O., Llimós, J. and Ponsá, S. 2021. Integrated solid-state enzymatic hydrolysis and solid-state fermentation for producing sustainable polyhydroxyalkanoates from low-cost agro-industrial residues. Food and Bioproducts Processing. 126: 334–344.

Martínez-Olivo, A.O., Durán-Castañeda, A.C., López-Cárdenas, F.G., de Jesús Rodríguez-Romero, J., Sánchez-Burgos, J.A., Sáyago-Ayerdi, S.G., et al. 2023. Advances in ultrasound-assisted extraction of bioactive compounds (antioxidant compounds) from agrofood waste. pp. 217–246. *In*: Lorente, D., Duarte Serna, S., Betoret, E. and Betoret, N. (eds). Advanced Technologies in Wastewater Treatment. Elsevier.

Mathias, T.R.D.S., Fernandes de Aguiar, P., de Almeida e Silva, J.B., Moretzsohn de Mello, P.P. and Camporese Sérvulo, E.F. 2017. Brewery waste reuse for protease production by lactic acid fermentation. Food Technology and Biotechnology. 55(2): 218–224.

Mazlan, N.A., Samad, K.A., Yussof, H.W., Saufi, S.M. and Jahim, J. 2019. Xylooligosaccharides from potential agricultural waste: Characterization and screening on the enzymatic hydrolysis factors. Industrial Crops and Products. 129: 575–584.

Mehmood, T., Nadeem, F., Bilal, M. and Iqbal, H.M. 2021. Recent trends on the food wastes valorization to value-added commodities. pp. 171–196. *In*: Khan, A., Jawaid, M., Pizzi, A., Azum, N., Asiri, A.M. and Isa, I.M. (eds). Advanced Technology for the Conversion of Waste into Fuels and Chemicals. Woodhead Publishing.

Mensah, J.K. and Twumasi, P. 2017. Use of pineapple waste for single cell protein (SCP) production and the effect of substrate concentration on the yield. Journal of Food Process Engineering. 40(3): e12478.

Milea, Ş.A., Aprodu, I., Mihalcea, L., Enachi, E., Bolea, C.A., Râpeanu, G., et al. 2020. Bovine β-lactoglobulin peptides as novel carriers for flavonoids extracted with supercritical fluids from yellow onion skins. Journal of Food Science. 85(12): 4290-4299.

Mohamad, N., Ramli, N., Abd-Aziz, S. and Ibrahim, M.F. 2019. Comparison of hydro-distillation, hydro-distillation with enzyme-assisted and supercritical fluid for the extraction of essential oil from pineapple peels. 3 Biotech. 9: 1–9.

Moirangthem, K., Zaky, A.S. and Tucker, G.A. 2022. Microwave subcritical water pre-treatment and enzymatic hydrolysis of geographical identification (GI) tag Indian black rice (Chakhao Poireiton) straw for fermentable sugar production. Biofuels. 13(7): 815–822.

Montibeller, M.J., de Lima Monteiro, P., Stoll, L., Tupuna-Yerovi, D.S., Rodrigues, E., Rodrigues, R.C., et al. 2019. Improvement of enzymatic assisted extraction conditions on anthocyanin recovery from different varieties of V. vinifera and V. labrusca grape pomaces. Food Analytical Methods. 12: 2056–2068.

Moreira, M.D., Melo, M.M., Coimbra, J.M., Dos Reis, K.C., Schwan, R.F. and Silva, C.F. 2018. Solid coffee waste as alternative to produce carotenoids with antioxidant and antimicrobial activities. Waste Management. 82: 93–99.

Mottaleb, M.A. and Sarker, S.D. 2012. Accelerated solvent extraction for natural products isolation. Natural Products Isolation. 75–87.

Murphy, J.N., Hawboldt, K. and Kerton, F.M. 2018. Enzymatic processing of mussel shells to produce biorenewable calcium carbonate in seawater. Green Chemistry. 20(12): 2913–2920.

Nadar, S.S., Rao, P. and Rathod, V.K. 2018. Enzyme assisted extraction of biomolecules as an approach to novel extraction technology: a review. Food Research International. 108: 309–330.

Nag, S. and Sit, N. 2018. Optimization of ultrasound assisted enzymatic extraction of polyphenols from pomegranate peels based on phytochemical content and antioxidant property. Journal of Food Measurement and Characterization. 12: 1734–1743.

Naik, B., Goyal, S.K., Tripathi, A.D. and Kumar, V. 2019. Screening of agro-industrial waste and physical factors for the optimum production of pullulanase in solid-state fermentation from endophytic Aspergillus sp. Biocatalysis and Agricultural Biotechnology. 22: 101423.

Nakthong, N., Wongsagonsup, R. and Amornsakchai, T. 2017. Characteristics and potential utilizations of starch from pineapple stem waste. Industrial Crops and Products. 105: 74–82.

Ng, H.S., Kee, P.E., Tan, G.Y.T., Yim, H.S. and Lan, J.C.W. 2020a. Surfactant as an Additive for the Recovery of Potent Antioxidants from Garcinia mangostana Pericarps Using a Polymer/Salt Aqueous Biphasic System. Applied Biochemistry and Biotechnology. 191: 273–283.

Ng, H.S., Kee, P.E., Yim, H.S., Chen, P.T., Wei, Y.H. and Lan, J.C.W. 2020b. Recent advances on the sustainable approaches for conversion and reutilization of food wastes to valuable bioproducts. Bioresource Technology. 302: 122889

Nouri, M., Khodaiyan, F., Razavi, S.H. and Mousavi, M. 2016. Improvement of chitosan production from Persian Gulf shrimp waste by response surface methodology. Food Hydrocolloids. 59: 50–58.

Nurdalilah, O., Teoh, Y.P., Ooi, Z.X. and Sam, S.T. 2018, September. Comparative study on the extraction of bioactive secondary metabolites from pomelo and pineapple peels extract. *In*: IOP Conference Series: Materials Science and Engineering. 429(1): 012040. IOP Publishing.

Oliveira, A.M.B., Viganó, J., Sanches, V. L., Rostagno, M. A. and Martínez, J. 2022. Extraction of potential bioactive compounds from industrial Tahiti lime (Citrus latifólia Tan.) by-product using pressurized liquids and ultrasound-assisted extraction. Food Research International. 157: 111381.

Padhi, S., Sarkar, P., Abedin, M.M., Sirohi, R., Singh, S.P. and Rai, A.K. 2023. Microbial Enzymes and Metabolites in Health Care: An Overview. pp. 3–23. *In*: Sirohi, R., Rai, A.K., de Souza Vandenberghe, L.P. and Parameswaran, B. (eds). Microbial Enzymes and Metabolites for Health and Well-Being. CRC Press.

Paramos, P.R., Granjo, J.F., Corazza, M.L. and Matos, H.A. 2020. Extraction of high value products from avocado waste biomass. The Journal of Supercritical Fluids. 165: 104988.

Paritosh, K., Kushwaha, S.K., Yadav, M., Pareek, N., Chawade, A. and Vivekanand, V. 2017. Food waste to energy: an overview of sustainable approaches for food waste management and nutrient recycling. BioMed Research International. 2017.

Parmar, I. and Rupasinghe, H.V. 2013. Bio-conversion of apple pomace into ethanol and acetic acid: enzymatic hydrolysis and fermentation. Bioresource Technology. 130: 613–620.

Patchimpet, J., Sangkharak, K., Eiad-ua, A. and Klomklao, S. 2021. Thermoseparating aqueous two-phase system for lipase recovery and partitioning from Nile tilapia viscera: Biochemical properties and effect of ultrasound. Journal of Molecular Liquids. 331: 115721.

Pattnaik, B., Sarangi, P.K., Jena, P.K., Sahoo, H.P. and Bhatia, L. 2022. Production of phenolic flavoring compounds from sugarcane bagasse by Lactobacillus acidophilus MTCC 10307. Archives of Microbiology. 204: 1–8.

Pereira, D.T.V., Zabot, G.L., Reyes, F.G.R., Iglesias, A.H. and Martínez, J. 2021. Integration of pressurized liquids and ultrasound in the extraction of bioactive compounds from passion fruit rinds: Impact on phenolic yield, extraction kinetics and technical-economic evaluation. Innovative Food Science and Emerging Technologies. 67: 102549.

Pereira, J.F., Freire, M.G. and Coutinho, J.A. 2020. Aqueous two-phase systems: Towards novel and more disruptive applications. Fluid Phase Equilibria. 505: 112341.

Phongthai, S., Lim, S.T. and Rawdkuen, S. 2016. Optimization of microwave-assisted extraction of rice bran protein and its hydrolysates properties. Journal of Cereal Science. 70: 146–154.

Phukon, L.C., Abedin, M.M., Chourasia, R., Sahoo, D., Parameswaran, B. and Rai, A.K. 2020. Production of xylanase under submerged fermentation from Bacillus firmus HS11 isolated from Sikkim Himalayan region. Indian Journal of Experimental Biology (IJEB). 58(08): 563–570.

Phukon, L.C., Chourasia, R., Padhi, S., Abedin, M.M., Godan, T.K., Parameswaran, B., et al. 2022. Cold-adaptive traits identified by comparative genomic analysis of a lipase-producing Pseudomonas sp. HS6 isolated from snow-covered soil of Sikkim Himalaya and molecular simulation of lipase for wide substrate specificity. Current Genetics. 68(3–4): 375–391.

Picot-Allain, C., Mahomoodally, M.F., Ak, G. and Zengin, G. 2021. Conventional versus green extraction techniques—a comparative perspective. Current Opinion in Food Science. 40: 144–156.

Poyraz, Ç., Küçükyıldız, G., Kırbaşlar, Ş.İ., Ciğeroğlu, Z. and Şahin, S. 2021. Valorization of Citrus unshiu biowastes to value-added products: an optimization of ultrasound-assisted extraction method using response surface methodology and particle swarm optimization. Biomass Conversion and Biorefinery. 1–11.

Prabhakar, P., Mukherjee, S., Kumar, A., Kumar, S., Verma, D.K., Dhara, S., et al. 2023a. Optimization of microwave-assisted extraction (MAE) of key phenolic compounds from pigeon pea (Cajanus cajan L.), their characterization, and measurement of their anti-diabetic and cytotoxic potential. Journal of Food Measurement and Characterization. 1–24.

Prabhakar, P., Mukherjee, S., Kumar, A., Kumar, S., Verma, D.K., Dhara, S., et al. 2023b. Optimization of MAE for Carica papaya phytochemicals, and its in silico, in vitro, and ex vivo evaluation: For functional food and drug applications. Food Bioscience. 102861.

Pradhan, D., Jaiswal, A.K. and Jaiswal, S. 2022. Emerging technologies for the production of nanocellulose from lignocellulosic biomass. Carbohydrate Polymers. 119258.

Puértolas, E. and Barba, F.J. 2016. Electrotechnologies applied to valorization of by-products from food industry: Main findings, energy and economic cost of their industrialization. Food and Bioproducts Processing. 100: 172–184.

Radenkovs, V., Juhnevica-Radenkova, K., Górnaś, P. and Seglina, D. 2018. Non-waste technology through the enzymatic hydrolysis of agro-industrial by-products. Trends in Food Science and Technology. 77: 64–76.

Rahimpour, S. and Taghian Dinani, S. 2018. Lycopene extraction from tomato processing waste using ultrasound and cell-wall degrading enzymes. Journal of Food Measurement and Characterization. 12(4): 2394–2403.

Reshmy, R., Philip, E., Unni, R., Paul, S.A., Sindhu, R., Madhavan, A., et al. 2023. Bioethanol—A promising alternative fuel for sustainable future. pp. 179–196. *In*: Gupta, V.K., Tuohy, M, Ramteke, P. Nguyen, Q. and Bhat, R. (eds). Valorization of Biomass to Bioproducts. Elsevier.

Ricarte, G.N., Coelho, M.A.Z., Marrucho, I.M. and Ribeiro, B.D. 2020. Enzyme-assisted extraction of carotenoids and phenolic compounds from sunflower wastes using green solvents. 3 Biotech. 10: 1–11.

Rifna, E.J., Misra, N.N. and Dwivedi, M. 2021. Recent advances in extraction technologies for recovery of bioactive compounds derived from fruit and vegetable waste peels: a review. Critical Reviews in Food Science and Nutrition. 1–34.

Rivero-Pino, F., Espejo-Carpio, F.J. and Guadix, E.M. 2020. Bioactive fish hydrolysates resistance to food processing. Lwt. 117: 108670.

Rojas, L.F., Zapata, P. and Ruiz-Tirado, L. 2022. Agro-industrial waste enzymes: Perspectives in circular economy. Current Opinion in Green and Sustainable Chemistry. 34: 100585.

Romano, R., Aiello, A., Pizzolongo, F., Rispoli, A., De Luca, L. and Masi, P. 2020. Characterisation of oleoresins extracted from tomato waste by liquid and supercritical carbon dioxide. International Journal of Food Science and Technology. 55(10): 3334–3342.

Romano, R., De Luca, L., Aiello, A., Rossi, D., Pizzolongo, F. and Masi, P. 2022. Bioactive compounds extracted by liquid and supercritical carbon dioxide from citrus peels. International Journal of Food Science and Technology. 57(6): 3826–3837.

Rosa, M.E., Ferreira, A.M., Neves, C.M., Almeida, M.R., Barros, R., Cristovão, A.C., et al. 2023. Valorisation of red beet waste: one-step extraction and separation of betalains and chlorophylls using thermoreversible aqueous biphasic systems. Green Chemistry. 25(5): 1852–1864.

Saad, N., Louvet, F., Tarrade, S., Meudec, E., Grenier, K., Landolt, C., et al. 2019. Enzyme-assisted extraction of bioactive compounds from raspberry (*Rubus idaeus* L.) pomace. Journal of Food Science. 84(6): 1371–1381.

Sadh, P.K., Duhan, S. and Duhan, J.S. 2018a. Agro-industrial wastes and their utilization using solid state fermentation: a review. Bioresources and Bioprocessing. 5(1): 1–15.

Sadh, P.K., Kumar, S., Chawla, P. and Duhan, J.S. 2018b. Fermentation: a boon for production of bioactive compounds by processing of food industries wastes (by-products). Molecules. 23(10): 2560.

Safdar, M.N., Kausar, T., Jabbar, S., Mumtaz, A., Ahad, K. and Saddozai, A.A. 2017. Extraction and quantification of polyphenols from kinnow (*Citrus reticulate* L.) peel using ultrasound and maceration techniques. Journal of Food and Drug Analysis. 25(3): 488–500.

Sagar, N.A., Pareek, S., Sharma, S., Yahia, E.M. and Lobo, M.G. 2018. Fruit and vegetable waste: Bioactive compounds, their extraction, and possible utilization. Comprehensive Reviews in Food Science and Food Safety. 17(3): 512–531.

Saha, N., Sarkar, B. and Sen, K. 2022. Aqueous biphasic systems: a robust platform for green extraction of biomolecules. Journal of Molecular Liquids. 119882.

Salimi, E., Saragas, K., Taheri, M.E., Novakovic, J., Barampouti, E.M., Mai, S., et al. 2019. The role of enzyme loading on starch and cellulose hydrolysis of food waste. Waste and Biomass Valorization. 10: 3753–3762.

Santana, Á.L. and Macedo, G.A. 2019. Effects of hydroalcoholic and enzyme-assisted extraction processes on the recovery of catechins and methylxanthines from crude and waste seeds of guarana (Paullinia cupana). Food Chemistry. 281: 222–230.

Santana, Á.L., Zabot, G.L., Osorio-Tobón, J.F., Johner, J.C., Coelho, A.S., Schmiele, M., et al. 2017. Starch recovery from turmeric wastes using supercritical technology. Journal of Food Engineering. 214: 266–276.

Santos-Martín, M., Cubero-Cardoso, J., González-Domínguez, R., Cortés-Triviño, E., Sayago, A., Urbano, J., et al. 2023. Ultrasound-assisted extraction of phenolic compounds from blueberry leaves using natural deep eutectic solvents (NADES) for the valorization of agrifood wastes. Biomass and Bioenergy. 175: 106882.

Saravana Pandian, P., Sindhanaiselvan, S., Subathira, A. and Saravanan, S. 2022. A correlative algorithmic optimization study for an integrated soft computing technique in aqueous two-phase protein extraction from Litopenaeus vannamei waste. Biomass Conversion and Biorefinery. 1–15.

Sarkar, P., Abedin, M.M., Singh, S.P., Pandey, A. and Rai, A.K. 2022. Microbial production and transformation of polyphenols. pp. 189–208. *In*: Rai, A.K., Singh, S.P., Pandey, A., Larroche, C. and Soccol, C.R. (eds). Current Developments in Biotechnology and Bioengineering.

Scaglia, B., D'Incecco, P., Squillace, P., Dell'Orto, M., De Nisi, P., Pellegrino, L., et al. 2020. Development of a tomato pomace biorefinery based on a CO_2-supercritical extraction process for the production of a high value lycopene product, bioenergy and digestate. Journal of Cleaner Production. 243: 118650.

Seabra, I.J., Braga, M.E.M., Batista, M.T. and de Sousa, H.C. 2010. Effect of solvent (CO_2/ethanol/H_2O) on the fractionated enhanced solvent extraction of anthocyanins from elderberry pomace. The Journal of Supercritical Fluids. 54(2): 145–152.

Selvaraj, R., Ramachandra Murty, V., Varadavenkatesan, T., Sekar, R. and Ramesh, V., 2011. Aqueous two phase systems for the recovery of biomolecules – a review. Science Technology. 1: 7–16.

Sepúlveda, L., Laredo-Alcalá, E., Buenrostro-Figueroa, J.J., Ascacio-Valdés, J.A., Genisheva, Z., Aguilar, C., et al. 2020. Ellagic acid production using polyphenols from orange peel waste by submerged fermentation. Electronic Journal of Biotechnology. 43: 1–7.

Shahram, H., Dinani, S.T. and Amouheydari, M. 2019. Effects of pectinase concentration, ultrasonic time, and pH of an ultrasonic-assisted enzymatic process on extraction of phenolic compounds from orange processing waste. Journal of Food Measurement and Characterization. 13(1): 487–498.

Sharif, T., Bhatti, H.N., Bull, I.D. and Bilal, M. 2021. Recovery of high-value bioactive phytochemicals from agro-waste of mango (Mangifera indica L.) using enzyme-assisted ultrasound pretreated extraction. Biomass Conversion and Biorefinery. 1–9.

Sharma, P., Gaur, V.K., Kim, S.H. and Pandey, A. 2020. Microbial strategies for bio-transforming food waste into resources. Bioresource Technology. 299: 122580.

Sharma, P., Gaur, V.K., Sirohi, R., Varjani, S., Kim, S.H. and Wong, J.W. 2021. Sustainable processing of food waste for production of bio-based products for circular bioeconomy. Bioresource Technology. 325: 124684.

Sharma, V., Bhise, K., Bhagwat, P., Pillai, S. and Ranveer, R. 2023. Waste-based bioethanol production by using food, fruit, and grain wastes. pp. 281–298. *In*: Kuddus, M. and Ramteke, P.W. (eds). Value-Addition in Agri-food Industry Waste Through Enzyme Technology. Academic Press.

Shekhar, S., Prakash, P., Singha, P., Prasad, K. and Singh, S.K. 2023. Modeling and optimization of ultrasound-assisted extraction of bioactive compounds from Allium sativum leaves using response surface methodology and artificial neural network coupled with genetic algorithm. Foods. 12(9): 1925.

Shen, D., Kou, X., Wu, C., Fan, G., Li, T., Dou, J., et al. 2021. Cocktail enzyme-assisted alkaline extraction and identification of jujube peel pigments. Food Chemistry. 357: 129747.

Silva, S., Costa, E.M., Calhau, C., Morais, R.M. and Pintado, M.E. 2017. Anthocyanin extraction from plant tissues: a review. Critical Reviews in Food Science and Nutrition. 57(14): 3072–3083.

Silva, T.O.M., Nogueira, J.P., Rezende, Y.R.R.S., Oliveira, C.S.D. and Narain, N. 2022. Bioactive compounds and antioxidants activities in the agro-industrial residues of berries by solvent and enzyme assisted extraction. Food Science and Technology. 42.

Singh, S., Verma, D.K., Thakur, M., Tripathy, S., Patel, A.R., Shah, N., et al. 2021. Supercritical fluid extraction (SCFE) as green extraction technology for high-value metabolites of algae, its potential trends in food and human health. Food Research International. 150(Part A): 110746. https://doi.org/10.1016/j.foodres.2021.110746.

Singh, A., Singhania, R.R., Soam, S., Chen, C.W., Haldar, D., Varjani, S., et al. 2022. Production of bioethanol from food waste: Status and perspectives. Bioresource Technology. 127651.

Singh, R., Singh, P., Pandey, V.K., Dash, K.K., Ashish, Mukarram, S.A. and Kovács, B., 2023. Microwave-assisted phytochemical extraction from walnut hull and process optimization using box–behnken design (BBD). Processes. 11(4): 1243.

Soquetta, M.B., Terra, L.D.M. and Bastos, C.P. 2018. Green technologies for the extraction of bioactive compounds in fruits and vegetables. CyTA-Journal of Food. 16(1): 400–412.

Srihanam, P., Thonpho, A., Sanarat, P. and Motham, P. 2021. Phytochemicals and antioxidant activity in sugarcane (*Saccharum officinarum* L.) bagasse extracts. Suan Sunandha Science and Technology Journal. 8(2): 26–35.

Sufficiency, E., Qamar, S.A., Ferreira, L.F.R., Franco, M., Iqbal, H.M., and ilal, M. 2022. Emerging biotechnological strategies for food waste management: a green leap towards achieving high-value products and environmental abatement. Energy Nexus. 6: 100077.

Suman, G., Nupur, M., Anuradha, S. and Pradeep, B. 2017. Characterization of dairy waste and its utilisation as substrate for production of single cell protein. Journal of Biotechnology and Biochemistry. 3: 73–78.

Sumere, B.R., de Souza, M.C., Dos Santos, M.P., Bezerra, R.M.N., da Cunha, D.T., Martinez, J., et al. 2018. Combining pressurized liquids with ultrasound to improve the extraction of phenolic compounds from pomegranate peel (*Punica granatum* L.). Ultrasonics sonochemistry. 48: 151–162.

Taheri, M.E., Salimi, E., Saragas, K., Novakovic, J., Barampouti, E.M., Mai, S., et al. 2021. Effect of pretreatment techniques on enzymatic hydrolysis of food waste. Biomass Conversion and Biorefinery. 11: 219–226.

Tang, J., Wang, X., Hu, Y., Zhang, Y. and Li, Y. 2016. Lactic acid fermentation from food waste with indigenous microbiota: effects of pH, temperature and high OLR. Waste Management. 52: 278–285.

Tang, J., Wang, X.C., Hu, Y., Zhang, Y. and Li, Y. 2017. Effect of pH on lactic acid production from acidogenic fermentation of food waste with different types of inocula. Bioresource Technology. 224: 544–552.

Teles, A.S.C., Chávez, D.W.H., Coelho, M.A.Z., Rosenthal, A., Gottschalk, L.M.F. and Tonon, R.V. 2021. Combination of enzyme-assisted extraction and high hydrostatic pressure for phenolic compounds recovery from grape pomace. Journal of Food Engineering. 288: 110128.

Tena, N. and Asuero, A.G. 2022. Up-to-date analysis of the extraction methods for anthocyanins: principles of the techniques, optimization, technical progress, and industrial application. Antioxidants. 11(2): 286.

Toepfl, S., Mathys, A., Heinz, V. and Knorr, D. 2006. Potential of high hydrostatic pressure and pulsed electric fields for energy efficient and environmentally friendly food processing. Food Reviews International. 22(4): 405–423.

Toledo-Merma, P.R., Cornejo-Figueroa, M.H., Crisosto-Fuster, A.D.R., Strieder, M.M., Chañi-Paucar, L.O., Náthia-Neves, G., et al. 2022. Phenolic compounds recovery from pomegranate (*Punica granatum* L.) by-products of pressurized liquid extraction. Foods. 11(8): 1070.

Torres-León, C., Ramírez-Guzmán, N., Ascacio-Valdes, J., Serna-Cock, L., dos Santos Correia, M.T., Contreras-Esquivel, J.C., et al. 2019. Solid-state fermentation with Aspergillus niger to enhance the phenolic contents and antioxidative activity of Mexican mango seed: a promising source of natural antioxidants. Lwt. 112: 108236.

Torres-León, C., Chávez-González, M.L., Hernández-Almanza, A., Martínez-Medina, G.A., Ramírez-Guzmán, N., Londoño-Hernández, L., et al. 2021. Recent advances on the microbiological and enzymatic processing for conversion of food wastes to valuable bioproducts. Current Opinion in Food Science. 38: 40–45.

Tripathy, S., Verma, D.K., Thakur, M., Patel, A.R., Srivastav, P.P., Singh, S., et al. 2021. Curcumin extraction, isolation, quantification and its application in functional foods: A review with a focus on immune enhancement activities and COVID-19. Frontiers in Nutrition. 675. https://doi.org/10.3389/fnut.2021.747956.

Tripathy, S., Verma, D.K., Thakur, M., Chakravorty, N., Singh, S. and Srivastav, P.P. 2022. Recent trends in extraction, identification and quantification methods of Centella asiatica phytochemicals with potential applications in food industry and therapeutic relevance: a review. Food Bioscience. 101864. https://doi.org/10.1016/j.fbio.2022.101864.

Tripathy, S. and Srivastav, P.P. 2023a. Effect of dielectric barrier discharge (DBD) cold plasma-activated water pre-treatment on the drying properties, kinetic parameters, and physicochemical and functional properties of Centella asiatica leaves. Chemosphere. 332: 138901.

Tripathy, S. and Srivastav, P.P. 2023b. Encapsulation of Centella asiatica leaf extract in liposome: Study on structural stability, degradation kinetics and fate of bioactive compounds during storage. Food Chemistry Advances. 2: 100202.

Tsang, Y.F., Kumar, V., Samadar, P., Yang, Y., Lee, J., Ok, Y.S., et al. 2019. Production of bioplastic through food waste valorization. Environment International. 127: 625–644.

Tsui, T.H. and Wong, J.W. 2019. A critical review: emerging bioeconomy and waste-to-energy technologies for sustainable municipal solid waste management. Waste Disposal and Sustainable Energy. 1: 151–167.

Usmani, Z., Sharma, M., Karpichev, Y., Pandey, A., Kuhad, R.C., Bhat, R., et al. 2020. Advancement in valorization technologies to improve utilization of bio-based waste in bioeconomy context. Renewable and Sustainable Energy Reviews. 131: 109965.

Vardanega, R., Fuentes, F.S., Palma, J., Bugueño-Muñoz, W., Cerezal-Mezquita, P. and Ruiz-Domínguez, M.C. 2023. Valorization of granadilla waste (*Passiflora ligularis*, Juss.) by sequential green extraction processes based on pressurized fluids to obtain bioactive compounds. The Journal of Supercritical Fluids. 194: 105833.

Veloso, A.V., Silva, B.C., Bomfim, S.A., de Souza, R.L., Soares, C.M.F. and Lima, Á.S. 2020. Selective and continuous recovery of ascorbic acid and vanillin from commercial diet pudding waste using an aqueous two-phase system. Food and Bioproducts Processing. 119: 268–276.

Verma, D.K., Dhakane, J.P., Mahato, D.K., Billoria, S., Bhattacharjee, P. and Srivastav, P.P. 2018. Supercritical fluid extraction (SCFE) for rice aroma chemicals: recent and advance extraction method. pp. 179–198. *In*: Verma, D.K. and Srivastav, P.P. (eds). Science and Technology of Aroma, Flavour and Fragrance in Rice. Apple Academic Press, USA.

Verma, D.K., Mahanti, N.K., Thakur, M., Chakraborty, S.K. and Srivastav, P.P. 2020. Microwave heating: Alternative thermal process technology for food application. pp. 25–67. *In*: Srivastav, P.P., Verma, D.K., Patel, A.R. and Al-Hilphy, A.R. (eds). Emerging Thermal and Nonthermal Technologies in Food Processing. Apple Academic Press.

Verma, D.K. and Srivastav, P.P. 2020a. A paradigm of volatile aroma compounds in rice and their product with extraction and identification methods: a comprehensive review. Food Research International. 130: 1–33. https://doi.org/10.1016/j.foodres.2019.108924.

Verma, D.K. and Srivastav, P.P. 2020b. Bioactive compounds of rice (Oryza sativa L.): Review on paradigm and its potential benefit in human health. Trends in Food Science and Technology. 97: 355–365.

Verma, D.K. and Srivastav, P.P. 2022. Extraction, identification and quantification methods of rice aroma compounds with emphasis on 2-acetyl-1-pyrroline (2-AP) and its relationship with rice quality: a comprehensive review. Food Reviews International. 38(2): 111–162.

Vijayaraghavan, P., Vincent, S.G.P., Arasu, M.V. and Al-Dhabi, N.A. 2016. Bioconversion of agro-industrial wastes for the production of fibrinolytic enzyme from Bacillus halodurans IND18: purification and biochemical characterization. Electronic Journal of Biotechnology. 20: 1–8.

Wang, Y., Yuan, Z. and Tang, Y. 2021. Enhancing food security and environmental sustainability: a critical review of food loss and waste management. Resources, Environment and Sustainability. 4: 100023.

Wang, Z., Xu, B., Luo, H., Meng, K., Wang, Y., Liu, M., et al. 2020. Production pectin oligosaccharides using Humicola insolens Y1-derived unusual pectate lyase. Journal of Bioscience and Bioengineering. 129(1): 16–22.

Waqas, M., Rehan, M., Khan, M. and Nizami, D.A.-S. 2019. Conversion of food waste to fermentation products. pp. 501–509. *In*: Ferranti, P., Berry, E., Jock, A. (eds). Encyclopedia of Food Security and Sustainability. Elsevier.

Wu, R., Dong, Q., Che, Z., Wang, H., Cao, J., Cao, F., et al. 2022. Green and efficient simultaneous enrichment and separation of multiple valuable bioactive compounds from agricultural waste ginkgo biloba exocarp using a two-phase deep eutectic solvent system. ACS Sustainable Chemistry and Engineering. 10(50): 16958–16968.

Wu, W.C., Ng, H.S., Sun, I.M. and Lan, J.C.W. 2017. Single step purification of bromelain from Ananas comosus pulp using a polymer/salt aqueous biphasic system. Journal of the Taiwan Institute of Chemical Engineers. 79: 158–162.

Wu, Y., Ma, H., Zheng, M. and Wang, K. 2015. Lactic acid production from acidogenic fermentation of fruit and vegetable wastes. Bioresource Technology. 191: 53–58.

Xie, Y., Guo, J., Li, W., Wu, Z. and Yu, Z. 2021. Effects of ferulic acid esterase-producing lactic acid bacteria and storage temperature on the fermentation quality, in vitro digestibility and phenolic acid extraction yields of sorghum (*Sorghum bicolor* L.) silage. Microorganisms. 9(1): 114.

Xiong, X., Iris, K.M., Tsang, D.C., Bolan, N.S., Ok, Y.S., Igalavithana, A.D., et al. 2019. Value-added chemicals from food supply chain wastes: state-of-the-art review and future prospects. Chemical Engineering Journal. 375: 121983.

Xu, G., Zhao, J., Shi, K., Xu, Y., Hu, H., Xu, X., et al. 2023. Trends in valorization of citrus by-products from the net-zero perspective: green processing innovation combined with applications in emission reduction. Trends in Food Science and Technology. 137: 124–141.

Xue, H., Tan, J., Li, Q., Tang, J. and Cai, X. 2021. Ultrasound-assisted enzymatic extraction of anthocyanins from raspberry wine residues: process optimization, isolation, purification, and bioactivity determination. Food Analytical Methods. 14: 1369–1386.

Yadav, M., Joshi, C., Paritosh, K., Thakur, J., Pareek, N., Masakapalli, S.K., et al. 2022. Organic waste conversion through anaerobic digestion: a critical insight into the metabolic pathways and microbial interactions. Metabolic Engineering. 69: 323–337.

Yan, Y., Qin, L., Gao, J., Nan, R. and Gao, J. 2020. Protein extraction and sludge dewatering performance of ultrasound-assisted enzymatic hydrolysis of excess sludge. Environmental Science and Pollution Research. 27: 18317–18328.

Yang, G., Tan, H., Li, S., Zhang, M., Che, J., Li, K., et al. 2020. Application of engineered yeast strain fermentation for oligogalacturonides production from pectin-rich waste biomass. Bioresource technology. 300: 122645.

Yang, L., Guo, Z., Wei, J., Han, L., Yu, Q.L., Chen, H., et al. 2021. Extraction of low molecular weight peptides from bovine bone using ultrasound-assisted double enzyme hydrolysis: Impact on the antioxidant activities of the extracted peptides. Lwt. 146: 111470.

Yang, X., Lee, J.H., Yoo, H.Y., Shin, H.Y., Thapa, L.P., Park, C., et al. 2014. Production of bioethanol and biodiesel using instant noodle waste. Bioprocess and Biosystems Engineering. 37: 1627–1635.

Yu, Q., Wang, F., Baroutian, S., Zhang, Y., Wang, Z., Yuan, Z., et al. 2023. A deep eutectic solvent binary-phase system for alkaloid extraction from Chinese herb Evodia lepta residue and its mechanism. Journal of Cleaner Production. 398: 136645.

Yu, X., Li, M., Yagoub, A.E.A., Chen, L., Zhou, C. and Yan, D. 2021. Switchable (pH driven) aqueous two-phase systems formed by deep eutectic solvents as integrated platforms for production-separation 5-HMF. Journal of Molecular Liquids. 325: 115158.

Yusoff, I.M., Chua, L.S. and Taher, Z.M. 2023. Valorization of fruit waste from Cucurbitaceae family: Profiling of phytoconstituent of Benincasa hispida and Citrullus lanatus rinds using ultrasound-assisted extraction. Food Bioscience. 51: 102190.

Yusree, F.I.F.M., Peter, A.P., binti Zulkifli, N.A., Nor, M.Z.M., Basri, M.S.M., Mokhtar, M.N., et al. 2022. Towards green recovery of β-amylase from slurry of sweet potato (Ipomoea batatas) of VitAto variety via liquid biphasic system. Sustainable Chemistry and Pharmacy. 25: 100579.

Zain, N.A.M., Aziman, S.N., Suhaimi, M.S. and Idris, A. 2021. Optimization of L (+) lactic acid production from solid pineapple waste (spw) by rhizopus oryzae NRRL 395. Journal of Polymers and the Environment. 29: 230–249.

Zhang, Q.W., Lin, L.G. and Ye, W.C. 2018. Techniques for extraction and isolation of natural products: A comprehensive review. Chinese Medicine. 13: 1–26.

Zhang, Y., Lei, Y., Qi, S., Fan, M., Zheng, S., Huang, Q., et al. 2023a. Ultrasonic-microwave-assisted extraction for enhancing antioxidant activity of dictyophora indusiata polysaccharides: The difference mechanisms between single and combined assisted extraction. Ultrasonics Sonochemistry. 95: 106356.

Zhang, Y., Wang, X., Zhu, W., Zhao, Y., Wang, N., Gao, M., et al. 2023b. Anaerobic fermentation of organic solid waste: Recent updates in substrates, products, and the process with multiple products co-production. Environmental Research. 116444.

Zhu, J., Kou, X., Wu, C., Fan, G., Li, T., Dou, J., et al. 2022. Enhanced extraction of bioactive natural products using ultrasound-assisted aqueous two-phase system: Application to flavonoids extraction from jujube peels. Food Chemistry. 395: 133530.

Index

P

R

S

T

V

W

Y

About the Editors

Prof. Sudhir Pratap Singh is a Professor of Industrial Biotechnology at Gujarat Biotechnology University, Gandhinagar, Gujarat, India. The focus of his research is gene mining, biocatalyst engineering and characterization to produce high-value rare sugar and functional biomolecules, such as D-allulose, trehalose, turanose, prebiotic oligosaccharides, resistant starch, etc. Dr. Singh has published over 80 research articles in SCI journals. He has edited 10 books, published by international publishers. The scientific contributions of Prof. Singh have been recognized by several prestigious scientific honours, e.g. TATA Innovation Fellowship, DBT, Govt. of India (2024), Associate Fellowship, Indian Science Academy (2024), Fellow, National Academy of Agricultural Sciences, India (2023), Prof. SB Chincholakar Award, BRSI (2023-24), Gandhian Young Technological Innovation Award to the team, SRISTI (2019), Professor Hira Lal Chakravarty Award, Indian Science Congress Association, DST (2018), SBS-MKU Genomics Award, BRSI (2016-17), and Young Scientist Award, International Bioprocessing Association (2016-17).

Dr. Jyoti Singh Jadaun, PhD, is an Assistant Professor in Dayanand Girls' P.G. College, affiliated to CSJM, University, Kanpur since 2017. She earned her PhD degree from Council of Scientific and Industrial Research (CSIR)- Central Institute of Medicinal and Aromatic Plants, Lucknow, India. She pursued her Post-Doctoral research as SERB-NPDF at DBT-Center of Innovative and Applied Bioprocessing, Mohali, India. She has carried out research in various domains of environment sustainability. Her keen interests are to develop the process for sustainable production of biopolymers, green technology to remediate the heavy metal contaminated sites and excursion of plants endophytes to formulate biocontrol agents. She has published 11 research articles, 05 review articles and 14 book chapters of international repute and 03 patents in her credit. Her research contribution are well cited with h-index of 15.

Dr. Amit Kumar Rai is Scientist D at the DBT-National Agri-Food Biotechnology Institute, Mohali, Punjab, India. He has served as Scientist at DBT-Institute of Bioresources and Sustainable Development, regional centre, Sikkim, India from 2012 to 2023. His current research focus is on fermented foods, microbial biotransformation, bioactive peptides, and enzymes for production of value-added products. To his credit, he has filed five patents and 126 publications including 90 publications in SCI Journals and Edited/Authored 4 books and author 32 book chapters with >3800 citations. His performance has resulted in the bestowment of the Associate Fellow 2023 from *Indian National Science Academy* (INSA), Young Scientist Award 2022 from International Bioprocessing Association 2019, *Young Scientist Award 2016* from Association of Food Scientists and Technologists (India), *Young Scientist Award 2015* by Association of Microbiologist of India, *Selected Member* of National Academy of Science, India - 2022, *AU-CBT Excellence Award 2009* by the Biotech Research Society of India, and *Award for Excellence in Food Biotechnology 2016* from Society of Applied Biotechnology. He is also member of Scientific Panel of FSSAI.